# Oil on the Brain

## The Discovery of Oil and the Excitement of the Boom in Northwestern Pennsylvania

Armstrong, Butler, Clarion, Venango Counties

*Third Edition*

by
Gary S. McKinney

Thanks to
Sally Jordan Reed
For Editing and Formatting

Layout and Design by
Kari McConnell



**THIRD EDITION**


First Printing December 2003
Second Printing September 2004
(Revised and Expanded Edition)
Third Edition © January 2008

Library of Congress Control Number: 2007941514

ISBN-13: 978-0-9793772-2-8
ISBN-10: 0-9793772-2-6

Printed in the United States of America

Published by
MECHLING BOOKBINDERY
PRINTING & BINDING
1124 Oneida Valley Road - Rte. 38
Chicora, PA 16025
1-800-941-3735
www.mechlingbooks.com

# FORWARD

Many seasons have passed since the historic oil boomtown of Antwerp City suddenly sprang to life in northwestern Pennsylvania. Colossal oaks, maples, and hardy pines now lend their exposed, grizzled roots, and cast-off leaves to cover the evidence of those days gone by.

Lush pastures framed with fence rows, small wooded lots, and a criss-cross of bubbling streams have now surrendered to the quietness of the forest. The old farms are gone and the old log homes and barns have disappeared along with the hardy men and women who owned them. A man's farm was his most prized possession, next to a good workhorse, an iron plow, and his wife. If a man took care of his farm, the farm took care of the family. The winters were long and hard; the family stocked their shelves and bins in the cellar, pantry, and cookhouse. Hard work made the difference between life and death.

But eventually came the progression of oil well drilling from Titusville and the Colonel Drake well in 1859 to the Parker's Landing oil fields. Drilling along the Allegheny River at Foxburg, Parker's Landing, Karns City, Petrolia, Lawrenceburg, and up the Clarion River to Grass Flats were all leading the "wildcatters" to Antwerp City and the new oil territories yet unknown.

This history reveals how it took many discoveries, wells, towns, inventions, and people willing to "take a chance," in order to open the oil fields in Clarion, Armstrong, Venango and Butler counties.

Northwest Pennsylvania has thousands of acres of abandoned oil fields. The pump houses, powers, engines, boilers, derricks, well-pulling poles, and old wood oil tanks have rotted away. Engines and boilers have long ago exploded. Pipe and shackle lines rust into the earth. The scrap drives of World Wars I and II have taken their toll. Nature has slowly returned with orchards of thorn apple, large stands of maple, cherry and oak, and finished the job with wild rose and jagger bushes. Old well roads and foot paths are now only an occasional rut in the woods. Cut stones that once held the poor driller's shanty, or maybe the finest oil town hotel, have been stolen away to now hold a younger man's dream. The rows of old time cellars are now shallow depressions filled with the debris of years and are the fading evidence of our glorious oil boom towns.

Perhaps the most famous Clarion County boom town was Antwerp City. The "old girl" exploded into a short life and created much excitement for all parties involved. People from all parts of the oil region tried to pump new life into her but, alas, she was gone. This is a history of those early days of oil.

TO THE AFFLICTED.

# OIL ON THE BRAIN

*The Yankees boast that they make clocks*
*Which "just beat all creation:"*
*They never made one could keep time*
*With our great speculation.*
*Our stocks, like clocks, go with a spring,*
*Wind up, run down again:*
*But all our strikes are sure to cause,*
*"Oil on the brain."*

*Stock's par, stock's up,*
*Then on the wane:*
*Ev'rybody's troubled with*
*"Oil on the brain."*

## SONG AND CHORUS.

BY

## EASTBURN.

3½

O. DITSON & CO., BOSTON.

New York: S. T. GORDON.......Boston: O. DITSON & CO.



ELECTROTYPED BY L. JOHNSON & CO., PHILADELPHIA.

### A song written in 1864 by Joseph E. Winner titled "I've Got Oil On The Brain," goes like this:

The Yankees boast that they make clocks, Which "just beat all creation;"
They never made one could keep time, With our great speculation.
Our stocks, like clocks, go up with a spring, Wind up, run down again;
But all our strikes are sure to cause "oil on the brain."

Chorus: Stock's par, stock's up, Then on the wane;
Everybody's troubled with "oil on the brain."

There's various kinds of oil afloat, Cod liver, Castor, Sweet;
Which tend to make a sick man well, And set him on his feet.
But ours a curious feat performs; We just a well obtain,
And set the people crazy with "oil on the brain."

There's neighbor Smith, a poor young man, Who couldn't raise a dime;
Had clothes which boasted many rents, And took his "mip" on time.
But now he's clad in dandy style, Sports diamonds, kids, and cane;
And his success was owing to "oil on the brain."

Miss Simple drives her coach-and-four, And dresses in high style;
And Mister Shoddy courts her strong, Because her "Dad's struck ile."
Her jewels, laces, velvets, silks, Of which she is so vain,
Were bought by "Dad" the time he had "oil on the brain."

You meet a friend upon the street; He greets you with a smile,
And tells you, in a hurried way, He's "just gone into ile."
He buttonholes you half an hour; Of course you can't complain,
For you can see the fellow has "oil on the brain."

The lawyers, doctors, hatters, clerks, Industrious and lazy,
Have put their money all in stocks, In fact, have gone "Oil Crazy."
They'd better stick to briefs and pills, Hot irons, ink and pen;
Or they will "kick the bucket" from "oil on the brain."

Poor Mrs. Jones was taken ill; The doctors gave her up.
They lost the confidence they had; In lancet, leech, and cup.
"Affliction osre long time she bore," Physicians were in vain;
And at last expired of "oil on the brain."

There's "Maple Shade," "Excelsior," "Bull Creek," "Big Tank," "Dalzell,"
And "Keystone," "Star," "Venango," "Briggs," "Organic," and "Farrell,"
"Petroleum," "Saint Nicholas," "Cornplanter," "New Creek Vein."
Sure, 'tis no wonder many have "oil on the brain."

This book is dedicated

to

***The Afflicted with***

# *Oil on the Brain,*

those men who ate, drank,
and slept oil.

*From left: Charles Vensel (driller and oil producer, oldest person to enlist in the Spanish-American War); Melissa Vensel (daughter of Joel and Eve Fink); John J. Ashbaugh; Hannah Vensel Ashbaugh. Photograph taken at Ashbaugh home.*

## THANKS BOYS

Some things in life seem so strange. Anyone who has written their family history, looked for a long-lost and forgotten grave, or searched for that missing piece of information, so important to the completion of the story, will understand these thoughts.

Either I have been very lucky in compiling my facts and forgotten stories, or I've had assistance from many helpful, friendly but long since departed friends.

My thanks to all those "old timers" who have helped in keeping up my spirits, especially my good friends Charles Vensel, Joel Fink, and John J. Ashbaugh. Job well done, boys!

– *Gary S. McKinney*

# Oil Discovered in Clarion County And The Marcus Hulings Antwerp No. 1

Ever since I was a little boy, I've heard tidbits of verbal history mixed with tales about local oil history and the ghost town of Antwerp City.

I had also seen extensive volumes written about Colonel Drake, Titusville, Pithole, Rouseville, Petroleum Center, and the Venango oil fields, but came to realize that the development of Clarion County's oil district also played a very important role in the history of oil in Northwestern Pennsylvania.

## First Oil Well in Clarion County

In the year 1810, Colonel Thomas Watson drilled a salt well at the mouth of Deer Creek, where it empties into the Clarion River. The Colonel's two sons, Abraham and James, used a "spring pole" to drive the well down. After some time, the supply of salt water failed and they drilled deeper. Late one night as they drilled by torch light, the tools dropped into a void and almost immediately the well took fire destroying the "spring pole" and surrounding buildings. The mysterious yellow fluid was flammable when brought into contact with fire. Salt water was refined into salt and used in both the preparation and preservation of food. With the salt water polluted, the enterprise was abandoned. Native Indians skimmed oil from springs but the Colonel unintentionally drilled the first oil well.

As the years passed by, other owners of the Colonel Watson property continued to collect the salty light colored fluid. Mr. Job Packer took a sample to Pittsburgh for testing, but was told it lacked good qualities. The property was bought by David Whitehill and the nuisance hole was plugged. Later the well was unplugged when Jacob Hahn, Charles Hahn and Jesse Thompson leased the property in 1860. They had heard of the Drake well and the excitement in the Oil Creek Valley and were determined to create their own excitement. The Civil War broke out and oil was temporarily forgotten.

After the war, other speculators tried their hand in the oil business using "spring poles" and crude pumping apparatuses around Deer, Piney, Canoe and Beaver Creeks. After a few barrels of petroleum had been collected, their high hopes of discovery and wealth were soon gone.

## Richland Township's First Attempt at Obtaining Petroleum

1860 was the year that Isaac Neely, William Neely, Daniel Bostaph, Alexander Wilson, Henry Barr, Edward Murray and others started putting down a well on the Isaac Neely farm two miles east of Petersburg. Again the news of the discovery of oil at Titusville fueled the excitement. The tools were very crude and the drilling was slow. The "spring pole" was used and later a horse was hooked up to ease the work.

The Civil War broke out and the drilling was stopped. After the war, a new set of tools was secured and again the drilling resumed. It must be noted that work on this well was done only after all the farm work was completed, but alas the tools were stuck and lost.

A small amount of gas was found but no oil, and finally with much disappointment the venture was abandoned as a failure. This is the first well ever put down in Richland Township specifically in search for petroleum. Twelve years later this farm proved to be rich in oil

*Drilling a well by using a spring pole.*
*Sketches in Crude Oil, McLaurin*

and many successful wells were drilled there.

Also in 1860, Thomas McConnell, W. D. Robinson, Smith K. Campbell, and Colonel J. B. Finlay formed the Foxburg Oil Company. They purchased two acres of land on the west bank of the Allegheny, near Tom's Run, from Elisha Robinson and began to put down a well. After reaching a depth of 460 feet, an accident occurred bringing a halt to their drilling. Before they could resume, the Civil War broke out and thoughts of continuing the well were abandoned. Later, those same men bought 100 more acres from Mr. Robinson for $50 an acre, a price considered far above its value. A 30-acre tract in turn was sold for $20,000 to a group of Philadelphia men who formed the Clarion and Allegheny Oil Company and put down a well, striking oil in October 1865, the first producing well of that region. Thus was introduced the Parker oil field, the largest and richest known. Significantly, the original well which had been abandoned with the outbreak of the war was eventually proven to be dry territory and had it been so determined originally, would have marked the beginning and the end of oil discovery in that district.

## Fox And Fink

In the fall of 1865, Samuel M. Fox, of the Fox Farm, and Joel Fink, of St. Petersburg, had drilled their well No. 1 on the east side of the Allegheny River on the "Reed" tract. The oil was of good quality, but small quantity. "In those days of Pithole with its 500 and 1,000 barrelers, a well which pumped four or five barrels was deemed almost worthless, and this received little attention. However, this well drilled by Samuel Fox and Joel Fink is considered Clarion County's first staying well of illuminating petroleum. In the summer of 1866, Fox and Fink put down well No. 2, also a small producer. No. 3 was drilled in 1867 and yielded little. In 1869 Joel Fink moved the rig down river from the old stone house, which now only sports its chimney, and on September 30, 1869, No. 4 came in at 15 barrels.

*Railroad siding for loading tank cars at Graham's Landing in Perry Township, Clarion County, directly across the Allegheny River from Parker's Landing*
*Courtesy Drake Well Museum*

In May of 1869 in the heat of the Parker's Landing excitement, a well was struck in the gulch at the mouth of the Clarion River, where it empties into the Allegheny, known as Graham's Landing. The Graham's Landing Oil Company was owned by R. L. Brown, Simon Truby, W. H. H. Piper, William Robinson, and Robert Crawford. It was soon followed by a 15-barrel per day well struck by Duncan Karns on the James Pollock farm on the "heights" back from the river. This in turn was followed by the "Buckeye," a 40-barrel well below Graham's Landing. Attention was then directed northward toward Foxburg where a few wells of small importance had been put down on the Fox Farm.

This created some excitement and soon a number of wells were underway. These included the "Mead Well," the "Elephant," and the "Island Queen." The Galey No. 1 was drilled on Grass Flats, a short distance up from the mouth of the Clarion River and hit in August 1871. Others on the Grass Flats ranged from 30 to 80 barrels.

## Parker City
## From Allegheny River Landing
## To Important Oil Center

## Parker's Landing

When the oil fields around Parker's Landing boomed, they brought together a most motley crowd. Parker's Landing, just like all oil towns, produced a very mixed crowd. It also produced many rich farmers, who for years before had barely scratched a living from the soil. The Parker, Robinson, Fox and other families in the area were rich enough to live a good life without oil, but when the "black gold" was forced on them, they very meekly accepted it. The three Robinson brothers were among the richest both before and after oil hit big. Each had a large farm and every acre was good oil territory.

*Old Times in Oildom* by George W. Brown, dated 1909, is a good source of oil history. Mr. Brown owned and operated a lumber yard in Parker City.

*View of Parker's Landing taken from Devil's Rock in 1870. Note the large house boat in the photograph is believed to be Ben Hogans Palace.*
*Venango County Historical Society*

*Main Street, Parker's Landing, 1870*
*Courtesy Drake Well Museum*

*Main Street, Parker's Landing, looking North from the Fuller House*
*Courtesy Drake Well Museum*

*Ferry crossing the Allegheny River at Parker's Landing. The Phillips House and the Allegheny Valley Railroad office are Across the river.*
*Courtesy Drake Well Museum*

*View from Devil's Rock with Graham's Landing below and Parker's Landing across the Allegheny River*
*Courtesy Drake Well Museum*

*Toll booth for ferry on the Clarion County side of the Allegheny River and view of Parker's Landing*
*Courtesy Drake Well Museum*

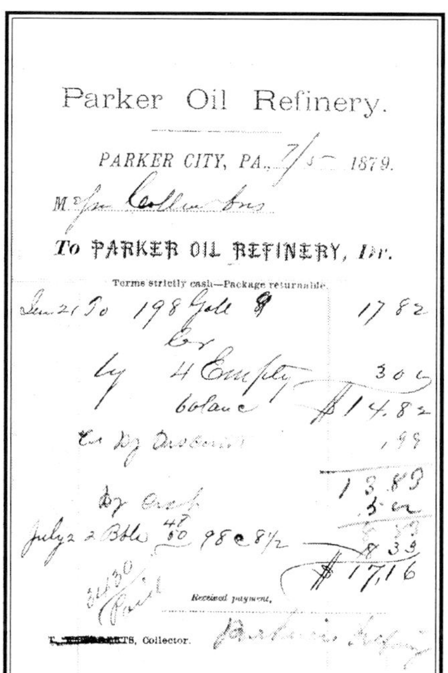

Parker Oil Refinery.

PARKER CITY, PA., 7/5 1879.

To PARKER OIL REFINERY, Dr.

Terms strictly cash—Package returnable.

Received payment,

Collector.

Parker Oil Refinery receipt from 1879

*"Mt. Vernon," the first well drilled on "the bluff" above Parker*
*Courtesy Drake Well Museum*

*The "Eagle" well above Parker's Landing*
*Courtesy Drake Well Museum*

*This elevator carried people from downtown Parker City up to the Bluff.*

## Clarion River

The oil rich Clarion River valley was known by the Indians as "Tobeco," a corruption of the Indian words "tupeek-hanne," meaning a stream that flows from a large spring, and the waters were known for its clearness and sweetness. As early as 1809 the river was called Toby or Stump Creek by early pioneers. Later two surveyors, Daniel Stanard and David Lawson, were camped along the banks of the river, and while lying in their tent listened to the sounds of the distant ripples. Stanard made mention that the waters sounded like a "distant clarion," thus by 1819 the name Clarion River was first used officially.

Many reports were recorded years ago about giant fish in the Clarion River. A fish in one case was reported to be "longer than the canoe I was in." Every year sturgeon worked their way out of the Allegheny River and up the Clarion River from the mouth in order to spawn. With oil wells and coal mines polluting the once clear waters, the giant sturgeon came no more.

The waters of this beautiful river flow through Clarion County toward its mouth but in reverse the discovery of oil starts at Clarion's mouth and works its way up the river to the Grass Flats, then jumps to Antwerp City, and onward throughout the Clarion Field. The following history of Stump Creek Island begins this progression.

## The Island Oil Field

Long Tree, or more commonly called Stump Creek Island, is the island laying between the Fox farm on the east side of the Allegheny River and the Fox and Robinson farms on the west. The island is opposite the mouth of the Clarion

River, one mile up river from Parker's Landing. Stump Creek Island proved to be very profitable to the owners of the oil wells located on the narrow strip of oil territory. While hundreds were drilling for oil on hilltops and in valleys and the earth was sending forth rich streams of wealth, there were those whose energy and capital induced them to locate a spot which had been utterly neglected.

***Wells on Stump Creek Island at the mouth of the Clarion River***
***Courtesy Drake Well Museum***

Who would have thought of drilling wells on an island in the river? At high water, the island is completely submerged. The surface, even at low water, is little above the level of the river, and hence to rig derricks and sink wells upon such territory would, to say the least, be assuming a great risk. The Allegheny River is known for terrible ice floods and rapid current, and everything in its path is submerged and swept away. To sink wells upon such an unprotected spot and invest many thousands of dollars with the prospect of everything being destroyed was assuming an immense risk, but well-known operators O. E. Shannon and William Hartley entertained the idea of drilling wells on the island. They were not of that class to be easily alarmed and in the fall of 1869 they put down the first well, called the "Point," which came in at 100 barrels a day.

John H. Galey, a businessman from Parker's Landing, was looking for a piece of the oil action and was offered the lease on Stump Creek Island for $10,000. Mr. Galey decided that $10,000 worth of oil could be pumped from the "Point" well, so he borrowed the money in Pittsburgh and bought the island lease. A second well, the "Island Queen," was put down by J. H. Galey, Dower and Co. in the summer of 1870, was torpedoed three times, and was making 30 barrels. Within 18 months, John Galey had made over $125,000 profit over and above his original investment. To further show the extensive operations on this small island, the following wells were drilled:

3rd and 4th. "Contest," double rig owned by William Hartley and Co. and both torpedoed once producing together 40 barrels.

5th. The "Island King," owned by William Hartley, John H. Galey and Co., was torpedoed on September 11, 1870 and averaged 70 barrels.

6th. The "Island King No. 2," owned by Hartley, Galey and Co. was 90 feet deep. They were very unfortunate, having to drive three strings of pipe 40 feet.

7th. The "Weller" well, owned by Weller, Crawford, John Morgan and Son, commenced drilling on October 14, 1870, was 40 feet deep and drove 45 feet of pipe.

8th. The "Douglass" well owned by O. E. Shannon and Co., of Emlenton, down 680 feet.

9th. The "Harry" well, owned by O. E. Shannon and Co.

10th. The "Ark" well, owned by O. E. Shannon and Co. "The derrick is up and will soon be in operation." (July 29, 1871)

In order to prevent these wells from being totally wrecked and swept away by ice and high water, E. O. Shannon spent nearly $4,000 of his own money to ensure the safety of the equipment. Large wooden cribs were built on the extreme upper end of the island to break the force of the ice. The derricks, engines, tanks, etc., were placed upon a frame work of logs, thus to a great extent rendering them more secure. One wooden crib was 130 feet long by 12 feet high and the other which shielded the "King No. 1" well was 50 feet long by 10 feet high. The only building on the island was a boarding house for drillers run by a Mr. Blair.

I have driven past this island for many years and never realized the commerce and oil excitement that had happened there in days passed by. Now tall walnut trees grow and deer, bear and copperheads work the silent wells.

## Oilmen's Journal, March 10, 1876

This article was discovered in a Parker City newspaper, *The Journal*, dated March 10, 1876. This article shows us how a man's lot in life can change overnight, especially in the "oil patch."

"ARRESTED – John H. Galey was recently arrested by officer McCandless on warrants issued by Stoughton, Duffy, and Alex Mitchell, charged, we understand, with attempting to leave the country for the purpose of defrauding his creditors. He was taken before the court on Tuesday, but the attention of that tribunal was so much taken up with the consideration of applications for license that no other business could be attended. Consequently the hearing of Mr. Galey was postponed. He has for years been one of the most prominent and extensive oil operators of the Butler, Clarion, and Parker oil regions. Whether he is guilty of the charges preferred or whether his creditors were unnecessarily alarmed remains to be seen. He will remain in the custody of the officer until the day of the hearing."

In this year 2002, 126 years later, Mr. Galey is, as far as we know, still in the custody of the officer. — Author's note (See page 289)

## Marcus Hulings

"About this time operators began to see the fallacy of the theory that oil would be found only in river bottoms and rigs began to climb the hillsides." (*History of Clarion County*, Davis, 1887 page 351). The pioneer who pushed developments back from the Clarion River to St. Petersburg was Marcus Hulings. In the fall of 1871 the discovery of the famous oil well, "Antwerp No.1," was drilled by Hulings and started the oil excitement that totally and completely engulfed St. Petersburg and surrounding areas.

*Marcus Hulings*
*Courtesy Drake Well Museum*

"Marcus Hulings, a leader in the world of petroleum, was born October 7, 1823, near Phillipsburg, Centre County, and began his career as a producer in 1860. For some years he had been a contractor and builder and he turned his practical knowledge of mechanics to good account. His earliest oil venture was a well on the Allegheny River above Oil City, for which he refused $60,000. To be nearer the producing fields, he removed to Emlenton and resided there a number of years. The Hulings family had been identified with Venango County from the first settlement, one of them establishing a ferry at Franklin a century ago. Prior to that date the family owned and lived on what is now Duncan's Island, at the junction of the Susquehanna and Juniata Rivers, 15 miles northwest of Harrisburg. Marcus was a pathfinder in Forest County and opened the Clarion region. He leased Clark & Babcock's 6,000 acres in McKean County and drilled hundreds of paying wells. Deciding to locate at Oil City, he built an elegant home on the South Side and bought a delightful place in Crawford County for a summer residence. His liberality, enterprise and energy seemed inexhaustible. He donated a magnificent hall to Allegheny College, Meadville, aided churches and schools, relieved the poor and was active in political affairs. Besides his vast oil interests, he had mines in Arizona and California, mills on the Pacific Coast, and huge lumber tracts in West Virginia. Self-posed and self-reliant, daring yet prudent, brave and trustworthy, he was one of the grandest representatives of the petroleum industry." (*Sketches in Crude Oil,* McLaurin, 1896).

Time has now come to give Marcus Hulings the credit and respect he most certainly deserves, here in Clarion County. He died January 26, 1892 and is buried beside his wife, Margaret, in the Grove Hill Cemetery, Oil City, Pennsylvania.

*Believed to be Marcus Hulings's Antwerp Well #1*

Marcus Hulings' grandfather was Michael Hulings of France. "Each summer he would skim a quart or two of 'earth-oil' from a tiny pond, formed by damming a bit of the creek, the fluid serving as a liniment and medicine. This was the small beginning of one whose relative and namesake, two generations later, was to rank as a leading oil millionaire." Michael Hulings died August 9, 1797 at the age of 27 years. His wife, Massar, died February 9, 1813 aged 67 years. (*Sketches in Crude Oil*).

Marcus Hulings was well known not only for his ability to open new oil regions, but also for his interest in local communities. In Emlenton, Pennsylvania, the Hulings Hose Company was named in honor of Marcus Hulings and was organized on the 17th of September, 1877. The newspaper in Emlenton was established in 1859 and was a very shaky venture for years. In the mid-1870s, Marcus Hulings and his partner in journalism, D. D. Moriarty, bought the *Emlenton Times.* This co-partnership lasted about one year and "Mr. H," as he was called, became sole owner. Drilling wells was more important then than publishing a newspaper and ownership was turned over to his son, A. A. Hulings. The name was changed to the *Emlenton Register* and had a weekly printing of 700 papers every Friday. As of 1879 the paper was still a paying enterprise.

On June 17, 1875, the first meeting of the Emlenton, Shippenville, and Clarion Railroad was held with Marcus Hulings as a member of the board of directors. This narrow gauge railroad was first built from Emlenton to Turkey City, a distance of seven miles. Then in December, 1876, a contract was given to extend on through the oil fields to Edenburg (now Knox), and shortly thereafter it was further extended to Clarion. This railroad did much business and handled both freight into, and oil out of, the expanding Clarion fields. The Emlenton, Shippenville and Clarion Railroad ran a distance of 30 miles throughout the oil fields. Stations were built at Emlenton, Turkey City, Monroeville, Pickwick, Jefferson, Beaver City, Edenburg, Elk

City, Shippenville, and Clarion. The 1879 *History of Venango County, Pennsylvania*, published by J. A. Caldwell, states "this railroad runs through a fine agricultural, lumber and coal region and will always prove a fine investment even should the Clarion District fail in the production of oil."

During the short time "Mr. H." lived in Emlenton, he was an elder of the Presbyterian Church there. The church was organized on January 13, 1858, and the first church was built in 1859. The present brick church building was built in 1874 and dedicated August 27, 1875.

Marcus Hulings followed each new oil field and thus lived in many different places. It is known that he built a house and lived in Antwerp City prior to living in Emlenton.

## Marcus Hulings and Jonathon Watson

Jonathon Watson's association with the petroleum industry goes back to the beginning of the industry. Jonathon came to Titusville, Pennsylvania, in 1845 when the town was just a crossroads in the woods. Mr. Watson was a hard-working lumber man and ran a very successful lumber business. When Colonel Drake came to Titusville in 1859 with his plans to drill for oil, Jonathon Watson stated, "I have no faith in the project myself, but I'm going to help him through with his venture." After the Drake Well came in, he hastened to secure or lease lands all along Oil Creek. The lumber business quickly turned to oil and at eight o'clock on that historic Monday morning, Jonathon Watson quickly leased 350 acres on the oil rich McClintock farm and drilled the third well to strike oil in the Oil Creek Valley. Watson's wells hit big, the riches poured in, and in time he became the Pennsylvania region's first oil millionaire. Mr. Watson pushed the edge of the oil territories miles ahead of the producing wells and had much luck in opening new regions. He was a believer in spiritualism and an Oil City clairvoyant helped him locate some of his best wells. On one well, Mr. Watson asked for a 300-barrel-a-day well. The clairvoyant picked a good spot and the well came in at 330 barrels. I guess that would make me a believer too!!

*Jonathon Watson (1819-1894), oil's first millionaire. He would sell everything, accumulating almost three million dollars by 1864, and retire to Rochester, only to find the lure of the oil industry too powerful to ignore. He would return to Titusville, build a "story book" mansion, and go on to lose his fortune. Courtesy Drake Well Museum*

The more a student of Clarion County oil history researches the information, the more we see how important the Titusville connection actually was in the development of our fields. Marcus Hulings and Jonathon Watson were friends and Watson bought in 1871 a three-sixteenth interest in the Antwerp No.1 well and its 20-acre lease in Clarion County. Mr. Watson accumulated a fortune of about 3 million dollars, but due to dry holes, bad management, marital problems, and bad investments, his fortune was gone. He died in 1894, broke. In the end he was as dependent upon financial assistance as Colonel Drake had been before him. He is buried in Woodlawn Cemetery in Titusville, Pennsylvania.

The following is an agreement between Marcus Hulings and Jonathon Watson. Through oil men such as Hulings, Watson, and others, Antwerp City and the Petersburg oil fields began to expand.

## Agreement

Know all men by these presents that I, Marcus Hulings of Oil City, Venango County and state of Pennsylvania, for value received by my full satisfaction, I did sell and convey for J. S. and S. McCauley, their heirs and assigns, Three sixteenth interest in a lease for oil purposes, of twenty acres of land of John J. Ashbaugh situated in Richland Township, Clarion Co. and state of Pennsylvania with one well fully equipped for pumping. The royalty on said lease granted from one tenth to one quarter of the oil in certain portions of the land as in said lease set up.

And whereas J. S. and S. McCauley for the consideration of the sum of two thousand dollars by money paid to them by Jonathan Watson of Titusville, Crawford Co., Penna.. did sell to said Watson his heirs and assigns the three sixteenths interest in said lease and well not the production from the time the well was first tubed and transferred to him my agreement for the same as of date October 28th 1871 when the money was paid which agreement was surrendered to one by Mr. Watson and my agreement given to him to make the proper transfer to him for said interest dated November 27th, 1871. That for in consideration of the foregoing I do for myself my heirs executors and assigns bargain and convey and by their presents have bargained, sold, and conveyed to said Jonathan Watson his heirs and assigns the full three sixteenths part and interest in a lease from John J. Ashbaugh to me, my heirs and assigns for the term of fifteen years with one well now pumping of said lease known as the Antwerp Well No. 1. Fully and properly paid with the same interest on the production of the working interest of said well Antwerp No. 1 from the commencement and subject to all conditions of said lease, as working the same oil proportion by said Watson interest form the time the well began to produce or was tubed said J. S. and S. McCauley having paid me in full for said interest before selling the same to said Watson.

In witness whereof I have hereunto set my hand and seal the first day of December AD 1871." Signed Marcus Hulings. Signed in the presence of S. G. Bayne. Signed Jonathan Watson.

The foregoing is recorded on page 395 of leases in Clarion County Courthouse, Clarion, Pennsylvania, dated February 13, 1872.

## Sand Pumpings

(Reprinted from *Petroleum Monthly*, page 70 Vol. I, 1870-72, Oil City, Penna.)

St. Petersburg and its immediate surroundings, which is just now creating a great deal of excitement among oil men, is the scene of the principal oil development in the oil region. It is located in Richland township, Clarion county, and about two miles from Foxburg, the nearest station on the A.V.R.R. It is reached by a stage line, which runs in connection with all up and down trains. The town of St. Petersburg, before oil was struck in its vicinity, only numbered about a dozen houses, which were owned by the well-to-do farmers who tilled the surrounding farms. The country is fertile and very productive, presenting as fine rolling farming land as can be found in the State. The town now numbers about 1,000 inhabitants, mostly oil men and the employees on the wells. There are two good hotels in the place, several groceries, hardware and dry good stores, restaurants, etc. It is a busy, thriving little town, which is growing fast, and will probably double its population in the next three months.

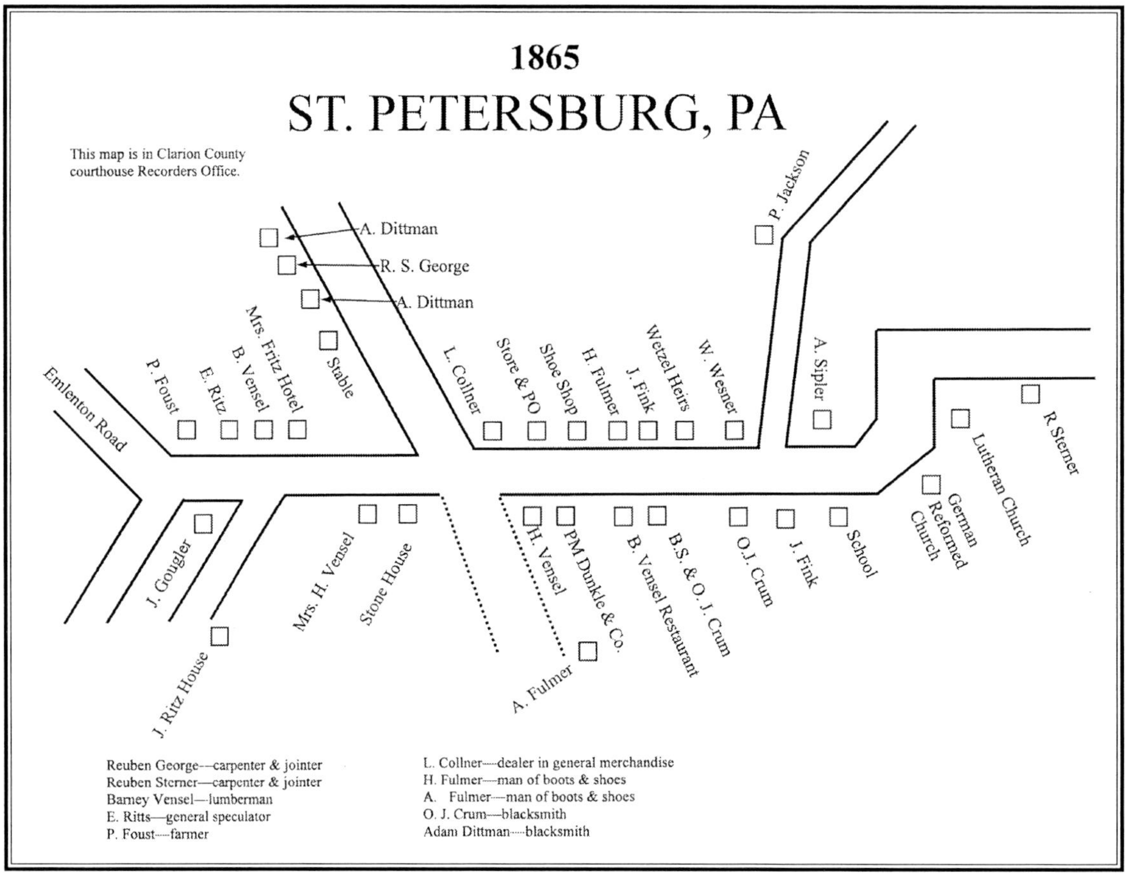


***1865 map of St. Petersburg***

The oil developments, so far, are principally on the south, eastern and northeastern sides of the town. This field is an extension of the Parker's Landing oil fields. The belt is supposed to run from the islands at the mouth of the Clarion river in a northeasterly direction along the banks of that river as far as Grass Flats, when it veers off more to the north. The wells at Foxburg are thought to be on the northern edge of the belt, and the Huling's well at Keating's Furnace on Turkey Run, which proved a dry hole, on the southeastern boundary of the belt.

Commencing at Foxburg and the territory lying between that place and the Clarion River, we will give the farms and tracts as they come in rotation from that point along the belt to the termination of the development.

On the Fox Estate, which comprises some 1,200 acres, lying around the station of Foxburg, there are a number of wells, of which Messrs. Fertig & Hammond have four, producing an average of ten barrels per day, each, on a twenty-acre lease. Those owned by the heirs of the Fox Estate are averaging about 15 barrels each per day.

The Rupert farm is next to the Fox Estate to the northeast, and has 8 or 10 wells, averaging ten barrels each.

On the Whitling farm G. R. Harris has ten wells in different stages of drilling.

The David Shoup farm adjoins the Whitling. On this the Palmer well is producing sixty barrels, although when first struck it pumped 150. David Young has a ten-barrel well on this farm, which two are the only producing wells. There are no wells drilling.

On the Collner farm is the Moriarty well, one of the best in the district, which has, for the past three months, produced 80 barrels per day, with no variation from the first day of its completion. Fleming & McKay have a 40-barrel well on the same territory. On the Barney Vensel tract, same farm, is the Frothingham well, now down to the third sand, with 500 feet of oil in the hole. Ralph & Riddle have a hundred-barrel well on the Johnas Fink tract, same farm.

*Hiram Neely. In 1807 the Conrad Moyer family settled on the E. Ritts farm, the second family to settle in that area. About the same time, William Porter settled on the land now known as the Hiram Neely farm. On this farm Richard Peters, the Revolutionary Secretary of State, had a small house and lived by himself for many years. Hiram Neely's farm was rich in petroleum.*

Grass Flats commences with the Galey tract, and is, so far, the most productive quarter of the Petersburg region. Walker & Bros. Have a well on the Galey tract, which is doing thirty barrels. Hall & Andrews also have a well on the same tract, producing 15 barrels. The Galey wells Nos. 1 and 2, are doing about 15 barrels each. Galey well No. 1, known as the Downer well, was the first well on Grass Flats, and, in fact, the first well in this region, having been completed about the first of last August. This well was the cause of the rise of St. Petersburg. There was a number of other wells on the same tract, among which is the Galloway well, east of that owned by Hall & Andrews, which started on 75 barrels, but is now doing only 35 barrels. The Parsons Bros.' Well is doing about 60 barrels per day, notwithstanding the report of its producing 100 barrels.

On the Keating farm is the Derousse well, pumping about 50 barrels. There is another small well on the upper end of the Keating farm, which is the farthest well east, and is about six miles from Parker's.

On the Daniel Ritts farm, which is north of the Keating, Siggins & Blanchard struck a good well on Monday (22d), which is flowing at the rate of 150 barrels. The Jackson, Vensel & Shoup well, on the same farm, has been pumping on an average of 30 barrels per day for the last ten days.

Bly & Andrews have six or seven wells drilling on the line of the C. Ritts and Ashbaugh farms, one of which was tubed yesterday, the 23d, and is pumping at the rate of 65 barrels. On the C. Ritts farm is the Nottingham well No. 1, which is doing 50 barrels. No. 2 is drilling. Patrick's well is producing ten barrels.

The Ashbaugh farm is the next to the north, and has only one producing well, the Hulings, or Antwerp well, which was finished in September, and has been producing 75 barrels per day since then. There are about 20 wells going down on this farm.

To the east is the Eli Ritts farm, on which a number of wells are drilling. Several will be completed in a few days.

The Neely farm adjoins the Eli Ritts on the east. The Neely well, owned by Amer & Nicholson, is located on this tract.

Coming back toward Foxburg, north of St. Petersburg and the above farms, the first is Collner farm No. 2. Adjoining this is the Foust farm, on which is a number of good wells. The Goss & Gould well is doing about ten barrels. The Big Indian, or Hamilton well, was struck in August, and produces 20 barrels a day. The Evans well, of which we have made mention, is on this farm, and is pumping something over 75 barrels per day. This well is a little south of St. Petersburg.

There are about 100 wells (or about one-quarter the whole number drilling in the oil region) in different stages of drilling at present in this region, many of which will be completed in ten or fifteen days, and as they are some distance form the producing wells, the extent and capacity of the territory will be then fully demonstrated.

## Development of the Petersburg Oil Field

The old village of Petersburg is located one mile above the Clarion River about two miles from Foxburg station on the Allegheny Valley Railroad, and was never disturbed by an oil excitement until the Hulings well began to flow on the Ashbaugh farm. News of this strike spread and was soon followed by a flight of operators from Titusville. These men began to invest around the Hulings property and in a few days the prices of land rose from the old conventional rates for farming purposes to $1,000 per acre. (*Titusville Morning Herald*, November 15, 1871)

Overnight, Petersburg went from a small village with a few buildings, log homes, and one general store, to a booming, almost out-of-control city. It took about ten years to understand the pattern of the oil route to be discovered from Titusville down through the Allegheny River Valley to the new Petersburg field. Many of the old founding families of this area went overnight from struggling farmers to wealthy oil people. Some of these families were the Collners, Ashbaughs, Finks, Vensels, Fousts, Shoups, Logues, Rittses, Keatings, and Neelys.

## Daniel Ashbaugh Farm

In the oil days, the Ashbaugh name was very well known in the St. Petersburg area. The Daniel Ashbaugh family came to this area in the early 1800s and was the first family to settle here. The original land owned by the Ashbaughs consisted of approximately 400 acres and was bought from Judge Richard Peters, for whom Petersburg was named. The 400 acres were eventually divided into four farms, each owned by an Ashbaugh heir. These farms are just east of St. Petersburg in Richland Township, Clarion County. The farms, in a row, are known today as:

First, the Whitling farm, later the C. W. Biery farm, which in 1903 consisted of 80 acres. This farm was sold by Daniel Ashbaugh on August 6, 1844, to George Whitling. George cut wood and made charcoal which he sold to the Keating blast furnace. It is said that George owned the first threshing machine in Clarion County.

Second, the John Ashbaugh farm, passed on to his son, Henry.

Third, the Adam Ashbaugh farm.

Fourth, the Jacob Ashbaugh farm, now owned by Gary S. McKinney, author of this book and local historian. John J. Ashbaugh bought this farm from his father, Jacob Ashbaugh, for $1,000 in 1870. Jacob and his two wives are buried in St. Petersburg. Jacob was born in 1800 and died in 1889.

The following is the deed from Jacob Ashbaugh to his son, John J., dated September 7, 1870. Little could anyone have imagined that so much excitement was about to explode on the old farm. I've always said that it was "a good thing that oil was discovered because they would have finally starved to death trying to farm it." Daniel Ashbaugh is buried in the Union Cemetery, St. Petersburg, Penna.

## Agreement

Article of Agreement – Jacob Ashbaugh and John J. Ashbaugh

Article of agreement made and concluded this seventh day of September AD one thousand eight hundred and seventy (1870) between Jacob Ashbaugh of the township of Richland and county of Clarion and state of Pennsylvania of the first part and John J. Ashbaugh his son of same township county and state of the second part. The party of the first part doth hereby sell all his right interest and claims in deed to a certain piece of land situated in Richland township Clarion county and state of Pennsylvania being part of land of his father, Daniel Ashbaugh bounded on the north by Adam Ashbaugh south by Charles Ritz and Danial Ritz east by Eli Ritz and west by German reformed church property containing about eighty-seven acres. In consideration the party of the second part has paid unto the said Jacob Ashbaugh the sum of one thousand dollars (1000.00) the receipt whereof is hereby acknowledged as soon as possible the party of the first part will make a good and sufficient deed as to the time performance of all this agreement both parties have here unto set their hand and seals day and date above written.

Signed: Jacob Ashbaugh; John J. Ashbaugh

Witness: C. Klotz, Justice Peace; Albert Vensel

Received this date of John J. Ashbaugh one thousand dollars being the consideration money in full this 7[th] day of September 1870.

Witness: C. Klotz, Justice Peace

This farm has been owned by Daniel Ashbaugh, Jacob Ashbaugh, John J. Ashbaugh, Howard Ashbaugh, Bill Cunchula, Ben Pesevento and presently, in 2003, by Gary McKinney. All deeds and information can be found in the Clarion County Courthouse, Clarion, Pennsylvania.

The John J. Ashbaugh farm is the closest to St. Petersburg and the house and barn were actually within the borough limits. On this farm and in particular the 20-acre lease is where the Antwerp No. 1 was drilled and the city of Antwerp was constructed.

## St. Petersburg Business Directory
## 1877 *Clarion County Atlas*

Arnold, P., Druggist
Aggers, H. C., Contractor and Producer
Ashbaugh, J. J., Oil Operator
Ashbaugh & Soliday, Teamsters
Ayers, Henry, Laborer
Abbinson, Jos.
Ashbaugh & Vensel
Aldine, A. A., Merchant
Ashner, A. J., Oil Operator
Aiken, Jerome, Hotel Keeper
Adams, F. G., Grocer
Barclay, F. H., Oil Producer and Real Estate Agent
Blakslee Bros., (J. B. Blakslee, C. F. Blaklee, W. Z. Blakslee) General Machinists and Dealers in Boilers and Tubing. Pipe Line and Oil Well Suplies at reduced prices.
Butt, G. W., Oil Producer
Beals, H., Butcher. Dealer in Fresh Meats of all kinds. Shop nearly opposite Chalfar.
Best, John, Carpenter
Brewer, Leroy B.
Beardsley, E. C.
Berine, Wm., Oil Operator
Bond, Robert, Butcher
Burnapp, G. W., Machinist
Bryan, Robert
Brown, Thomas
Bush, John E., Laborer
Bayne, S. G., Oil Operator

Blair, S. Q., Contractor
Blair, John A., Teamster
Burk & Hall, Grocers
Burk, M. M., Grocer
Brainard, C. A., Laborer
Brice, John, Laborer
Barger, L. F., Bookkeeper
Brocket, T. H., Oil Operator
Blackall, James
Bliley Bros., Tinners
Cauley, D. M., Oil Operator
Cortelyor, John, Agent
Collner & Bros., H., Wholesale and Retail Dealers in Dry Goods, Notions, Millinery, Hats, Caps, Flannels, Shawls, Socks, Car pets, Oil Cloths, Trunks and Satchels.
Brick Block, Main Street, St. Petersburg
Collner, C., Merchant
Collner, J., Merchant
Cramer, L. B.
Chappell, Henry, Tailor
Chalfant & Graff, Hardwaremen
Collner, Louis
Collins, Wm., Carpenter
Coulter, R. C., Oil Operator
Cook, Smith, Oil Operator
Colwell, W., Pumper
Cruthers, Alex., Driller
Coulter, J. E., Clerk
Dutton, Thos., Operator and Producer
Dunbar, I. C., Carpenter
Dittman, Adam, Blacksmith
Derolph, John, Carpenter
Dunning, O., Carpenter
Dinsmore, John, Miner
Dunbar, Robert, Constable
Derouse, William, Pumper
Derouse, Henry, Oil Operator
Dower, Peter, Oil Operator
Dittman, Jr., John, Pumper
Dorn, S. B., Physician
Demaux, ___, Clerk
Davis, L. A., Carpenter
Dooley, Philip, Policeman
Edinger, S. J., Oil Producer
Ewing, ___, Carpenter
Eno, H. W., Stonemason
Eno, J. H., Stonemason
Fink, Joel, Oil Producer
Foust, Jr., Philip, Oil Producer and Operator of Coal Works
Finley, W. P.
Foust, Solomon
Fullerton, H. R.
Fulmer, Henry, Pumper
Foust, Benjamin
Foster, Wm., Physician
Ferris, M. B., Carpenter
Foust, Samuel, Teamster
Foust, Joseph, Farmer
Fitzpatrick, Edward, Blacksmith
Foust, Sr., Philip, Farmer
Ginley, J. W., Agent
Griffin, Thomas & Co., Manufacturers and Repairers of Boilers. Shop opposite the Monitor House.
Guffey, J. M., Oil Operator and Producer
Geer, L. R., Harness Maker
Goldsborough, H. C., Oil Buyer
Gerard, J. B., Oil Operator
Godfrey, A. O., Wagon Maker
Gibner & Carter, Merchants
Goodmiller, A., Tailor
George, West, Bookkeeper
Hill, B. B., Druggist
Hardison, Jas. H., Oil Operator
Hurlbut, H. S., Carpenter and Contractor
Howell, P.
Hanshaw, Marion, Carpenter
Houston, Robert, Carpenter
Hill & Tyler
Herron, M., Merchant
Harwick, Frank
Hawk, John
Haslett, James, Grocer
Hoffman, E. H., Grocer
Hugus, Simon, Oil Operator
Hagan, Simeon, Pumper
Hardison, Simeon, Oil Operator
Humphrey, T. K., Miner

Hatch, J. W., Shoemaker
Houder, John, Helper
Hamilton, D. G, Carpenter
Hagan, W. F., Tool Dresser
Harley, Daniel, Teamster
Hays, N.
Herron, L. S.
Jones, David, Agent for Safety Valve. Head quarters Monitor House
Johnston, John
Johnston, R. H., Pumper
Johnston, D.
Jackson, Peter
Johnston & Wilson
Kisley, M.
Komer, Theo., Oil Producer
Kratzer, Paul, Oil Producer
King, Charles, Hotel
Kratzer, Paul, Oil Operator
Keating, John, Oil Producer
Kane, Patrick, Laborer
Kerr, Geo, Tool Dresser
Kennard, Robert, Blacksmith
Kelley, Peter, Teamster
Lawson, Miss Maggie, Milliner and Dress-maker, three doors below Post Office
Leiblick, Sr., Geo., Baker
Lobough, Samuel, Teamster
Loyd, Thomas, Hotel Keeper
Luke, McClaine, Miner
Luke, Albert, Teamster
Levegue, Joseph, Painter
Lockwood, A., Horse Trainer
Little, C. W., Wholesale and Retail Dealer in Watches, jewelry and Diamonds. Pure fine Toilet Articles.
Mahan, J. W., Proprietor New York House. This house, being refitted anew, guests will find a hearty welcome. Table the best the market can afford
McCutcheon Bros., Grocers
McGee, J. J., Contractor
McSweeney, H., Attorney-at-Law
Morgan, P. W. Opera House. In connection with the above is an excellent Hotel and Billiard Parlor. Guests will be promptly attended to. Rates Moderate.
Mathews, G. W.
McElwain, J. B., Merchant
Moore, C. A.
Messinger, Chancey, Lumber Dealer
Mason, A. F., Oil Operator
Munroe, Wm., Oil Operator
Ming, Wm., Restaurant
Maxler, J. A., Oil Operator
Mathias & McCalley, Pumpers
Miller, James
Mackey, L. J. W., Driller
Mahl, Paul, Carpenter
McCamet, J. G.
Myers & Aiken, Hotel
Minich, Charles, Hackman
Neely & Scott, Printers
Nolder, John, Tool Dresser
Neely, R. H., Laborer
Nevil, Jacob
Nash, M. C.
Obet, Andrew
Orr, Robert
Place, J. M.
Phillips, David, Blacksmith
Powel, Wm., Boot and Shoemaker. Men's Fine Boots and Ladies' Fine Shoes a specialty. Shop opposite Monitor House
Painter, M. L., Oil Producer
Perkins' Heirs, F. A.
Painter, Israel, Oil Operator
Peacock & Roof
Robinson, James M., Oil Well Cleaner
Reichart, C., Justice of the Peace and Surveyor
Ryan, John W., Oil Operator and Real Estate Agent
Ritts, Elias, Oil Producer, Farmer and Stock
Raiser. All oil wells on farm producing.
Ritts, J. V., Cashier of St. Petersburg Savings Bank

Roschy, Frederick, Manufacturer of Carriages, Buggies, Spring and Farm Wagons. Repairing neatly done. Shop west end of town, near Monitor House.
Riggall, Robert. Monitor House. Being refitted and refurnished, we wish to call the attention to the public that we are prepared to accommodate guests comfortably. Good stabling attached.
Runkle, A., Teamster
Ritts, Sr., Heirs, John
Ritts, Charles, Oil Operator
Robinson, James, Gauger
Rugh, M. J., Oil Operator
Ruffner, Geo., Shoemaker
Rouley, M. L., Driller
Ritts, Harry, Teamster
Ratell, Joseph, Laborer
Shoemaker, L. L., Principal of school
Shulenberger, Rev. W. C. B., Pastor of Reformed Church
Springer, J. M., Superintending Oil Wells, Elk City
Siggins, L. W., Contractor
Smith, J. G., Driller
Sargent, G., Proprietor of Sargent's Dining Hall and Restaurant. Best brands of Cigars and Wines at the bar.
Smith, H. C., Oil Producer
Skeel & Beebe, Manufacturers of Skeel Folding Clothes Rack, Mops and Washboards. For sale by the dozen or half-dozen.
Address the above at Hayfield, Crawford Co.
Shoemaker, G. W., and Morison, A. H., Dealers in Fine Groceries, Provisions, Fruits, Cutlery, Wood and Willow /Ware. Store opposite St. Petersburg
Stoltz, J. J., Practical Boot and Shoemaker. I am prepared to do any kind of work. I keep a large stock of my own work on hand. Waterproof Boots a specialty. Seed bags, etc. etc. Shop East End, nearly opposite Lutheran Church
Sith, A. H., Pumper
Sipler, Chas., Laborer
Sterner, Reuben, Carpenter
Sipler, C. M., Constable
Smith, Wm.
Sheerer Bros., Merchants
Shell, E.
Smith, John
Shoup, Miles, Carpenter
Shoup, John
Samuels, Charles, Miner
Snyder, Frank, Blacksmith
Schoener, Julius, Baker
Silverburg, D., Merchant Tailor
Soliday, A., Oil Operator
Stofer, Samuel, Blacksmith
Shoup, John D.
Snow, J. H.
Terney, M.
Truby, J. H.
Taylor, L. F., Painter
Trace, A. W.
Taylor, J. P., Carpenter
Thompson, John W., Agent for Celebrated
Estey Organs, Decker & Bros. Pianos, and Haines Bros and Fisher Pianos; also, Sheet Music, Music Books and Pic-
ture Frames. Instruments tuned on short notice. Instruments sold on monthly payments.
Thompson, C. H., of firm of Thompson & Riggall, proprietors of Monitor House
Urban, John, Laborer
Vensel, C. H., Oil Producer
Vensel, Jacob
Vensel, Barney, Oil Operator
Weatherbee, E. H., Blacksmith. Horse shoeing a specialty. Shop opposite Monitor House
Wilson, S. E., Wholesale Dealer in Foreign and Domestic Liquors.
Watt & McElwaine, Dealers in Hardware and Oil Well Supplies
Wise & Gilger, Manufacturers of Boots and Shoes. Men's Calfskin Boots and Ladies' Fine Shoes a specialty. Repairing done on short notice. Shop opposite Chalpant & Graft's Hardware Store.

Wise, L. B., Laborer
Wilson, A.
Wing, W. O., Bookkeeper
Weeter, Levi, Teamster
Westerman, Louis
Wing & Wilber
Whitling, Daniel, Jeweler
Wachter, H. C., Oil Operator
Wilson, Robert M.
Whitling, F.
Walley, Thomas
Westner's Heirs, Wm.
Yozer, Frank A., Printer
Youst, C. K.

***This lithograph, dated 1877, is from Caldwell's Clarion County Atlas. This was the home and barnyard on the John J. Ashbaugh farm. Note the oil derrick in the vicinity of Antwerp City. During the oil excitement, John J. Ashbaugh caught a transient trying to climb through the window to the right. This home burned about 3:00 in the morning sometime about 1965. The Ashbaughs sold the farm to Bill and Violet Cunchula, who in turn sold to the Swarthout family. The farm is now owned by the Gary McKinney family.***

## It Will Never Last

Charles H. Cramer, later a hotel proprietor in Pittsburgh, left the Butler field to drill the Antwerp well, in which he had a quarter interest. James M. Lambing, for whom he had been drilling, jokingly remarked, "When you return 'broke' from the wildcat well on the Ashbaugh farm I will have another job for you." It illustrated the ups and downs of the oil business of the 70s to note that, when the well was completed, Lambing had met with financial reverses and Cramer was in a position to give out jobs on his own hook. (*Sketches in Crude Oil,* McLaurin, 1896)

People flocked to St. Petersburg by the hundreds and then by the thousands until that sleepy little village became the largest town in Clarion County. St. Petersburg's "Hay Day" was probably between 1872 and 1882. There were years when "the oil flowed and the good times rolled," followed by years of just "hanging in there." The years between 1882 and 1910 were years of slow to fast decline. As the oil dried up, people and businesses moved on to the next field.

*St. Petersburg, 1872*
*Main Street, St. Petersburg, looking West from Church Street.*
*Second building from left was The Progress newspaper.*
*All this burned in the fire of February 25, 1873*
*Courtesy Barbara L. Hajel*

## Fires in St. Petersburg

The first fire occurred October 10, 1872, at eight o'clock in the morning. It broke out at the east end of the town at the house of Mr. Patton, who had rented it to four families. There were nine houses burned, two torn down, and one rolled over the hill.

The following account of the October 1872 fire is taken from the *Titusville Morning Herald* of October 11, 1872.

## "Destructive Fire at St. Petersburg – Thirteen Buildings Burned – Loss $8,650"

About half-past six o'clock this morning our citizens were alarmed by a cry of fire from the extreme eastern end of the town and in less than half an hour one thousand people were on the ground, and all that could were actively engaged in rendering assistance. The chance of saving any of the houses near by was slight, as the town has no facilities for fighting fire, not even water. The following is a list of the property destroyed, with names of owners and amount of loss, as near as could be ascertained:

L. D. Patton house where the fire broke out; occupied as a dwelling by L. D. Patton, A. Layman, Widow Pratt, and E. Hasley. Total loss $2,600; Patton's loss $1,300, insured in the People's Company of Philadelphia.

Next west on the street was the Travelers' Home, owned by L. P. Rogers of Foxburg, and occupied by J. Ratell as a dwelling; Roger's loss $650; Ratell's loss on household goods estimated at $75; amount of insurance unknown.

Two story building owned by J. Rishell and occupied by S. T. Taylor as a dwelling; Loss $315; no insurance.

D. Gottschalk, tin store and dwelling, Loss $900; insured for $500 in the Republic of Philadelphia.

Dwelling of widow Ritts, Loss $200, no insurance. This building was torn down.

Widow Sesston's dwelling, Loss $100, no insurance. This was left standing but partially torn in pieces.

W. H. Whitting, oyster saloon and dwelling. Loss $400; no insurance. This building was torn down.

*Fields of Derricks – Clarion County farms went from pasture fields to oil fields overnight. This photograph represents a typical scene in the Petersburg Field. Actual photograph is Red Hot, Venango County. Courtesy Drake Well Museum*

Next east of Patton was the bakery and dwelling of George Sible. Loss $1,000; no insurance.

Wm. McCarty, carpenter shop and dwelling upstairs. Loss $500; no insurance.

John Malloney, shoe shop and dwelling. Loss $600; no insurance.

J. Rena, ice cream saloon and dwelling house owned by J. J. Ashbaugh; total loss $350; no insurance

Unoccupied building, owned by W. H. Steffey; loss estimated at $200.

On the opposite side of the road, one building owned and occupied as a dwelling by R. Slarnes, was burned; loss $800, no insurance.

A meeting has been called and a fire and hook and ladder company will be formed at once.

*Cornfield Turned Oil Field – This photograph represents that the David Shoup farm might have looked like in the summer of 1873. The Shoup farm was typical of the tranquil, rural settings which burst forth with oil wells. The well in the foreground is being pumped, the shed at the left housing a steam boiler, an engine and a small vent pipe. Starting up the hill, one can see a typical oil field house with two children in the doorway. An outhouse is in the back. Actual photograph is the Shaffer farm. Courtesy Drake Well Museum*

## Making Fun

The October 14, 1872 issue of the *Herald* newspaper reported:

*The Progress* in its report of the recent fire says: On the floor of one of the houses was heaped some five or six bushels of apples. Soon after the building caught fire the proprietor came in with a wooden bucket and commenced to fill it with withered and squashy apples. When told to leave the apples go he insisted that they cost him "fivf and swantzig cent." At this time the house was in flames. Another went through a room containing many valuable pieces of household furniture, and carried out the dishpan. Two men went into a building and after effectually overturning the breakfast table, proceeded to carry out a hot cooking stove. They did, too. Had there been at the breaking out of this fire ten barrels of water in the vicinity, it could have been checked. We have called the attention of the public to this matter frequently, and must attribute the loss to the indifference of the people.

This account of the aftermath of the fire appears in *Sketches in Crude Oil*: The behavior of some people in a crisis is a wonderful puzzle, sometimes funnier than a pig circus. At the St. Petersburg fire, which sent half the town up in smoke, an old woman rescued from the Adams House, with a bag of money containing $400, was indignant that her fifty-cent spectacles had been left to burn. A male guest stormed over the loss of his satchel, which a servant had carried into the street, and threatened a suit for damages. The satchel was found and opened. It had a pair of dirty socks, two dirty collars, a comb and a toothbrush! The man with presence of mind to throw his mother-in-law from the fourth-story window and carry a feather pillow down stairs was not on hand. St. Petersburg had no four-story buildings.

## More Fire

The *Titusville Morning Herald* of October 24, 1872, carried this article titled:

## "Another Fire—The New Water Works"

About six o'clock this morning the St. Petersburg hotel was found to be on fire, the origin presumedly a defective flue. The fire was speedily extinguished; although creating at the time a great deal of excitement. The recurrence of fires will hasten the building of the water works that have been decided upon, as all feel the insecurity of the town in case a fire should obtain any headway.

***Oil Field Fire (Harper's Weekly). A sketch of the fire scene by W. W. Goodwin captures the drama as spectators look on helplessly. This scene was an oilman's nightmare. If one well caught fire, the possibility of other wells catching fire was a real threat.***

The second fire broke out at the west end of the town, at half past six o'clock on the evening of February 21, 1873, starting in Fred Hepp's saloon, adjoining the opera house. In an instant the entire building was in flames, which shot high into the air, enveloping the opera house in a sheet of fire from end to end. The European House adjoining it took fire, also the Boston clothing house, Hall's restaurant, Seymour & Taylor's hardware store, Mat Colwell's billiard room, William Richard's barber shop, Michalisky's dry goods store, Wilson's dry goods store, Jamestown clothing store, Mr. Ritts's dwelling, Johringer's jewelry store, Wilcox & Butt's furniture store, J. H. Welsh's wholesale liquor store, post office, and Smith Cook's new residence.

On the opposite side of the street the fire commenced with the Savings Bank, spreading eastward to Barney Vensel's office and residence, thence to the store of R. W. DeHaven, and to St. Petersburg Hotel, then jumped across a wide street and caught Lewis Collner's residence, then leaped two hundred feet to D. Herron's office, taking the entire row of buildings to *The Progress* office. By the noble efforts of the fire department and the aid of thousands of people who flocked from Antwerp, Turkey City, Richmond, and Foxburg, the fire was extinguished, but not until it had destroyed nearly the entire business part of the town, together with many fine residences. In all, about 40 houses were destroyed. The loss amounted to $200,000. The burnt portion of the town was rapidly rebuilt. So great was the business activity that new lumber was hauled upon the dying embers. (Read Fire Article, February 25, 1873, page 171)

*Mr. and Mrs. Barney Vensel. Mr. Vensel was an oil driller and producer.*

## Fred Hepp

Concerning the fire of February 21, 1873, the following appears in *Romance of American Petroleum and Gas*, Derrick Publishing Co., Oil City, Penna., Vol. 1, pp. 155-156, 1911:

On the night of February 21, 1873 fire broke out in the saloon of Fred Hepp, St. Petersburg, Clarion county, shortly after 10 o'clock. At three o'clock on the morning of Washington's Birthday the town was a mass of blazing and smoking ruins. More than 500 families were homeless on a bitterly cold winter day. Such is the summary of a typical oil town conflagration. The flames from the little groggery communicated to the big Cramer Opera House adjoining. There was no fire fighting apparatus in the town and no adequate water supply. There had been too much haste in building the town and developing the surrounding oil territory to give thought to protection from fire. The Opera House burned like tinder. A high wind from the west drove the flames up the main street and they traveled at amazing speed. Bucket brigades worked with frenzied energy, willing hands carried out household goods, merchandise, anything portable and worth while. But the fire increased in fury and spread in area. An attempt at midnight to make a gap by tearing down a corner building only resulted in strewing the debris across both streets for the flames to cross easily. Finally the destruction was arrested by tearing down two buildings, covering the roofs of the next with wet salt and curtaining the sides with quilts, kept saturated with brine. Three score of buildings were saved, but $200,000 worth of property had been destroyed in four hours. The men from Antwerp, a mile away, were heroes in the battle against the flames, and each of them took home with him some of the shivering, homeless sufferers. Rebuilding had begun on the evening of the twenty-second. September 3rd, less than seven months afterward, Antwerp burned, with the exception of three buildings and was never rebuilt. The men of St. Petersburg returned the winter favor, with less success. The loss of the smaller town was about $50,000.

The destruction of St. Petersbutg was blamed on Fred Hepp. Refer to February 25, 1872 edition of the *Morning Herald* on Hepp's troubles.

Another fire occurred in April, 1881. Several temporary houses, most of them vacant, were burned, but as no important buildings were included, the loss was consequently not heavy. In May, 1884, the small schoolhouse and the Presbyterian Church were burned. They were not rebuilt. (*History of Clarion County* by A. J. Davis, 1887)

## The Fire Fiend

Fire breaks out in the St. James Hotel – Refer to article later in this book, April 24, 1873. The last destructive fire occurred in 1892, when lightning struck the opera house.

## Wickedest Town in Oildom

Antwerp City was located on the west side of the ravine, "Antwerp Hollow," through which runs a small stream called Myers Run. The town was built over the rounded top of the hill, cleared by the axe of an early German settler. Thus sprung up in a night the magic town of Antwerp City. Antwerp City started with a cluster of pumping wells surrounded by dry holes. Farms and stony tracts of scrub timber sold at fabulous prices. John J. Ashbaugh leased the part of his farm where Antwerp City would grow and was paid well over $80,000 with more monies promised. New streets were laid out with Main Street being the road which is now Route 58 east toward Richmond. Other streets were laid out up and down the rolling hill. The new main street with its busy throng and many stores, saloons, and hotels stood upon the ground which but a fortnight before was a field of pine stumps.

Men of every description, rich and poor, honest and dishonest, stupid and intelligent, invaded the new town. Antwerp City became a town of over 2,500 in population. A wooden boardwalk was built from St. Petersburg down the J. J. Ashbaugh Hill and out the flat to Antwerp City so people could move more easily without all the mud.

Antwerp City was built solidly on both sides of the hill and along the flat or "Pike." St. Petersburg newspapers of 1872 have advertisements of several hotels in Antwerp City, such as "The Mortimer House and "The Antwerp City Hotel." There are also ads for two large hardware stores and many drug stores, eateries, and groceries. (*St. Petersburg Centennial History Book*, 1972)

The *Derrick* reported that Antwerp City was the "wickedest town in oildom," where gambling dens flourished with open doors and packed houses. Drunken brawls were an every-day occurrence. There was too much oil excitement to think of discipline. Restaurants sprung up with earth floors, sky roofs and two boards laid side by side for tables. Businesses of all kinds flourished in groves by the roadside. The old "Pike," which for years had carried the family wagon on its way to church or the mill, now sustained two continuous lines of vehicles moving in opposite directions and bearing all manner of freight.

Along the roads were wooden troughs built to collect water for thirsty horses wherever there was a spring. Drillers and their families lived in small shanties which were nothing more than small wooden shacks that were located on the oil lease near a well. They were double-boarded with tarred burlap roofs, a small stove to cook on, and lit with kerosene lights. It is important to note that there were many accidents, burns and deaths due to the fact that the oil for the lights had not been refined properly. Too much gas left in the oil or kerosene caused lamps to explode. Strict measures were set in place and refiners paid strict attention to them.

## Sad Accidents

Many men were killed, burned to death, drilling wells because the drill bits would cut through the cap rock and release the pressure of natural gas. This natural gas shot up the well under pressure. At this point an explosion occurred and drillers were burned to death. Victor Gretter was one of the spectators of the oil flowing over the derrick. The waste suggested to him the idea of the oil saver, which he patented. (*Sketches in Crude Oil,* McLaurin, 1896)

Sad accidents happened before drillers learned how to manage a flowing oil well with casing in it. At Frank Fertig's well, Antwerp, a man was burned to death. The burning of the Shoup & Vensel well at Turkey City cost three lives and led to an indignation meeting at St. Petersburg to protest against casing. Danger from its use was soon removed by victor Gretter's invention of the oil saver. Gretter, a small, dark-haired, dark-eyed man, lived at St. Petersburg. He was an inventive genius and a joker of the first water. His oil-saver doubtless saved many lives by preventing gas and oil from escaping when a vein was tapped, and coming in contact with the tool dresser's fire in the derrick. (*Sketches in Crude Oil, page 256, McLaurin*).

***Titusville Morning Herald,* November 23, 1871**, "Frightful Accident – Narrow Escape From a Horrible Death":

On Tuesday last a driller named A. B. Rockwood employed on the "Wizard" well near Antwerp, met with a fearful accident, which resulted in a narrow escape from instant death. He was drawing the tubing from the well, and while in the act of arranging the rope on the bullwheel, his right hand glove was caught under the rope and his fingers entangled. In this position he was drawn over the wheel, which made from twenty to thirty revolutions before the machinery was stopped. Mr. Rockwood was then extracted by his fellow workmen, and though dreadfully bruised from the effects of the accident, no bones were broken. He may be confined for some time, but he is exceedingly grateful at his wonderful escape from a shocking death.

In the spring of 1872 the Armstrong well was completed and began flowing over the top of the derrick. While flowing furiously, there came an explosion like a clap of thunder, and inquiring eyes saw the rig in flames, all one mass of seething, boiling fire. The unfortunate driller and tool dresser were on the spot and ran with their clothes on fire until willing hands snatched the burning raiment from their bodies. The burning well, shooting a column of flaming oil a hundred feet high, falling again in drops of liquid fire, was a scene of horrible beauty, but the two poor fellows who suffered by it, lay on beds of untold anguish for months and only recovered through the untiring efforts and care of those who nursed them. A few months later, the Fertigs well was burned in the same manner and the driller, poor fellow, lost his life after nine days of terrible suffering.

***Titusville Morning Herald,* May 7, 1872:**

The boiler on well No. 2, Ritz Farm, Petersburg, exploded last Friday forenoon. The boiler and engine house was shattered to atoms. The engine, derrick and tools remain intact. Peter McMahon, the tool dresser, was slightly injured in the back, having been knocked down by the concussion. No one else was hurt. The boiler was an old one, and the engineer in charge states that he carried 70 lbs. of steam and two solid cocks of water when the explosion took place. Damage $700.

***Titusville Morning Herald,* August 6, 1872**, "Horrible Accident in St. Petersburg District – Three Men Burned to Death – Full Particulars":

The people of St. Petersburg and vicinity were thrown into a state of great excitement early yesterday morning by the intelligence that a fatal accident had taken place on the Filman Farm, near Turkey Run. The news flew over the country and caused intense excitement, and hundreds of people abandoned whatever work they had in hand and hastened to the spot. The extent of the disaster was unknown, but all feared the worst. It seems that five men were at a well on the Filman farm, either as drillers or lookers-on, when the tools struck through the cap rock,

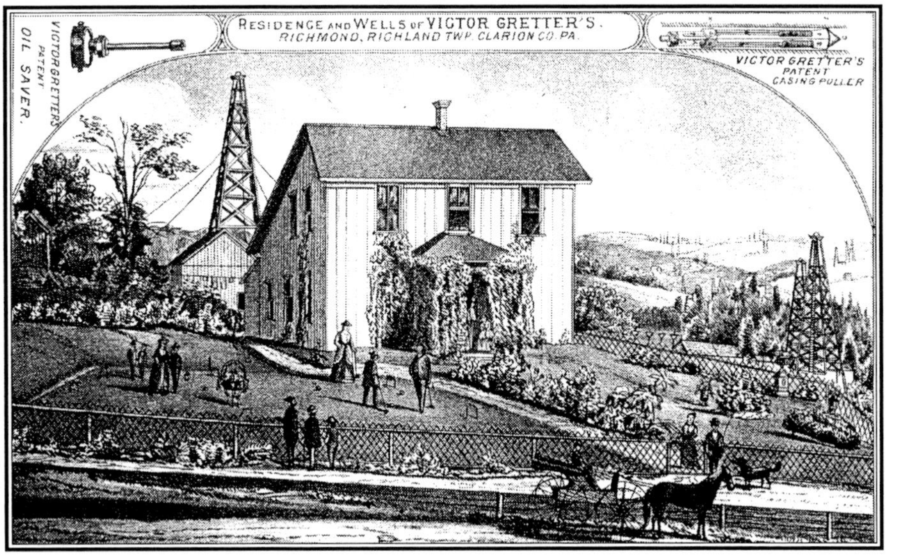


***Residence and wells of Victor Gretter in Richland Township.***
***1877 Caldwell's Clarion County Atlas***

and an immense volume of gas was liberated from its long imprisonment and rushed with great force up through the well and almost instantly communicated with the fire under the boiler and exploded with great force, enveloping the engine house and derrick. All this was the work of an instant, and the men working at the well had no time for flight, but were caught in the burning gas and three of them, Nelson Vensel, Michael Shoup, and Alexander McGee, were burned to death. Martin Frank was dangerously and John Malthorn slightly burned. The services of Doctors J. H. Sutherland and Cyrus France were at once procured and all was done for the sufferers that was possible under the circumstances. The accident cast a gloom over the place. Groups of men were seen discussing the cause and the means of preventing such occurrences in the future. It is conceded that a driller's life is often in danger, and he may be hurled into eternity without a moment's notice.

***Titusville Morning Herald*, September 9, 1872**, "An Oil Well On Fire – An Unknown Man Possible Burned – Foxburg, Sept. 7":

The derrick and engine house of the Edinger well, located near Antwerp, and in the St. Petersburg district, and owned by J. W. Irwin and others, of Petroleum Center, burned last night. No one connected with the well was injured, but it is reported that a man who was seen near there with a lantern and who, it is supposed accidently fired the well, is missing and fears are entertained that he may have been burned. The well, which is a very good one, was burning at last accounts.

***Titusville Morning Herald*, September 24, 1872**, "Explosion and Fire at Antwerp – Narrow Escape from a Large Conflagration":

At nine o'clock last night the boiler at the Coleman well, on the John Ashbaugh farm, owned by Robert Bryan and others, exploded and immediately the derrick and engine house were wrapped in flames. The noise of the explosion was heard for some distance around, and brought to the ground in less than five minutes a crowd of people variously estimated at from five hundred to one thousand in number. The Antwerp City Water Works, situated about one thousand feet from the fire, furnished a line of hose, and an abundant supply of water, which combined with the exertions of the crowd present soon extinguished the flames. Had their action not been prompt, the fire would have communicated to the tank containing two hundred and fifty barrels of oil, and had this caught it would have been impossible to have saved over a dozen adjacent wells and several dwellings. The engineer was severely, but probably not fatally, burned. It is supposed that he was asleep at the time of the explosion.

***Titusville Morning Herald*, October 18, 1872**, "A Burned Rig the Penalty of Pumping During Suspension":

Foxburg, Oct. 17 – This morning about three o'clock a well owned by S. N. Dunlap, on the Jamison farm, was destroyed by fire – the cause of which is unknown to many. The party was requested to shut down with others but flatly refused, and has now concluded he will give the carpenters a few days work and will not start up the well until he gets through, which will be after the expiration of the thirty days suspension. A few neighbors who have been stealing heads at night have resolved that it is of no use to pump it by night, lest the fire fiend might serve them in the same manner. The well being off the public road quite a distance, there is no doubt but that the owner thought his actions were unobserved. If he did he should reverse his opinion.

## Crime in Antwerp

In Antwerp, the first buildings were erected late in 1871 and in the spring of 1872, and in 60 days had the company of 4 groceries, 3 hotels, innumerable saloons, telegraph office, schoolhouse and 200 dwellings. Its over-all style was summed up by a victim of a poker game in the expressive words: "If you want a smell of brimstone before supper, go to Antwerp!" The town of Antwerp was said by the *Derrick* to have "some of the meanest drinking and gambling dens to be discovered this side the domains of his Satanic Majesty." (*Sketches in Crude Oil,* John McLaurin)

This information explains the moral status of Antwerp City and was taken from a statement dated December 18, 1872: With the lack of law, a shortage of police officers and the overwhelming numbers of people flocking to the area for new opportunities and work, Antwerp City was a pretty rough place. Punishment was dealt with quite swiftly in certain cases, in others quite slowly and in others the culprits left town. The office of the Justice of the Peace was located on East Main Street and was always a busy place. Oil men, lawyers, and businessmen were lined up across the porch, down the stairs, and out on the wooden boardwalk that lined both sides of Antwerp City's Main Street, all waiting to get their business completed. The building held the office of the Justice, which was located in the front of the building. Living quarters for his family were located on the second floor. The temporary "lock up" was located in the rear of the building. If a prisoner needed special security or was sentenced to jail for more than a few days, he was sent to the "Iron Jail" in St. Petersburg.

The following list of criminal actions for the year 1872 shows us someone was busy breaking the law:

| | | | |
|---|---|---|---|
| Malicious mischief | 4 | Public Drunkenness | 18 |
| Resisting an officer | 6 | Keeping gambling house | 3 |
| Selling liquor against the law | 18 | Nuisance | 4 |
| Assault and battery | 64 | Forcible entry | 6 |
| Robbery | 6 | Embezzlement | 5 |
| Desertion | 1 | False Pretenses | 11 |
| Larceny | 12 | Keeping house of ill-repute | 9 |

| | | | |
|---|---|---|---|
| Vagrancy | 7 | Assault and battery with intent to kill | 4 |
| Damaging property | 6 | Assault and battery with intent to Commit rape | 1 |

Civil actions entered for the year 1872 in Antwerp City: 165 (*The Golden Flood, Sin Among The Derrick*, page 226)

An old saying states that "time marches on," which it certainly does. When it came to Antwerp City, "crime marched on." In the fall of 1873 the criminal element was reported to have started a fire in the office of the Justice. Antwerp City went "hellward" and crime marched on.

Antwerp City would have been more important but for its nearness to St. Petersburg. Even the disastrous fires in 1872-73 could not prevent ranking St. Petersburg with the best towns of Oildom. Stages from Foxburg were crowded until the narrow-gauge railroad furnished improved facilities for travel. Schools, churches, hotels, newspapers, one bank, and an opera house flourished. The Pickwick Club was a famous social organization. The Collner, Shoup, Ritts, Vensel, Neely, Fulmer and Ashbaugh farms and Grass Flats produced 3,000 barrels of oil day. Oil was 5 to 6 dollars and business strode ahead like the wearer of the Seven-League Boots. Now the erstwhile busy town is back to its pristine quietude and the farms that produced oil have resumed the production of corn and grass. (*The Golden Flood, Sin Among The Derricks*, page 226)

The course of operations extended past Richland Furnace, up and beyond Turkey Run City, a dozen miles from the mouth of the Clarion River. Good wells on the Ritts and Neely farms originated Richmond, a small place that fizzled out in a year. The Irwin well, a mile further, flowed 300 barrels in September of 1872. The gas took fire and burned three men to death. The entire ravine and contiguous slopes proved desirable territory, although to the east of the famous Slicker farm, for months was second only to St. Petersburg as a frontier town. It had four stages to Foxburg, a post office, daily mail service and two passable hotels. George Washington, a newspaper reporter who took a hack at a cherry tree, might have preferred walking to the drive over the tough cut-up roads that led to and from Turkey Run City. The wells averaged eleven hundred feet, with excellent sand and loads of gas for fuel. Richard Owen and Alan Cochran, of Rouseville, opened a jackpot on the Johnson farm, above town. Wells lasted for years and his nook of the Clarion district could match pennies with any other in the business of producing oil. (*The Golden Flood, Sin Among The Derricks*, page 226)

Other "Boom Towns" that sprang up between Antwerp City and Edenburg were Richmond, Keating Furnace, Turkey Run City, Monroeville, Blairs Corners, Triangle, Jefferson City, Pickwick, Forest Home, Mongtown and Beaver City.

## Name That Town

In my search to uncover the history of Antwerp City, I often have asked the question, "Where did the name for this oil boom town originate?"

President Abraham Lincoln appointed A. W. Crawford, a resident of Emlenton, Pennsylvania, as U. S. Consul at Antwerp, Belgium, in 1861. Mr. Crawford, representing Venango County's oil interests, pushed the importation of oil into European markets.

This information comes from statements made by Mr. Crawford. In the early to mid-1860s printed in the reports of the Commercial Relations of the United States with Foreign Nations for

*Dr. A. W. Crawford, October, 1861. The first shipment of oil abroad was introduced into the Netherlands by Dr. Crawford, who was Consul at Antwerp. – Hand Book of Petroleum, P. C. Boyle*

the year ending September 30, 1861, page 176, Mr. Crawford could see great potential in marketing petroleum products and felt it was part of his duty to, "Let the oil springs and wells of the Allegheny Valley profitably let their lights shine in this direction."

Again on May 26, 1865, Mr. Crawford wrote that "petroleum has been the principal article of import from the United States and Antwerp, Belgium still takes the lead of the European markets in this branch of trade. The imports of this article this year have doubled those of the preceding, and are as follows: In 1863, 115,000 barrels; in 1864, 230,000 barrels; increase of 115,000 barrels." The preceding information shows that Mr. Crawford was quite influential in understanding the importance of Northwesten Pennsylvania crude oil. He saw the potential uses of oil and its profit-making abilities in the European market. His U. S. Consul duties were done from his office in Antwerp, Belgium, and this is where I will try to tie things together.

Before and after A. W. Crawford returned to the Emlenton area, he was good friends with Marcus Hulings, the well-known oil producer and driller. Mr. Hulings at one time lived in the Emlenton area and followed the oil fields. Marcus Hulings was well known and "well healed."

In the fall of 1871 while drilling on the Ashbaugh farm near St. Petersburg, the famous Antwerp No.1 came in big. Overnight Marcus Hulings had opened the Petersburg oil territories and laid out the new oil boom town of Antwerp City named after Antwerp, Belgium and in respect of his old friend, Mr. A. W. Crawford, consul to Antwerp and promoter of petroleum.

*Emlenton Penna., looking up river from Fullerton.*

***Emlenton in 1872. Emlenton was an established river community at the southern tip of Venango County when the oil excitement started. The covered bridge across the Allegheny was erected in 1856 and went out in the flood of 1883.***

*A Sketch of the John Crawford Family who settled near Emlenton, on the Allegheny River, in 1797*: Dr. Alexander W. Crawford, son of Alexander and Mary McMillen Crawford, settled in Allegheny Township, Butler County. He served two terms in the State Legislature and was appointed by President Lincoln in 1861 as Minister to Belgium. He was located at Antwerp and was reappointed for a second term. During his second term, his wife died and he resigned his appointment and brought her remains to be buried in the Allegheny Cemetery, at Pittsburgh. His second wife, Mrs. Jennie G. Walker, whom he married in 1872, was a daughter of the Hon. Jacob Black of Shippenville, Clarion County, Penna.; died Nov. 28, 1915, in Buffalo, N. Y. Her remains were laid away in the same plot with his first wife. When he reached Antwerp, continental Europe had no cheap illuminant, wealthy people using rape seed oil for light and the poorer folk going to bed in the dark. Not a glimmer could be seen outside the towns of the most populous region on earth. A few gallons of American refined petroleum, regarded as a curiosity, had been sent to Antwerp prior to the Doctor's arrival. He guaranteed against loss August Schmitz & Sons, a leading firm, who ordered a shipment of forty barrels of kerosene from New York, being the first oil ever imported into continental Europe. It was hard to sell the stuff in the face of deep-seated prejudice, but the firm ultimately got rid of its supply at a profit of forty francs. The oil won its way at length and 1,500,000 gallons were marketed the next year. Dr. Crawford returned home from abroad in 1866, and died July 14, 1895 at Emlenton and was buried in Allegheny Cemetery at Pittsburgh.

## Drilling, Torpedoes, and Nitroglycerine

The following information about the cost of drilling an oil well in 1870 was taken from the *Titusville Morning Herald* of December 8, 1870:

"The expense of putting down wells varies a little with the locality, and is generally in proportion to the depth, the pebble or oil bearing rock, being found all the way from six hundred feet to eleven hundred feet from the surface. Most wells are put down now-a-days with large casing, six to six and a half inches diameter. This is for the purpose of shutting off the water previous to drilling in the oil rock. The use of this casing entails an expense of from $1,000 to $1,200. In place of the old driving pipe which used to be driven to the rock, a conductor as it is termed, made of four pieces of two-inch plank, prevents the caving in of the gravel. A six or six and a half inch hole is then drilled, until it reaches below the water courses, after this the casing with seed bag, water packers, or cups attached, is let down until the water is cut off. This effectually prevents the fresh water reaching the oil-bearing crevices. The drilling now proceeds with a reduced bore of about five and a half to six inches in diameter, to the required depth, after which the tubing can be immediately let down, and pumping commenced at once. Salt water usually accompanies the oil, but unfortunately the oil does not always accompany the salt water. Small yielding wells generally give off gas sufficient to supply half the requisite fuel for working, while large producing wells furnish enough gas to run, not merely the engine which pumps them, but several others besides. Great economy is found in the use of one large boiler, to supply steam to several wells at once. The power of the engines vary from ten to twelve horse-power each. The expenses of rig and machinery are as follows:

*A driller on duty*

| | | |
|---|---|---|
| Rig | $ 500 to | $ 600 |
| Driving Pipe | 200 to | 300 |
| Engine | 600 to | 1,000 |
| Casing, 6 inch | 500 to | 1,000 |
| Tubing | 350 to | 400 |
| Sucker Rods | 100 to | 125 |
| Tools & Sundries | 400 to | 500 |
| Cost of drilling | 1,200 to | 2,000 |
| Fuel | 250 to | 300 |
| | $4,100 to | $6,225 |

These figures cover the average cost of shallow and deep wells without tankage. In some cases, however, they are put down cheaper, and in others where accidents occur, such as caving in, tools sticking, etc., they cost more money."

One method of locating a likely spot for an oil well was by use of the divining rod. The following account is from "Contemporary Accounts 1859-1872", *Early Days of Oil,* Paul H. Giddens, 1947, pages 296 and 297.

*Blacksmith's forge, inside of derrick, with perspective view of anvil, bellows, bull wheels, brake, and cast off with cord. The broken rope in the foreground is the bull rope. The sand pump is under the bull wheel. This driller is dressing the drill bit so that it had a cutting edge. This required a fire, an anvil, a couple of sledge hammers, and some muscle. They unscrewed the bit from the stem, put it in a forge, heated it to a proper cherry-red, and placed it on the anvil. With sledge hammers, they dressed the bottom of the bit, which had the cutting edge on it, out to gauge so that it would just go inside the hole they were drilling. After they tempered it in what was called the slack tub, which was filled with water and generally some salt, they put in back on the stem.*

*In this photograph taken near Pleasantville by John Mather in 1874, four men pose with some tools used for drilling. (From left: wrench, hammer and pipe wrench, drill bit, and a bailer.) Courtesy Drake Well Museum*

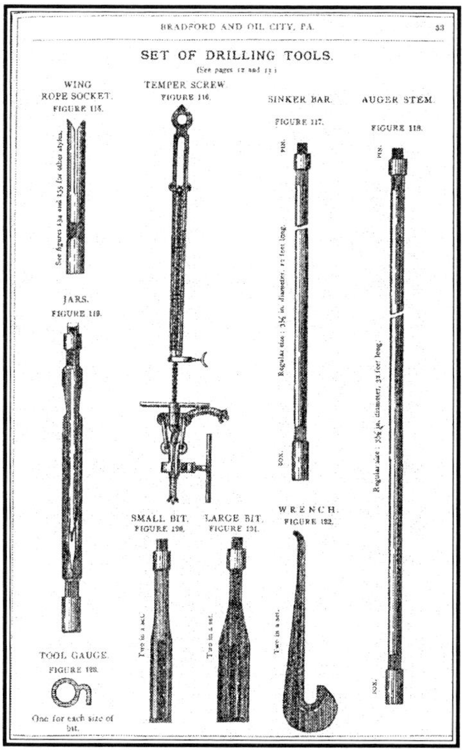


*A set of drilling tools.*

*Nitroglycerine wagon*
*Courtesy Drake Well Museum*

*A nitroglycerine wagon explosion in which Adam Cupler was killed near Titusville. Courtesy Drake Well Museum*

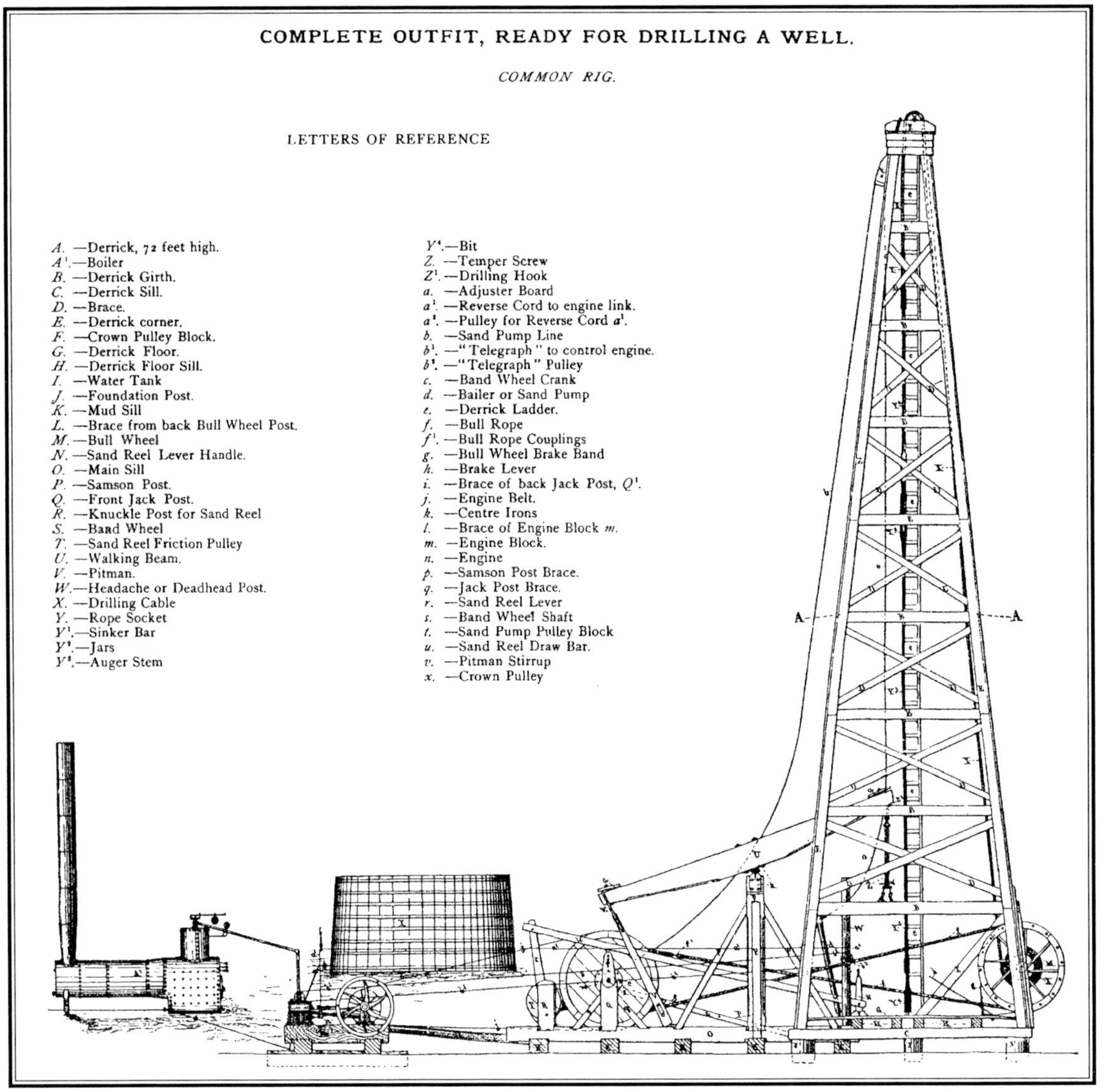


***Complete outfit, ready for drilling a well***

# Divining Rod

The use of the divining in the locating of oil wells is now being generally practiced by those mining for petroleum. The mode of operation is exceedingly simple. The expert uses a forked twig of witch hazel. This he takes by the two prongs, or ends, walks slowly over the ground where he intends to prospect, holding it in a horizontal position. The theory is, that in passing over a vein of oil the existence of the same is indicated by the action of the twig thus held, as the larger end points down toward the ground. By this mode of operating experts claim that they cannot only discover the existence, but can trace the vein of oil in any direction, locate the spot where it is most desirable to drill an oil well with great certainty of success. Many wells located by this means have proved to be good producing ones. We are not a believer in the efficacy of the divining rods, as indicating with any degree of certainty the existence of a productive oil vein. But as many of the most practical oil men have practiced this mode of locating their wells, even if they don't believe in it, their opinions are entitled to respect. The sticks used will only operate in the hands of certain persons, and with these only in places where oil exists beneath the surface. At least so it is alleged by the experts. The reason given for the vibration and turning of the divining rod is this: The party operating being possessed of an undue amount of positive electricity, this electricity acts upon the sticks as above described and has consequently some affinity with minerals beneath the earth's surface. Now the sum and substance of the whole matter is simply this: As there are no certain indications by which to locate an oil well that will prove productive, it is just as well to try the divining rod. The investment is generally moderate. And this view is the one taken by most of those boring wells.

## Roberts Torpedo

Colonel E. A. L. Roberts on November 15, 1864, filed an application for a patent to use a torpedo as a way of increasing the production of oil wells. The first public test of this torpedo was at the Ladies Well on the Watson Flats near Titusville, Pennsylvania. Many oil men claimed they were the first to use a torpedo type device, and it took many years before Roberts was granted a patent.

There are three reasons that three-fourths of all oil wells drilled do not produce. If you miss the oil veins, you have a problem; if paraffin or other materials plug the flow through the vein, you have a bigger problem; and if you have a "dry hole" and had absolutely no idea why, you, at this point in time, were broke!

A torpedo consisted of a tube about four feet long, which was made to fit the bore of the oil well. The tube was filled with black powder or nitroglycerin with a wire tied to the tube. The torpedo was carefully lowered into the well until it was secured at the desired level. An iron weight, also called a "follower," is attached to the wire and dropped. When the "follower" hits the cap, an explosion occurs. The following diagrams show how a torpedo explodes.

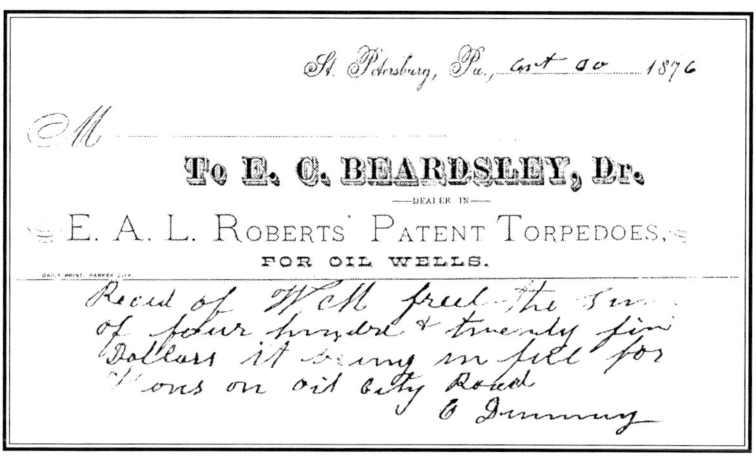
St. Petersburg, Pa., Oct 30 1876
M
To E. C. BEARDSLEY, Dr.
DEALER IN
E. A. L. ROBERTS' PATENT TORPEDOES,
FOR OIL WELLS.
Rec'd of Well [illegible] the sum of four hundred & twenty five Dollars it being in full for [illegible] on Oil City Road
C. Dunning

*Robert's Torpedoes receipt of 1876*

## Deadly Mistake

The Roberts torpedoes were exploded by dropping a weight on fulminating caps inserted in the top. W. H. Harper secured a patent on one to be exploded by caps placed at the bottom, whereas the Roberts type had the caps at the top. It was exploded by means of a plunger, passing through an inner tube, which was jerked up forcibly by the line used to lower the shell. W. H. Harper, who received a patent for his improvement in torpedoes, went to his doom on Turkey Run at Keating's Furnace, two miles from St. Petersburg, in 1875, when one of these torpedoes failed to explode. Harper pulled it up, emptied the shell, carried it far from the derrick, moved the plunger to see why it did not work, and an explosion followed, driving pieces of tin-shell deep into his flesh and through his body. Harper lived nine days with a hole entirely through his lungs, "big enough to insert two fingers in it," to quote Dr. France, the St. Petersburg surgeon who attended him. (*Sketches in Crude Oil*, page 397, and *Northwestern Pennsylvania*, page 538)

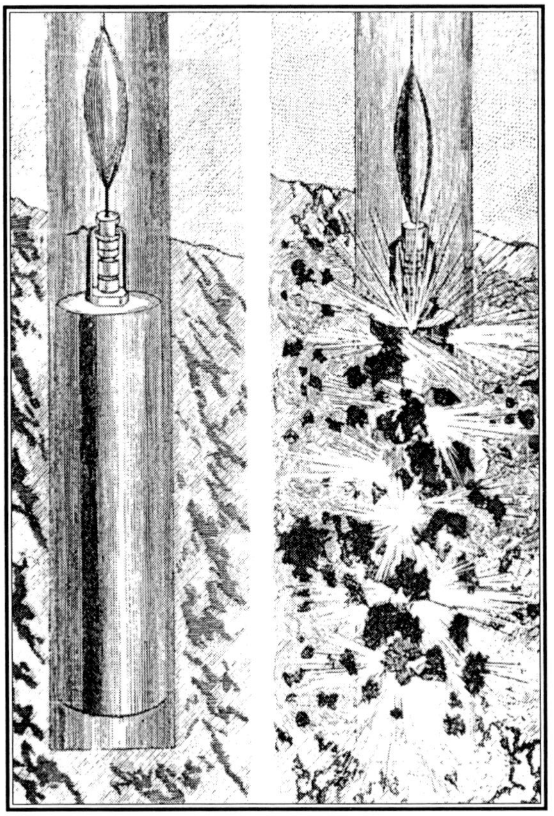

*Before Explosion* *After Explosion*

*The Roberts Torpedo*

## Blown To Atoms

James Barnum, a "shooter" in the Clarion County oil fields for the Roberts Company, was killed on February 23, 1876, a bitterly cold day, as he was transferring 300 pounds of nitroglycerine from a factory near Edenburg to his magazine near St. Petersburg. Probably benumbed by the cold, he let something drop when he went to open the magazine, which contained about 200 pounds of the explosive. The concussion was felt five miles away, broke hundreds of windows and rocked houses to their foundations. A huge iron safe was imbedded in a cave dug into the hill and acted as the repository of the explosives. Barnum had tied his team to a small tree and must have been taking the cans from the wagon to the safe. Both horses lay dead and disemboweled. The biggest piece of the luckless agent would not weigh two pounds. One of his ears was found next morning, a half-mile away. The few remains were collected in a box and buried at Franklin. Oddly, Mr. Barnum had been one of the first persons on the scene when, two years earlier, William Pine had been "blown to atoms" by some mishap of his nitroglycerine cargo between Shaw Farm and Rouseville. (*Sketches in Crude Oil*, page 396, and *Northwestern Pennsylvania*, page 542)

Prosper Wellman was a Civil War veteran who came to St. Petersburg in 1876. Here he embarked in the oil business and also the manufacture of nitroglycerine. In his new undertaking he met with a fair amount of success until one fateful day in 1902. He was blown up at Little Bridge near Grass Flats, and is buried in the cemetery at St. Petersburg. His son was Dr. Wellman, local well-known doctor.

Business in St. Petersburg kept steady pace with the growth of the population. The hotels and boarding houses were frequently unable to accommodate all the transients, and people were often forced to seek shelter for the night under the boardwalks or in barns. No less than 50 business places existed, which included not only food and clothing stores, but the various enterprises necessary to the development of the oil industry. For a time, four oil companies had offices in the town, and an oil exchange was located there. The social, educational, and spiritual interests grew in accordance with the industrial progress. St. Petersburg now is essentially a residential community with a population slightly over 500.

## Life In and Around Antwerp

Today, when a person thinks of the northwestern Pennsylvania oil fields, a word that one might use is "primitive," but that is not entirely correct. The following are a few of the goods and services available in Antwerp City in 1872. What you read just might surprise you.

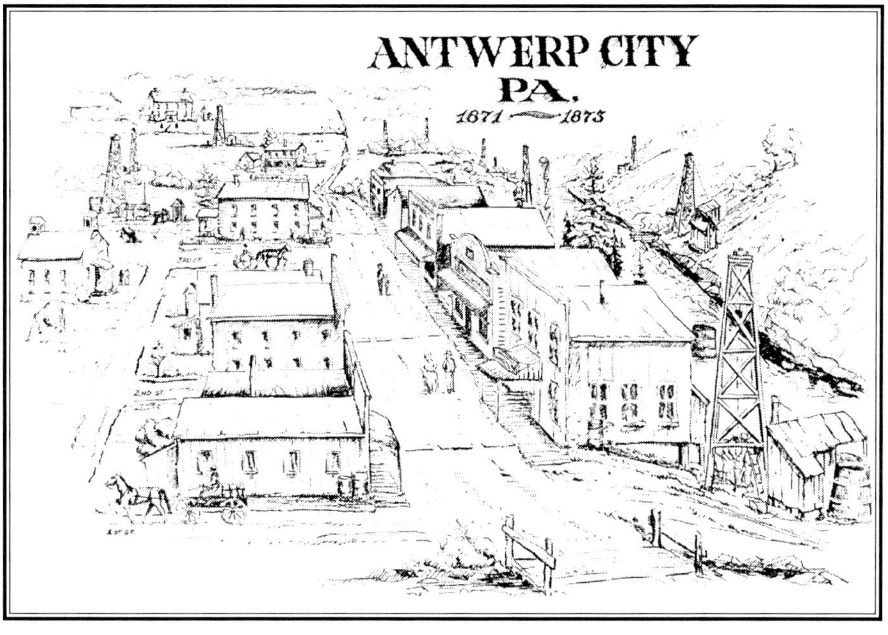


*Sketch of Antwerp City 1871-1873*

## Wash House

First of all, there was the much-needed wash house. Wash houses were quite necessary due to the fact that clothes do get dirty. Oil work is greasy and certain business women seemed to have just the right touch to get things clean. Washing work clothes was hard work, but with the right tools, such as a wooden wash tub, scrub brush, lye soap, and a good wash board, the job was made easier. Most wash houses prided themselves on just how clean they could get the laundry and word got around as to who washed clothes the best. The money paid was usually much needed and this gave many women an income.

## Barber Shop

Barber shops were another business that gave the oil men a sense of being civilized. Barber shops were usually located in or near boarding houses and hotels. One particular barber shop in Antwerp City advertised "hair cutting done in the latest styles." A good shave, shampoo, and a hot bath made new men out of many. Most shops had the familiar red and white candy cane striped barber pole out front.

The barber shops in Antwerp City had a room in the rear which contained a tin or wooden tub. After a good shave and haircut, the oil pumper could get a hot bath for 15 cents. Add another dime to that and he could have the privilege of using water no one else had used.

## Boarding House

There were numerous boarding houses in Antwerp City. Boarding houses rented rooms or only a sleeping space by the day or by the week. Some nights when beds were scarce, the boarders slept side by side on the floor. Fights broke out often over invasion of other men's space or someone snoring too loudly. Rooms were usually inexpensive and boarding houses were only for sleeping, no food was served, nor was there a bar or livery stable for horses. If the unlucky could not find a room for the night, the only alternative was to sleep under a building, inside someone's shed or barn, and if things got real tough, in someone's outhouse or beneath a boardwalk. Many unwanted boarders (or should they be called intruders), were marched from a shed or barn at the point of a loaded gun by its owner. Farmers in the Antwerp-St. Petersburg area were especially on the lookout for overnight guests in their barns. Fire was of great concern in those days and farmers were very careful not to have a careless transient smoking in a barn or shed full of hay. Henry Ashbaugh owned a farm about one mile from downtown Antwerp and many a night he chased away unwanted guests from his barn.

*Henry and Susanna Ashbaugh*

In much contrast were the hotels of Antwerp City to the "flop houses" or boarding houses scattered throughout the so-called city. The hotels of Antwerp City were located on main street, and were much more luxurious with restaurants, bars, and their own personal livery stables for its patrons.

*The hotels were full and oil men slept in chairs.*

## Antwerp Hotel

The following is an advertisement from the July 10, 1872 edition of the St. Petersburg *The Progress* newspaper:

Antwerp City Hotel – Mrs. A. Mackey, Proprietress, Formerly of the Sheridan House, Scrubgrass. This house is now open for the reception of guests, having been refitted and enlarged. I can now accommodate the public with as good rooms fare as any house in the Oil Regions. My table is supplied with all the substantial and luxuries of the season. My rooms are well furnished and attentive. Good liquors and cigars are always found at the bar. Good stabling, livery, and barber shops connected. Hacks running to all trains on the A.V.R.R. at Foxburg. This house is centrally located one mile from the Turkey Run Oil Field, and three miles from Foxburg. The patronage of the public is respectfully solicited. Mrs. A. Mackey

*(Left) The Oil Seeker – He seeks and finds.*
*(Right) The Oil Seeker – He seeks and don't find.*

A lease between Marcus Hulings and Mrs. A. Mackey, owner and proprietress of the Antwerp City Hotel on main street, is for lots No. 15 and No. 16 on which the hotel was built. Leases were sold for the building lots in Antwerp City and eventually all leases reverted back to J. J. Ashbaugh.

## Agreement

Article of Agreement made and entered into this 20th day of November AD 1871 by and between M. Hulings, party of the first part, of Emlenton, Penna. and sub-leaser of a certain piece or parcel of land on the J. J. Ashbaugh farm, Richland Twp., Clarion Co., of Pennsylvania and Mrs. A. Mackey formerly of Scrubgrass, Scrubgrass Twp. Venango Co. and state of Pennsylvania of the second part witnesseth. Let it be known that all of that certain piece or parcel of land leased by M. Hulings on farm property of John J. Ashbaugh and known and described as

Antwerp City Richland Twp. Clarion Co. state of Pennsylvania. That for and in consideration of the party of the second part has paid unto the said Marcus Hulings the sum of six hundred dollars in hand paid to the party of the first part the receipt where of is hereby acknowledged the said party of the first part hereby agree to lease to the party of the second part their heirs and assigns lots No. 15 and No. 16 according to a plot or survey made by Marcus Hulings for a term of fifteen years. The party of the second part does not have the sole and exclusive right to bore or dig for oil on said lots. Lot No. 15 and lot No. 16 are to be for the sole and exclusive purpose of building the Antwerp City hotel. Signed by M. Hulings and Mrs. A. Mackey in the presence of J. C. Patterson, J. P. – On this 19th day of June AD one thousand eight hundred and seventy-two, Marcus Hulings did appear before me one of the Justices of the peace in and for said county. J. C. Patterson, J. P., Antwerp City, Pennsylvania. Signed John J. Ashbaugh, Marcus Hulings, Mrs. A. Mackey.

## Mortimer House

Another fine hotel in Antwerp City was the Mortimer House. This advertisement is from the August 17, 1872 *The Progress* newspaper:

"The Mortimer House, Antwerp City, Clarion Co., Penna., has just been opened for the accommodation of the public. This house is first class in every respect, and its accommodations for guests are unequaled in the oil region. The best brands of cigars and the finest liquors at the bar. W. S. Mortimer and Co. proprietors."

***A typical oil town***
***Courtesy Drake Well Museum***

Another advertisement for the Mortimer House states that it has the "Finest Sleeping Rooms" available and boasts, "Those of you who wish a cool glass of ale and hear a good song are invited to call and see for yourself." Also, billiard tables were available at the tap rooms and were usually in constant use. The hotel restaurants prided themselves in the fact that they served the finest fresh oysters, clams, mussels, and all types of fresh fish including rainbow trout.

## Undertaker

When an undertaker was needed in the area, funeral parlors were located in both St. Petersburg and Antwerp City. This advertisement dated July 10, 1872, comes from *The Progress*: "First Class furniture and bedding at popular prices wholesale and retail at Wilcox and Butts, St. Petersburg, Penna. Goods delivered free and undertaking promptly attended to." Another ad states: "Call on Giebner and Carter. Furniture and bedding of all kinds. Undertaking promptly attended to. Parlors in Parker City, St. Petersburg and Antwerp City. General agents for St. Johns Sewing Machines." Another undertaking firm was McLain and Gealy of St. Petersburg, "undertaking in all its branches."

Most burials were made in the St. Petersburg cemetery. Two things Antwerp City never had were a cemetery for the dead and a church for the living. Home lay-outs were the way funerals were held in the oil era days. The embalming was done in the home and the casket was kept at the deceased's home and friends visited there. After two days and before the body showed signs of deterioration, it was buried. Caskets were handmade of stained and varnished wood with a closed top, except where the face was. Here a piece of glass showed the face only. Silver plated thumb screws held the top on and usually a silver plated name plate was screwed to the top of the box at about the chest area. These read "Mother," "Father," "Daughter," "Son," or "Friend." The casket was placed in a wooden box and buried.

The grave was usually dug by family or friends, with a small marble marker marking the spot. With so many people coming to work in the oil fields, many people died with no known families. These people were buried in what is known as the "pauper's plot" in the St. Petersburg cemetery. These poor souls had no funeral, a very cheap wooden box donated by someone, and no marble marker. Today after 125 years, the number of people buried here and their names have long been forgotten and all that is left is an open area with no memorials in the center of the St. Petersburg Cemetery.

## Grocery Stores

Back then, just as now, the grocery store was a very important place in any community. Note the many different types of goods at the New York Grocery and Provision Store, Antwerp City: "We have the largest and most complete stock of groceries and provisions in Clarion County and offer the greatest inducements to the public. We keep the best tea, dried, and canned fruit, cheapest sugar, and flour, eggs, lard, ham, potato, fresh dairy butter by the roll or tub. We also have a very large stock of Queensware, glassware, lamps, crockery, and tinware which we sell at the lowest market rates. Everyone who buys groceries or anything in our line, will find it to their advantage to call and buy from us. Thompson and Patterson, corner Third and Main Street, Antwerp City, Penna."

Competition is always good for business and this advertisement was in the July 10, 1872 edition of *The Progress* newspaper printed in St. Petersburg, which had a circulation of 1,500 papers a week: "Neither Grant or Greely buy at the Antwerp City Store, which has the most complete stock of goods in town consisting of groceries of all kinds, canned goods and fruits, produce, hardware, and tinware. Also a full line of notions, clothing, and gents furnishing goods at the lowest cash prices. J. H. Bethune."

Grant and Greely both individually owned their own stores in Antwerp City and were competition to "poor old" J. H. Bethune and his Antwerp City Store.

With the ups and downs of business in the oil fields and oil boom towns, it was always advantageous to have a good, positive attitude and the following shows the good humor in these advertisements:

**Go to the Smith meat shop and buy your Christmas poultry for 18 cents per pound, and first class beef cheaper than any other market in Antwerp City.**

**Don't Doubt. Cures guaranteed by using Vail's V. S. P. Bitters.**

**The man was hanged because he failed to visit Thompson and Patterson's immense Grocery establishment at Antwerp City**

**The "Driller's wife" is pleased with the fruit, vegetables, crockery and groceries of every kind at Thompson and Patterson's New York Grocery, Antwerp City, Penna.**

**This is for a shoe shop in St. Petersburg:**
**Ladies and Gentlemen**
**Horses and Mules Get their Shoes at Charley Bruel's**

**A Great Riot was experienced, because Thompson and Patterson failed to wait upon all their customers at once. Antwerp City is the place where they do business.**

**The "Wildcatter" said he was pleased with the fruit, nuts, and candy etc. which he bought at Thompson and Patterson's, Antwerp City, Penna.**

**The Hungry Man goes to Thompson and Patterson's to get ham, eggs, butter, flour, pork, beans, salt, fish, etc. and be pleased, they are at Antwerp City.**

**Machine Shops**

**This ad is from the St. Petersburg *The Progress* newspaper dated August 1, 1872. The Blakslee Brothers started their oil well supply and machine shop in the upper fields and moved operations to St. Petersburg and Antwerp City when the Petersburg fields opened in late 1871.**

**Oil Supply and Hardware**
**Blakslee Machine and Tool Headquarters for oil well supplies**
**Tubing, Casing, Steam and Gas Pipe,**
**Working Barrels and Valves, New Bedford Drilling and Plain Laid Lines**
**In fact, everything needed in drilling and equipping an oil well.**

**Monitor engines on hand and for sale by**
**Blakslee Bros.**
**Corner of Machine Street and Main,**
**Petersburg**

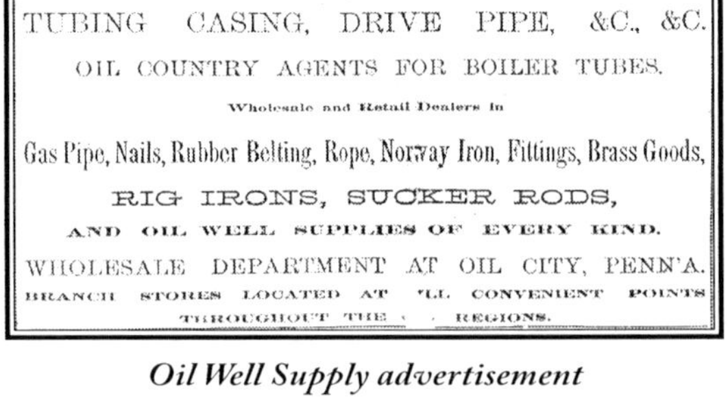


*Oil Well Supply advertisement*

Another Blakslee Brothers ad reads: "Hardware and house furnishing goods, heating stoves and cook stoves and ranges in endless variety. Horse shoes and horse nails. Wagon makers stock of all kinds, wheels, springs and axles. All materials necessary to build a wagon or buggy. We have a first class shop in connection with our establishment where all kinds of tin, sheet iron or copper work is manufactured and repaired. No trouble to show goods. Satisfaction guaranteed. Come and see us. Blakslee Bros., corner of West Main Street and Mill Street. Also at Antwerp City on Third Street in rear of Antwerp City Hotel."

## *Titusville Herald* Sold Here

Crawford and Palm's News Room and Cigar Store was always a very busy place. Each day's edition of local papers showed yesterday's statistics on who, how much and where new oil strikes were found. Oil speculators were eager to lease new ground with the hopes of "hitting the Big One." Not only were local papers sold here, but magazines from places such as New York City, Boston, and Pittsburgh. The *Clarion Democrat, Clarion Republican, Titusville Herald,* and the *Oil City Derrick* were much sought after and studied.

Crawford and Palm kept a steady flow of news from the front, to be reported in the *Titusville Herald.* This newspaper provided a wealth of information to the author.

**Buy a new Mansard cook stove of Porter, Wickham and Co. and bake good bread**

**Fresh oysters always to be found at Patterson Grocery, Antwerp City, Across the street from Mrs. Mackey's Hotel. Choice XXX, 30 cents a can.**

*Well in the Clarion County oil field*
*Venango County Historical Society*

All types of tobaccos were available, including sniffing tobacco, chewing tobacco, and smoking tobacco. One particular cigar store proudly proclaimed that it sold to its clientele, "the fine men of the oil region," the finest imported "Havana" cigars from Cuba and also the smooth smoking good tasting "Syracuse Prides."

The Westmoreland House of Antwerp City was managed and owned by proprietor Simon Hugus. This hotel was located about half way down main street, on the right-hand side, heading toward Turkey City. Its advertisements mention nothing about how nice the rooms were or how reasonable rates were; also nothing is said about the stabling or livery available, but "boy-oh-boy" could you get a rare seasonal scotch, London Porter, stout, genuine cognac, brandy, and "Old Tom Gin." Thank goodness they also had available ice cool soda water, always on hand and dispensed with particular reference to the taste of the customers.

## Drug Store

On a little lighter side, the drug store owned by A. O. Blias provided for the buying public fancy goods and notions of all kinds with a good assortment of hair goods. Mr. Blias dispensed the purest drugs, medicines, and rare chemicals. He also carried a complete assortment of trusses, shoulder braces, canes, supporters, elastic stockings, bandages, and surgical instruments. And finally, last but not least, all types of toilet articles, soaps, cosmetics, combs, brushes, writing papers, blank books, and receipt and notebooks. Quite a line of products for a store to stock in the city of Antwerp. The finest wallpapers imported from England and France were also kept in stock. (Author's note, "Wow, this place was primitive.")

## Livery Stables

A rather humorous practice most livery stables used to keep horses calm and soothed was to keep a "Billy" goat in the livery barn. The strong smell of "Mr. Billy" was believed to be soothing to the horses.

## Hardware Stores

The hardware store was always an important part of any main street scene. An advertisement for a hardware business in Antwerp City reads: "Hardware, Hardware. Remember Robert Bryan has the only gas pipe cutter in this oil region running by steam that will cut thread on 5 5/8 and 6-inch casing. He is also prepared to manufacture sheet iron work such as smoke stacks. Breechan and Co. are dealers in Boston Tubing and Casing, Meadville Rig Irons and New Bedford Cables. Stores at East Foxburg and Antwerp City, Penna." Robert Bryan sold other items such as paints, oils and varnishes, nails, screws, hinges, and ropes. Some of the finer things for sale were colored and painted glass lamps, tinware, crockery, silver place settings, and to hold your food, dinnerware of every description, fishing tackle of all kinds, rifles, powder, caps, and shells.

Another business in town was Wolf Hardware Store. This store sold oil well supplies, nails, ropes, tubing, pails, tallow, etc. It is said that at a meeting of local oil men in the Wolf Oil Supply Store, Victor Gretter introduced his new invention, the oil saver, which was well received by the oil industry.

Jesse M. Smith of Edenburg reported in his journal that he came out of the Civil War broke and worked at Pithole. In 1872 he located at the new City of Antwerp and worked at Wolf Hardware Store on Main Street and was still broke.

## Attorneys

Antwerp City rounded out its services available by having many attorneys at law who not only counseled their clients but also made collections throughout the oil region. Real estate agents sold properties both in town and out. Physicians, surgeons, mechanical and surgical, dentists and veterinary surgeons were here to take care of both human and animal needs. First-class surveyors always on hand.

## Blacksmiths

There were blacksmiths and machine shop services available. The machine shops either made or sold every type of tool or machine needed to produce oil. Drilling tools, fishing tools, tubing, well casing, gas pipes, working barrels, valves, brass and iron fittings, belts, steam engine packing, boilers and sand pumps were available. Blakslee Bros. Machine and Tool had shops in both St. Petersburg and Antwerp City. Chalfant and Graff, Watt and McElwaine, T. Griffin, boiler maker, and S. W. Carpenter sold oil hardware

## More Places to Spend Your Money

Lumber yards and planing mills sold all kinds of dressed lumber, siding, flooring, roof shingles and even pickets for your fence. Large assortments of pre-made doors, window frames, sashes and glass helped building houses easier and at a lesser cost. C. A. McCafferty, dealer of lumber, bills sawed to order, Keating City.

## Tintypes

Millinery shops carried ladies' clothes in the latest styles, beautiful hats, shoes and underclothing. After buying a new suit of clothes and before getting them dirty crossing a rutted, muddy street, you had better get your picture taken at one of many photograph galleries in Antwerp City. Many young gents sent pictures to their best girl and even home to show the folks just how well they were doing out here in the "petroleum patch." One particular gallery reported they did "copying of old pictures as a specialty, current photographs done, also water, oil, ink and charcoal portraits. Your satisfaction we promptly guarantee," E. G. Chappell, Studies at Antwerp City and Oil City, Penna.

## Ice Cream

Who would ever believe that in a rough and tumble oil town back in the 1870s people could sit down and enjoy a nice cold bowl of ice cream? The Ice Cream Parlor or Ice Cream Saloon were actually a part of the Antwerp City main street scene. This is an advertisement that states the benefits of the ice cream saloon. "Our parlor is splendidly fitted up for gentlemen and ladies to spend an evening there, for they can obtain every kind of fruit, strawberries and ice cream of every flavor, also soda water, lemonades, and clean candies to our customers." Again, just how primitive can we get? *The Progress*, St. Petersburg

## Telegraph Office

Feed mills produced the different feeds needed by horses, cows, dogs, and other farm animals. Carpenters, house painters, and paper hangers were in much demand with so much new construction. There was an express office that received and shipped packages and crates throughout the oil region with prompt delivery. The telegrpah office sent messages throughout the oil region and stage coaches and delivery hacks left Antwerp City for all points in the oil region. Stage coaches left for Foxburg every hour on the hour.

## Letter to the Editor

Finally, with the growth of the oil boom town came those establishments of ill repute mostly looked over unless they became a real nuisance. This letter to the editor of the July 10, 1872 issue of the St. Petersburg *The Progress* lets "Wild Cat" express his views:

"Antwerp July 1, 1872. Allow me to congratulate you, Mr. Editor, in behalf of the citizens of this place for the noble stand you have taken in regard to the mysterious death and libel suit which occurred in your place a few days ago. Law abiding citizens can duly appreciate the worth of such a paper as *The Progress.* It is a public servant in bringing to justice the perpetrators of the crimes and misdeeds which will occur in such a center of excitement and travel as is St. Petersburg. Only too glad are we to know that 'adventures' and 'black legs' who have hither to and in other places given the oil business an unenviable and false reputation by their foul deeds will in this district meet with full exposure. Our town is cursed with the presence of a large number of 'star boarders' and 'guzzling shops.' Whiskey will dwarf a dog. Has it 'dwarfed' Antwerp? I pause for a reply. Signed 'Wild Cat.'"

## Roughest Town in Clarion County

No wonder the traveler told the old man to take another route if he expected to end up in Heaven!. Do not go through Antwerp!

This article was published in the *Oil City Derrick*, Wednesday morning, January 4, 1882:

## Early Days of the Roughest Town in Clarion Co.
## Shootings and Cutting Affrays – Every Other House a Saloon
## Fire Casualties – Delirium Tremens
## The Thirty Day Shut Down – Swept by Fire

In the early days of the oil development in Clarion County, Antwerp, one of the most remarkable and short-lived of oil towns, rose and fell, shot athwart the sky of history, meteor-like, and disappeared amid a blaze of light never that reappear.

During the year 1870, the festive producer, then more cautious than now, gradually marched forward from Foxburg, up the banks of the classic Clarion River, and when Marcus Hulings, now of Oil City and known to everybody in the oil regions, in August 1871, went nearly a mile in advance of operations and located a well on Ashbaugh run, about half a mile east of St. Petersburg, then a thriving village just beginning to bloom under the stimulus of the oil excitement, he was considered a venturesome "wildcatter."

October 31st of that year found the well completed and flowing at a wonderful rate. There was no pipe line connection near, and five 250-barrel tanks were set in rapid succession to save the valuable fluid, for oil was then $4.50 a barrel. One of the tanks was set at night, the men working by lantern light, one carrying a lantern round and round the tank, as the sledge and set-holder made their rounds, driving the iron bands tight.

The well continued for some time to put forth from 125 to 150 barrels of crude per day. Pipe line connections were made as soon as possible, and the income of the well the first month footed up $10,000. Mr. Hulings having a 20-acre lease on the farm on which the well was located, immediately laid out lots for a town in the approved style, and named the place Antwerp City, though there were few buildings except a boarding house in the town. Producers crowded forward and rigs were put up rapidly during that winter. Houses began to spring up, mushroom like, in the embryo city. Wells were completed and proved good producers, and the spring of 1872 found Antwerp perhaps the most promising town in the Clarion field. St. Petersburg had grown wonderfully, but Antwerp was soon a good second, with everything in favor of its outgrowing its rival. Half the houses were saloons, or places where intoxicating liquors were sold. Money was plenty, wages high, oil a good price, and everything was booming until the memorable Thirty Day Shut Down of that year, which had the effect of stopping the growth of the town completely. After the shut down was over, the oil ranged at four dollars again, the town took another leap, but, the rapidly declining price of crude put a damper on it and it virtually ceased to grow, though it continued to be the wickedest town in oildom. St. Petersburg had been chartered as a borough and provided with a police force which tended to drive the most lawless portion of humanity to the town at the foot of the hill. In the fall of '72 a large maison do joie was built in the east end of town, by a Dutchman, by contract, for parties from Petroleum Center. When it was completed the persons who had it built refused to pay for the work, and the contractor barred the doors and refused them admittance. Several loads of furniture arrived and were piled up outside as the Dutchman would not let it be placed inside until he was paid.

One night, however, the doors were broken open and the work of piling in furniture commenced. The contractor and his two workers hastened to the scene bent on ejecting the invading parties. They entered the house, loud words followed, and presently two shots rang out on the still night air, followed by a terrible racket. Men ran from all directions and were hastened on by a cry of "My God, I'm shot. I'll die. I'll die," pronounced with a strong Teutonic accent. Three men were seen to leap from an upstairs window and run rapidly up the hill back of the house. They proved to be the contractor and his two men, the former of whom, upon examination was found to have his coat, vest, and shirt perforated with bullet holes and the skin on his side was cut so blood ran profusely, the wound was not serious.

*This photograph is the Monoitor Well, located at the end of 1st Street, Antwerp City. The well was located on the John J. Ashbaugh farm and was named after the famous Civil War fighting vessel. The photograph was taken by B. F. Gilger, photographer, St. Petersburg, Penna. 1873.*

No one liked to venture in the house in midnight darkness to seek a man possessed of a good revolver which probably still contained several cartridges, and in the excitement the murderous villain slipped out and made off to a healthier latitude, as threats of lynching were freely made.

Soon after that the house changed hands, new proprietors came, the contractor was paid off, and a dozen demimonde of the worst type took up their abode there. Among others was one yelept "Crazy Nell," who "made Rome howl" frequently. She had a penchant for committing suicide, and next to self murder, the dearest desire of her heart seemed to be to murder someone else, especially when she had partaken freely of the flowing bowl. She never succeeded, however, in murdering anything, except the sleep of good citizens. This she did frequently by getting on a tear and requiring half a dozen fellows to hold her to prevent self-destruction or the destruction of a fellow creature. She, meanwhile, affirmed her determination to do something terrible in the loudest tones her lungs were capable of.

## Crazy Nellie

I would like all students of oil history to answer this big question, "What did Pithole City have that Antwerp City did not have?" Nothing. Pithole had that sweetie "French Kate" and Antwerp had "Crazy Nell." Both were well known, both knew their trade and both eventually moved on to their next assignment. The photo of Nellie Jane, "Crazy Nell," was given to me by Florence Ashbaugh Shingledecker, granddaughter of J. J. Ashbaugh, in 1975. Nell's last name was Robinson and her whereabouts after 1873 are not known. 2007 update on Crazy Nellie, see page 390.

***"Crazy Nell" – Hundreds of boarding houses cropped up in the oil regions during the boom of the 1870s. They not only housed oil lease workers, but also teamsters, carpenters, railroad crews and many others who thronged to the boom towns. Few records remain of these enterprises. "A woman named Nellie Robinson, residing at Seibert's boarding house on the alley near the Monitor well at the end of 1st street, Antwerp City, was arrested yesterday for running and operating a house of ill-repute. The case was heard before Justice Patterson during the afternoon, and bail was set at $100. A decision will be rendered tomorrow at two o'clock." The Progress, July 23, 1872***

## Curly Green

"Curly Green" held forth here in all his glory, running a gambling hall and saloon, and a worse hole would be difficult to find anywhere. Nightly a drunken crowd was gathered about his gambling tables, passing coins and bills about, swearing, drinking, quarreling, and fighting. One night the games were more than usually interesting, stakes were larger, drinks were more frequent and animosities were more bitter. Two men played at a table eagerly. One was losing regularly. His breath came hard, his eyes were dilated, hectic spots burned on his feverish cheeks as he saw his despair, when his adversary dropped a card from his sleeve. Immediately he leaped to his feet, but the other gambler was up as quickly. They clinched and struggled out on the street, unnoticed by the other players. The cheating gambler attempted to draw his revolver, but the other seized his wrist, and together they rolled on the ground until the loser got his cheating adversary pinned tight under him, when he whipped out a gleaming knife, raised it aloft and would have plunged it in the heart of the man beneath him but that several gentlemen passing seized his arm and prevented him.

A night that passed without three or four fights, or the week that did not witness one or two attempts at murder, were rare exceptions, yet, strange to say, the life of man was not taken by his brother in the whole history of the town.

A fellow one night got a touch of the "jim jams" and went out in the street where he lay down in the mud. He imagined he was drowning and all assurances to the contrary, notwithstanding, he gasped and gurgled like a drowning person. He turned livid in the face, struggled violently, and would have died, but that a bystander seized him and made believe he was pulling him from the water he imagined he was in. When they reached land, the lunatic from rum was considerably relieved and improved rapidly.

When the great devastating fire broke out in St. Petersburg in February 1873, it was heroic men of Antwerp who checked the progress of the flames, working like heroes, and fighting fire until water had to be constantly thrown on them to keep their clothes from igniting.

## Down Hill Side of Life

With the opening days of spring, in 1873, an exodus to newer fields set in, and the busy town of Antwerp City was on the down hill side of life, its feverish vitality gradually ebbing away. But, one night in July, or rather about three o'clock in the morning, the incendiary's torch did its fiendish work and the terrible cry of "Fire! Fire! Fire!" rang out on the still air. The devouring flames leaped from house to house with a gleeful energy, according well with the vigorous enterprise of the place, and with the dawn's early light the busy little hamlet perished, passing away with the night. Only a handful of houses were left, and for years two rows of blackened posts alone marked the spot where once stood the lively and wicked little town of Antwerp City.

The following article entitled "The Town of Antwerp" was written by Gary McKinney and printed in *The Progress News*, Emlenton, Pennsylvania:

Not many people realize that a town of over 1,000 people existed near St. Petersburg in the 1870s. This town was called Antwerp and was located about one mile east of St. Petersburg, Penna. The town was formed in 1871 and in a short time it had grocery stores, hotels, saloons, a schoolhouse, a telegraph office, livery stables, the Antwerp Pipe Line Co., and about 200 houses.

It was a tough fighting, hard drinking oil town as indicated by a loser in a poker game. Antwerp had no paved streets and the sidewalks were wooden planks.

A story is told that a man in a buggy overtook an old man walking along the muddy road between Turkey Run City and Antwerp City. Before asking the old man to ride, he said to him, "Where are you going?" to which the old man replied, "I am going to heaven and I have been on my way for 18 years." The man in the buggy replied, "Goodbye old man, if you have been traveling towards heaven for 18 years and have gotten no closer than Antwerp City, I'll take another route."

## Railroad Siding

It has been noted that a stagecoach made runs from Foxburg through St. Petersburg and Antwerp City each day, delivering passengers and mail. After the narrow gauge railroad was built near Antwerp Hollow, it made the moving of passengers and goods easier. The railroad built a siding on the east side of the ravine at Antwerp. The siding came off the main line on the eastern side of the curved railroad trestle. I've seen railroad ties torn out of the main road, Route 58, just across from the St. Petersburg Rod and Gun Club. Trains were loaded and unloaded there and then sent back to the main line to continue their journey. There exists a time schedule showing all the stops including the Ashbaugh farm in the 1940 *Clarion County Centennial History Book.* The trains ran every day except Sunday. Note page 123

## A Self-Respecting Ghost

A story I recall has to do with bumps in the night, souls long lost, and ghosts. Every self-respecting ghost town needs a self-respecting ghost. This story was first told to me by Charlie Rupert about 1964 when we were building the St. Petersburg Rod and Gun Club. I was just a young kid and we always liked to hear Charlie tell the story of the old doctor from Antwerp, whose spirit still wanders on those dark warm summer nights. Only on the nights when the moon creeps in and out of its dark cloudy cover, usually near the midnight hour, does the old doctor begin his eerie nightly rounds.

Located on the eastern end of the ghost town of Antwerp City, runs a dark hollow, with a small winding but gently flowing stream called Myers Run. Large old oaks, maples, and black cherry trees line the "brook" and tower up, holding the darkness from escaping into the night. In those early days, roads were almost non-existent and stream beds made fairly good paths. Now that the stage is set, we must be very quiet and sit back on this old stump on the banks of this old hollow and very, very quietly listen. As we listen we hear the leaves rustle and off to the left a strange noise seems to whistle, then moan. Just at this very moment, down the winding stream bed comes a slow-moving dark figure. You feel the "goose bumps" on your arms and the hair stand up on your neck. You strain your eyes through the darkened night and behold! There comes a small but beautiful black carriage, with a large coal black horse in the lead. There at the reins you see the old gentleman, stove pipe hat on his head, carefully guiding his rig over the rocky stream bed towards his destination at Keating City which was located at the mouth of the ancient run. Through the dark still night comes the clickety clack, clickety clack of hooves prancing over the stones along with the steely ring of the buggy's wheels as they glance from rock to rock. As we watch and listen in the frozen darkness, we turn to each other and look into each others' saucer-sized eyes. With no words spoken, away we run up the steep stream bank and out the Antwerp flat as fast as our feet can go toward the lights and safety of good old St. Petersburg.

Keep this story in mind and spend some dark warm summer evening in the Antwerp Hollow and I guarantee you, too, will be a believer.

## Mrs. Payne's Newsroom

Many women at the time took in wash to make a living, and there was always much laundry to do. A Mrs. Payne, widowed by an oil well accident in which her husband was horribly burned, did laundry seven days a week and with her hard-earned money had saved enough to open a small news store in Antwerp City. The arrival of the stage coach from Foxburg each day brought the daily newspapers from Titusville, Oil City, and Pittsburgh. These were sold both from her store and out on the streets by her son.

Across the street from the news store was the most prominent hotel in town, the Antwerp City Hotel. It had more class than the other hotels and boarding houses. It had a big porch with a railing and men sat out there and read the papers and exchanged the latest news.

## "Dutchmen" Stories Were Popular

A jolly Dutchman near St. Petersburg, who married his second wife soon after the funeral of the first, was visited with a two-hours' serenade in token of disapproval. He expostulated pathetically thus: "I say, poys, you ought to be ashamed of myself to be making all dish noise ven der vas a funeral here purty soon not long ago." This dispersed the party more effectively than a bulldog revolver could have done.

A girl just returned to St. Petersburg from a Boston high school said, upon seeing the new fire engine at work, "Who would evah have dweamed such a vewy diminutive looking apawatus could hod so much wattah!" (*Sketches in Crude Oil,* John J. McLaurin, 1896)

An honest Dutchman near Turkey City declined to lease his farm at one-fourth royalty, insisting upon one-eighth as the very lowest he would accept. He did not discover that one-eighth was not twice one-fourth until he received his first installment of oil, when he fired off the simple expletive, "Kreutz-millionendonnerwetter!"

## Pioneer Families And Their Stories

Northwest Pennsylvania held many dangers to the settlers as they traveled through the wilderness to their new homes. Some of the many life-threatening situations those early travelers faced were Indians, wild animals, lack of provisions (such as food, clothing, and medicines), snake bites, drowning due to many stream and river crossings, and the harsh Clarion County winters.

The Judge Richard Peters Tract and other land near to and surrounding the future village of Petersburg, was beginning to be settled by the pioneer families of Fink, Vensel, Shoup, Snyder, Moyer, Porter, and Ashbaugh. The families traveled in a wave of migration which began in 1806. These families began the task of clearing and settling those wild and forested lands. Farms were surveyed and home sites were chosen. Open-sided lean-to's were the first means of shelter with log cabins being built later. Food and shelter were the first priorities before winter set in. Small gardens and limited crops were planted immediately after arrival to insure a food supply for the upcoming winter.

About 1815 the Shoup family came to this area and bought land near the Daniel Ashbaugh farm in Richland Township on the outskirts of the future site of Petersburg. Little did Solomon and Elizabeth Shoup know that the farm they chose would bring their son, David, a fortune in oil wealth. Their farm was located near the oil town of Antwerp City.

The following information comes from a family history, "Yearning to Breathe Free," the German migration about the Shoup family of St. Petersburg, by Sara Anson Vaux and Kenneth Lester Shoup, 1977. This story is known by very few people and what follows shows the heartaches a pioneer family could face.

In 1806 the Daniel Ashbaugh family traveled in a wagon as far as the Clarion River. The family then proceeded on foot, carrying their household goods with them. They carried but little food and clothing, and "the want of proper food and goods caused great suffering." In their travels they were compelled to camp under trees and eat bread made of old wormy flour mixed with life-saving fresh water and baked on the coals. For Solomon and Elizabeth Shoup, these hardships climaxed years of backbreaking work, beginning with their original removal from Bucks County 25 years before. Through accident and exhaustion this was to be the Shoup family's final migration. The Shoups entrenched themselves firmly in the Allegheny Clarion Valley and their spirits have never left it.

The legend persists that the Shoups arrived in Richland Township near the future village of Petersburg in the fall of 1815. Wearily they climbed to the top of a hill overlooking Daniel Ashbaugh's land, as Daniel had done a few years before. They built a "shanty" of poles and clapboards, open on the sides, with slight protection against the wind, snakes and rain. Above their heads towered two ancient willow trees, ten feet across and 400 years old. After the first night on their new farm, the Shoup family awoke the next morning, a brilliant fall countryside greeting them; a sea of red, orange, and gold stretching across the valley to the hillsides beyond the Clarion and Allegheny River valleys. Even today, the view from Solomon Shoup's hill is awe inspiring. The family "fell in love with the view" and decided to end their journey.

The family group grew and prospered on the farm, raising corn, oats, wheat, and buckwheat. When oil was discovered by Marcus Hulings on the Ashbaugh farm in the hollow below their farm in 1871, the Shoup farm was leased and the David Shoup family gained much wealth from oil.

David Shoup was Solomon and Elizabeth's eighth child of ten. David took over the farm and built a fine large house on this very hill, commanding a view of Antwerp City and the surrounding oil fields. Solomon's family consisted of Jacob born 1788; Catherine 1789; John 1790; Elizabeth 1793; Susan 1795; next child unknown; Sara 1798; David 1800 who died 1889; Jonathan 1804; and Solomon 1805. David first married Rosina Shakely, then Lavina Goughler.

## Family Helps Family

After many years of hardships, clearing fields, building homes and then moving on, always looking for something better, Solomon had finally found his home. The future was filled with opportunity and Sol looked forward to a long and prosperous life. By this time he had two married sons, several married daughters, grandchildren, and at home were Sara 19, David 17, Jonathan 13, and Solomon 2.

In these early days people helped each other, whether in planting crops, harvesting grain, or building log homes, barns and out buildings. In those days a man's handshake was a man's word. In the fall of 1817 Sol's friend and neighbor, Daniel Ashbaugh, decided that the barn he built shortly after he had settled his farm in 1806, was not large enough and a new log barn was to be built. All during the summer of 1817 logs were cut and hewn in proper lengths and widths. Rocks were cut and squared for the foundation. Daniel decided that after crops were harvested in the fall there would be time to erect the barn.

On the day of the barn raising the sun made the colors of the changing leaves most beautiful and the chill in the air reminded the farmers that winter would be just around the corner. Many farm families had gathered that fine fall day, the Neelys, Finks, Delos, Ashbaughs, Bests, Shoups, Porters, Heeters, Vensels, Snyders, and Moyers, all looking forward to hard work, along with good company, story telling, the latest news, and lots of good food cooked up by the hard-working farmers' wives. Children ran and played, chasing each other as they played hide and seek. In the shade under the large white oak near the Ashbaugh's log house were two large wooden barrels full of cold water with a tin dipper hanging from each container. It was up to the older girls to watch the smaller children and to keep the water barrels full.

Work went well on the Ashbaugh log barn as each man knew his job. Slowly the beams stacked together. By 4 o'clock in the afternoon work had progressed to the point where the roof sheeting was being nailed to the round roof rafters. Solomon Shoup had worked hard all day and was well liked by all his neighbors. He decided the men needed help with those first layers of hand-made wooden shingles. In his time he had roofed a good many buildings and the first rows had to be put on straight and square in order to make a rain and wind tight roof. Ladders were set up and they transported men to different parts of the unfinished structure. Solomon Shoup climbed a ladder which stuck up above the barn roof. No one knew just what happened at this point but just as he reached the top, he lost his balance. Through the sound of the hammers, the chatter of conversations, and the bark of a dog, a short shrill scream was heard, a cry for God's help and a terrible thud as Sol Shoup's lifeless body lay on the ground. Sol had somehow lost his balance, falling on a pile of cut stone and left-over hewn timbers. The resulting broken neck caused his instant death.

Solomon's death was a great loss to the farm community and each man, woman, and child took to their grave the terrible memory of that fateful fall day.

On January 14, 1818, Captain Henry Neely and Jacob Vensel filed a bond for $500 in the death of Solomon Shoup, with Henry Best and George Berlin signing as sureties. The bond entitled the men to administer the estate since there was no will. Solomon's friends and traveling companions were on hand to settle his last affairs as they also had accompanied him for years through the wild and unfamiliar wilderness. David Shoup took over the responsibility of the farm. David is buried in the St. Petersburg Cemetery with three small stones on the lot inscribed "Father," "Mother," and "Sister." The David Shoup house was bulldozed down in about 1983 and the farm was strip mined for coal. C. & K Coal Company of Clarion sold it to the Clawson family.

## Richland Township

Richland Township, Clarion County, was first formed in 1806 when it was part of Venango County. It was divided between Venango County and Clarion County in 1839, the larger section going to Clarion and the smaller to Venango County.

The earliest settled portion of the township was the region about the mouth of Ritchey's Run. This stream derives its name from the Ritchey family, of which the first representative here was James Ritchey. He was from Westmoreland County and the date of his settlement here is placed as early as 1796. He located on the Clarion County side of Ritchey's Run, a mile and a half from its mouth. The maiden name of his wife was Jane Kennedy, and she was the mother of thirteen children, five sons and eight daughters.

It is supposed the name Richland for the township was based on the name of this early pioneer family as well as the quality of the land. This name was retained in Venango County's township and carried over into Clarion County. Ritchey's Run forms a boundary line between Venango County and Clarion County at this point. Richland Township was later divided up to form parts of Salem, Beaver, and Licking Townships.

## Richard Peters

Saint Petersburg was at first called "Petersburg" in honor of Judge Richard Peters of Philadelphia who once owned the town site. When the post office was established in 1862, the postal authorities, to avoid confusion with Petersburg in Huntington County, added the prefix "Saint" in imitation of the name of the former capital of Russia.

Richard Peters was Secretary of the Board of War in 1779 and throughout the remainder of the American Revolution. In point of fact, if not name, Richard Peters was our first Secretary of War, and by his energy, enterprise and resourcefulness in securing funds, supplies and munitions, he contributed a small measure to ultimate triumph of the Colonial Patriots. He was Judge of the U. S. District Court of Pennsylvania for 36 years.

St. Petersburg is situated in the western part of Clarion County on a ridge rising between the Allegheny and Clarion Rivers. Judge Peters donated five acres to the village for church and cemetery purposes. When the village was incorporated as a borough, February 23, 1872, Barney Vensel was chosen the first burgess of the town, and a council comprised of G. C. Fink, Joel Fink, D. S. Herron, Nelson Vensel, C. H. Wilton, and Daniel Whitling were elected. To assist in keeping law and order, a 24-hour police force of eight to ten men was employed.

## John J. Ashbaugh

The following information on John J. Ashbaugh shows the varied and interesting aspects of his life, from his early years in this area's wilderness, to his service in the Civil War, to oil producer, to respected town citizen.

When J. J. came home from the Civil War after being a prisoner at the infamous Andersonville Prison Camp and the prison at Florence, South Carolina, he was going to go back to farming the family farm. Typically this consisted of plowing, raising cattle, feeding chickens and harvesting the fields and gardens. It seems there were always chores to do such as fixing barn roofs, repairing fences, and looking after the horses. The work horses were the farm's modern day bulldozers and tractors. The Ashbaughs always had good teams of work horses and strong, sleek carriage horses. J. J. did what most young men did by marrying Hannah Vensel, sister of Charles Vensel, and raising a family. Then everything began to change after the discovery of oil on his farm in 1871. John J. went from farmer to oil producer almost overnight.

## A Final Farewell To Drake Well

In 1876 the centennial of the United States was celebrated in Philadelphia. This was called the Philadelphia Centennial Exposition and was well attended by the people of the northwestern Pennsylvania oil regions. Many people now had more money and could afford some of life's finer things. They rode the railroad to Philadelphia in some of the finest passenger cars available at that time. Great steam engines pulled the passenger cars and many oil families went on a trip of a lifetime with their new-found wealth. This story I find very interesting and it comes from the book *Sketches in Crude Oil* by John J. McLaurin, 1898, page 74.

Someone decided that one of Northwest Pennsylvania's oil region's contributions to the Philadelphia Centennial Exposition would be to send something that was known to everyone, historically important, and hopefully interesting to the visitors of the great 1876 Exposition. Much thought was given to just what their display would be. Finally it was agreed upon that

*Drake Well, Philadelphia Exposition, 1876. Drake's original engine house and derrick burned October 6, 1859, after "Uncle Billy" Smith used a lantern to look down the well. It was rebuilt in November. In 1875, a Mr. Millard of Titusville was hired to dismantle and load the derrick on a flat car to be transported to Philadelphia. Courtesy Drake Well Museum*

what could be more reprsentative of the oil country than Colonel Drake's original engine and pump house and derrick. We are all so familiar with the replica known today. The Drake well engine house was dismantled board by board, piece by piece, and labeled for its reconstruction in Philadelphia. John J. Ashbaugh of St. Petersburg and Thomas O'Donnell of Foxburg were in charge of the project from start to finish. The original engine had already been stolen or sold piecemeal. Everything was shipped by rail, accompanied by these two men and a crew. Unfortunately, it seems that the whole display must not have been what interested people and the display was mostly unsuccessful. After the Exposition closed, the Drake well display was carted away and burned. This act of final destruction to us oil history buffs was very unfortunate.

The following information was handwritten by John J. Ashbaugh on his Civil War Service. The spelling is left as he wrote it:

"I enlisted January 4th, 1862 in Co H, 103 P. V. This company was raised here in Clarion Co. by Capt. J. G. Mackey of Clarion and G. W. Kelly of Philadelphia and from here we went to Washington City. From there to Fortress Monroe and then started on the Peninsuler Campaigne. We were at the siege of Yorktown, Battle of Williamsburg, Va., Chiekatomig, Seven Pines or Fair Oaks, and Seven days fight. After the seven days fight we landed at Harrison Landing. From there we marched back to Fortress Monroe. Then went to Suffolk, Va. We made seven raids to Black Water River while at Suffok. We went with an expectation to Goldsborough from Newbern. The first battle on the raid was at Kingston. We 450 men in our regiment at this time and lost 80 killed and wounded in this battle. Next was Whitehall and then Goldborough where we burned a railroad bridge. Then retreated back to New Berne. From there they sent our brigade to Plymouth, N. C. Plymouth is on Roanoke R. 7 miles form the mouth. We wer at Plymouth about a year when on the 17th 1864 of April we wer attacked by a force

*Empire Transportation Company's exhibit at 1876 Philadelphia Centennial Exposition*

of rebels estimated form 10 to 1500 (our force numbering about 2500) on the 20th of April we had to surrender and made prisoners of war and took us to Andersonville Prison in Georgia on the 12th of Sept. 1864. They took 11000 of us prisoners out of Andersonville and took us to Florence Prison, S. C. and on the 7th of Dec. same year I was paroled with 10 or 1100 others and delivered to our lines in Charleston Harber. From there we came on a boat to Annaplis, Maryland to the Parole Camp. We all got a 30 day furlough. Eugene Widel and I came home together. We stretched our 30 day furlough a few days. When we started back and got to Pittsburgh went to the Ticket Office to buy a ticket to Annapolis. We had to show our furlough to some high much of the Provost Guards. He seen our furlough was run out. He put us in the Guard House. From there they took us to Harrisburg and put us in the guardhouse there from there to Philadelphia guardhouse was our hotel again. From there to Baltimore. Shure enough we were put in the guardhouse again. From there they took us to Fort Federal Hill a distribution camp. We stayed there about two weeks. Then they took us to the parole camp at Annapolis to the officer in command. He ask us when we were paroled we told him. He said go to your quarters that was the end of our furlough bussiness. After we were exchanged they send us to Roanoke M. C. We stayed there until we were discharged Jun 25 1865. Thirty six of my company went in Andersonville Prison and 18 came out a live (and I was one of these). When I was taken prisoner I had about $250.00 in money. I think that help me to see the stars and stripe again. Flour was selling at one dollar a lb. Beans one dollar a qurt Salt 25 ct a spoonful. Eggs $4.50 a dozen I paid two dollars for a slice of ham 6 inches long and abot 3 inches wide and as thin as could be cut. I seen them 6 men hung in Andersonville on the 11th of July for robing and murdering in camp. I wa a prisoner 235 days. John J. Ashbaugh, Company H, 103rd Pennsylvania Veteran Volunteer, St. Petersburg Penna. Jun. 20th."

The old Civil War veterans always marched in the parades here in St. Petersburg. On Memorial Day after the parade and memorial service, all the Civil War veterans' graves in the St. Petersburg Cemetery were decorated. The Civil War veterans here belonged to the "Look Out Post" which was a pre-runner to the modern day American Legion. The last Civil War veteran in St. Petersburg died in 1935. He was John D. Murphy, the music teacher at the St. Petersburg Public School. He is interred in the St. Petersburg Cemetery. Murphy was a driller and started out at Pithole.

*German Reformed Church, St. Peterburg, built with oil money. The steeple was 167 feet from the ground to the top. It was torn down in 1974.*

## Stay Off That Motor

One of the last wells to be pumped on the J. J. farm was on the hillside just below the workshop near the house. It seems that they had an old natural gas engine that ran the pump which pumped oil from the well. When starting this natural gas engine, J. J. would step up on the flywheel and by pushing down on the flywheel would turn the engine over. At this point hopefully it would fire and then start to run. This one particular day the old motor was being cantankerous and many attempts to start the motor failed. J. J. figured he would give it one more chance. After adjusting the gas flow to the motor, once more he crawled up on the flywheel and gave it just one more spin. With all his weight he jumped down on the spoke of the old wheel and "BANG" at the very next moment the "old girl" backfired spinning the big flywheel backwards, which instantly propelled J. J. across the room and almost through the roof of the old pump house. He hit the rafters in the roof so hard several were broken in two. This accident half killed the poor old guy.

## Ring The Bell

Many families in this area looked to the Reformed Church in St. Petersburg for their spiritual guidance. The church bell rang whenever a member passed away. As the funeral procession neared town, the giant bell began to toll. The J. J. farm is located at the east end of town with quite a view, looking east toward the Alum Rock area. A member of the church was situated in the yard so that when the funeral procession came into view, he would run to the church and the bell would begin to ring. Each family had their own pew in the church and every Sunday no other family would sit in another family's pew.

## Wash Day

Each week one day was set aside as wash day. On that day the women of the family washed, dried, and ironed clothes all day. From Mrs. Ashbaugh's yard she could see Mrs. David Shoup's home and surrounding yard. The two women were competitive, but very good friends, and as each wash day rolled around one would try to out do the other. It seems that for years one lady would try to get her wash out on the line to dry before the other. This was a great game for them and many times the women had their wash out drying even before the sun came up.

The Shoup farm was stripped of its coal and now the Jamie Clawson family lives there.

## J. J.'s Obituary

In J. J.'s later years he developed cancer and that proved to be his demise. The story was that a chicken had pecked him on the arm which started the problem. During his last days he had a male nurse to take care of his needs. Bill Morgan was this nurse and lived in St. Petersburg until he passed away. The following is his obituary, August 31, 1932.

"Funeral services in memory of John J. Ashbaugh, veteran resident of St. Petersburg, were held at the family home at 2 o'clock Friday afternoon. Rev. W. F. Ginder, pastor of the St. Petersburg Reformed Church, officiated. Burial with military honors was made in St. Petersburg Cemetery in charge of the American Legion. The deceased was 89 years of age and a veteran of the Civil War. His entire life was lived at St. Petersburg with the exception of the time spent in military service. Mr. Ashbaugh enlisted Jan. 4, 1862 in Company H, 103$^{rd}$ Regiment, Pennsylvania Volunteers, in the Army of the Potomac as a private and drummer boy, and was honorably discharged from the service June 25, 1865. He participated in all the battles, skirmishes and picket duty in which his regiment was engaged. He was a participant in the siege of Yorktown, the Battles of Williamsburg, Fair Oaks, Malvern Hill, Blackwater, Va., Williamstown, Kingston, and Plymouth, N. C. He was captured at Plymouth and confined in the Andersonville, Ga., and Flor-

*John J. Ashbaugh with his Civil War drum*

ence, S. C. prisons for seven months and twenty-one days, and finally exchanged December 11, 1864. Mr. Ashbaugh was actively engaged in civic affairs in St. Petersburg for a number of years. He served as burgess for two terms, and was at various times a councilman and borough assessor. Surviving are the following children: W. A. of Butler, H. W. and F. C. of St. Petersburg, Charles B. Ashbaugh of Wilkinsburg, and Mrs. R. M. McQuillen of Rochester, N. H.; seven grandchildren; four great-grandchildren and an aged sister, Mrs. Sarah Shoup, 92, of St. Petersburg, also survive. He had been a member of St. Peter's Reformed church at St. Petersburg for 73 years and served as deacon and elder for several terms. He was a member of the Grand Army of the Republic for many years."

Mr. Ashbaugh often said the only reason he survived the American Civil War was he had the good luck of being transferred from the terrible conditions of Andersonville prison camp to the better conditions at the prison camp at Florence, South Carolina He told his son, Howard, that if ever he should have a little girl, please call her Florence. Florence Ashbaugh was born and lived in the St. Petersburg area her entire life. I always enjoyed talking to Florence and listening to her many stories.

## Pipe Line Transportation Of Oil

Prior to 1865, in the upper fields, the transportation of oil through pipe lines had failed due to leaking joints, pumping problems and vandalism. Finally, Mr. Samuel Van Syckle laid a two-inch line from Pithole to Miller Farm, a distance of six miles.

The mechanical success of this line soon caused the matter to be taken up by others. New pipe companies' lines were extended over the upper fields and with the opening of the lower belt, lines were built here.

## Clarion County's First Oil Pipeline

This is just another example of the Titusville connection to our Clarion County oil field development.

On October 19, 1869, an agreement was signed between Gus R. Harms, of Petroleum Centre, M. C. Martin, of East Foxburg, and Samuel Fox. The purpose of this agreement was to engage in the business of transporting petroleum from the wells to the railroads. This pipeline was operated under the name of Harms and Martin. Mr. James H. Snow was the superintendent for this line. Note Page 297.

## Agreement

For and in consideration of the sum of one dollar in hand paid, the receipt of which is hereby acknowledged, I Samuel M. Fox of Richland township, Clarion County, Pennsylvania, do hereby sell and assign, and by these presents, set over to M. C. Martin and Gus R. Harms, their heirs and assigns for the period of five years from the date hereof, the exclusive right of way over my land in Perry township, Armstrong County, Pennsylvania, and my lands in Richland township, Clarion County, lying near the mouth of the Clarion river for the purpose of piping, transport and shipping petroleum oil over and across said lands for shipment on cars; provided always that the rate of transportation charged by said M. C. Martin and Gus R. Harms, shall never exceed twenty-five cents per barrel. Should said rate at any time be exceeded, this sale and transfer shall have its legal termination; otherwise, I bind myself, my heirs and assigns, to protect said M. C. Martin and Gus R. Harms, in exclusive possession of said right away for said term of five years; neverless the said S. M. Fox reserves for himself and his lessees on the east side of the river, the right to ship oil of their own production, by car or otherwise, as they may see fit, and to lay a pipe for that purpose. In witness whereof I hereunto set my hand and seal this 19th day of October, 1869. Signed Samuel M. Fox.

The other part of the agreement is as follows: In consideration of Samuel M. Fox having granted us the right of way through certain lands of his, for the purpose of piping petroleum, we agree to lay a pipe across the Allegheny River, and to extend the same through his lands in Perry township, Armstrong County, within two months from the date hereof, and to lay pipe to the tanks of his lessees as soon after their wells become productive as conveniently practicable, and to allow him a drawback of five cents per barrel on his oil piped over the river, unless our general charge should be reduced to fifteen cents per barrel, when the drawback shall only be two and a half cents. We also agree, in case the oil passed through our pipes shall amount to five hundred barrels per diem, to build an iron tank of not less than eight thousand barrel capacity, for its reception, and at the end of every four months, to divide among our customers, pro rata, any excess of oil that we may have received over and above what we may have given credit for. Witness our hands and seal this 19th day of Oct. 1869. Signed M. C. Martin and Gus R. Harms.

**Monitor engines on hand and for sale by Blakeslee Bros. Corner Of Machine Street and Main, Petersburg**

## The Mutual

*Mutual Pipe Line Office, Foxburg*

As more wells were drilled and the oil transportation business picked up, Martin and Harms were joined by James Bishop and C. Myer, under the name of "Mutual Pipe Line Company." This partnership had its headquarters at Foxburg and the date was late 1870.

Wells were being put down at a rapid rate and by July 20, 1872, there were 233 producing wells and 106 drilling wells in the Petersburg (third) district. By the middle of September 1872, the producing wells numbered 300, with hundreds more drilling.

# Pipe Line Companies

In 1872 the Antwerp Pipe Line Company entered the field and in the same year a line was laid to Oil City by way of East Sandy called the Sandy Pipe Line Company. In 1874 the Emlenton Pipe Line began gathering oil. This line ran from Turkey City to Emlenton and later was extended on to Oil City. The name was later changed to The American Transfer Co. and became the nucleus of the United Pipe Line System. Mr. Daniel O'Day was general manager and James H. Snow, Superintendent.

***Rail transportation of oil. This photo could have been taken at Fullerton, showing the loading of wooden tank cars. Amos and James Densmore, oil brokers at Millers Farm in Venango County, devised these tankers in order to ship more oil by rail. Each tank held between 42 and 45 barrels of petroleum. Later, steel tank cars were developed. Actual photo is Gregg's Switch, Gregg Farm, Venango County. – Mather/Drake Well Museum***

Early in 1872 the Empire Transportation Co. took control of the Mutual Pipe Line Co. headquartered at Foxburg. Mr. Charles P. Hatch of Titusville was president and Alfred W. Smiley of Foxburg was Superintendent. Under Smiley, the Mutual Pipe Line grew. After the Empire purchased The Mutual, they also purchased the Union Pipe Line at Parker and the Atlantic Pipe Line Company. Foxburg and the mouth of the Clarion River were delivery points for the Mutual and Parker and opposite the mouth of Bear Creek were the Union's. In 1876 the Union Pipe Line extended its line to Butler.

In 1875, the "Atlantic" began piping oil to Sligo but soon afterwards changed its drop off point to Foxburg. Alfred Smiley was known throughout the oil region. He knew and was a friend of John Wilkes Booth, wrote books on the history of oil, lived, died and is buried in the cemetery at Foxburg.

When Edenburg became a center of oil activity, all of the Clarion County pipe line companies centered in Edenburg. These included the Sandy, United, Antwerp, Oil City, American Transfer and the Atlantic and as a result of the competition between companies, carrying rates became ruinously low, dropping to as low as five cents per barrel. This competition was good for the producers but almost put the transfer companies out of business. This all changed on March 12, 1877, when all the pipe line companies were consolidated with the "United Pipe Line" which was owned and controlled by our old friend, John D. Rockefeller and his Standard Oil Company. The central office for the "United" was in Oil City and later evolved into the "National Transit Company."

On January 1, 1878, Alfred Smiley, due to reorganization, lost his job. Smiley now went to work on the Fox Farm Pipe Line. This line served mainly the Fox Farm and a few surrounding farms. This independent line lasted a little over two years. On April 29, 1880, William Logan Fox died and shortly thereafter the Fox Farm line was absorbed into the United Pipe System.

## Antwerp Pipe Line Company

As the Edenburg field opened up, Antwerp City became a stepping stone toward the next magical oil gusher. Between Antwerp City and St. Petersburg, thirty rigs were underway in November 1871. In a short time every available foot of ground in a 45 degree line between St. Petersburg and Turkey Run City was in the hands of eager speculators.

By the time the Petersburg area had its oil excitement, transportation of oil was somewhat better organized than when the upper fields were opened. Oil was still hauled in trains, on wagons, in barrels, or on boats on the river, but the pipelines were now facilitating the transportation of oil to the tank yards and refineries.

***Loading steel tank cars Venango County Historical Society Franklin, Penna.***

## New Oil Town

One of the largest pipeline companies in the area was the Antwerp Pipe Line Company, which had its offices in Fullerton, Pennsylvania, located on the Allegheny River, and on Main Street in Antwerp City and St. Petersburg. Its lines were laid throughout the Petersburg and Turkey City oil districts to collect oil and pipe it to Fullerton. Fullerton was named after Ira Fuller, oil producer. "Half Way Town" was located above Fullerton. Here they had large wooden or iron tanks to hold the oil until it could be pumped into cars at the Fullerton railroad stop on the Allegheny Valley Railroad, later the Pennsylvania Railroad. This stop was half way between Foxburg and Emlenton, along the Allegheny River. From Fullerton, oil was shipped by train to the refineries. The town or stage coach stop called "Half Way Town" was half-way between Kittanning, Pennsylvania, and Meadville, Pennsylvania. In the late 1790s and early 1800s, this was a major route. My friend, John Chechak, was born in the old hotel before it burned in the 1930s.

***Oil Line Pump Station – This photo represents what a Clarion County pump station consisted of. Large boilers, engines and pumps forced oil through miles of steel pipe lines from the wells to the loading dock. The large storage tanks held surplus petroleum. The Antwerp Pipe Line Company had a lease on the John J. Ashbaugh farm for a pump station. The small boy near the coal pile is newsboy J. M. Place, who later became a well-known newspaperman in the oil region and St. Petersburg's postmaster. Actual photo is Tidioute Pipeline Company near Pleasantville, Venango County. – Mather/Drake Well Museum***

### Don's Doubt. Cures guaranteed by using Vail's V. S. P. Bitters

From an article in the *Venango Spectator* of June 20, 1872, comes the following news item: "Fullerton now has eight families and two fine hotels. It is a receiving and shipping point for the St. Petersburg Oil District. The Antwerp Pipe Line Company has been completed to Fullerton from the Antwerp-Petersburg fields. The daily run is about 800 to 1,000 barrels. The Antwerp Pipe Line Co. has already built one 10,000 barrel wooden tank and another has just been started. The hauling of oil in barrels was largely eliminated by a system of storage tanks at wells and feeder pipe lines to the main lines."

***Titusville Herald*, April 1872:** The Antwerp Pipe Line is being laid from Petersburg district to Spruce Run, on the Allegheny river, between Emlenton and Foxburg. The pipe line and tanks will be completed and in working order in about one week. Ground is being sold and leased at the terminus of the line for building purposes, the land being admirably adapted for the purpose, a thriving town will no doubt suddenly spring up, which is to be named Fullertown. Work has already been commenced on a large hotel, and other buildings will no doubt soon be under way.

***Titsuville Herald*, April 1872,** "New Oil Town": Fullerton is the name of a new town being laid out by Mr. Fuller, late of Oil City, and Mr. M. Hulings, of the South Side. It is situated a short distance from Foxburg on the Allegheny Valley Railroad. Streets are being laid out and town lots are selling rapidly. Mr. Fuller intends erecting a large hotel to be called the Fullerton House, and to make other permanent improvements. A pipe line is at once to be laid from Antwerp City to that point, and the great portion of oil produced in the Petersburg region will be shipped via Fullertown instead of Foxburg.

## Certificate of Incorporation of the Antwerp Pipe Co.

This is to certify that Ira C. Fuller, M. Hulings, H. M. Hughs, C. F. Blakslee, James B. Blakslee, William Z. Blakslee, E. C. Bradley, and R. Richardson being tenants in common of the lands and property hereinafter mentioned and described and thus desire to organize the Antwerp Pipe Co. for the transportation and storage of petroleum in the counties of Venango, Warren, Forest, Armstrong, Clarion, Butler, Crawford, and Erie in the Commonwealth of Pennsylvania have associated themselves together under the provisions of an act entitled, an act relating to corporation for mechanical, manufacturing, mining, and quarrying purpose affirmed the eighteenth day of July A. D. 1863 and the supplements thereunto as extended in the transportation and storage of petroleum by the supplement to said act appeared on the twelfth day of March A. D. 1872.

**Name**

That the name of such company is desired to be formed is the Antwerp Pipe Company.

**Purpose**

That the purpose for which the company is organized is the transportation, storage, shipment, insurance of petroleum and to lay down construct and maintain pipes, tubing, tanks and such other decisions or arrangements for carrying on of said business as may be necessary.

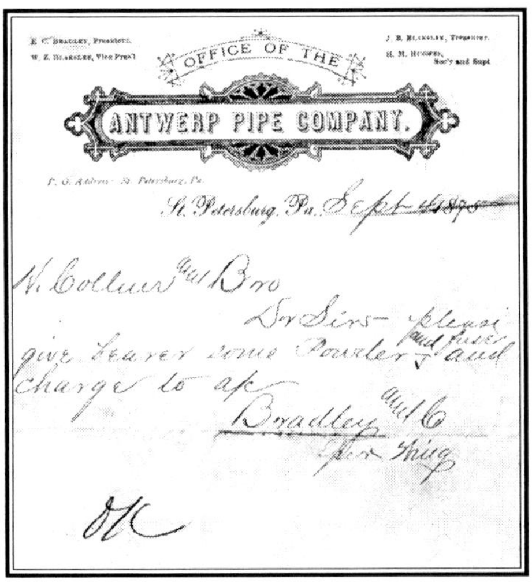


*Antwerp Pipe Company document, 1875*

**Place**

That the place under which said corporation is desired to be established is at Fullerton on the Allegheny River in the county of Clarion at which place is located the principal offices of said company and from which it is desired to lay and construct there main or principal lines of pipes for transporting oil into from the county aforesaid to the said Allegheny River.

**Real Estate**

That the lands and property of the company thus to be formed as described as follows: All that certain property or tenement and lot of land situated in Richland Township, Clarion County and state of Pennsylvania, bounded and described as follows: Beginning at a Black Oak on the bank of the Allegheny River at a corner of lands of James L. Colgan, thence by said Colgan's land North fifty eight degrees, East twenty eight perches to a post on the road to Emlenton thence by said road North fifty three and a quarter degrees West thirteen perches thence North fifty seven and one quarter degrees West Thirty six perches North sixty one and one half degrees West sixteen perches thence North seventy one degrees West twenty two and three tenths perches and thence North fifty three and one half degrees West six and three tenths perches to the corner and distances being along said road. Thence North seventy six degrees West four perches to a Black Oak on the bank of said Allegheny River. Thence down said river to the corner and distances there of thence thirty six degrees East ten perches South forty three degrees East seventeen perches South fifty degrees East forty two perches South fifty one degrees East sixteen perches to the place of beginning, containing ten acres more or less, and being part of Warrant 5728 recorded in the name of Samuel M. Fox.

All that certain piece or tract of land situated in Richland Township, Clarion County, and state of Pennsylvania, it being lot No. three (3) containing five rods front on Allegheny River and extending thirty one rods reference to map or survey will more fully explain Bounded on the East by Amos Goughler, West by Allegheny River, North by lot No. 4 and South by lot No. 5 same survey being part of a larger tract of land conveyed by Hannah Fox, Samuel Fox, and Mary his wife, by Hunsicker by three deeds dated the twenty first day of November A. D. 1854 and recorded in the office for recording deeds in said County of Clarion recorded in Book 5 page 48 in the 25$^{th}$ day of May A. D. 1865 and conveyed by Martin Hunsicker and Elvira his wife to James L. Colgan by deed dated Oct. 5 A. D. 1864 and recorded in Book I Vol. 9 pages 151 and 152.

Also that certain lot or piece of ground situated in Richland Township, Clarion County, and State of Pennsylvania, being five rods front on Allegheny River and extending a distance of ninety one rods, being lot No. 4 on plan of lots made out by James Colgan in reference they to will more fully explain bounded on West by Allegheny River, North East by Amos Goughler, South by Lyon Shot being part of larger tract of land conveyed by Hannah Fox and Samuel Fox and Mary his wife to said Hunsicker by three deeds dated the first day of April A. D. 1854 and recorded in Deed Book Vo. 5 page 48 and conveyed by Martin Hunsicker and Elvira his wife to James L. Colgan deed dated Oct. 5, 1864 and recorded in Book I Vol. 9 pages 151 and 152.

Also all that certain piece or tract of land situated in Richland, Clarion County and state of Pennsylvania being lots numbers one, two, five, six, seven, eight, nine, ten, eleven, twelve, and thirteen each lot contains one acre front on Allegheny River and extending back thirty one rods, reference by a map of plan in hands of James L. Colgan will more fully explain, bounded East by Amos Goughler, West by Allegheny River, North to the lot of Hunsicker, South by Lyon Short being part of a larger tract of land conveyed by Hannah Fox and signed the tenth day of

Nov. 1854, recorded in Deed Book No. 5 page 48 and by Martin Hunsicker to James L. Colgan deed dated Oct. 5, 1864, recorded in Book I, Vol. 9, pages 151-152.

**Personal Property**

The personal property consists of about 95 miles of iron pipe laid and in operation from Fullerton to St. Petersburg, Antwerp City, Penna., and vicinity in the county of Clarion, Penna.

Platform and back for loading cars

Iron Tank No. one. Ten thousand barrels

Iron Tank No. two. Twenty thousand barrels

Seven wooden tanks

Three boilers and two engines

Two oil wells with fixtures and machinery

One lease described as follows: situated in Clarion River Northside No. 6 in Blair Farm will be the same as said leases ten rods square bounded on the Clarion River on the South.

Also one lease on the J. J. Ashbaugh Farm in Clarion County, Penna. for pump station

Also one lease of ten acres on the Keating Farm, Clarion County, Penna.

That the real and personal property sum owned by us and belonging to said company is of the value of Eighty Thousand Dollars and constitutes the capital block of said company. That said capital will be divided into eighty shares of One Thousand Dollars each. That the number of said shares of stock and their place of residence and number of shares is as follows:

Ira C. Fuller, Fullerton, Clarion Co., 30 shares, $30,000

M. Hulings, Emlenton, Venango Co., 10 shares, $10,000

H. M. Hughes, Franklin, Venango Co., 5 shares, $5,000

C. F. Blakslee, Jos. B. Blakslee, and Wm. Z. Blakslee, St. Petersburg, Clarion Co., 20 shares $20,000

E. C. Bradley, 3 shares, $3,000

R. Richardson, Tarr Farm, Venango Co., 5 shares $5,000

The above information on the Certificate of Incorporation of the Antwerp Pipe Line Company was found in the Clarion County Courthouse, Clarion, Penna. in 1995, Title Book 2, page 249, October 12, 1872.

"There are a few gravity pipe lines, but generally the oil is forced through the lines by pumping. The capacity of each line is about 1,500 barrels in every twenty-four hours... The present cost of the pipe used is about 30 cents per foot and the average cost of the lines, including pipes, tanks, pumps and boilers, is about $1,500 per mile. A large share of the production is purchased at the wells by the companies and then transported on their own account. Many of the large producers, however, prefer marketing their own oil, and employ the pipe companies to transport it to the railway. The charge varies form 20 to 30 cents per barrel, according to distance. In such cases, the oil is usually pumped into the company" tanks, and from these, 42 gallons are delivered for each 43 gallons received at the well. One gallon per barrel was deducted for wastage." *The Early and Later History of Petroleum with Authentic Facts in Regard to its Development in Western Pennsylvania*, J.V. Roger, 1873, Co. Philadelpha, Penna.

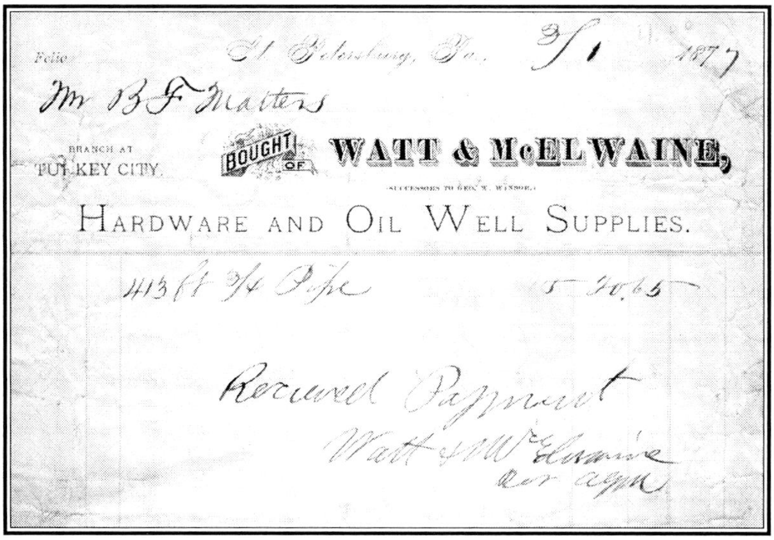
St. Petersburg, Pa. 3/1 1877
Mr B F Matters
BRANCH AT TURKEY CITY
BOUGHT OF WATT & McELWAINE,
HARDWARE AND OIL WELL SUPPLIES.
413 ft 3/4 Pipe
Received Payment
Watt & McElwaine

*Watt & McElwaine, Turkey City, recipt of 1877*

## Driving Too Fast

About five o'clock yesterday afternoon, Mr. F. I. Johnson was driving on Main Street, Antwerp City, and had occasion to stop near the oak trees. Samuel Rupert was not far behind and came along with a lively gait. Mr. Rupert unfortunately did not observe that Mr. Johnson had stopped, and before he had the opportunity to deviate from his course, he ran into Mr. Johnson's cutter, utterly demolishing it, and precipitating Mr. Johnson to the ground, who, however, luckily escaped with a slight injury to his back. No further damage was done. – Feb. 9, 1872, *The Progress*, St. Petersburg.

## John D. Rockefeller

The following is from the 1940 *Clarion County Historical Centennial Book*:

Marcus Hulings drilled the famous 'Antwerp' well, the first oil well in this territory on the Ashbaugh farm at the east end of town in 1871-1872. St. Petersburg became historically important. Almost overnight the population rose from 100 to 10,000 making it the largest town in Clarion County. Of the noted personalities who then visited and transacted business in the town, probably the most important was John D. Rockefeller, who organized among the local oil men the Antwerp Pipe Line Company. Joel Fink was a good friend, and many times invited him to the Fink House for dinner and business. Joel Fink from St. Petersburg was an oil producer and distributor of crude oil.

*John D. Rockefeller in 1872*
*Courtesy Drake Well Museum*

John Davison Rockefeller was born in Richford, Tioga County, New York, on July 8, 1839. John D. was brought up in Cleveland, Ohio, graduated from high school and a business college course. He then took a job as bookkeeper for a produce commission house. Rockefeller became interested in the uses of crude oil and helped steer his employer into the business of refining kerosene from Venango County petroleum in 1862.

John D. realized from the very beginning that there would be a great need for businesses that could take the raw crude oil and turn the black-green liquid into a finished product. The small refining company was called Rockefeller, Andrews and Flagler and by 1867 John D. was the major owner of the firm. A refinery needs oil and the Venango County oil territories were producing oil in the quantity never before seen. Rockefeller made many trips into the "Upper Oil Territories" which included Titusville, Pithole, Rouseville, etc., and then beginning in the 1870s the "Lower Oil Territories" which included places like St. Petersburg, Antwerp City, Edenburg, Parker's Landing, Foxburg, Elk City, and Petrolia. John D. formed alliances with local oil producers, such as Joel Fink of St. Petersburg, Clarion County, pipe line companies such as Antwerp Pipe Line Company, small independent refining companies and railroads. All were in his plan to send oil to his refineries. In January 1870, the name of the firm was changed to the Standard Oil Company of Ohio, which grew to control at least 70% of all refining facilities in the United States. In those years when the Standard Oil Company had a "strangle hold" on the oil industry, Rockefeller earned the name "robber baron."

Titusville, Pennsylvania, and the Drake Well started the flow of oil that led to the rise of John D. Rockefeller and his Standard Oil Company. The Parker's Landing, Lawrenceburg and Petrolia fields, followed by Antwerp City and the Hulings Antwerp No. 1, opened up the Clarion County fields which added to the Rockefeller oil producing reserves. From these humble beginnings "oil on the brain" went far and wide, east and west, north and south, and continent to continent. Oil related products have even gone from planet earth to the vast expanses of outer space.

## Ida Tarbell

Ida Tarbell, a world-class reporter and writer, grew up in Titusville during the early days of oil. Her father was an oil man and had bad dealings with the Standard Oil Company and in later years Miss Tarbell's writings and her History of the Standard Oil Company, helped in the break-up of the company by the U. S. Supreme Court in 1911. As early as 1895, Rockefeller began to turn the daily management over to his associates.

In later years John D. gave hundreds of millions of dollars away to good and worthy causes. He was well known for carrying a pocket full of shiny new dimes to give small children he met. John Rockfeller was a strong Baptist and abstained from the evils of smoke and drink. He died May 23, 1937 in Ormond Beach, Florida, just short of 100 years old.

## St. Petersburg Directory For 1872

This business listing verifies that the people living in St. Petersburg at the time were not "natives," but oil men and merchants who had transplanted here to reap the benefits of the oil boom. The spelling has been corrected when known to be incorrect.

Adams House, King & Hook propr's. Main
Alabaugh W., driller, bds. Main st
Arnold Geo. P., clerk at Donaldson's
Armstrong T., contractor, bds Main st
Armstrong Jas., driller, bds Main st
Atleburger J. R., clerk, Main st
Balliet A., propr St. Petersburg Hotel
Barr John, driller, bds Main st
Bayne S. G., opr. Bds. St. Petersburg Hotel
Beals Hiram, meat market
Beals & Bro., meat market
Beals James F., meat market
Bennett Jas., stoves, tinware and house fur goods
Bingham W. C., blacksmith helper
Black A. R., opr. Bds. St. Petersburg Hotel
Blakslee Bros., machinist, blacksmiths and oil well supplies
Blakslee C. F., of Blakslee Bros.
Blakslee W. Z., of Blakslee Bros.
Blakslee J. B., of Blakslee Bros.
Booth A. J., porter St. Petersburg Hotel
Booth M. S., clerk, St. Petersburg Hotel
Bonnell B. F., opr. Bds. St. Petersburg Hotel
Bould Jas. W., opr. Bds St. Petersburg Hotel
Boyce C. C., tinner, bds Main st
Boyce Robt. Manager for James Bennett
Boyce Wilson S., tinner
Boyle Jno., carptr, bds St. Petersburg Hotel
Bryan R., stoves and hardware, Main st
Brownell Jno., tank building, bds Main st
Brenen James, blacksmith helper
Butt G. W., of Wilcox & Butt
Butt H. H., cigars and tobacco, Main st
Burnap G. W. machinist
Cahrenger John, watch maker
Caller T. W., carpr, bds St. Petersburg Hotel
Caldwell M. G., billiard room
Campbell M., driller, bds Main st
Carney J., driller, bds Main st
Chalfant, Graff & Co., dealers in hardware, oil well supplies etc., Main st
Charlton James, machinist
Chisholm D., carpr, bds St. Petersbug Hotel
Christy Jno. A., of J. B. Donaldson & Co., res Oil City
Clingan G. W., livery stable, bds St. Petersburg Hotel
Collner H. & Bros., dry goods, groceries, etc., Main st
Collner H., of H. Collner & Bros., Main st
Collner Saml., of H. Collner & Bros., Main st
Collner Jos., of H. Collner & Bros., Main st
Collner C., clerk, Main st
Collner Lemuel, clerk, Main st
Cochran J. A., contractor, bds Main st
Coleman Jno., supt at McDonalds livery
Connelly James, blacksmith helper
Cunningham Mrs. H. M., millinery Main
Cunningham R. D., dollar store, cigars, tobacco, Main st
DeHaven R. W. & Co., dealers in clothing and cigars
DeHaven R. W., of J. R. W. DeHaven & Co., bds St. Petersburg Hotel
Donaldson J. B., of J. B. Donaldson & Co., bds St. Petersburg Hotel
Donald Jno., rig builder, bds Main st
Donaldson J. B. & Co., drugs & medicines, Main st
Dorn S. B., of Hopkins & Dorn
Duffey Jno., shoemaker, Main st
Dunbar A., porter, Monitor House
Dutt R., supt., St. Petersburg Hotel
Dunning O., contractor, bds Main st
Eshenour A. J., oil operator, office Main st upstairs, bds Main st
Eshnour Andrew, oil operator, bds Main st
Farner J. T., oil operator, bds Main st
Fertig & Milligan, oil producers, office Main Street upstairs
Fertig Frank, of Fertig & Milligan, Main st., bds St. Petersburg Hotel
Finley J., stage driver, bds St. Petersburg Hotel
Fiscus J. D., driller, bds St. Petersburg Hotel
Felming & Uncapher, billiard room, Main st
Fleming Jno., of Fleming & Uncapher, Main st
Flisher J. B., oil operator, bds Main st
Flisher & Farner, oil operators, office Main street up stairs

Four Geo. H., proprietor Monitor House
Foster J. E., carpenter, bds Main st
Frankenstein S., clothing & gents furnishing goods
Frack S., clerk, St. Petersburg Hotel
Fruenthal L., of M. Goldsmith & Co
Fulmer Henry, boarding house, Main st
Galbraith & mcClure, restaurant, Main st
Galbraith Wm. Of Galbraith & McClure, Main street
Galloway Jno., operator, bds St. Petersburg Hotel
Gaincy Mat, driller, bds St. Petersburg Hotel
Geer E. N. & Co., harness, etc., Main st
George Bros., flour and feed
George W., of George Brothers
George Theod., of George Brother
George J. M., of George Brothers
George J. M., oper., bds St. Petersburg Hotel
Gibbs, Russel & Co., branch office
Gibson D., driller, bds Main st
Gilfillan Jas., groceries, Main st
Gipps ___, bds Main st
Goss A., oper., bds St. Petersburg Hotel
Goldsmith M., watches and jewelry
Goldsmith, M. & Co., hotel
Graham Wm., clerk, Wing, Wilber & Co
Groves J., contractor, bds Main st
Graham Thos. S., stage driver, bds St. Petersburg Hotel
Green M. W., livery stable, bds St. Petersburg Hotel
Hallock I. W., blacksmith
Hall A. D., of Hall Bros., bds Main st
Hall H. D., of Hall Bros., bds Main st
Hall Bros., restaurant, opp Petersburg Hotel, Main st
Haver A. S., atty at law, bds St. Petersburg Hotel
Hazlett John, laborer, bds Main st
Hazlett Jas., restaurant, Main st
Hazlett John, carpenter, bds Main st
Hazlett Wm., carpenter, Main st
Herron D. S., attorney at law
Henry T., bookkeeper, bds Main st
Hoaks H. E., carpenter
Hopkins T. W., of Hopkins & Dorn
Hopkins & Dorn, physicians & surgeons, Main st
Huston Robert, carptr, bds St. Petersburg Hotel
Jack Matthew, pipe cutter, bds Main st
Johnson F. I., drugs & medicines, Main st
Johnson D. C., carpenter, bds Main st
Johnson J., shoemaker, Main st
Jones E., carpenter
Keating Hugh, oper, bds St. Petersburg Hotel
Kearney P., lumber dealer, bds St. Petersburg Hotel
Kibbe C. A., of Wochter, K. & Co
Killgore Jno., carpenter, bds Main st
King & Hook, props Adams House
Koch Max, of Sonneborn & Koch, Main st
Lawson T. J., blacksmith
Latin James, waiter, Main st
Levi Alfred, clothing, etc.
Levans A., driller, bds Main st
Lewis Ira, at McDonalds livery
Lewis Ira, livery stable, bds St. Petersburg Hotel
Little J. J., clerk, bds Main st
Lehman N., boots and shoes
Lehman C., clerk
Lugardts William, cook, Main st
Matters Wm., hackman, bds St. Petersburg Hotel
Mathew Jas., rig builder, bds Main st
Mayer Bros., jewelry and cigars, Main st
Mayer L., of Mayer Bros.
Mayer H., of Mayer Bros.
Mayer L., jeweler, bds St. Petersburg Hotel
Mayne S., laborer, bds St. Petersburg Hotel
Mecer R., driller, bds Main st
Milligan Groves, of Fertig & milligan, Main
Mong Wm., groceries and confectionery
Monitor House, Geo. H. Fourl proprietor
Morrison E., blacksmith helper
Morow G. W., driller, bds Main st
Naylor Thos., engineer
Nocker & Roschi, bakery & grocery
Northup B., oil operator, bds Main st

Obite And., oil operator
O'Donnel Ja., clerk at Welsh's
Palmieter Louis, driller, Main st
Patton E. H., painter, bds Main st
Place J. M., postmaster & news depot
Place Marvillo J., assistant post master
Place Jas., of Place Bros., bds Main st
Place N. W., of Place Bros., bds Main st
Phillips David, blacksmith
Phillips Geo., New York bakery, Main st
Pike Albert, carpenter, bds Mainst
Quigley G. B., operator, bds St. Petersburg Hotel
Quigley B., hackman, bds St. Petersburg Hotel
Reecher Isaac, meat market
Rea Albert, waiter, bds Main st
Ramsey J., driller, bds Main st
Retchel W. C., barber, bds St. Petersburg Hotel
Reyburn J. H., oil operator
Reynolds W. H., surgeon dentist, Main st
Retcher W. C., barber
Robinson Jno., butcher
Robinson Bros., gaugers, Parker, St. Petersburg & Petroleum Centre
Robinson William
Robinson James
Runkel Anthony, of Shiler & Runkel
Roscoe James, machinist
Ryan B. H., driller, bds Main st
Seymour, Taylor & Co., hardware and oil well supplies
Seymour C. B., of Seymour, Taylor & Co
Seiple A., butcher
Share & Cooperman, dry goods, Main st
Share Doritz, of S. & Cooperman
Shaler Jno., livery stable, bds St. Petersburg Hotel
Shanner ___, driller, bds Main st
Shannon E. O., attorney at law
Shearer W. E., of McKee & Shearer
Shell ___, carpente, bds Main st
Sheffield Geo., of Seymour Taylor & Co
Shirley Jos., clerk, bds Main st
Shiler & Runkel, livery and feed stable
Shiler J., of S. & Runkel
Shull M., carpenter, bds Main st
Silverberg H., clothing, etc.
Silverberg D., oil operator
Sill Geo., opr., bds St. Petersburg Hotel
Sloan A. R., bookkeeper, Blakslee Bros.
Sonneborn & Koch, cigars and gents fur goods
Sonneborn Saml., of S. & Koch
Steffee William, clerk, bds Main st
Stevens S. D., opr., bds St. Petersburg Hotel
Stewart, T., meat market
Stewart J.
Stewart S. S. of Wachter, Kibbe & Co
Sloughenhopt Jas., laborer, bds Main st
St. Petersburg Hotel, A. Balliet prop., R. Dutt superintendent
Taylor Captain, hardware, bds St. Petersburg Hotel
Taylor Lee, hardware, bds St. Petersburg Hotel
Taylor A., clerk, Seymour, Taylor & Co
Taylor C., of Seymour, Taylor & Co
Urmson W. H., machinist
Uncapher A. J., of Fleming & Uncapher, Main st
Vogan John, blacksmith
Vensel Barney, land agt, Main st upstairs
Wachter, Kibbe & Co., cigars & tobaccos, and gents furnishing goods
Wachter H. C., of W. Kibbe & Co.
Wachter & Kibbe, real estate agents, opp St. Petersburg Hotel
Watt Chas., bookkeeper, Seymour, Taylor & Co
Watt Chas., hardware, bds St. Petersburg Hotel
Welsh Jas., eating house, Main st
Welsh J. H., wines, liquors & tobaccos, Main st
Whitling Dan'l., cigars, tobaccos, etc., Main st
White Saml., carpenter, bds Main st
Wilcox & Butt, furniture and undertakers, Main st
Wilcox S. J. of Wilcox & Butt, bds

St. Petersburg Hotel
Wilber M. O., of Wing, Wilber & Co.
Wilber W. C., merchant, bds St. Petersburg Hotel
Wilson Robert, cigars, tobacco, etc.
Wilson G. J., billiard hall, Main st
Wilson T. C. & G. W., dry goods, clothing, etc., Main st
Wilson G. W., of T. C. & G. W. Wilson
Wilmot H. H., carptr, bds St. Petersburg Hotel
Wilson Chas., of Chalfant & Graff, bds Main
Wing, Wilber & Co., groceries
Wing, A. K., of Wing, Wilber & Co.
Wing W. O., of Wing, Wilber & Co.
Woodford N. S., oper., bds St. Petersburg Hotel
Wollf Saml., W. V. tel oper., bds Main st
Wyman A. L., barber

## Ritts and Rockefeller

John Vensel Ritts was born in 1852 at Petersburg, the son of Elias Ritts, prominent citizen and "general speculator." Mr. Ritts, prior to 1890, was a bookkeeper and cashier at the St. Petersburg Savings Bank. In 1882, Mr. Ritts married Irene C. Blakslee, daughter of W. Z. Blakslee of Blakslee Bros. of St. Petersburg. Mr. Ritts was in the oil and gas business from the discovery of oil in Clarion County. Elias Ritts owned considerable acreage in Richland township, Clarion County, and all proved to be excellent oil territory. J. V. Ritts, in 1890, established the Butler County National Bank in Butler, Pennsylvania, and accumulated vast holdings which brought him great wealth. Mr. Ritts was a pioneer in the early days of oil discovery in Clarion County and created personal contacts with all the early oilmen that remained throughout life. John V. Ritts was an intimate and personal friend of the late John D. Rockefeller and such other prominent figures in Standard Oil as Dan O'Day and Joseph Seep. Once, when John D. Rockefeller was confronted with a monopoly law suit, Mr. Ritts promptly went on his bail bond and kept Rockefeller out of jail. Mr. Ritts regularly visited Mr. Rockefeller at his Florida residence and reminisced "Old Times in Oildom." Mr. Ritts's granddaughter loved to swim in Rockefeller's pool. J. V. Ritts died in Butler, Penna., June 24, 1937 and is buried at the North Cemetery. (*Butler Eagle*, June 24, 1937)

*John V. Ritts*
*Clarion County History 1887*
*A. J. Davis*

# Jack Cribbs

The first Chief of Police, Jack Cribbs, was hired by the newly formed town council of St. Petersburg in 1872. At the time of the "oil boom," crime was running rampant with law and order utterly disregarded. John W. "Jack" Cribbs wrote of his travels by writing the *Adventures of Jack Cribbs*, Volumes 1, 2 and 3. Cribbs supposedly "cleaned up" many an early oil town but said one of his toughest arrests was the "Big Dutch Man" at Antwerp City.

Jack's life-long career of fighting crime and dangerous felons came to an abrupt end. The following information comes from the 1968 supplement to the Clarion County History book, page 81. Jack Cribbs, noted Clarion County detective and author of several hair-raising books of his experiences, died of burns received when his Monroe Township home burned. He was hurt while trying to save his house and valuables. The year was 1915. These three books are very informative and interesting to the student of early Clarion County history.

## The Resurrected Dutchman

Taken from *The Adventures of Jack Cribbs*

John Dearolph, the High Constable, called me one evening and said there had been a man killed at Headtown, or Antwerp, that some Dutchman had been shot. We doubled quick and were soon there, about a half mile from St. Petersburg. When we arrived, we found the street crowded with people, probably a thousand or more, gathered around the building where the shooting occurred. The crowd yelled at me not to enter, that there were two or three desperate men in there and they swore that they would kill the first man that would enter. I never stopped

*An early view of Edenburg*

only long enough to inquire whether the men had been seen leaving the building. I called out to Dearolph to come on. He did. We went straight through the building and found the house entirely deserted, not a soul in it. I then inquired from the crowd where the dead Dutchman was. They pointed out the house. I proceeded at once, and thought that I could get a clue if the man was still living. I went in and found a man lying in bed. No one else in the room. He was moaning and groaning. I went up to him and at once recognized him as a man I took for a tank builder. I inquired where he was hit. He replied, "Mine Got I am a dead man, I can't talk, where is the doctor? I am a dead man."

He was groaning and moaning. I at once uncovered him, taking the bed clothes off him. I said, "Who shot you? Tell me at once, do you know? Can you identify them? Speak out." He says, "Oh, my Gott, I am a dead man."

"Tell me where you are hit."

"Oh mine Gott," and he put his hand across his bowels. I examined him and found only a slight scratch across his stomach merely enough to draw a few drops of blood. I took hold of the fellow and raised him out of bed and I convinced him that he was not hurt, only scared, but it required some work to do this for the Dutchman, Hans, had fully made up his mind that he was fatally shot and it looked to me like a case of being scared to death. The fellow might have never rallied from the scare had I not raised him up, stood him on his feet and made him walk around the room to fully convince him that he was not hurt but only scared almost to death. But he never told me who his assailants were.

## Hollenback and Barber His Prisoner

Taken from *Adventures of Jack Cribbs*

Another sketch that is somewhat amusing occurred in 1872. A gentleman called on me while I was acting policeman and said his name was Hollenback and that he was from the upper oil country, that he wanted a man arrested, and that I had been highly recommended to him for the job. I informed him that I was busy and had no time, that I was paid for looking after my duties as a policeman. I then inquired of him if he had any track of this man. He replied that he had, that he was in Headtown. I informed him that the place was only a short distance away and that the original name was Antwerp, but that the town was called Headtown for the reason that a man could at any time get a fight there and a head thrown in. He told me that the fellow he wanted was Hank Barber. I at once told him that I knew Hank Barber and I would try to get him. The gentleman told me that he had been Barber's bondsman and the latter had skipped him and that he wanted to make sure of Barber. He said that if I would arrest him and lock him up, he would take him back O.K.

I had met this man, Barber, and heard that he was quite an actor, and I informed Mr. Hollenback that I would arrest Barber as soon as possible and advised him that when I would get Barber and lock him up, that he had better put a watch on him. He thought not. I went to work and the next evening I had Mr. Barber in the lock up. I then reported to Mr. Hollenback and he paid me for my services. Again I cautioned him to put a watch on his man and to secure a good man to accompany him on his return trip. Hollenback replied that he had brought iron along and that he would land him all right and that he had examined the lockup and it would hold the Devil himself secure. He said to me that he would furnish Mr. Barber with his supper and breakfast himself; that he had been a landlord for years and well knew how to cater to the wants of guests. He then requested me to be present at the supper that he would furnish his guest, saying that he would get him the best the town could furnish.

I was at that time stopping at the St. Petersburg Hotel. Mr. Balliett was the proprietor. It was a first-class hotel. Mr. Hollenback was also stopping at the same hotel, and we arranged to get his prisoner a good supper. I was on hand promptly with key to the lock up and to unlock the cell door. Mr. Hollenback entered with his big tray containing a supper fit for a prince. Mr. Barber certainly enjoyed it and did justice to it, and thanked Mr. Hollenback for the supper. Mr. Hollenback replied, "There is nothing too good for you. I am delighted to see you. You were lost, but now you are found. I am delighted to have the pleasure to attend you and escort you back. I have gone to great trouble to get you and I intend to hold you, my friend."

Then Barber replied, "There is many a slip 'twixt the cup and the lip."

"Well," said Hollenback, "you slipped me once and I am satisfied you will never get out of here, and when I get those on you (pulling out a pair of hand cuffs) you will be safe enough with me."

Mr. Hollenback bade Barber good night saying that he would call early with a breakfast to match the supper and that after a good breakfast they would drive to Foxburg and take the train on the A.V.R.R. for home. Mr. Barber replied: "The trains are not worrying me in the least but I am thinking now of the good breakfast."

Mr. Hollenback replied that he would be on hand with as good a breakfast as the town would afford.

Mr. Hollenback requested me to be on hand in the morning to let him in to attend to the wants of this bird.

Well, the morning came, and with it came Mr. Hollenback, telling me that he had ordered breakfast and that he would be around in a few moments. The time came and we proceeded to the lockup, and when I unlocked the front door, Lo and behold you could see daylight through the lockup. Mr. Hollenback was there with his tray, loaded with eatables, and yelled, "Barber, Hello Barber," no reply came.

"My God," says Hollenback, and down fell the platter, dishes and all. We went in and examined it and found that Mr. Barber and two common drunks had skipped and from the looks of the jail they must have had outside help. Mr. Hollenback had to return home without his man and he was a much disappointed man.

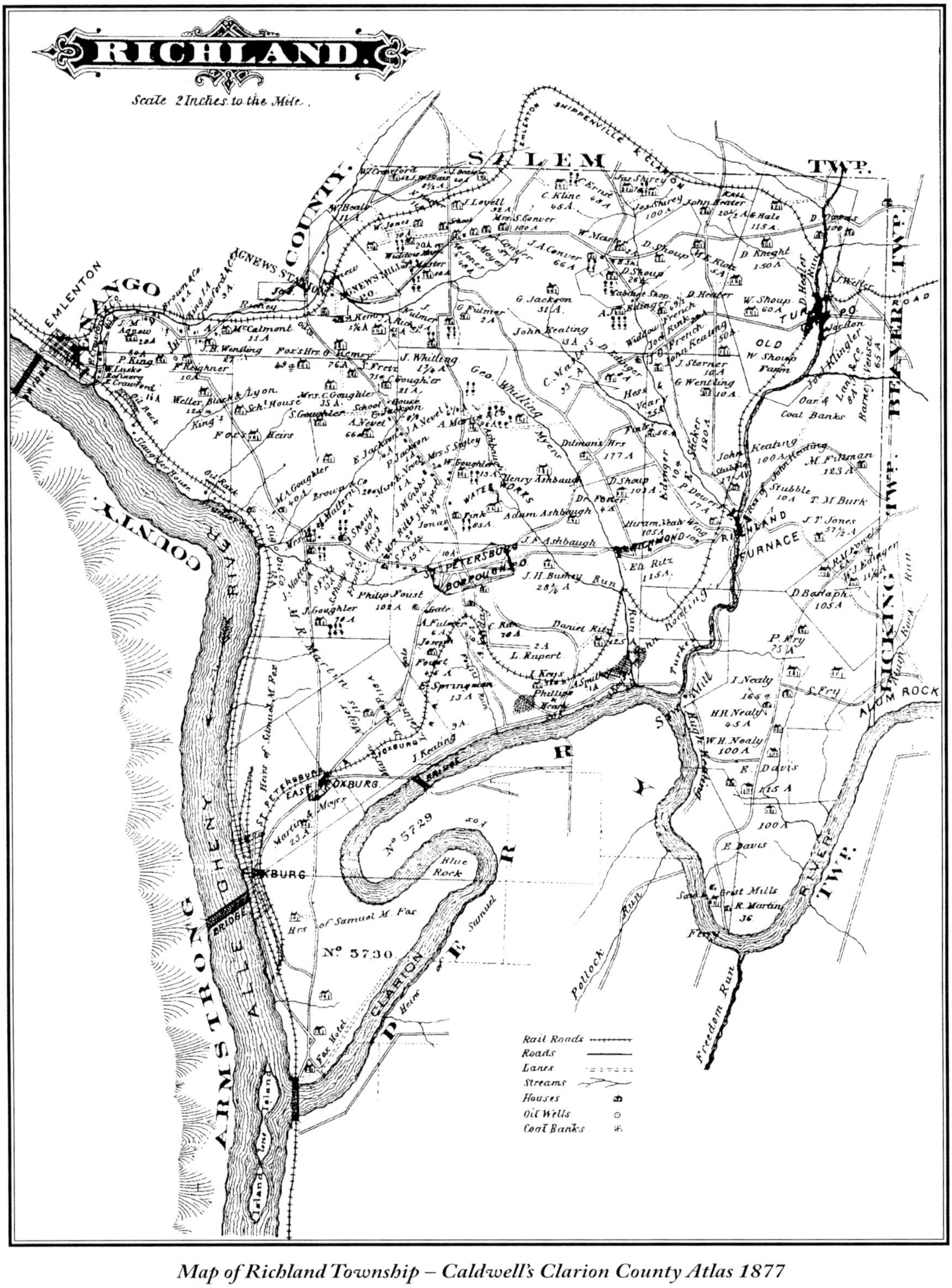


*Map of Richland Township – Caldwell's Clarion County Atlas 1877*

## Richland Township Directory of 1872

Spelling has been corrected when known to be incorrect

Agnew James
Amsler Peter
Arbuckle J. B.
Ashbaugh Adam
Ashbaugh Henry
Ashbaugh Daniel
Ashbaugh Jacob
Ashbaugh Abraham
Ashbaugh John J.
Barkley Augustus
Beals Jacob
Beals Hiram
Beals William
Beals Levi
Beals James
Best Simon
Bostaph Daniel
Bostaph Samuel
Brant Harmon
Brice James
Bryon Robert
Bryon Lowen
Chears Thomas Sr.
Chears Thomas Jr.
Chrisman Adam
Collner Samuel
Collner Harrison
Collner Levi
Collner Lewis
Conver Jacob
Conver Aaron
Coulter J. E.
Crawford William A.
Crum Oliver
Culp Samuel
Curtis J. A.
Dansen Mate
Davis Ebinger
Dearolph John
Devore John
Dittman David Sr.
Dittman Adam
Dittman David Jr.
Dittman John
Dittman Martin
Don Leper
Donken William
Dower Peter
Dunkle B. B.
Edinger Wm. F.
Edinger George
Edinger Wm. H.
Edinger Samuel S.
Edinger A. J.
Edinger Daniel
Edinger Henry
Edinger Samuel J.
Edinger Andrew
Edinger Charles
Edinger John
Fertig Charles
Fillman Jonathan
Fillman Michael
Fink Joseph
Fink Joel
Fink Jonas
Foust Joseph
Foust Philip
Foust Samuel
Foust Benjamin
Foust Solomon
Foust Wm.
Frazier George
Frank Martin
Frets Solomon
Frederick Wm.
French Samuel G.
Fritz Joseph
Fullmer George
Fry Samuel
Fry Paul
Garner Henry
Goughler Amos
Goughler Samuel
Goughler Joshua
Goughler Henry
Goughler Charles
Goughler Jacob
Goughler Wm.
Gilbert David
Gilger Jonas
Graham John
Hale George
Heeter Daniel
Heeter John J.
Heeter S. B.
Heeter Amos
Irvin Miles
Jackson George
Jackson Peter
Jackson Andrew
Jamison J. T.
Johnston Peter
Johnson Adam
Jones Wm.
Jones George Jr.
Jordan Adam
Keating John
Keating M. L.
Keter Peter
Kinny Henry A.
Kinny George
King Peter
Kline Charles
Klinger Joel
Klotz Charles
Knight Daniel
Knight Emanuel
Laughner James
Laughner Newton
Lovil Jesse
Lowers Jacob
Lowers Anthony
Malarkey Thomas
Martin Samuel
Martin Richard
Master George
Master Charles
Master Nathan

Mathorn John
Mathorn George
Miller John
Mitchell William
Monroe W. H.
Morgan Morgan
Morgan John Sr.
Morgan John Jr.
Morehead Alexander
Myers Jacob
McCamon Alexander
McCrea H. A.
McFerson Albert
Neely William
Neely Isaac
Neely Hiram
Neely Richard
Neely Harison
Neely A. M.
Neely F. J.
Nevil J. A.
Nevil Amos
Nevil Jacob
Parmer A. S.
Powell J. N.
Reighner Jeremiah
Reighner Franklin
Richison Albert
Ritts Elias
Ritts John H.
Ritts Daniel
Ritts Charles
Ritts John Sr.
Ritts Elias (of J)
Ritchey Alex
Rupert Louis
Rupert Wm.
Rupert Jeremiah
Shell Emanuel
Shicker Arnst
Shirey Rudolph
Shirey Samuel
Shirey Saml. A.
Shirey James
Shoemaker D. O.
Shoup David Sr.
Shoup Wm.
Shoup James Sr.
Shoup David Jr.
Shoup David (of John)
Shoup Barney
Shoup John Jr.
Shoup John Sr.
Shoup James W.
Shoup Jacob
Shoup Michael
Sigley George
Sitelr Alex
Slaugenhoupt John
Smith Campbell
Sterner John
Sterner Reuben
St John T. C.
Stobble Charles
Stoble Frederick
Tippery Andrew
Tippery Samuel
Vensel Nelson
Vensel Charles H.
Vensel Barney
Vensel Alfred
Watson H. C.
Way Benjamin
Way John
Weaver G. W.
Wentling Joshua
Wentling Geo.
Wesner I. M.
Wesner Wm.
Whitling John
Whitling George
Whittling Saml
Whittling Daniel
Whittling Wm.
Williamson John

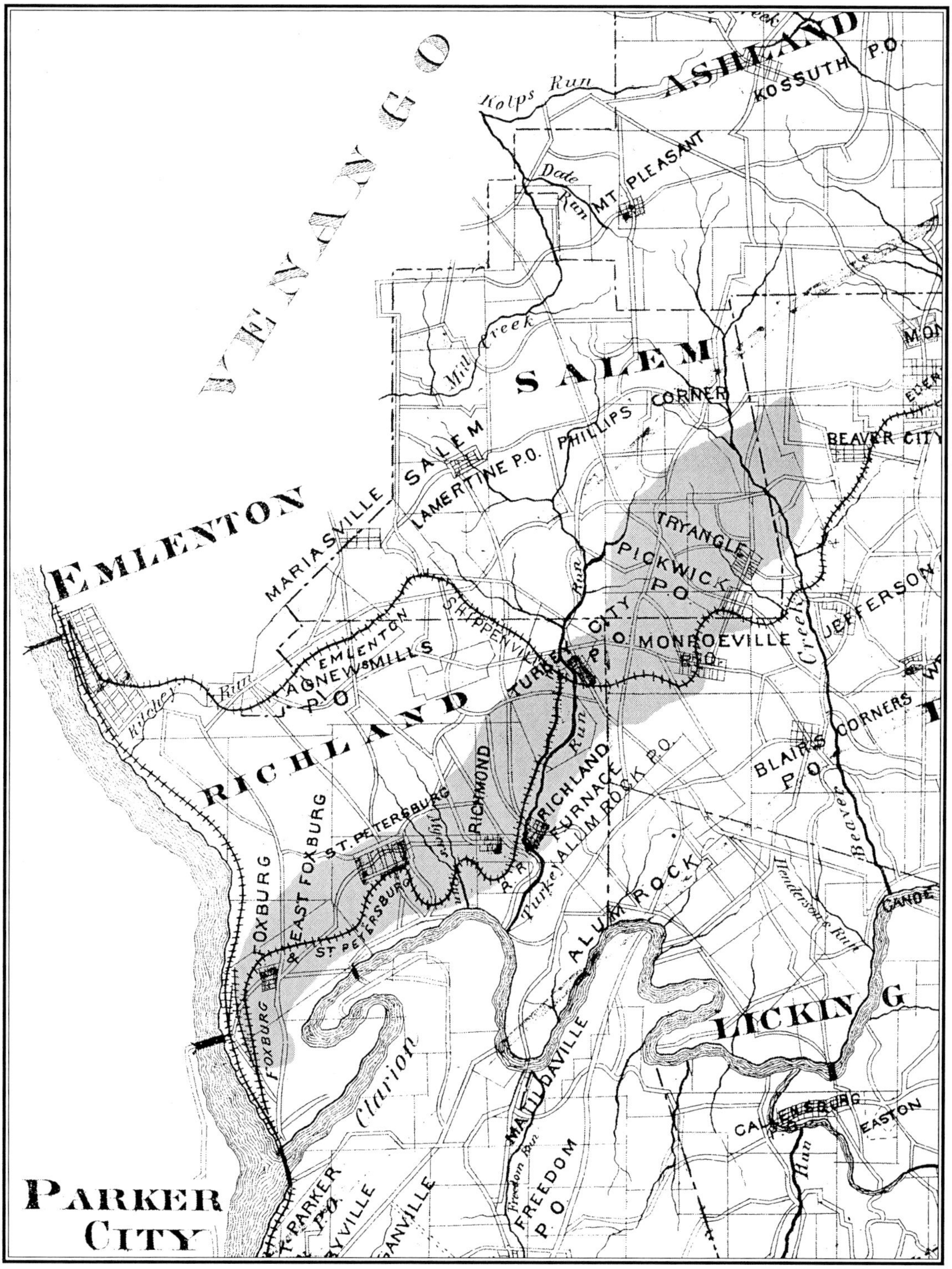


*This lower belt began at or near Clarion on the Clarion River and extended a distance of 43 miles to and terminating at St. Jo in Butler County. The entire width of the sand rock does not exceed three miles and because the productive area is so vast, a rapid development was due. This map of the oil belt is from the 1877 Caldwell's Clarion County Atlas*

*Residence of Elias Ritts on left. St. Petersburg Savings Bank on right. 1877 Caldwell's Clarion County Atlas*

*Elias Ritts*
*Clarion County History 1887*
*A. J. Davis*

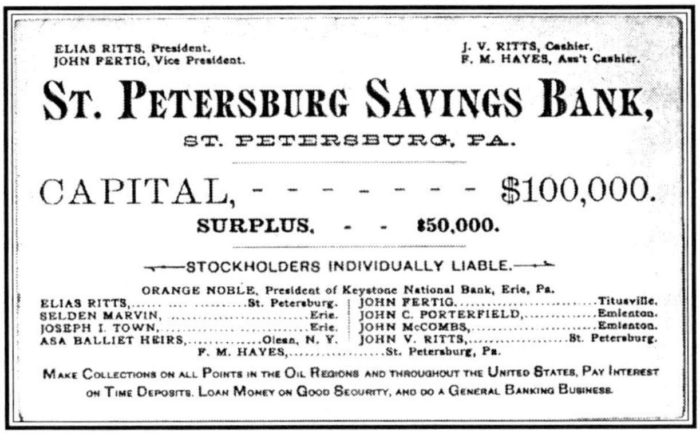


*St. Petersburg Savings Bank advertisement*

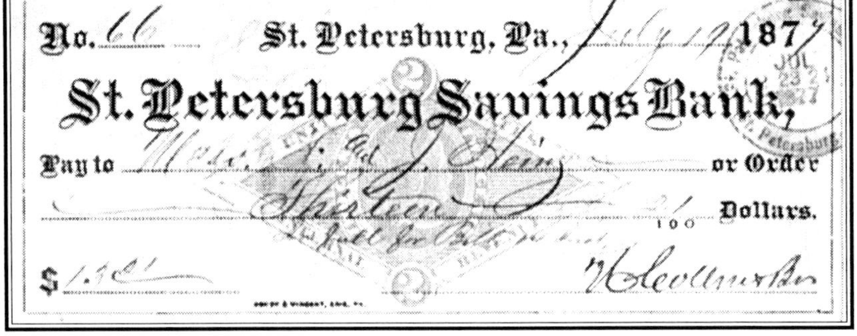


*St. Petersburg Savings Bank check*

## Elias Ritts

In the early days of oil excitement near the mouth of the Clarion River, Elias Ritts became an oil operator and producer. Upon the discovery of petroleum in territory contiguous to St. Petersburg, the necessity of banking facilities soon became apparent and resulted in the St. Petersburg Saving Bank in 1872. Its organization was effected by the election of Hon. John W. Hammond of Erie, president, Hon. John Fertig of Titusville, vice president, and Charles Horton of Erie County, cashier. Within a year Mr. Horton died and J. V. Ritts became cashier and general manager. As business increased, additional banking houses were established at Foxburg and Turkey City. At the banks' heighth, they were handling in excess of one million dollars a day. The St. Petersburg Bank peaked out by 1878 and was finally closed in 1911.

## Those Titusville Boys

The following newspaper articles come from the *Titusville Morning Herald*. Many of the newspapers published in this area during the oil boom era have through the years been lost and destroyed forever. The *Titusville Morning Herald* has provided a wealth of information on the beginning of the oil industry in Clarion County. I spent much of the winter of 2001 at the research library of the Drake Well Museum. These articles show the progression of oil discoveries in the lower field.

A very important fact that I need to get across to the readers of this history is how our lower field was explored and drilled. This field was developed by a great number of oil producers, wildcatters and drillers from the Titusville, Petroleum Center, Rouseville, Oil City, and Parker's Landing areas. These men talked oil, ate oil, and slept oil. Their capital and expertise of the oil business is what opened up our fields. Not only did they have a need to expand oil producing territories, they also expanded their bank accounts. Little old Clarion County farmers owned the land and those Titusville boys rolled up their sleeves and dug in. It was a time of great excitement, all created by "Oil on the Brain."

## Oil Field Progression

**AUGUST 1871**

A well on Grass Flats some 5 miles above the mouth of the Clarion, owned by John H. Galey and others, finished and starts at 60 brls/day but declined to 30.

The Chestnut Shade, on the Anchor farm, east side of Allegheny River opposite Parker's Landing, finished, a 15 brl. Producer.

A well on the Fox farm near Foxburg railroad station starts at 15 brls.

The Clarion Democrat notes that a farmer in Richland Township, Clarion County, refused an offer of $1,000 an acre for his farm and that by reason of the oil excitement in that neighborhood, land worth $50 per acre for farming purposes is commanding $1,200 per acre.

A well on the Grass Flats, Clarion River, owned by I. C. Fuller, R. P. Crawford, R. L. Brown and F. Jackson, finished and is a 25 brl. Producer.

**SEPTEMBER 1871**

H. A. Speer & Co. complete a well near the mouth of the Clarion river doing 40 brls/day.

The Palmer well on the David Shoup farm near St. Petersburg, owned by Palmer, Bovard & Hill, finished and is a 25 barrel producer.

The Peach Orchard well, Galey farm, near Clarion river, owned by H. Spears and others, drilled 52 feet in the third sand and started at 25 brls/day.

**OCTOBER 1871**

*Petroleum Centre Record* notes an increasing activity in the oil operations in the vicinity of St. Petersburg.

A well on Ashbaugh Run near St. Petersburg on land of John J. Ashbaugh owned by Marcus Hulings and others, starts at 150 brls/day.

The Pioneer well on the Clarion River near the mouth of Turkey Run, owned by D. L. Kistler and others, finished and is a 10 brl. Producer.

A well owned by S. M. Kier and others on the Joseph Foust farm between St. Petersburg and the Grass Flats, started at 35 brls/day.

Much excitement along the Clarion.

Wells Nos. 1 and 2 on the Grass Flats owned by J. H. Galey and others, finished each yielding 10 brls/day.

*The rough road from East Foxburg to Grass Flats*

Marcus Hulings' Antwerp well near St. Petersburg increased to 250 brls/day.

Three hundred wells reported under way along the Clarion River between Alum Rock and the Allegheny River.

A party of civil engineers start out from Parker's Landing to establish a grand belt theory connecting the Allegheny Valley oil regions with those of West Virginia and Tennessee

The Moriarity well on the Collner farm near St. Petersburg completed and pumped 150 barrels the first 24 hours.

Excitement running high along Turkey Run, Clarion County.

Owners of the Collner farm near St. Petersburg refuse $1000 an acre for the property.

Hulings' Ashbaugh or Antwerp well still flowing at 75 brls/day.

## From Parker's Landing, October 30, 1871

Editors *Morning Herald*

Business here at Parker's Landing continues active and progress is the order of the day. New buildings are going up rapidly and the oil development throughout the surrounding districts is astonishingly on the increase.

We hear of a new well at Petersburg called the "Boss", whether named in honor of "Boss" Tweed is not stated, but it will be a good thing if it has struck the Tammany "belt."

Yours Truly, " Sand Pump Line"

## Oil News, November 3, 1871

A well was struck on the Traver's track near Foxburg adjoining the Nolan farm last Saturday. Doing 35. Owned by M. C. Martin and G. R. Harms.

Several wells are being put down to test the territory in Knox township. One on the land of Ellison and Co., and others whose names we have not learned.

## Oil News, St. Petersburg, November 6, 1871

Editors *Morning Herald*

The "Neely" well on the Neely farm, one quarter of a mile east of the Hulings well, penetrated the oil rock on Saturday last, and yesterday was flowing through the casing at a rate of 50 barrels per day.

The "Pleasantville" well on the Eli Ritts farm, was tubed Saturday, and is pumping about fifteen barrels. The proprietors of this well, Goss Bros., Goal, Heart, Hicks, and Johnson, have a large stock of territory around this well and will commence the drilling of more at once.

A new well was struck on the H. Neely farm on Friday last. The well is flowing at the rate of 35 barrels per day with much gas and power. It is owned by Messrs. Patterson and Dickey, M. L. Lockwood, and W. H. Nickolson of Petroleum Centre, Keny Bros. of Columbia Farm, and Armor and Raymer of Parker's Landing.

Yours Truly, "Sand Pump Line"

## Oil News, November 13, 1871

The Parker's Landing district, and the territory in the vicinity of St. Petersburg near the mouth of the Clarion River, are the points where the most extensive operations are being conducted, not less than 200 wells now going down in that section, and with favorable indications for the success of most of them.

## *Emlenton Sun*, November 14, 1871

To have some idea of the amount of developing that is being done, it is only necessary to say that from a single point on the road between Petersburg and Foxburg there are 16 derricks to be seen where only one stood one month since and at a point on the Adam Ashbaugh farm we counted where 20 rigs were going up on the farms of John Ashbaugh, Daniel Ritts, Charles Ritts, David Shoup and David Dittman.

A new well on the Hiram Neely farm has a fair prospect of being equal to any in the regions if reports are to be relied on. It was reported that from last Saturday till Monday there were some 50 barrels of oil saved. The well was not finished drilling at latest accounts.

*Main Street, St. Petersburg, 1876, taken from roof of St. Petersburg Hotel*

*Main Street, St. Petersburg – Adams House, Oil Exchange, and Opera House*

Coal is produced in great abundance here, some of the mines being on oil producing farms, and coal can be had for fifteen cents per bushel, delivered at the wells. The wells in the immediate vicinity of St. Petersburg go through a rich vein of coal.

*Main Street, St. Petersburg. Bank on left, telegraph office in center, hotel on right*

## Sketch of Petersburg

Petersburg, November 15, 1871
Editors *Morning Herald*

As a brief sketch of the new El Dorado down the "river" at Petersburg may not be interesting to your readers. I shall endeavor to give you some information regarding it.

The old village of Petersburg was a one-horse place on the Clarion River, about 2 miles from Foxburg Station, Allegheny Valley Railroad, and was never disturbed by an oil excitement till the Hulings well began to flow on the Ashbaugh farm some three weeks since. News of this strike spread up the "creek" and was soon followed by a flight of operators from Titusville consisting of Messrs, Watson, Cody, Hall, Andrews, Blake, Bayne and Fertig. These gentlemen began to invest round the Huling property and in a few days the prices of land rose from the old conventional rates for farming purposes to $1000 per acre, at which figure Messrs. Watson and Bayne made a purchase of twelve acres on the Ritts farm, while Messrs. Andrews and Bly gave $13,500 for twelve acres adjoining. Leasing then became brisk, and the smallest coin heard of in the way of bonus was $1000. This state of things lasted but a short time, as a break took place in the market, and leases went down to par at the close. The "bulls" however had another card to play in the "Neely" well, which was drilling about half a mile ahead of the Hulings, which they claimed would bring leases up to the opening price, but they were doomed to disappointment, as when this well was completed last Friday, it produced but 20 barrels. However, with the Hulings doing 75 and the Neely 20, this promises to be the most reliable in the country, as the sand rock is of a fine quality in both wells. Some fears were entertained after the striking of the Huling that it might affect the price of oil by opening up a large field, but these thoughts were soon dispelled after the "Neely" well commenced pumping.

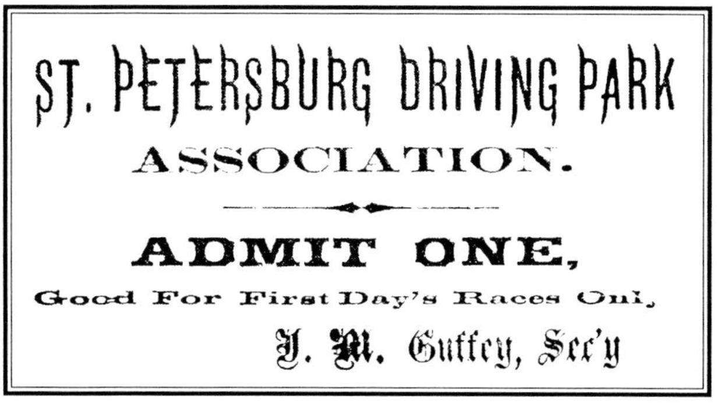


***St. Petersburg Driving Park admission ticket***

"Petersburg is reached by stages from the railway and is being rapidly built up, the new buildings consisting of hotels, stables, stores, machine shops and boarding houses.

Mr. Wing is selling lots as high as $1000 from his recent purchase, while Mr. G. J. Sherman, of Titusville, is starting a bank in Mr. Vensel's house. We understand his "greenback trap" is on its way from Herring's.

The Balliet House is headquarters and it takes one to brave a thousand shocks to get through its densely crowded rooms. Three "bar keeps" preside at the bar from early morning till late at night. Mr. Balliet does all he possibly can to accommodate his patrons, but notwithstanding this fact, we noticed ten noblemen courting "Morpheus" on the coca nut matting around the office stove at four in the morning.

The new city of Antwerp is just started on the banks of Ashbaugh Run. This city is a continuance of Petersburg. The first lots were sold at auction, ranging from $200 to $260 per lot. The city is in rather a crude condition, nary a board having been erected at this writing, but preparations are being made to out rival her sister city of Petersburg. Active operations are commenced on the under-mentioned leases, but as ninety days must elapse before they can get their first well down (the depth of drilling being over 1,000 feet), matters must remain in status quo till that time.

On the Ritts farm, Andrews lease, four wells are going down.

Fezzin's lease, two wells are going down.

Watson and Bayne, two wells.

Wyble, two wells.

Fertig, two wells.

On the John J. Ashbaugh farm, Hulings and Co. are drilling five wells.

Wilkinson, two wells.

Watson and Bayne, and others, three wells.

Messrs. Hall, Patterson, and Dickey are about to drill one in the vicinity of the Neely farm, while Mr. Hulings is putting a test well on Turkey Run on Judge Keating tract. Everything taken into account, we believe Petersburg and its surroundings to be in the most desirable point for operations found within the present year, and shall be happy to report further progress as the plot thickens.

Yours truly, "Sand Pump Line'"

## Oil News, November 15, 1871

*Wells along Clarion River near Grass Flats Courtesy Drake Well Museum*

The well on Hiram Neely's farm, a short distance from Keating's at Petersburg, was about six feet in the third sand at the time of my visit, and was flowing through the casing while they are drilling.
The gas is very strong. The tank was about half full of oil and everything indicates a large well. There is much excitement in that quarter over this strike, as it brings the development a long distance eastward. The "Antwerp" well, between the "Neely" well and Petersburg, still yields between 80 and 100 barrels a day. These are the only wells completed in that locality, though just west of Petersburg and about the Grass Flats there are several producing wells yielding about six to 75 barrels a day.

Both sides of the Clarion river, at the Grass Flats, are lined with new derricks. Land ranges in price from $200 to $1000 per acre, according to location. Land owners near the big wells ask $200 an acre bonus and one-fourth royalty on leases.

## Strike at Parker's Landing, November 20, 1871

The *Derrick* says: A new well was struck at Parker's Landing on Tuesday, belonging to S. W. Morehead and F. Ripley of this city and Sam Woods of Petroleum Center. It has not been tested. When the tools struck the third sand a volume of gas escaped so great that all the fires in the vicinity had to be put out. The tools were still in the well yesterday, with about a thousand feet of oil in the hole. It is expected to be a very large well.

The "Moriarty" well on the Lewis Collner farm, Petersburg, is pumping 70 barrels a day. The "Kerr" well on the Forest farm is pumping 20 barrels daily.

The "Hulings" well, with several acres of a lease on the Ashbaugh farm near Petersburg, has been sold for $80,000. The well has been yielding from 75 to 80 barrels daily for some time.

The "Wizard" well owned by Messrs. Woods, Ripley and Morehad, is now flowing between 50 and 60 barrels a day.

The Hiram Neely farm near Petersburg is leased on the following terms: For wells averaging 8 barrels, one-sixth royalty; for wells averaging more than eight barrels, one-fourth royalty and $200 bonus.

## Oil News, November 22, 1871

Our townsman, Mr. Victor Gretter of Titusville, has taken a lease on the Moffatt farm, and has the engine house and rig up for a new well. (*Petroleum Center Record*)

(Note: Gretter followed the excitement to Petersburg, see page 26.)

## Oil News, Parker Oil Field Production, November 23, 1871

Out of 48 districts, Parker's Landing stands at the head, the daily production being 3,152 barrels, or 370 barrels more than double the amount of any other district. (*G. W. B. Erie Dispatch*)

The "Hiram Neely" well, near Petersburg, is not thoroughly tested, as the sand still cuts the cups, but it is actually producing 40 barrels a day and indicates a much larger production.

The towns of Foxburg, East Foxburg, Petersburg, Antwerp City and several others in Richland township are springing up like mushrooms, but with a prospect of much longer life.

A new well on the Fox land, on the Clarion river below the Grass Flats, has just been tested and is producing about 60 barrels. The well on the Collner farm is now yielding about 80 barrels per day. A well on the Shoup is nearly as large and the "Ashbaugh" well is larger still. These, with several wells producing from 10 to 20 barrels per day and no dry holes, establish Petersburg as the best oil field now known. The sand rock is thick, giving indications of a rich and lasting territory.

A new well belonging to Galey and others near the old Galey well on the Grass Flats, Clarion River, was tubed and tested on Thursday and is now producing 40 barrels per day. (*Clarion Democrat*)

## Oil News, December 20, 1871

The "Porterfield" well at Emlenton is down to the third sand and has every indication of proving a paying one. The flow of gas is very strong.

There are several new wells in progress and if they should prove to be "A No. 1," it will give impetus to this place that will place it in the front rank. Emlenton is but three miles from Petersburg on the direct belt.

## Oil News, December 21, 1871

The following Rousevillers, Bradley and Chambers, Deming and Co., and the George Bros., have wells going down on the St. Petersburg territory at present. Messrs. Bradley and Co. are the fortunate possessors of an extensive tract of first class territory near Petersburg.

## Oil News, Petersburg and Foxburg Oil District, December 21, 1871

The Fox Estate is erecting a large new hotel at Foxburg, which will be completed about the first of January. Mr. Sterrett, of the Phillips house at Parker Station, has leased the hotel and will open it as soon as completed. The A. V. R. Company is putting down more side tracks and switches, and erecting a new station house and as soon as the hotel is opened, the cars will stop at that point for meals instead of Parkers. Foxburg is becoming quite a town; it already boasts of two hotels, a large hardware and dry goods store, a bank, an extensive machine shop, livery stable, lumber yard, feed store, groceries, eating houses, etc. and a large number of comfortable and tasty dwelling houses. Matthews, the cashier of the new bank, is a thorough and clever business man. Mr. Palmer and his two gentlemanly sons are superintending the immense Fox Estate, and managing a number of good oil wells of their own. S. S. Marshall, a live man, runs the lumber yard of D. W. Beers and Co. J. E. Coulter, formerly of Clarion, is conducting the immense hardware and dry goods establishment of Porterfield and McCombs. A. Vandersall and E. Wilson each keep groceries and restaurants.

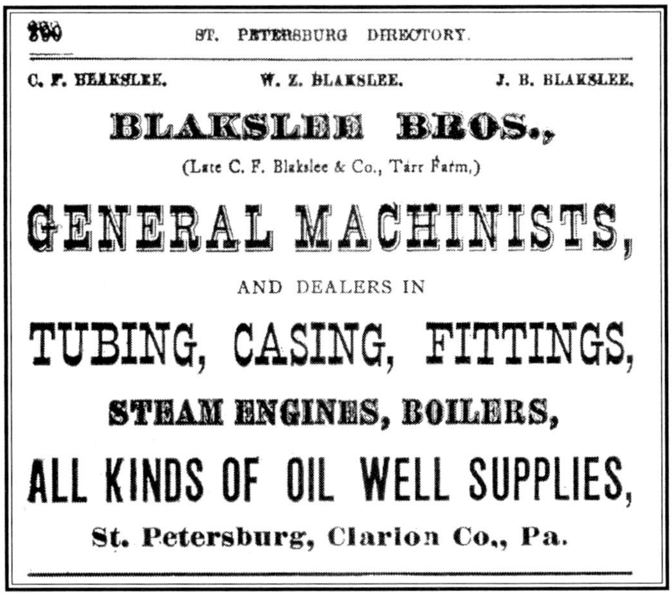


*Blakslee Brothers advertisement*
*Arters Oil Field Directory 1872*

East Foxburg – On the road east, towards Petersburg, about one mile from Foxburg, is situated the town of East Foxburg. It is on what was formerly the Rupert Farm. The Finley House, built and kept by Harris Finley, is a favorite stopping place where good meals, comfortable beds, and every reasonable accommodation is afforded. Panton and Nolan, the well known oil men, are the proprietors of the farm and reside here. Parker and Dennison have a large grocery, provision and feed store, nearly opposite the Finley House. There is a telegraph office, two hardware stores, a restaurant, shoe shop. This place is surrounded by oil wells. The Finley House is heated throughout by gas from the big gas wells near. Martin, Harm and Co. own land adjoining on which there are several good wells. A new strike on the Collner farm between St. Petersburg and East Foxburg was recently made by James Panton. It is producing about 35 barrels of oil a day. M'Cullen and Sanderson, on the Shoup farm adjoining the Collner farm, made a big strike on the 12th. The well is yeilding some 65 barrels a day.

*Lumber raft tied at mouth of Turkey Run on the Clarion River, getting ready for a trip to Pittsburgh. Note the evergreen and "scarecrow."*
*Venango County Historical Society*

St. Petersburg – Two miles east of Foxburg was a village previous to the past summer with one store, a hotel, blacksmith shop, and a few other small establishments and dwelling houses. Since the strikes made in the neighborhood last August, it has grown to be a town of large dimensions, being nearly a mile long, and building up so rapidly that it is impossible to enumerate the various establishments and businessmen. The following might be mentioned as some of the leading establishments: J.

*Clarion River at Alum Rock*

B. Donaldson and Co.'s drug store, Chalfant, Graff and Co.'s hardware store, Seymour, Taylor and Co.'s hardward store, Blakslee and Co.'s machine shop, M. Goldsmith's hotel, Asa Balliet's hotel, J. M. Place, news dealer and postmaster, Hiram Beals, meat shop, George Phillip's baker, Henry Fulmer's boarding house, Hopkins and Dorne, physicians, J. Birch, tank builder, Wm. H. Whittling, photographer, W. H. Steffee's restaurant, Collner's store, Bennett's hardware store, Johnston's drug store, Wing's grocery, and a great many other establishments, old, new, and getting ready.

There is a new well on the Fink lot, north of Petersburg, going about 10 barrels. The territory between Petersburg and the Grass Flats is well dotted over with derricks. A new strike has just been made in this locality by Frank Fertig, doing 12 barrels. Two new strikes have been made near the Grass Flats, on the Palmer lot by Galey, Soult and Dower. A new strike was made last week at the lower end of Grass Flats, on the south side of the Clarion on the Fox land, yielding 75 barrels a day. There are a large number of producing wells along the Clarion river at this point, and all holding their own, and some increasing their production.

Antwerp City – This town was laid out by Marcus Hulings and is being built up on the Ashbaugh farm, one mile east of Petersburg, where the celebrated Hulings well is located which for several weeks has steadily 80 barrels of oil a day. Twenty-five new derricks are up within site of the town.

About one mile further east is located the Neely well, which has been producing largely for several weeks, but at the time of our visit had fallen off somewhat, but this was considered only temporary. On the road, near this point, P.Graham and Co. have opened a large dry goods store. A number of new derricks are up and drilling and about commencing in this locality. A short distance east, near Keating's Furnace, on the Stubble lot, a well is almost completed. The most of the land for some distance further east and north has been purchased or leased and wells are being commenced. North of Petersburg and all the way to Salem, new wells are going down. The coming winter and spring will see a large amount of this county developed as oil territory. Pioneers are going ahead leasing and preparing to put down wells in Beaver, Salem, Ashland, Licking, Toby, Madison and other townships, and at various points further up the Clarion river. (*Clarion Democrat*)

## Foxburg Business Directory
### *1877 Clarion County Atlas*

Answers, E., Laborer
Armor, J. C., Teamster
Banter & Bursell, Dealer in lumber of all kinds, and all orders promptly attended to. Oil well lumber a specialty
Benzino & Co., Manufacturers of oil well tools
Craig & Son, Wm., Foxburg, Gunsmiths. All kinds of guns made to order and repaired. Ammunition of all kinds on hand. We would call special attention to our rifles and shot guns, made together, one barrel above the other, with sight and triggers to both stocks, with the same set of triggers. Also, to our celebrated Quarter StockLong Range Rifles. Call and examine our goods.
Carpenter, T. H., Oil Producer, Armstrong Co.
Croken, J. H., Carpenter
Culp, H., Driller
Dorrance, G. K., Pumper
Doule, J., Pumper
Dunkle, J. F., Proprietor of hotel at the mouth of Clarion River
Elliott, J. A., Drayman
Edwards, W. H., Foxburg Iron Works. Machinist and general blacksmith. Well tools manufactured; also, repairing promptly attended to.
Fox Estate, 2000 acres of oil territory
Fox, E. C., Shoemaker
Furn, J., Laborer
Fay, S., Pumper
Gregory, Wm. S., Foxburg Hotel
Gibson, A., Wagon Maker and General Blacksmith. All work promptly attended to on short notice. Rates low.
Grant, Dale & Co., Producers
Gray, F., Miner
Hare, J. V., Driller
Hely, J., Gauger on Union Line
Hely, D., Oil Producer
Healy, T., Oil Producer
Healy, J., Gauger
Himtsburg, ___, Laborer
Harvey, C., Pumper
Hunter, J. R., Teamster
Hate, G., Laborer
Jack, John, Proprietor of Boarding House
Kerr, H., Notary Public and Carpenter
Kightlinger, J. D., Oil Producer
Kerr, Joel, W. H. Harper & Co's Producers Self-Suspended Patent Renovator for oil wells. No other torpedo can excel it. This we claim without fear of contradiction. This torpedo cannot be exploded without reaching the bottom of well, then raising it up six inches; then it can be exploded every half inch from then to top of well, or can be drawn out in safety. We are liable for all damages. Orders by mail or telegraph solicited. Address Joel Kerr, Foxburg, Penna.
Kightlinger, G. A., Carpenter
Kennerdell & Bro., Groceries and Produce
Koos, Philip, Butcher
Layman, G., Laborer on Pipe Line
Martin, M. B., Oil Producer
Manderville, M., Dealer in flour, feed and coal
McGuire, E. T., Operator
Mancel, G. W., Pumper
Mull, D. E., Proprietor of the Foxburg Livery, board, sale and feed stables
Marshall, Geo., Farmer, Butler Co., Penna., 80 acres. Post Offce at Emlenton, Penna.
Morgan, M., Oil Producer and Farmer
Murphy, M., Laborer
Mercer, J., Laborer
Morgan, D., Laborer
McGee, J. T., Pumper
Millit, J., Pumper
McGee, J., Pumper
Mooney, S., Oil Producer
Oday, F, Carpenter
O'Donnell, T., Real Estate Agent
Patterson, J. C., Justice of the Peace
Pollock, O., Oil Producer, Perry Township. 250 acres
Price, Austin, Proprietor of feed store

Painter, J. M., farmer and producer, 195 acres
Parson, J., Gardner
Riggle, J., Pumper
Rois, W. H.,Engineer
Smith, G. W., Pumper
Stanford, A., Driller
Tippery, A., Oil Producer and Farmer
Templeton, W. D., Bridge Keeper, Armstrong Co., Penna.
Terrill, W., Keeps a good supply of teams on hand for hire. Rates moderate.
Truby, S., Laborer
Tume, E., Laborer
Tippery, S., Producer and Farmer
Wright, Simpson & Co., Proprietors of the Boiler and Machine Shop. Repairing of all kinds done on short notice.
Weaver, W., Contractor and Producer
White, M., Contractor
Wood, F. B., Teamster
York, W., Shoemaker

## Oil News, New Well in The Petersburg District, December 28, 1871

The *Pittsburgh Commercial* says: "The Riddle well struck the sand on Saturday morning, and immediately commenced flowing; shortly after she took fire and the driller, Perry Ross, was badly burned. After the fire was extinguished, she continued to flow with the tools and cable still in the well. She has the appearance of making the largest well in the Petersburg district. This well is owned by G. W. Ralph and C. E. Riddle of Parker and Wm. D. Moore and H. Campbell of Allegheny City."

The oil well owned by Porterfield, McCombs and others near Emlenton is flowing oil at intervals, so that drilling had to be suspended in the night, as the quantity of oil and gas makes it dangerous for any fire to be near. This well is looked on by operators with much interest, as it will open another large field. Emlenton has had several small wells, but not large enough for a general excitement.

## Letter from Petersburg Oil District, St. Petersburg, Clarion County, January 1, 1872

Editors *Morning Herald*

A letter from this "red hot" section of the oil regions may not be deemed unworthy of a place in your columns. Times are decidedly lively yet, although the excitement has become considerably toned down the past week or two owing to the fact that a great many of the operators and transient population have "gone out to America," to spend the holidays at home in a civilized manner.

A few lines descriptive of what Petersburg was and is may be of interest to *Herald* readers. The town is located about two and a half miles northward of Foxburg station on the Allegheny Valley Railroad, and previous to the discovery of oil here comprised about a dozen houses. Few towns can exceed this place in beauty of location. It contains two churches, in which both German and English services are held, one being Reformed and the other Lutheran. The people were nearly all descendants of Pennsylvania Dutch settlers from east of the mountains, and were mostly of a thrifty and industrious character, although considerably old fogyish. An irreverent resident remarked to the writer that the natives were all Democrats of the old school, many of the kind which regularly vote yet for old Hickory Jackson. He also said there was a large congregation in his neighborhood named Best, all of whom were born *at* Best's church, were married *in* Best's church, and finally died and went to h—l *from* Best's church.

The inhabitants of St. Petersburg, although not mingling much with outside barbarians, were much less old fogyish in their notions than the above community. The oil has changed the old town vastly already, and in case success still attends development, it is only in its infancy. At present the town is stretched along the main road fully half a mile long, and large numbers of lots are sold out or leased, upon which buildings will be erected in the spring. When the excitement was at its height, lots 40 by 100 feet brought from $300 to $1000, and other lots in good location, 30 by 40, are rented out for a term of five years at a rental of $200 per annum, merely for the ground.

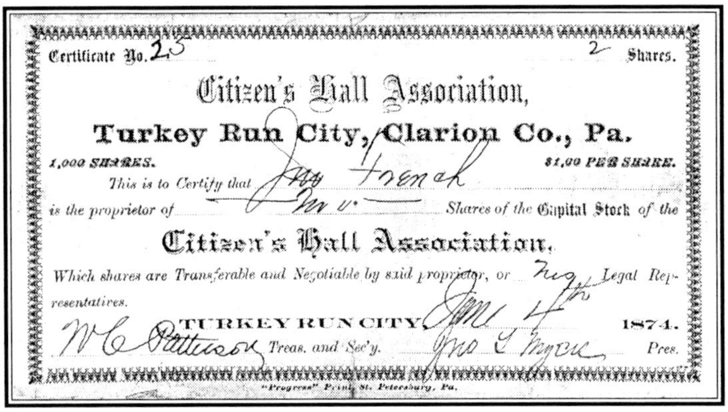
Certificate No.          Shares.
Citizen's Hall Association,
Turkey Run City, Clarion Co., Pa.
1,000 SHARES.          $1.00 PER SHARE.
This is to Certify that          is the proprietor of          Shares of the Capital Stock of the
Citizen's Hall Association.
Which shares are Transferable and Negotiable by said proprietor, or          Legal Representatives.
TURKEY RUN CITY          1874.
Treas. and Sec'y.          Pres.
"Progress" Print, St. Petersburg, Pa.

***Mr. French was an active driller and producer in the newly opened fields. He bought nine shares in the Citizens Ball Association at the oil town of Turkey Run City.***

Nearly all kinds of businesses are represented now by one or more houses, and most kinds are overcrowded already and will not more than pay expenses until spring at least. Hotels, boarding houses and saloons, or buildings intended for them, make up nearly one-half of the town, and in consequence of the competition good board and lodging can be had hereabouts as reasonably as up on the creek. A few weeks ago all the beds about town were occupied about three deep and many had to lie on floors or go to Parker's for beds.

There will be a host of applicants, at the January term of court, for license to sell liquor, but whether they get license or not, they yet do a rushing whisky trade at any rate. When the one hotel monopolized most of the business, it was a rare thing to see the bar when it was not crowded by thirsty suckers viewing the ceiling through their glasses—whisky glasses of course. Mr. Whitling purchased the St. Petersburg Hotel shortly before the oil excitement broke out here for $4,000, held it for some three months, and sold it to Mr. Balliet for $6,000, who yet runs it and derives therefrom a rich harvest of stamps. It is stated that steps were taken some time ago to have St. Petersburg made a borough, and it is to be hoped that the effort may be successful, as at times roughs, rowdies, cutthroats, etc., seem to be the ruling class and settle their little difficulties in their own way. Hardly a day passes that from one to a dozen of the riff raff have not "heads put on them" by some of their equally "whiskified" companions. Mr. Balliet has been compelled, on more than one occasion, for the safety of himself and employees and his house, to close his bar when his constituents became more than usually lively or, as they would say, were having a more than usually good time.

A large number of wells are going down or preparing to begin drilling as soon as other, considered test wells, are completed. Quite a number are nearly down and a great many more are far advanced toward completion and will be completed in January.

Several other wells near by are down, some of which will be completed and tested this month. The following is a list of some of them: Pete Dower & Co., and Finley & Co. have each wells down about 900 feet. Dower had the misfortune to have his engine house burned down lately, and drilling on his well was suspended a week or so in consequence, but he is at work again. The fire was the work of the "dead beats" who numerously infest this place and who, during the cold nights, flock to and make themselves at home in the engine houses near town during the absence of the drillers on Saturday and Sunday nights. The George Brothers have a well located near the west end of town, which is down about 400 feet and cased. .The drilling on this well as was the case with many others, was greatly delayed in consequence of a lack of water during the very cold weather.

The owners of the Gosse well are erecting a new rig and will soon be drilling their No. 2. This company has a lease of 50 acres adjacent to their well. The original Gosse well is yet steadily doing from 15 to 18 barrels per day.

Blakslee & Co., formerly of the Tarr farm, is drilling a well about a mile west of town, on Spruce Run, and had just struck the sand when the "jars" broke and are not yet removed from the well. They thought that they had struck the "sand" in this well nearly a month ago, but it proved to be only the "stray sand" which has so often mystified experienced drillers from the up creek oil regions. Had the contractor not met with an unaccountable amount of bad luck by the breaking of tools and plugging the hole, this well would have been completed a month ago.

A new rig was erected just in the rear of the St. Petersburg Hotel, and the drillers would soon have been at work but the storm of a week ago leveled derrick and engine house with the ground, which necessitates considerable delay.

A well, located just beyond the northern end of town in a ravine, which has been drilling since last August, was completed on Friday last at a depth of nearly 1,100 feet.

The well on first beginning to pump started off as if it was going to be a very large well, but something soon got wrong with the valves and it ceased pumping entirely. There is hardly a doubt but it will be a paying well, although not likely a large one. Finley Bell & Co., who are extensive operators here, are the owners of this well.

An immense number of derricks are erected and the drilling commenced on many wells on territory adjacent to the famous Huling well. A well near the latter is down through the second sand, from which an immense volume of gas issues, the roar of which can be heard at a considerable distance. The old well is yet about holding its own, doing from 70 to 80 barrels daily. It is reported that the owner of the Huling well has sold it, the prices given as the prices received for it vary all the way from forty to one hundred thousand dollars.

The great depth to which it is required to sink the wells here (about 1,100 feet) and the hardness of the rock requires about three months to drill them. The developments now in progress hereabouts are scattered over a vast extent of territory and will fully decide whether St. Petersburg is to be the center of the great oil producing district now known, or whether it has already seen its best days and will go down as rapidly as it went up. But its prosperity seems to be assured, for almost daily is chronicled the striking of large or at least good paying wells in its vicinity. One of the exceptions is the Gates & Co. well, located just below and east of the town, which is only doing about 200 barrels per day. As a good well was expected, both from the nature of the territory and its location with regard to the other large wells, the failure has caused operators in the vicinity to be considerably less sanguine of bit strikes. The well has been torpedoed already, but with no beneficial result, and some operators even doubt that the well was drilled into the "third sand."

The Deming and McBride well, located near by on a ten acre lease, is only down about 400 feet, although drilling has been in progress upon it for several months already. Mr. Campbell, of Rouseville, is the contractor, and has long been noted "on the creek" as a rapid and successful driller of wells, but in the present instance, he has had his work of drilling greatly retarded and often suspended by a miserably defective rig, and also by striking a vein of fresh water and being compelled to draw the casing and drilling deeper before casing again, and meeting with many other mishaps calculated to vex the driller's pious soul.

"Antwerp City," which is located near Huling's lease, has grown to be quite a place and is still growing rapidly and providing that half of the wells now going down near it are paying ones, it will become quite a brisk town in six months from the present time.

Operation on Turkey Run are brisk, many wells are considerably advanced toward completion, but few, however, will be tested within twenty days. Mr. Huling tested a well last week on Turkey Run and it is reported as being very nearly a dry hole.

Several good strikes were made on territory southeast of this place last week. One of them owned by Riddle, Ralph & Co., on the sand first being struck the flow of oil and gas took fire, and burned the derrick down leaving the cable and tools in the well. The derrick has been erected again and drilling resumed. It will doubtless prove to be one of the largest wells of this section.

The Walker Brothers of Rouseville are the lucky owners of a 50 barrel well struck last week. They have a five acre lease and propose putting another well down thereon as soon as practicable.

Those latter strikes are on territory about three-quarters of a mile from St. Petersburg and near the Mariaty well, which has been pumping about three months, making from 60 to 70 barrels per day.

On Saturday last a fine strike was made on a lease adjoining the Deming and McBride lease, which is at present pumping form 50 to 60 barrels and increasing. This well is owned by Mr. Frothingham, an operator well known in oil circles upon the creek.

## Need New Shipping Point

"There is a great deal of complaining heard here in consequence of the delay and annoyance in receiving any shipments by railroad. Cars and freight are often weeks, and often

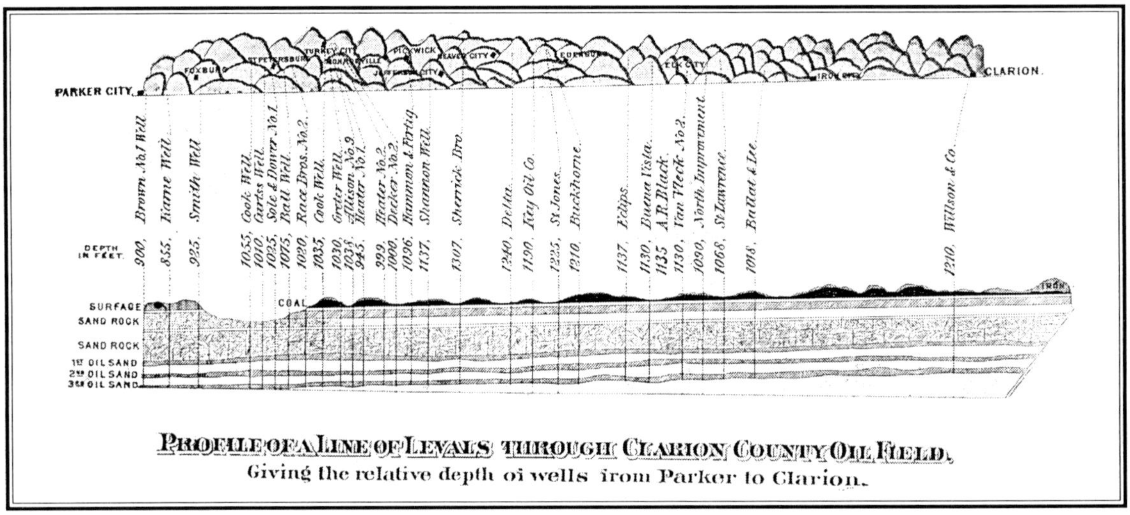

Profile of a Line of Levals through Clarion County Oil Field, Giving the relative depth of wells from Parker to Clarion.

*1877 Clarion County Atlas*

*"Shooting" a well*
*Courtesy Drake Well Museum*

*Wells along the Clarion River looking towards the mouth of the Clarion. Note that the bridge can faintly be seen at upper center, just under the tree branch.*
*Courtesy Drake Well Museum*

nearer a month, on the way between this and Oil City, before the party for whom the shipment is made receives his goods. The accommodations for receiving freight at Foxburg station are about as bad as they well can be, and many are having their goods come by way of Emlenton station. The space between the hill and the river is so narrow at Foxburg that it can never be made a good shipping point for the amount of freight that comes for the Petersburg oil regions. There is talk of opening a new road from this place to a point on the railroad above Foxburg, which would be shorter than any now in use. It is to be hoped the project may be carried into effect at an early day."

Yours Truly, "Sand Pump Line"

List of United States Patents Issued to
Western Pennsylvania Citizens for the Week Ending January 9, 1872
Patent Office, Washington D. C.
122,654 torpedo for oil wells
E. A. L. Roberts, Titusville, Penna.

## *Clarion Republican*, January 12, 1872

It is difficult to set up engines and drill oil wells in mid-winter, consequently few new wells are commenced at this season of the year, and the work on those that are drilling progresses slowly. A great many new wells, however, throughout Clarion county, are being started, even at this inclement season. But when spring opens up, experienced oil men tell us there will be a great rush and excitement in this section, as recent developments and present appearances satisfy operators that Clarion territory will prove more remunerative than any other the coming season.

## Oil News, January 12, 1872

The well owned by A. Evans Jr., near St. Petersburg on the Foust farm, struck the sand on Wednesday morning and in two hours the hole filled with oil. When the drill had penetrated the sand to a depth of ten feet, commenced flowing, and made it necessary to extinguish the fires. The well is expected to do about 75 barrels.

## Oil News, January 15, 1872

From the *Clarion Republican*, January 13: At the Ellison well in Knox township the parties are drilling rapidly. The well east of town is progressing and the parties will commence drilling shortly.

We learn that on Saturday last a well belonging to Ralph and Co., between Foxburg and St. Petersburg, reached the third sand and struck a large gas vein. The gas took fire in the engine house and severely burned the driller, a man named Lewis. He was going to the engine house to put out the fire at the time, but the gas got there before him. He was taken to Parker for treatment.

## Largest Gas Well in the Oil Region, January 18, 1872

On Thursday of last week, probably the largest gas well in the entire oil region, or the world for that matter, was struck in the Petersburg oil district. The well was struck about four miles beyond Petersburg on the Kronser farm, and is owned by Mr. Richardson, of Tarr farm. At a depth of 900 feet, and while the workmen were engaged in drilling, the tools struck a vein of gas of such immense force as to lift them, together with about four hundred feet of large casing, the cable, dirt, stones, water, mud, etc., clear out of the hole on dry land. The sound of the gas as it came from the hole, our informant states, was "like the rush of many waters," and could easily have been heard more than a mile away. As a matter of course the men had to flee for the lives "before the destroying agent," and it was many hours before the most venturesome of them dared approach the well. Fortunately the volume of gas, so immense was the force, was carried in a straight line to a great height in the air, hence the danger of fire catching from the engine was avoided. Our information estimates that gas enough is escaping from the hole to light two or three cities of the size of Buffalo or Rochester. If the gas could be taken and utilized, there would be power enough to run twenty engines with ease and then not exhaust the supply. The engine and boiler are now being removed to some distance from the well, and an effort will be made to conduct it off in pipes so that drilling can be resumed. This well is the most wonderful freak of nature that has yet been discovered in the Pennsylvania oil regions, surpassing even the celebrated "Burning Well" struck in East Sandy in 1865. The owner anticipates a big well.

Adjoining this lease, and within eight rods of the gas well, our townsmen, Messrs. Howe and Cook, D. O. Fisher, S. P. Armstrong, and James H. Smith, have a lease of 50 acres upon which they intend to commence developing at once. *Morning Herald*

## Oil Matters, January 19, 1872

Mr. Harry Derousse, of Petroleum Center, struck a 75-barrel well on the Keating farm, southeast of Petersburg, last week.

## Antwerp City, January 21, 1872

While Nelson Vensel was engaged in pumping here on the Marcus Hulings "Antwerp No. 1" well on the John J. Ashbaugh farm, he met with an accident Thursday by which his left thumb was almost cut off. It is feared it will have to be amputated. On this same well, Thomas Morgan had the misfortune of having his face painfully burned by steam last Friday afternoon.

*Nelson Vensel, driller, at a flowing well on the J. J. Ashbaugh farm. Mr. Vensel died in August 1872 in a fire on the Michael Fillman farm near Keating Furnace.*

## *Titusville Herald*, St. Petersburg Oil Field, January 24, 1872

St. Petersburg and its immediate surroundings, which is just now creating a great deal of excitement among oil men, is the scene of the principal oil development in the oil region. It is located in Richland township, Clarion county, and about two miles from Foxburg, the nearest station on the A. V. R. R. It is reached by a stage line, which runs in connection with all up and down trains. The town of St. Petersburg, before oil was struck in its vicinity, only numbered about a dozen houses, which were owned by the well-to-do farmers who tilled the surrounding farms. The country is fertile and very productive, presenting as fine rolling farming land as can be found in the state. The town now numbers about 1,000 inhabitants, mostly oil men and the employees on the wells. There are two good hotels in the place, several groceries, hardware and dry good stores, restaurants, etc. It is a busy, thriving little town, which is growing fast, and will probably double its population in the next three months.

The oil developments, so far, are principally on the south, eastern and northeastern sides of the town. This field is an extension of the Parker's Landing oil fields. The belt is supposed to run from the islands at the mouth of the Clarion river in a northeasterly direction along the banks of that river as far as Grass Flats, when it veers off more to the north. The wells at Foxburg are thought to be on the northern edge of the belt, and the Hulings well at Richland Furnace on Turkey Run, which proved a dry hole, on the southeastern boundary of the belt.

Commencing at Foxburg and the territory lying between that place and the Clarion river, we will give the farms and tracts as they come in relation from that point along the belt to the termination of the development.

On the Fox Estate, which comprises some 1,200 acres, lying around the station of Foxburg, there are a number of wells, of which Messrs. Fertig & Hammond have four, producing on an average of 10 barrels per day each, on a 20-acre lease. Three owned by the heirs of the Fox estate are averaging about 15 barrels each per day.

The Rupert farm is next to the Fox Estate to the northeast and has 8 or 10 wells, averaging 10 barrels each.

On the Whitling farm, G. R. Harris has ten wells in different stages of drilling.

The David Shoup farm adjoins the Whitling. On this the Palmer well is producing 60 barrels, although when first struck it pumped 150. David Young has a ten barrel well on this farm which two are the only producing wells. There are no wells drilling.

On the Collner farm is the Moriarty well, one of the best in the district, which has, for the past three months, produced 80 barrels per day with no variation from the first day of its completion. Fleming & McKay have a 40 barrel well on the same territory. On the Barney Vensel tract, same farm, is the Frothingham well, now down to the third sand, with 500 feet of oil in the hole. Ralph & Riddle have a hundred barrel well on the Joel Fink tract, same farm.

Grass Flats commences with the Galey tract and is, so far, the most productive quarter of the Petersburg region. Walker & Bros. have a well on the Galey tract, which is doing 30 barrels. Hall & Andrews also have a well on the same tract, producing 15 barrels. The Galey wells Nos. 1 and 2 are doing about 15 barrels each. Galey well No. 1, known as the Downer well, was the first well on Grass Flats and, in fact, the first well in this region, having been completed about the first of last August. This well was the cause of the rise of St. Petersburg. There was a number of other wells on the same tract, among which is the Galloway well, east of that owned by Hall & Andrews, which started at 75 barrels, but is now doing only 35 barrels. The Parsons Bros.' well is doing only 60 barrels per day, notwithstanding the report of its producing 100 barrels.

On the Keating farm is the Darousse well, pumping about 40 barrels. There is another small well on the upper end of the Keating farm, which is the furtherest well east, and is about six miles from Parkers.

On the Daniel Ritts farm, which is north of the Keating, Siggins & Blanchard struck a good well on Monday, 22nd, which is flowing at the rate of 150 barrels. The Jackson, Vensel & Shoup well, on the same farm, has been pumping on an average of 30 barrels per day for the last ten days.

Bly & Andrews have 6 or 7 wells drilling on the line of the C. Ritts and Ashbaugh farms, one of which was tubed yesterday, the 231, and is pumping at the rate of 65 barrels. On the C. Ritts farm is the Nottingham well No. 1, which is doing 50 barrels. No. 2 is drilling. Patrick's well is producing ten barrels.

The Ashbaugh farm is the next to the north and has only one producing well, the Hulings, or Antwerp well, which was finished in September and has been producing 75 barrels per day since then. There are about 20 wells going down on this farm.

To the east is the Eli Ritts farm, on which a number of wells are drilling. Several will be completed in a few days.

The Neely farm adjoins the Eli Ritts on the east. The Neely well, owned by Arner & Nicholson, is located on this tract.

Coming back toward Foxburg north of St. Petersburg and the above farms, the first is Collner farm No. 2. Adjoining this is the Foust farm on which is a number of good wells. The Goss & Gould well is doing about 10 barrels. The Big Indian, or Hamilton well, was struck in August and produces 20 barrels a day. The Evans well, of which we have made mention, is on this farm and is pumping something over 75 barrels per day. This well is a little south of St. Petersburg.

There are about 100 wells (or about one-quarter the whole number drilling in the oil regions), in different stages of drilling at present in this region, many of which will be completed in ten or fifteen days and as they are some distance from the producing wells, the extent and capacity of the territory will be then fully demonstrated.

Coal is produced in great abundance there, some of the mines being on oil producing farms, and coal can be had for fifteen cents per bushel delivered at the wells. The wells in the immediate vicinity of St. Petersburg go through a rich coal vein.

A savings bank is to be started in a few days by Messrs. Fertig & Hammond. It will be a branch of the Foxburg bank.

From the *Clarion Democrat*: A letter from Perry township says that W. H. H. Piper finished a well on the farm of James Pollock last Saturday, which started off about 25 barrels a day. It is on the run below the coal bank.

## Oil News, January 31, 1872

There will not be any wells down on the upper Clarion for a month or two, as drilling is now only about commenced.

A large gas vein was struck in the well drilling at Troutners in Salem township, and in a short time a strike is expected there.

The well on the Weeter farm near Salem is progressing favorably.

A new well on the Ashbaugh farm, one mile east of Petersburg, near the celebrated Hulings well which for several months has been pumping 80 barrels a day, reached the third sand last week and a large vein of oil was struck, filling up the hole, and water was poured in to prevent it from flowing. Unfortunately, the jars broke and left the tools fast in the well and thus it stood at last accounts. It is thought this strike will equal the Hulings.

## *Oil Man's Journal*, January 31, 1872, reported in *Titusville Herald*

A drilling well, owned by McFarland, Kier and others on the Collner Farm near St. Petersburg, struck a vein of gas at a depth of 800 feet on Wednesday night last, which flowed into the furnace, took fire and entirely destroyed the engine house and derrick and damaged the machinery to some extent.

A new well owned by Henry Derousse and others, Keating farm above the Grass Flats on the Clarion River, commenced pumping first on Wednesday last flowing oil at the rate of over 150 barrels per day. On Thursday it was still pumping and it was estimated generally by those who witnessed its operations that the production for the day would not be much short of 150 barrels. This well, thus far, proves the largest in the Petersburg district.

**"E. K. Thompson's Dandelion Mandrake pills will cure sick headache pain in the side and back."**

## Oil News, Lower Allegheny, February 6, 1872

We clip the following from the *Clarion Democrat*: Messrs. James and Crane have commenced drilling their new well on Blyson Run, and are preparing to start two or three more in the same locality.

A well on the Daniel Ritts farm between Petersburg and the Clarion river was struck last week, and has since been producing from 100 to 150 barrels of oil a day. This is now about as large a well as any in the whole oil regions.

The new well near the Hulings well, on the Ashbaugh farm, is doing from 40 to 50 barrels a day.

Preparations are being made to put down two new wells near Callensburg, one near Kingsville, and one a short distance east of the Clarion borough. This spring will witness lively times in this county.

There is a new well on the land of Joseph Foust, owned by Philip Foust and others just southwest of Petersburg, which has been producing something like 80 barrels a day for the last two weeks.

The *Emlenton Sun* says: The well at Kings' Mill is claimed to be doing 10 barrels per day. This well was drilled through a large casing, but struck a heavy vein of salt water in the second sand, which kept the hole full, making no improvement over the usual "wet hole" system, except the fact that it would lighten the pressure on the "cup" or "seed bag." Instead of putting a seed bag on the tubing, the owners had the well tubed the same as if the water had been shut off, the consequence is that the well has been pumping a large amount of water with a good quality of oil, some estimating the yield at 50 barrels. The stream of water is still increasing in size and is no doubt hindering the production of oil.

The well that is being drilled by Wm. Agnew, on the Agnew farm, will not be finished as soon as was supposed some time since, as the large flow of gas hinders the well from being drilled at night, and working only through the day will necessarily take about double the time that it was expected to finish it.

There are rigs up on the Samuel Knause, William Jones, Jesse Lovell and Alex Ritchey farms which when finished will thoroughly test the Ritchey's Run territory.

The well on the Jonas Fink farm, and the two on the George Whittling farm, will be finished in two or three weeks if no serious ill-luck should happen.

There is a well being drilled on the Chas. Goughler farm, situated between Ritchey's Run and Petersburg. The rig is up but drilling had not commenced at latest accounts.

The gas well on the Troutner farm was at last accounts about 925 feet.

From Oil Man's Journal Report in the *Herald:*

A well owned by Palmer and others on the Shoup farm, near Petersburg, was torpedoed on Saturday last, and its production increased to 100 barrels per day.

A well owned by A. Solliday and others on the Shoup farm east of Petersburg has been drilled to the third sand and is flowing oil at intervals in large quantities. The indications are that it will prove a large well.

A well on the D. Ritts farm, southeast of Petersburg, was sunk through the third sand and tubed Wednesday last, there was a good sand found, and notwithstanding it was drilled wet, a good show of oil made its appearance during its drilling process. This well is owned by S. M. Keir, A. W. Murry and Mr. Wilson of Parkersburg. It is thought it will prove one of the best wells in the district after being thoroughly tested.

The Segans well on the Ritts farm, southeast of Petersburg which reached a production of 100 barrels after a few days pumping, bursted its seed bag recently which will give it a serious back set, and may in fact spoil it to some extent for all time to come.

Well No. 2, owned by A. Evans and others on the Grass Flats, is in the third sand, makes a good show and promises to prove a good one to come.

## February 9, 1872

About five o'clock yesterday afternoon, Mr. F. I. Johnson was driving on Main Street, Antwerp City, and had occasion to stop near the oak trees. Samuel Rupert was not far behind and came along with a lively gait. Mr. Rupert unfortunately did not observe that Mr. Johnson had stopped, and before he had the opportunity to deviate from his course, he ran into Mr. Johnson's cutter, utterly demolishing it, and precipitating Mr. Johnson to the ground, who, however, luckily escaped with a slight injury to his back. No further damage was done.

**"If you have a cough or cold, try McDonald's Tolu Cough Syrup"**

February 12, 1872 – Total daily production from all oil districts
For the month ending January 31, 1872 is 17,826 bbls.

## From the Monthly Petroleum Report, February 12, 1872, *Titusville Morning Herald*

The point at which the greatest development is in progress is at Petersburg, which, in our table or production, is included under the head of "Parker's Landing and vicinity." Fully one-third of the whole number of wells in progress of drilling in the oil region during the month under review, are located at that point. The first well struck in that locality was in August 1871, it was on the Grass Flats, and proved a good well. Since that time up to the close of the year, not much was done in the way of development, but about the first of January 1872 active operations commenced, and now the number of wells is being daily increased, while those completed have, with few exceptions, proved large wells.

## Oil News, Petersburg, February 12, 1872

A well on the Clarion River, some distance below the mouth of Turkey Run, owned by Speer, Ginder and others, has been drilled into the third sand and presents an excellent appearance. While the drilling process was being continued, oil had flowed from the hold to such an extent as to enable the workmen to save 100 barrels. A big well is expected when the pumping process is commenced.

The well on the Adam Ashbaugh farm, owned by B. Vensel and others, northeast of Petersburg, has reached the third sand. The well being drilled through large casing, immediately after striking the sand filled several hundred feet with oil, thus giving unmistakable evidence that the sought after treasure exists in abundant qualities in the particular locality in which the drilling process is going on. Mr. Vensel and his associates will no doubt be the fortunate owners of a big well.

The well owned by Segans and others, on the Dan Ritts farm, in which the seed bag was bursted a week or more since, has been repaired and started up. It is reported as again yielding 100 barrels per day. This well is located about one mile southeast of Petersburg.

We glean the following from the *Emlenton Sun*: The new well of Andrews, Bly and Co., on Ashbaugh Run, is doing 40 barrels a day; several other wells will be completed within a week or two.

The well on the Daniel Ritts farm, near the eastern line, continues to pump 75 to 80 barrels of oil per day.

## Oil News, Lower Allegheny, February 15, 1872

The following items regarding Petersburg we clip from the *Clarion Democrat*:

The lumber for a derrick is on the ground on lands of Sloan and McMasters, about a mile southeast of Clarion borough. Capitalists from the Venango oil region are about to put down a well.

A heavy gas vein was struck last week in a well about half way between Gas City and Lineville.

A well owned by R. L. Neely and others, on a lot leased from John H. Galey on what is known as the Fulmer Farm between Petersburg and the Clarion River, is now some twelve feet in the third sand. The hole was drilled dry and has filled 900 feet with oil and indicates a large well.

A well on the John Ashbaugh farm in the woods, to the north of the road, just east of Petersburg, owned by Vensel and others, is nearly completed with a good show. A number of other wells about Petersburg will be down soon and strikes are becoming almost a daily occurrence.

Leases have been taken on the Isaac Neely and Paul Fry farms, some three miles east of Petersburg, and preparations are being made to put down two wells. A number of other wells on lands of Neely and others in the same neighborhood are to be put down, and lively times are anticipated as the spring opens up.

The Stubble well, on Turkey Run, which was considered a dry hole, has been drilled deeper and on Monday last was pumping oil at the rate of three barrels a day.

The well noticed last week on the Daniel Ritts farm between Petersburg and the Clarion River is still producing about 100 barrels of oil a day. It is owned by Keir, Murray and Wilson.

The Speer well, on land formerly belonging to the same farm, was tested last week and is claimed to be from a 100 to 150 barrel well. It is owned by Speer and others.

Jackson well No. 2, on the southeast side of the Clarion River, on the Smith lot, was completed last week and is rated as a 60 barrel well.

A well owned by Soliday and others, on the farm of Eli Ritts, Sr., lying immediately north of the John Ashbaugh land, on which the celebrated Hulings well is located, was tested last week, and started off large, and although its production has decreased, the indications are that it will be a good well.

Marcus Hulings and others associated with him have taken several leases in the neighborhood of Callensburg and are about to operate. They are erecting a derrick on the land of Solomon Shirey in Licking Township, about one mile east of Callensburg.

## Oil Matters, Lower Allegheny, February 21, 1872

The *Emlenton Sun* says:

The Vensel well on the Dittman Run, mentioned last week, is doing 25 barrels.

The well on the Peter King farm was torpedoed last Monday and is claimed to be doing about 16 barrels daily.

The Crawford and Haldman well is still being pumped, but as yet they failed to get oil in paying quantities.

Andrews, Bly and Co.'s No. 1 Ashbaugh Run was increased from 30 to 60 barrels per day. No. 2 was tubed on Tuesday and is doing fully 100 barrels.

The well that is being drilled by Wm. Agnew, on the Agnew farm, is nearly finished with over 260 feet of oil in the hole and twice as much gas as needed for running the machinery.

The new well northeast of Sandy is still being drilled. They are 42 feet in the sand, and when last heard from, had a sand pump stuck; this well has been drilled wet, which of course materially interferes with a speedy completion.

There are several new derricks up on the line of the "belt" from Petersburg to Sandy, leaving Sandy some two miles to the west; several wells are in process of drilling, and the operators all seem very sure that they are on the spot.

The Black farm owned by McKinney and Nesbit, Fertig and Hammond and others, has not proven as good as expected, as there have been several very small wells gotten, some of which will scarcely pay the owners.

If all the territory that is being developed in this section should prove one-half as good as the operators expect, there would be more oil produced within six months than has ever been produced at any one time, but there is no doubt but that the visions of greasy fortunes that have arisen before the vision of these operators will vanish like the shades of night before the morning sun, at least the majority of them.

A well located south of St. Peterburg, owned by John H. Galey, Richard Neely and others, commenced pumping on Saturday of last week. It threw out a large amount of oil when first started up, but finally settled down to 15 or 16 barrels per day.

The Speer wells No. 1 and 2 on the D. Ritts farm, located on the river below the mouth of Turkey Run and yielding about 100 barrels per day each. One of these wells does not yield quite 100 barrels per day but the other exceeds this amount. No. 2, the last one sunk, further up the Clarion, is the best well. No. 1 has been pumping for about three weeks, No. 2 for about 10 days.

The well on the Logue farm, south side of the Clarion river, continues to yield 35 or 40 barrels per day.

Jackson well No. 2, south side of the Clarion River, is in the third sand and making the greatest kind of a show. Owners and others look for one of the largest wells in the entire district when the pump is put in operation.

*Bird's eye view of Foxburg 1876*

The Hulings well near Keating's Furnace, Turkey Run, which was thought to be dry is now yielding three or four barrels per day.

A well on the Ritts farm near Ashbaugh Run, above the old Hulings Antwerp No. 1 well, has been pumping for some days, averaging a daily production of about 50 barrels.

The Soliday well is now pumping and yielding a goodly amount of oil.

The well owned by B. Vensel and others on the H. Ashbaugh farm has been tubed and pumping for some days.

P. R. Bohlen of Memphis, Tenn., and Powell Neely, of Armstrong county, have a rig up on the Isaac Neely farm, boiler and engines on the ground, and are about ready to commence drilling. This territory is a little in advance of the development and it was thought for a time that it lay too much to the southeast to be properly on the oil belt. The striking of the Speer well on the Clarion River, the Jackson well on the south side of the Clarion, and the well on the Logue farm, all point to this territory as being properly in the field.

Mr. W. W. Dewey, an extensive operator at Fagundas City, is now here and has the lumber on the ground for an oil well on Trout Run, about half a mile east of Clarion borough.

A correspondent writing from the Petersburg oil field says: Speers and company tubed a new well at the Grass Flats last Thursday, which is pumping 120 barrels a day, and they have a well on the next lease above that is flowing 150 barrels a day. These wells are on leases 10 rods wide, fronting on the Clarion river, and running back 40 rods on the John H. Galey tract, formerly part of the Daniel Ritts farm. James Boyd and others are drilling a well on one of these leases.

R. H. Neely is tubing a well today on the Galey and Hartley tract (adjoining the D. Ritts farm on the south) with good prospects, there being 900 feet of oil in the hole.

The *Tidioute Commercial* says: The new King well on Ritchie Run, in the Petersburg district, is only doing six barrels, so we are informed by a gentleman who returned from there yesterday.

## February 28, 1872

Letter to Editor of *Titusville Morning Herald*, February 28, 1872 - A producer wants to stop Sunday work to observe our God, and to drill fewer wells. St. Petersburg, Penna. Feb. 26, 1872

## Letter from Parker's Landing to the editors of the *Herald*, March 5, 1872:

Business is dull, the cause agitating the whole oil country is the selfish, diabolical plot of the Southern Improvement Company, which is against the interests of the oil region.

The ice on the river is rapidly breaking up, and in another week, if the present weather continues, it will be clear enough to ship oil by water to Pittsburgh or below.

The town of Foxburg, a few miles above here on the A. V. Railroad, whose visitors to the Petersburg district leave the train, is growing very fast. A large number of new buildings have been completed during the last two weeks, and the town is assuming the air of a thriving business place. The oil depot having been found to be inadequate to the business, a large new one has lately been finished, which is quite an ornament to the place.

The Southern Improvement Co. consists of railroad companies and several Cleveland and Pittsburgh refining companies entered into a combination to refine and ship the entire product of the oil region and control the price of crude and control shipping rates.

## March 7, 1872

Mass meeting at St. Petersburg, Pennsylvania, to fight the Southern Improvement Co. – Special to the *Herald*

St. Petersburg March 7 evening. A rousing mass meeting is now in session here. They pledge themselves a unit with the oil producers, no matter what the action may be. Committee A. R. Black, E. Barton, Frank Fertig.

## Oil Matters, March 8, 1872

St. Petersburg, March 7, 1872 – Another new well on the D. Ritts farm, near Huling's No. 1, commenced pumping on Saturday evening and is flowing at a rate of 75 barrels, owned by J. A. Pauly and H. D. Calderen, both deserving a good well for their energy and perseverance in this business.

***Railroad trestle across Bear Creek near Parker***
***Venango County Historical Society***

## March 11, 1872

New Well – Messrs. Richardson, Ormsby and Thayer struck their first well on Saturday, which is producing at a rate of 40 barrels per day. It has been named the "North Improvement Well," and is on the Ashbaugh farm, about six rods from the Huling's well.

## Petroleum Producers Union, March 14, 1872

The territory forming the Pennsylvania Petroleum field shall be divided into sixteen districts. The third district consists of Foxburg, Petersburg, Clarion River, both sides, Emlenton and Ritchie Run.

## Meeting at St. Petersburg, March 18, 1872

Mass meetings of operators and producers were held at St. Petersburg on Thursday and Friday nights, March 14th and 15th, at which much enthusiasm and determination to resist the "infamous combination" to the bitter end were displayed. The meetings passed resolutions heartily and fully endorsing the action of the general committee, and pledging cooperation in carrying out the plans of the same.

The following are the names of the local committee of the third district: J. E. Blake, Frank Fertig, David Schuyler, Joel Fink and S. G. Bayne. Many signatures were obtained to the pledge of the Producers Union.

An adjourned meeting will be held at the school house, St. Petersburg, on Friday evening next to organize a permanent oil exchange for the third district. Committee on Constitution etc. is expected to report. Let all attend.

## Letter from Foxburg to editor of *Morning Herald*, March 24, 1872

As I am daily reading the *Herald* and seeing articles from different localities near here, but nothing said about Foxburg, I thought I would send you a few lines and let your numerous readers know what it was four months ago and what it is now. At that time it only contained some six or eight dwellings, a boarding house, T. & J. Richardson's blacksmith and machine shop, store and post office. This composed about the whole of the town. Since then there has been built about 25 buildings, some private dwellings, others for groceries, restaurants, etc. The railroad company has built a fine large depot. The bank which started here some time ago has built a fine building, which, when finished, will be, perhaps, the fanciest building in the place. T. & J. Richardson have enlarged their premises and are doing a large business being well acquainted with the wants of the oil region and, also, first class workmen themselves know how to do their work in that manner that gives general satisfaction to all. A want long felt here is at present supplied, that is, a good hotel, the traveling community and those wishing to stop a day or so here, would take the evening train to Parker's or other stations where comfortable lodgings could be had. The new, large and commodious hotel, which was opened a few weeks since by Mr. Sterret, who seems to be doing a good business, therefore those wishing to stop in Foxburg for a day or so will no doubt be entertained and cared for in that style and manner which becomes a first-class hotel.

The coming summer will, no doubt, be a busy one, as people are daily seeking lots to build upon, and as it is the main outlet to the Petersburg and Clarion Oil Region, it will, no doubt, for the future maintain that "goaheaditiveness" which for the past few months has made it what it is. Yours truly, I. Patterson.

## Letter to *The Independent*, a newspaper published in East Brady, Penna., March 25, 1872, "East Foxburg Oil Field"

East Foxburg is the name of a prospective town east of Foxburg, one mile from Petersburg. As towns are generally named in their infancy, and especially in the oil region, I christen this one. I have not seen anything in your paper for sometime in regard to this oil field and will, therefore, give you a brief account of it.

*Well near Bear Creek*
*Venango County Historical Society*

There are eight rigs up. Three wells are pumping an average of ten barrels each, the others are drilling. Considerable property has changed hands of late in selling and leasing, and the prospect for a large and paying oil field is good. Among the principle operators are Fertig & Hammond, two wells pumping; Soult & Dower, one well pumping and one drilling, down 700 feet; A. S. Palmer's well down 800 feet; Panton & Nowlan down 400 feet; Brice and McConnell have abandoned one well owing to bad luck, but commenced another in the same derrick and as they are men of ability and push, they will overcome all difficulty. H. B. C.

## March 28, 1872, Let's have some fun!

A correspondent tells the story of some young men who undertook to clean out the Hope well near Bear Creek Station. They swung the tools and let them in the well, but couldn't make the "hitch," so they got a gentleman to assist them to accomplish this feat. After drilling for two or three hours, they drew up the tools as they thought, but there were no tools on the rope and they commenced to "sand pump for them." Their success was not very good, and placing a bucket bottom upward over the hole, they departed from the oil regions.

The *Titusville Morning Herald* news agents in Antwerp City are Crawford and Palm, News Room.

## April 5, 1872

The condition of the producers today is somewhat different than what it was 30 days ago. They then all agreed to sell no oil to the agents of the Southern Improvement ring, and they have thus far maintained their position heroically and so little oil has gone into the hands of the enemy, that the Southern Improvement Co.'s refiners have been compelled to shut down for now.

The oil business problems stem from our ability to produce excessive amounts of oil over the demand. This pushes the price of oil per barrel down.

Let us stop all drilling and we may be safe enough in predicting that the price can be maintained at good paying figures for the next six months.

## This letter published in *The Independent*, an East Brady, Penna., newspaper dated April 6, 1872

St. Petersburg Oil Field – On Monday we visited the St. Petersburg Oil Field. Meeting with our good friend Cap. Chalfant, of Parker's Landing, we secured a fast horse each at Foxburg and started on our dusty pilgrimage. The road from the river to Foxburg was crowded with wagons hauling goods, people on foot, and many enjoying a ride in buggies or in coaches. Wherever we looked, new derricks were visible and drilling going rapidly. At the Cross Roads, Mr. H. Finley has erected a fine hotel and is doing a good business. A new town, East Foxburg, is springing up in its vicinity. We spent nearly an hour in St. Petersburg seeing the sights, and then started over the hill to the Clarion River to take a glimpse of the Grass Flats oil field that now looms up as an interesting spot in oildom. We suppose from the time we struck the top of the hill above Foxburg

*Wells near Bear Creek*
*Courtesy Drake Well Museum*

until we reached the Grass Flats, 100 new rigs were seen, some completed and pumping oil, others nearly down; again others only beginning, while everywhere were seen candidates seeking for oleaginous honors, and may reap wealth, or worse, disappointment.

Well, we passed over the hill, the captain and ourself, making close observations of men and things and chatting pleasantly of things past, present, and to come. And, by the way, we may just here incidentally remark that the Captain is one of the best traveling companions in the world, and of course we enjoy his jovial company. After a very dusty ride, we reached the margin of the Clarion River, where we found wells on every hand, new stores, dwellings, etc., the evidences that the handmaids of civilizations and business had been there before us and here in a deep wild gorge, overwhelmed by giant trees, men were seeking for the new element of commerce which is not only enlightening but lighting and enriching the world.

We found several wells pumping, among them Dower & Gaily's producing 15 barrels per day; the Spurr well 6 to 7 per day; the Jackson well 20 barrels, and the Palmer well 20 barrels. W. P. Finley is just testing his new well, with a splendid abow.

## Oil News, Petersburg District, April 16, 1872

The Antwerp Pipe line is being laid from Petersburg district to Spruce Run on the Allegheny river, between Emlenton and Foxburg. The pipe line and tanks will be completed and in working order in about one week. Ground is being sold and leased at the terminus of the line for building purposes; the land being admirably adapted for that purpose, a thriving town will no doubt suddenly spring up, which is to be named Fullertown. Work has already been commenced on a large hotel and other buildings will no doubt soon be under way.

Well No. 1 on the Fink farm, southwest of Petersburg, has been pumping for three or four weeks and is still averaging 50 barrels per day. This well is owned by Porterfield, Treat, Fuller and Jackson.

No. 2, on the same farm, owned by the same company, has been pumping about ten days and is estimated at a 50 barrel well.

A well was recently sunk by Porterfield, Treat and others on the Foust farm, southwest of Petersburg, which is yielding 65 or 70 barrels per day.

A well owned by F. Andrews and others on the D. Ritts farm was tubed on Thursday of last week and started at a production of 40 or 50 barrels. It is thought the daily yield will not sink below 20 barrels.

The Harrington well, on the Slicker farm southeast of Petersburg, has been drilled to the third sand. Oil and gas flows from the hole and a big strike is looked for.

The Geary well on land of Joseph Foust was tubed and commenced pumping on Saturday last starting, it is thought, at a production of 100 barrels per day. It will no doubt remain a good well. Geary, Chapen and Clark are the fortunate owners.

A well owned by Crooks and Dion was recently sunk on land of Galey & Hartley, opposite the fording, at Grass Flats, formerly part of Ritts farm. Up to Saturday last this well had been pumping one week and had averaged a production of 100 barrels per day.

A well being drilled near Antwerp City, owned by Fertig, Long and others, took fire, caused by a flow of gas and oil, on Thursday of last week. The rig was destroyed and a young man named Moses Johnston was severely burned. It is feared his injures may prove fatal.

It is said the Hulings well on Turkey Run, near Keating's Furnace, continues to improve and that its production at the present time is near ten barrels per day.

**These names need to be mentioned when talking about the progression of oil in Clarion County.**

- Petersburg
- Richmond, now Alum Rock
- Grass Flats
- Keating Furnace
- Turnip Hole
- Jefferson Furnace
- Keating City
- Turkey Run City
- Monroeville, named for President James Monroe
- Cropps Corners
- Edenburg – 1876 Borough organized; named after Garden of Eden; between 1877 and 1880 was ravaged by fire six times, most being the result of arson. The fire of October 13, 1878 was the most destructive with a loss of $400,000 and 170 buildings destroyed.
- Triangle City
- Shippenville
- Elk City – 1875 commenced by winter of 76-77 over 3,000
- Blairs Corners
- Jefferson City
- Wentling's corners, first called Pleasantville
- Beaver City
- Mongtown
- Moyers Corners
- Dog Town
- Pilgraheim
- Graham's Landing
- Phillip's Corners
- Salem
- Mehrtena
- Chestnut Ridge
- Slambang
- Callensburg
- Ritts Farm
- Fullertown
- Pickwick
- East Sandy
- Dahte City
- Mount Pleasant
- Antwerp City
- Fern City
- Cogley City
- Forest Home
- Foxburg
- East Foxburg
- Blue Rock, Clarion River
- Fox Estate
- Matildaville
- Parker's Landing
- Lawrenceburg
- Brady's Bend
- Pollock (Perryville)
- West Freedom
- Kossuth
- Clarion
- Ritts Station
- Loguetown
- Alum Rock
- Mariasville
- Nickleville
- West Monterey
- Buttertown
- Shady City, east of Edenburg
- Moyer City
- Knox
- Phillipston
- Haganville
- Paint Mills
- Canoe Canoe Riffle
- Kribbs Farm
- Agnew Mills
- Lamartine
- Mouth of Piney
- Valley
- Mouth of Deer Creek
- Churchville
- Upper Hillville
- Piney
- Easton
- Brady's Bend
- Reidsburg
- Sligo
- Callensburg

Hummel Farm
Phillips Corners
Curllsville
East Parker
Sassafrass Point
Tippecanoe Furnace
Eagle Furnace
Beaver Furnace
Keeversburg
Williamsburg
Black's Furnace
Millerstown
Iron City
Agnew's Station
Sarrah Furnace

## April 18, 1872

Titusville celebrates the grand re-opening of the European Hotel. It is one of the most convenient and elegantly furnished of its kind in the city. The first floor of the brick building is where the sample room is located, in the center of the first floor is the sitting room, and in the back is the dining room. The second floor is occupied by offices and the third and fourth floors are filled up in the most modern and elegant style for bedrooms and parlors.

Mr. Fred Hepp will appreciate in the sample room. (Note: Fred Hepp moved to St. Petersburg and a fire at his bar burned half of the town down. He was almost hanged! – See February 25, 1873, Fire)

## Oil News, April 20, 1872

A vein of oil was struck in the well now drilling on the Michael Reichart farm, about two miles and a half southwest of Callensburg. It is supposed to be in the second sand.

The James and Crane well on Blyson Run, mentioned last week as having struck a heavy vein of oil at 450 feet, is now down 550 feet. Another vein of oil was struck last Saturday, but since that time the oil has not interfered in the drilling and the work is progressing rapidly.

The well on the Widow Cooner farm between St. Petersburg and Salem, where the rig has been burned twice, is now pumping 15 or 20 barrels of oil a day, thus proving a good well, notwithstanding the many predictions to the contrary. - *Clarion Democrat*

## New Oil Town, April 24, 1872

Fullerton is the name of a new town being laid out by Mr. Fuller, late of Oil City, and Mr. M. Hulings of the South Side. – *Register*

## St. Petersburg *The Progress*, April 27, 1872

Antwerp City is the home of a lad who is fast making for himself the reputation of an amateur thief. A few weeks ago he stole some pocket knives and candy from the home of Mr. Marcus Hulings on East Main Street and last Friday while Mr. Hugus was at supper, he entered the pool hall building, taking therefrom some goods including candy. He was noticed coming out and the plunder recovered. He should be placed where his thieving propensity will be curbed.

## St. Petersburg *The Progress*, April 27, 1872

M. Shear and Brother, Wholesale and Retail Dealers in Dry Goods, Millinery Goods, Ladies and Gents Furnishings – Establishments on Main Street of both Antwerp City and our sister city, St. Petersburg, Penna.

## Spring Opening at Antwerp City

Ladies and gents furnishings have just been received from New York and Philadelphia, extensive purchases of spring and summer dry goods. White goods, lace curtains, curtain poles, and window shades. Notions and hosiery. Ladies and gentlemens fine shoes. Hats and caps, gents. Youths' and children's fine clothing, Brussels 2 and 3 ply and Ingrain carpets and floor oil cloths.

Making all their purchases direct from manufacturers for spot cash, thereby securing jobber's prices and discounts. This enables them to offer special inducements to their customers and they extend to all a cordial invitation to visit their stores and examine their stocks, and be convinced that they can save you money. M. Shear and Bro., Main Street, Antwerp City.

## Local Notices, April 29, 1872

Soda water, with choice fruit and cream syrups; also Congress, Vichy and Kissengen Waters on draught at the City Drug Store.

Clark & Moon are now dealing out to their numerous customers from their Frigid Soda Apparatus, a superior article of Soda Water to any before offered in this city. They have Puffer's Octagon Apparatus, the largest in the city, having four draught tubes for soda and mineral waters, and a double cooler, which insures a cool drink.

They have nothing but pure Fruit and Cream Syrups.

Twelve tickets for one dollar, each good for one glass of Soda or Mineral Water.

List of syrups: lemon, strawberry, coffee, pineapple, nectar, orange, vanilla, cream, sarsaparilla, ginger, raspberry, banana, catawba grape, morello cherry, blackberry, chocolate, wintergreen, orgent, sherbet, wild cherry, Dolly Varden, on to Buffalo.

Call and be convinced that the City Drug Store is the place to buy your Soda Water. Clark & Moon, proprietors. *The Progress*, St. Petersburg.

## Fire at Parker's - April 29, 1872

The reported fire at Parker's Landing last Saturday was confined to the notorious "bagnio" kept by Ben Hogan, and none of the respectable residents of that locality will mourn the loss. The building was located in the heart of the city, and the fire might have proved disastrous had it not been for the splendid execution of the fire department, who with their new engine promptly and effectively arrested the progress of the flames. *The Progress*, St. Petersburg.

*Information and lithographs are from "Early days of Oil", Paul H. Giddens, Princeton University Press, 1948*

*Ben Hogan, "The wickedest man in the world."*

*Hogan's "Floating Palace" in the middle of the Allegheny River at Parker City. Hogan operated a free-and-easy in the city but was forced to close when the townspeople burned him out. See newspaper article of April 29, 1872, "Hogan's Bagnos Fired." He bought a boat and put his gambling tables and girls aboard. Here, in the middle of the river, he could escape the jurisdiction of the authorities. In three years Hogan made $210,000 at Parker's Landing.*

*"French Kate," Hogan's business associate*

*Scene on board the Floating Palace*

## Municipal Cleanliness, May 2, 1872

Notwithtanding the tardy approach of warm weather, there is every indication according to various authorities of the summer being intensely hot, and the coming season very unhealthy, and conducive to the propagation of contagious and malignant disease. This city is an unusually healthy one, but the stagnant and putrescent condition of our streets, alleys, and surroundings last summer, and the prevailing diseases consequent thereon, is sufficient proof that we are not altogether impregnable in that respect. The abominable habit of throwing decaying animal and vegetable garbage in the streets, and in rear lots, should be checked at once, and every offence of the kind punished to the fullest extent which the ordinance allows, and in this at least there should be no respect of persons, no matter who they may be or how influential their position, every citizen violating the ordinance should be treated alike. There are thousands of cart loads of accumulated filth in the heart of the city, which should be removed before the hot weather arrives. Personal cleanliness and attention to diet, of course, have a direct relation to health, and there are few so ignorant as to be unaware of this fact. But no matter how cleanly each individual may be, if the great bulk of the population are compelled to breath a vitiated atmosphere arising from open lots and streets teeming with putrescent garbage, disease in various forms will most assuredly take root and spread among us. The Health Committee has a large amount of work before them, and it is to be hoped that the city will undergo a thorough cleaning within the next few weeks. The relation between municipal cleanliness and municipal health has been demonstrated everywhere by the most bitter experience, and cannot be over-estimated. Boards of Health, in order to effect anything, require to be cautious and stringent, without being tyrannical, in their measures, which is the only safe plan to guard against the spread of disease or the ravages of pestilence.

## Oil News, May 7, 1872

Boiler explosion – The boiler on well No. 2, Ritts farm, Petersburg, exploded last Friday forenoon. The boiler was an old one.

*Oil man's friends*
*Venango County Historical Society*

## Dog Killing – Antwerp City, May 9, 1872

There are several cases of hydrophobia reported in different parts of the county. The dog crop in this city is large this year and almost every street is teeming with worthless curs. We are not therefore surprised that dog killing with the hot weather has commenced early. Four defunct canines were found yesterday morning. It might be wise for owners of valuable animals to muzzle them when allowed to run at large. It must be noted that it is in direct violation of the criminal code of this state to poison dogs or other animals with a fine not exceeding $500 and confinement, not exceeding three years.

## Watch Your Step!

After leaving one of Antwerp City's many drinking establishments, a poor oil man was slowly making his way down a very dark boardwalk on Main Street. After passing through a gateway onto 2nd street, in the darkness hit his nose against a post.

"I wish that post was in hell," said he.

"Better wish it were somewhere else," said a bystander, "you might run against it again!"

*Oil Wells In The Woods* by John Christopher O'Day, M. D., Deposit, N.Y., The O Quaga Press, 1906, is an informative book on oil history.
The student of early oil will find this book very interesting.

**If you have a cough or cold, try McDonald's Tolu Cough Syrup**

**Go to the Smith meat shop and buy your Christmas poultry for 18 cents per pound, And first class beef cheaper than any other market in Antwerp City.**

## May 10, 1872
## Dog Poisoning

A valuable dog belonging to Mr. Thomas Stewart was poisoned last evening. Quite a crowd gathered round to witness the struggles of the dying animal. Mr. Stewart was offered fifty dollars a few days ago for this dog. This poisoning was quite unjustifiable, as the dog was harmless as a kitten and wore a collar with the owner's name on.

**Fresh halibut, fresh cod, haddock, flounder, whitefish and bass**
**Just received at the Fish Market at end of 2nd street, Antwerp**

## A Narrow Escape

On the night of the fire at Mr. Bronson's a gentleman residing in the neighborhood was roused by the alarm, and upon raising up in bed discovered the window of his chamber open and the outline of a human figure, which he supposed to be entering the room. Having a revolver under his pillow, he at once seized and fired it, and what was his horror at discovering that he had shot at his own wife. She had risen before her husband awoke and was looking from the window to learn the location of the fire. The ball from the pistol slightly grazed her right arm, and what narrowly proved a domestic tragedy, terminated in harmless fright.

## Blakslee Bros. Robbed, May 14, 1872

John Gilgrist, a young man about eighteen years of age hailing from Buffalo, was yesterday arrested by Officer Miller on a charge preferred against him for stealing boiler makers' tools from the shop last Friday. He was taken before Justice Patterson, pled guilty, and was sentenced to six months in the county jail and $50 fine.

## The St. Petersburg Oil District, May 14, 1872

*The "Nettie" – This well, on the J. J. Ashbaugh farm, was named for Nettie Ashbaugh, his young daughter.*

New strikes are almost a daily occurrence in the St. Petersburg region and nearly all prove good paying wells, there being a few dry holes found in hundreds of wells drilled there during the past year. The development is still spreading, and a number of new wells are being put down in Salem and Beaver townships.

The Harris well, near St. Petersburg, was sunk some depth into the third sand some days ago and commenced flowing at the rate of 250 barrels per day. The last report received from the well represents that it is still flowing over 200 barrels per day. The tools have not yet been removed from the hole.

Big wells have lately been struck on the Collner farm and other portions of the Petersburg district. – *East Brady Independent*

## Antwerp City Lock-Up, May 20, 1872

About half past eleven o'clock last night two prisoners escaped from the lock-up in the following simple manner: The door is fastened with a hasp and padlock. The hasp is fastened by bolts passing through the door to the interior. On the ends of the bolts are screws and nuts. All, therefore, that the prisoner had to do was to unscrew the nut, pull back the door, and they were free. One of the fugitives was the young man Victorine Snyder, charged with forgery, whose case we mention in another column; the other was a drunk and disorderly named Brennan.

## Oil Matters, May 20, 1872

*The Oilman's Journal* of May 11th says: The Armstrong well, on Turkey Run, Stubble farm, above the old Hulings well, has been pumping for some ten days yielding about 20 barrels per day.

A well on the Slickler farm, owned by Mr. Chambers and others, on the west branch of Turkey Run, is in the third sand and large quantities of oil are flowing from the hole as the drilling progresses. Owners and others expect a large well.

It is said that the well on the Collner farm, struck some weeks ago, still continues to yield near 200 barrels of oil per day. This is the largest well in the entire district.

The Blair well on the south side of the Clarion opposite the mouth of Turkey Run is in the third sand. The hole is filled to a considerable extent with oil while a goodly amount flows out as the drilling progresses. It is expected to prove a good well when the drilling is completed and the pump put in motion.

## Mad Dog, May 21, 1872

A mad dog made its appearance at Porter, Moreland & Co. about 9 o'clock yesterday morning. Most of the men involuntarily betook themselves to the roofs of the tanks, while the delirious canine snapped at those who remained. The animal was finally shot by Daniel Cummings, the foreman, who fortunately had a revolver close at hand.

## May 21, 1872

Foxburg, two and a half miles above Parker's Landing, is a lively and growing oil town. It was founded way back in the dim ages of six months ago, and as its chief produce was oil and greenbacks, its first institution was the Savings Bank, of which Mr. J. W. Hammond of Erie, is President, and our townsman John Fertig, Vice President. The Fox House is kept by Mr. Sterrett, the well known landlord of Parker's Landing. Foxburg is decidedly a town of moral ideas. Being on the land of Mrs. Fox, she has been very firm and strict with her tenants, having a great abhorrence of dram shops and billiard saloons. She will not tolerate them on the precinct, but rules them out in the conditions of all her leases. In spite of this lack of diversion and privations of city life, the town is said to grow and flourish, having some thousand and more inhabitants of good temperate and industrious habits.

*Wells along the Clarion river at Grass Flats*
*Courtesy Drake Well Museum*

## Oil Matters, May 27, 1872

Another large well has been struck on the southeast corner of the Dan Ritts farm, the production of which is about 125 barrels per day.

The Lee & Thomas well on the D. Ritts farm was completed last Thursday and has since yielded 10 barrels per day.

Verbach & Co., of Warren, Penna., have got a new well on their lease on the D. Ritts farm, which is pumping over 100 barrels per day. It filled two 250 barrel tanks in four days' pumping.

Chalfant, Graff & Co. have got a well on their store lot in Petersburg, pumping 30 barrels per day.

The Harris well is pumping 150 barrels per day.

Treat, Porterfield & Crawford have a new well about 20 rods southeast from the Harris well, which is doing 30 barrels per day.

A new well owned by Palmer & Co., directly south of the Harris and just a few rods distant, is pumping about 100 barrels per day.

Dr. Smith and others of Oil City are testing a well on the John Ashbaugh farm, which promises to be a 50 barrel well.

On the Adam Ashbaugh farm, Barney Vensel has a new well doing 40 barrels per day, is located a few rods south of his No. 1 on the same tract.

A 30 barrel well has been struck on the south bank of the Clarion opposite the mouth on Turkey Run. This is leading eastward and considerably off the belt.

Huffman & Crowel, of Warren, Penna., are testing a well on the Andrews & Bly twelve acre lot of the Dan Ritts farm, the show in this well has been very poor. – *Emlenton Sun*

Three or four new wells, some of them large ones, were struck last week in Richland township. One of them is on the Sliker farm, another in the Borough of Petersburg, and still another on the D. Ritts farm.

## Oil Matters, June 6, 1872

The various oil fields from Brady's Bend to St. Petersburg are very active. Development is going speedily forward and new and large wells are reported every day. Fairview, Butler county, appears to be the objective point. A large amount of property has changed hands and some splendid strikes made. We hope next week to have a good report from this locality.

St. Petersburg holds its own and success to the operator is the order of the day. This is splendid territory, and has the gift of continuance in a wonderful degree.

There was never as much activity as at present, and should the present price of oil be maintained, or should it go higher, the prosperity of this region will be without a parallel. – *East Brady Independent*

## June 8, 1872

Well No. 6 on the Foust farm, owned by Spencer and others, is a new strike, which is yielding twenty-five or thirty barrels per day. St. Petersburg, Penna.

*View from Emlenton along the Allegheny River looking towards Fullerton, the oil loading dock for the Antwerp Pipe Line Company*

## Important to Oil Operators

We would respectfully announce to Operators and Contractors, that we have purchased the right and title to and in the invention of Mr. Victor Gretter, known as the "Oil and Gas Controller." Signed Chalfant and Graff & Co., Main Street, Petersburg. *The Oil Field Record*

## Oil News, June 11, 1872

On Saturday last ex-Sheriff Neely informed us that oil had been struck in one of the wells he is putting down on the Isaac Neely farm in Richland township, east of Turkey Run. Oil was flowing through the casing and the prospects are for a large well when the drilling is finished. This well is on the eastern end of the belt, indicating the general course of the oil up the Clarion River.

The well on the Master farm near Monroeville, Beaver township, on Turkey Run, is down about 1,000 feet. A strong vein of gas was struck and the well is now throwing a heavy volume of salt water. – *Clarion Democrat*

## About Oil at St. Petersburg, June 15, 1872

St. Petersburg, June 12, 1872, Editor *Herald*:

There were four wells tubed on Monday and Tuesday, on D. Ritts farm, which commenced pumping as follows:

The Speers well, which caught fire while drilling in third sand, causing the death of Frank Levans and Robert Bowen, is good for 50 or 60 barrels.

*Jefferson Furnace near Jefferson City*
*Courtesy of Venango County Historical Society*

The Blanchard and Siggins well No. 2 starts off at 60 barrels and increasing.

Wetmore & E. Parchall well, on what is known as the three acres lease, D. Ritts farm, starts off at the rate of 70 or 80 barrels. A few more such wells will make the Daniel Ritts Farm one of the richest oil producing farms in the Petersburg district.

Respectfully yours, J. C. B.

P. S. – Since writing the above, this district has been visited by a powerful thunderstorm, the lightning striking the derrick of the Warren Oil Company, destroying one-half of the derrick, including tank house and three 250 barrel tanks, with 200 barrels of oil.

## June 17, 1872

Well No. 6 on the Foust farm, owned by Spencer and others, is a new strike, which is yielding 25 or 30 barrels per day.

## Oil Items, June 18, 1872

A new well owned by Kittanning parties was struck on the Meldren farm last week. It has not been tubed yet, but the indications are for a 40 barrel well.

M. Cool & Lockhard's well known as "Greely No. 1," on the E. Logue farm, east of the Clarion River, was tubed on the 28th of last month and was reported to us as doing 100 barrels of oil per day. It is since reported to be doing 20 barrels. – Petersburg, *The Progress*

Fleming & Salsbury No. 2, Collner farm, was torpedoed last Thursday when she commenced to flow with such force that she lifted a set of tools standing on a plank over the hole, and weighing 1800 lbs., clear off and set them to one side. She is now doing about 50 barrels. – *Volcano Lubricator*

For the last two months the entire section known as the Petersburg district has decreased in production. There are a number of new wells in process of drilling, but the principal excitement seems to be in the lower region. The Petersburg district is drained by two pipe lines, the Antwerp Pipe Line, the office of which is at Fullerton, and the Mutual Pipe Line of Foxburg. By these the oil is taken from the wells and brought to the storage tanks, from which it is shipped to its different destinations.

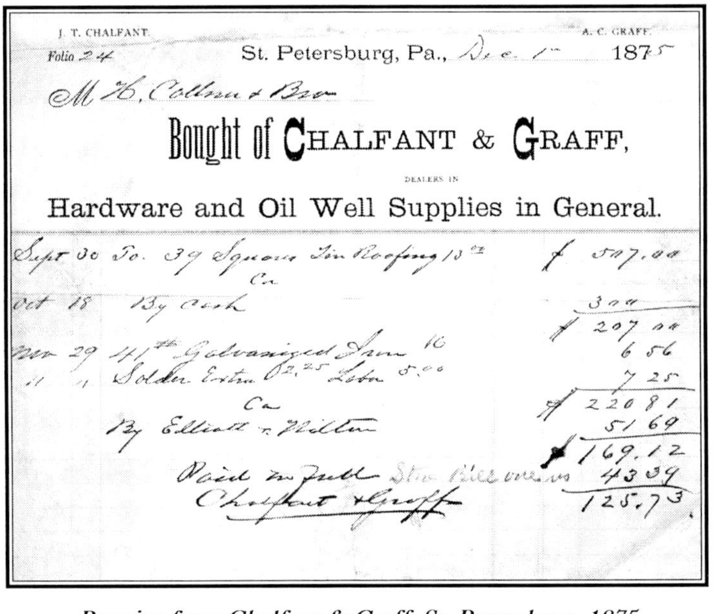
J. T. CHALFANT. A. C. GRAFF.
Folio 24 St. Petersburg, Pa., Dec 1 1875
M. H. Collner & Bro
Bought of CHALFANT & GRAFF,
DEALERS IN
Hardware and Oil Well Supplies in General.

| | | |
|---|---|---|
| Sept 30 | To 39 Squares Tin Roofing 10 oz | 507.00 |
| | Cr | |
| Oct 18 | By cash | 300 |
| | | 207.00 |
| Nov 29 | 41 lbs Galvanized Iron 16 | 6.56 |
| " " | Solder Extra 2.25 Labor 5.00 | 7.25 |
| | | 220.81 |
| | Cr | |
| | By Elliott & Hilton | 51.69 |
| | | 169.12 |
| | Paid in full Chalfant & Graff | 43.39 |
| | | 125.73 |

*Receipt from Chalfant & Graff, St. Petersburg, 1875*

**The Parker District** – The great Parker oil field now by far the most important of any in the region, producing, as it does, nearly one-third of the total production, embraces a very extensive range of country. From the Antwerp and St. Peterburg district to the McClimonds well at Fairview, is a distance of nearly sixteen miles.

## Oil News, June 24, 1872

A new well has been sunk on the Keating farm, called the Odd Fellows' well, which has been pumping since Tuesday of last week and is yielding 20 or 25 barrels per day. Taylor, Lithgow, Daugherty and others are the fortunate owners.

We clip the following from *The Progress* of the 15th inst.:

The Mason well, D. Shoup farm, on Turkey Run, is about down and there is every indication of a good well.

The Eudora well has been playing strange freaks. A few days ago she dropped from 130 barrels to 40 barrels within one hour, and for some time did not pump anything. The tubing was drawn by Messrs. Thompson, Gillespie & Co., and she was cleaned thoroughly, when she commenced to pump at the rate of 85 barrels per day. She has been steadily increasing since that time and is now going about 110 barrels per day.

## Oil News, June 26, 1872

The Elizabeth is the name of a new well sunk by Paul Neely and P. B. Bohlen on the Isaac Neely farm, east side of Turkey Run. The well has been pumping for some ten days and is regarded as a 25 or 30 barrel well. Mr. Isaac Neely owns a large tract of land in this vicinity which has heretofore been regarded as one of the best farms in Clarion county. Now that oil abounds in paying quantities under a portion of his land at least, it may be estimated at five times its former value. We congratulate the fortunate land owner and operators on their success. The striking of this well opens up new fields of operation as it is located some distance east of former development. – *Oil Man's Journal*

A well owned by Cook, Miniger and others, located on the Keating farm, commenced pumping about ten days ago. It is estimated as a 40 barrel well. It will certainly be good for this amount if it doesn't sink below it as others have done.

The Mason well on the Heeter farm, Turkey Run, above the furnace, has been pumping since the 5th of June, averaging a production of ten barrels per day.

A well owned by H. O. Robins and others, Keating farm on the hill above the Grass Flats, has been pumping since the 27th of June, yielding about ten barrels per day.

A well on the D. Ritts farm, a short distance southeast of Petersburg, owned by Plumer and others, has been pumping since the 28th of June and is regarded as a 40 barrel well.

A well was completed by Crooks & Dion on the D. Ritts farm, which is reported as yielding 40 barrels per day.

A well owned by Graham & Shoup, located on the Hiram Neely farm, commenced pumping on the 30th of June at the rate of about 40 barrels per day. – *Oil Man's Journal*

# Oil News, June 27, 1872

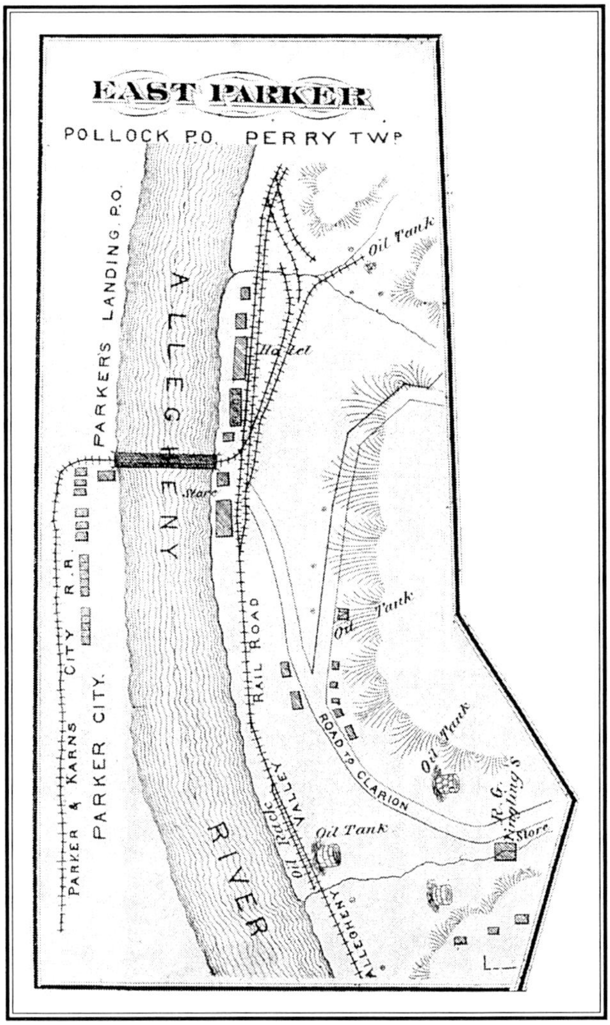


*Map of East Parker – Caldwell's Clarion County Atlas 1877*

Siggins No. 2, D. Ritts farm, is pumping 65 barrels per day.

The well on the I. Neely farm is pumping about 15 barrels per day.

Brice and others have a well on Grass Flats doing 25 barrels per day.

The Soult & Dower well which was torpedoed last week continues to pump about 70 barrels per day.

Mason is in the sand on the Shoup farm, Turkey Run, and the well is making a big show.

Ebon Crawford and others have another new well on the Foust farm doing 25 barrels a day.

The Hess well, on the Knight farm, is minus a sandy bottom. This well is about 50 rods too far west to touch the belt.

They are getting a good show at Berlins, on Beaver creek. They struck the third sand on Monday, which looks very favorable.

In the neighborhood of Jefferson Furnace, Beaver creek, Clarion County, there are four new wells going down. They are owned by Hammond and others.

McCloud is in the sand on the Keating farm. The well is located near the Neely farm line and directly east of Neely's house. Signed "Sand Middling"

The Vensel well on the John Ashbaugh farm in the woods opposite the churches (Petersburg) is doing fully 35 to 40 barrels per day.

On the Master farm, east of Turkey Run, there is a well down to the boulder which has an immense flow of gas, throwing the water at times as high as the derrick and with such force that it passes off in a kind of spray, representing some gigantic fountain bursting forth with a sudden gush of power unknown or seen.

This well is located on a belt passing through from the Grass Flats being N. 32 east. About one mile further on, on the same belt, Mr. Hulings has a lease of 90 acres on the Delo farm, upon which he has a well nearly down. And still further on about one mile and a half there are several more wells drilling and others being started,; here you strike the head waters of Beaver creek three miles north of the Clarion River. According to recent deveopment, this is undoutedly the course of the oil belt if it continues further than Ritchey Run, which fact is now plain.

## Mutual Pipe Company, Manager's Office, Foxburg, Penna., July 22, 1872

The public is hereby notified that on and after date all oil not removed from this company's tanks within ten (10) days after receipt by the company will be subject to extra charge, at the rate of one cent per barrel per day, from date of receipt until removal. Oil now in the custody of the company will be rated as received on the date of this notice.

Receipts or certificates are now ready for issue to all persons entitled to receive them for oil delivered to the company, and buyers of oil may at once require them. No orders will be received by the company after July 31st. The company has recently put in use a large amount of additional tankage; this tankage they must keep clear of all illegitimate accumulations, to enable them to move oil freely and promptly from the wells on their line. To secure this object and prevent the few from monopolizing tankage at the expense of the many, as well as to prevent serious detriment to the company's business, the above system has been adopted. Chas. P. Hatch, Manager

## Not Drunk, But Sleeping
## July 22, 1872

Last Sabbath, a man named Farnham, alias Barnum, a driller on the Hinkley and Barnsdell Lease, Curry Farm, thinking he would like a cool place to take a nap, went up on the top of a 250 barrel tank filled with oil, (the tank being partially covered with boards), lay down on the boards, and was soon dreaming of the fairy land. He suddenly awoke to find himself struggling for life, having rolled off into the tank of oil. Fortunately, a man named Charles Palmer happened to see the man fall and through his (Palmer's) exertions, he was rescued.

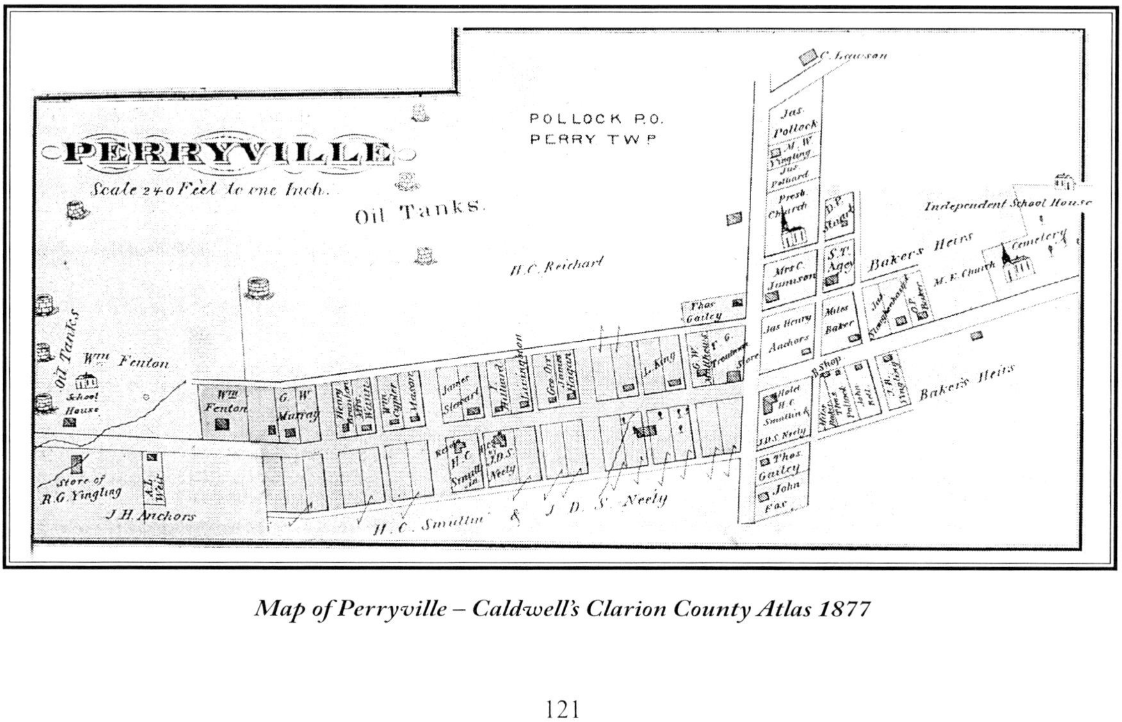


*Map of Perryville – Caldwell's Clarion County Atlas 1877*

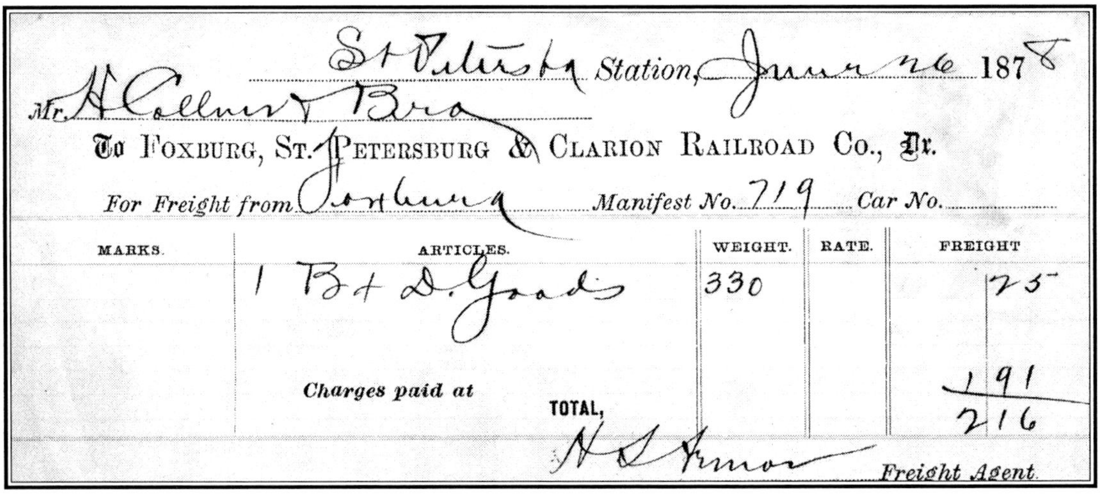

St Petersburg Station, June 26 1878

Mr. H Collner & Bro

To Foxburg, St. Petersburg & Clarion Railroad Co., Dr.

For Freight from Foxburg Manifest No. 719 Car No.

| MARKS. | ARTICLES. | WEIGHT. | RATE. | FREIGHT |
|---|---|---|---|---|
| | 1 B+ D. Goods | 330 | | 25 |
| | Charges paid at | | | 191 |
| | TOTAL, | | | 216 |

H S Armor Freight Agent.

*Foxburg, St. Petersburg and Clarion Railroad Company receipt, June 26, 1878*

*A depiction of the Allegheny Valley Railroad*

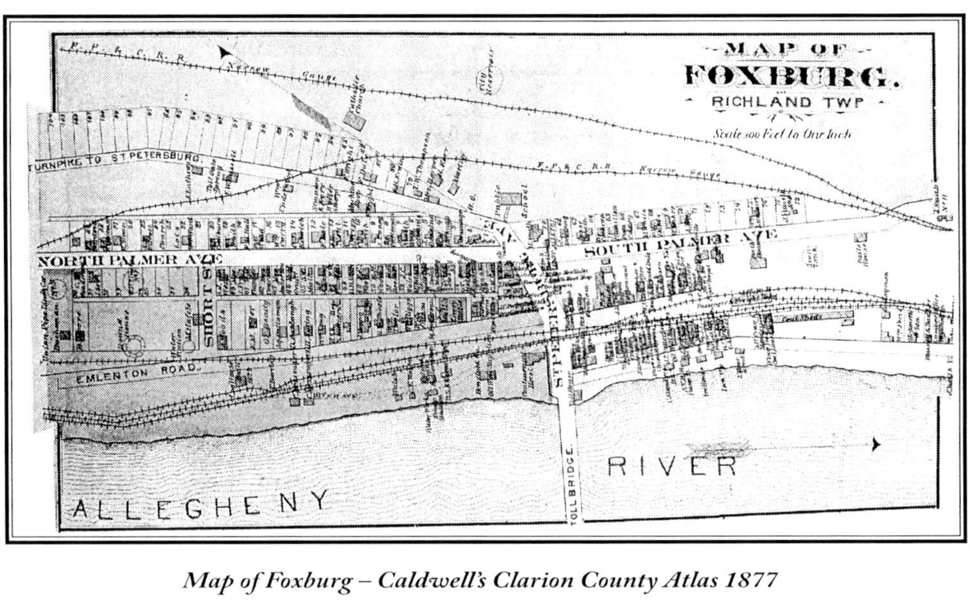


*Map of Foxburg – Caldwell's Clarion County Atlas 1877*

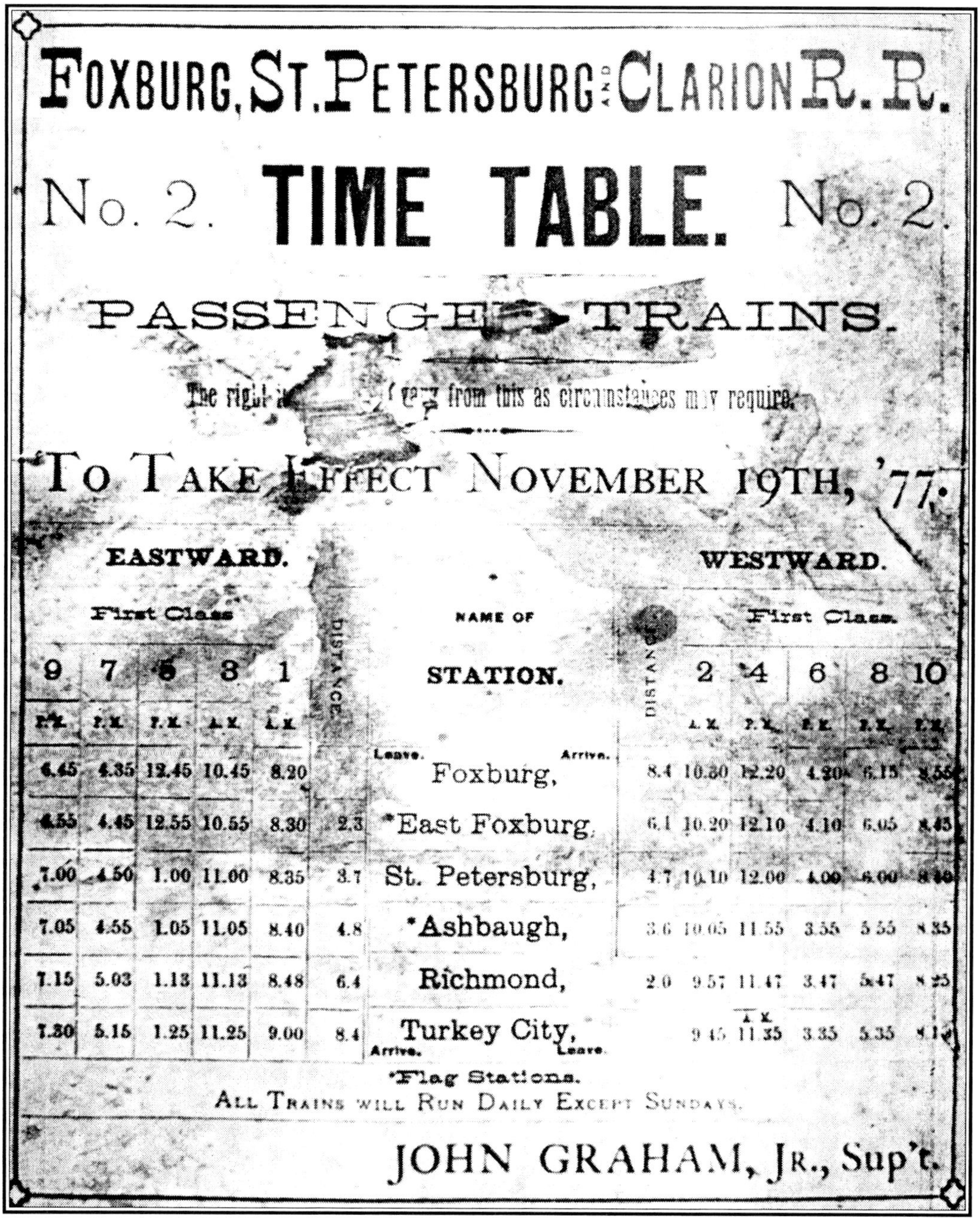

FOXBURG, ST. PETERSBURG AND CLARION R. R.

No. 2. TIME TABLE. No. 2.

PASSENGER TRAINS.

The right [illegible] vary from this as circumstances may require.

TO TAKE EFFECT NOVEMBER 19TH, '77.

| EASTWARD. First Class 9 | 7 | 5 | 3 | 1 | DISTANCE | NAME OF STATION. | DISTANCE | WESTWARD. First Class 2 | 4 | 6 | 8 | 10 |
|---|---|---|---|---|---|---|---|---|---|---|---|---|
| P.M. | P.M. | P.M. | A.M. | A.M. | | | | A.M. | P.M. | P.M. | P.M. | P.M. |
| 6.45 | 4.35 | 12.45 | 10.45 | 8.20 | | Leave. Foxburg, Arrive. | 8.4 | 10.30 | 12.20 | 4.20 | 6.15 | 8.55 |
| 6.55 | 4.45 | 12.55 | 10.55 | 8.30 | 2.3 | *East Foxburg, | 6.1 | 10.20 | 12.10 | 4.10 | 6.05 | 8.45 |
| 7.00 | 4.50 | 1.00 | 11.00 | 8.35 | 3.7 | St. Petersburg, | 4.7 | 10.10 | 12.00 | 4.00 | 6.00 | 8.40 |
| 7.05 | 4.55 | 1.05 | 11.05 | 8.40 | 4.8 | *Ashbaugh, | 3.6 | 10.05 | 11.55 | 3.55 | 5.55 | 8.35 |
| 7.15 | 5.03 | 1.13 | 11.13 | 8.48 | 6.4 | Richmond, | 2.0 | 9.57 | 11.47 | 3.47 | 5.47 | 8.25 |
| 7.30 | 5.15 | 1.25 | 11.25 | 9.00 | 8.4 | Arrive. Turkey City, Leave. | | 9.45 | A.M. 11.35 | 3.35 | 5.35 | 8.1[illegible] |

*Flag Stations.

ALL TRAINS WILL RUN DAILY EXCEPT SUNDAYS.

JOHN GRAHAM, JR., Sup't.

*Time Table of the Foxburg, St. Petersburg and Clarion Railroad*
*November 1877*

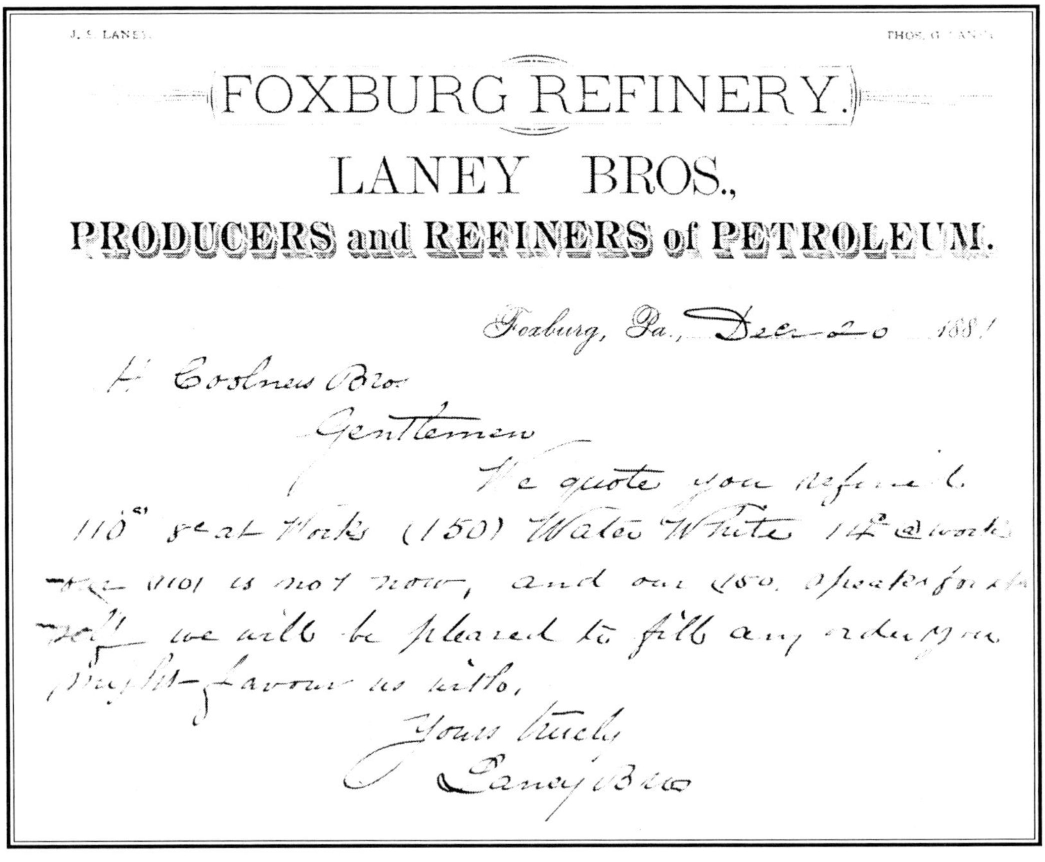

J. S. LANEY. THOS. G. LANEY.

FOXBURG REFINERY.

LANEY BROS.,

PRODUCERS and REFINERS of PETROLEUM.

Foxburg, Pa., Dec 20 1881

H. Coslner Bro

Gentlemen

We quote you refin'd 110° 8c at Works (150) Water White 11c @ works — our 110 is not now, and our 150 speaks for itself. We will be pleased to fill any order you might favour us with.

Yours truly

Laney Bro

*Foxburg Refinery document, December 20, 1881*

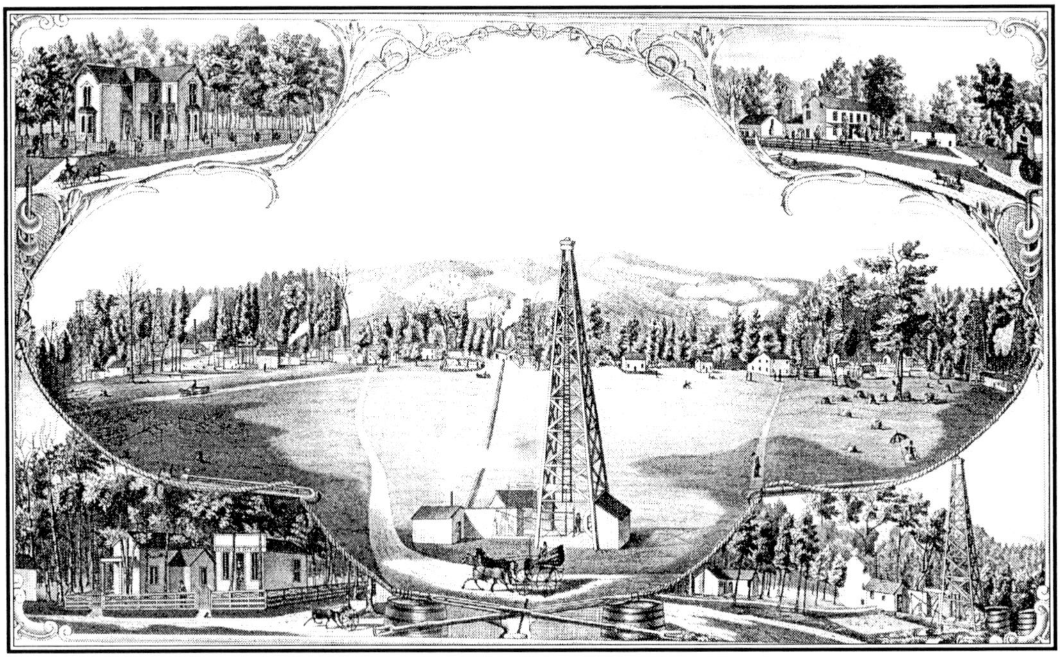

*Logue Oil Farm in Perry Township – Caldwell's Clarion County Atlas 1877*

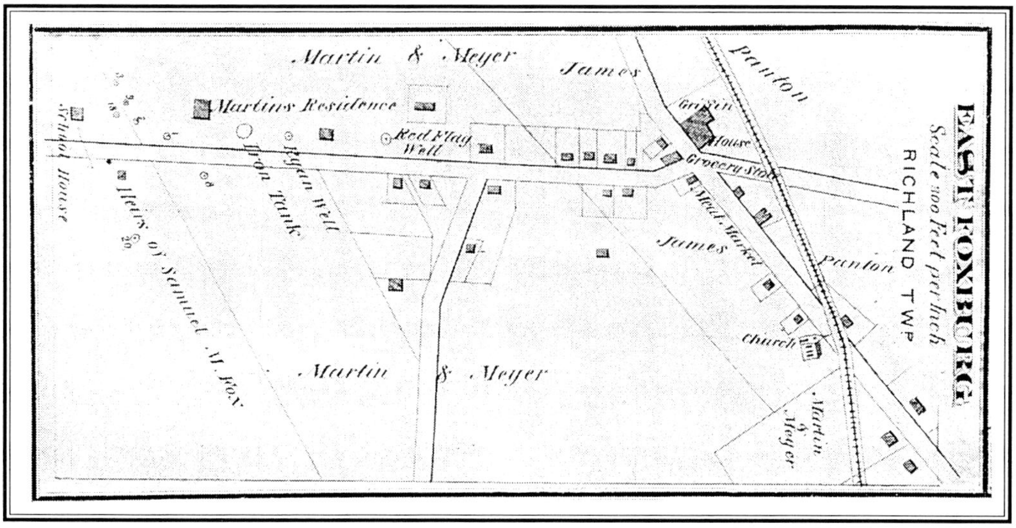


*Map of East Foxburg – Caldwell's Clarion County Atlas 1877*

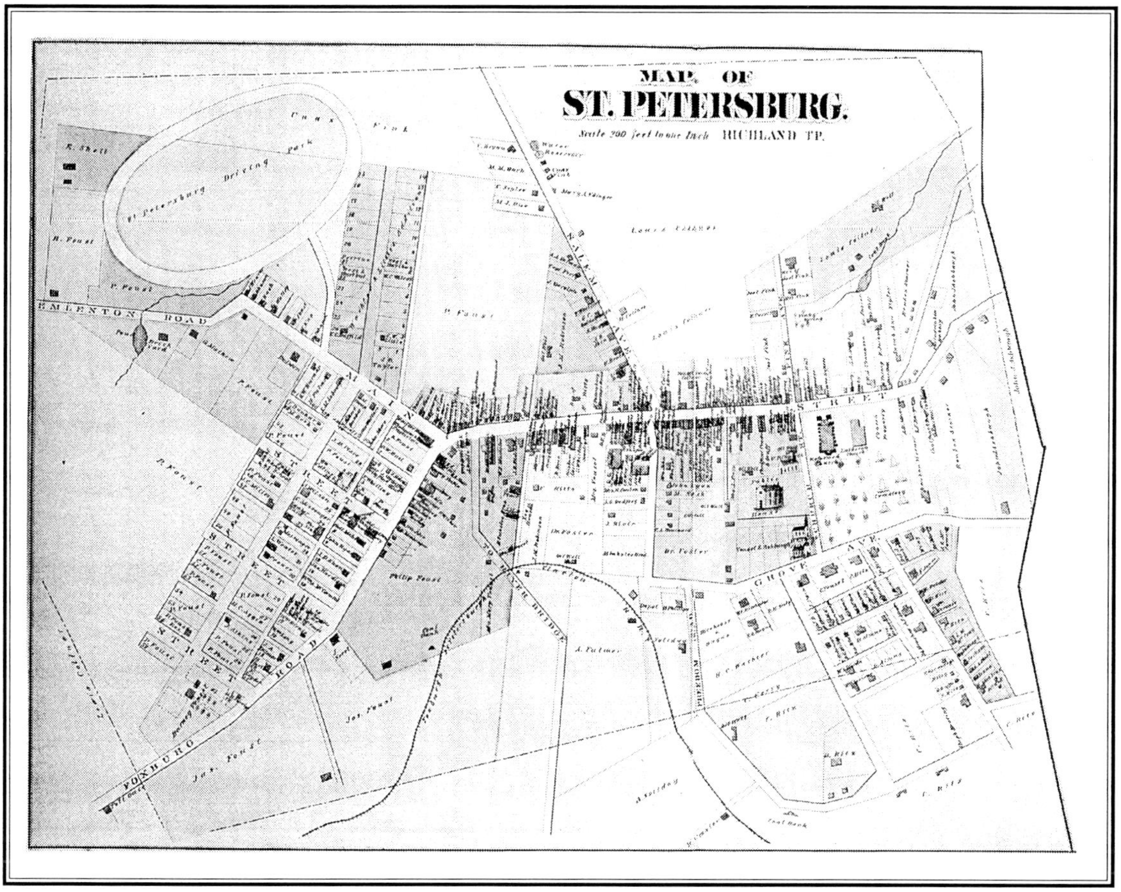


*Map of St. Petersburg – Caldwell's Clarion County Atlas 1877*

*Residences of J. B. and William Z. Blakslee, St. Petersburg – Caldwell's Clarion County Atlas 1877*

*Blakslee Brothers Machine Shop, St. Petersburg – Caldwell's Clarion County Atlas 1877*

*Residence of Joel Fink, St. Petersburg – Caldwell's Clarion County Atlas 1877*

*The Monitor House, St. Petersburg – Caldwell's Clarion County Atlas 1877*

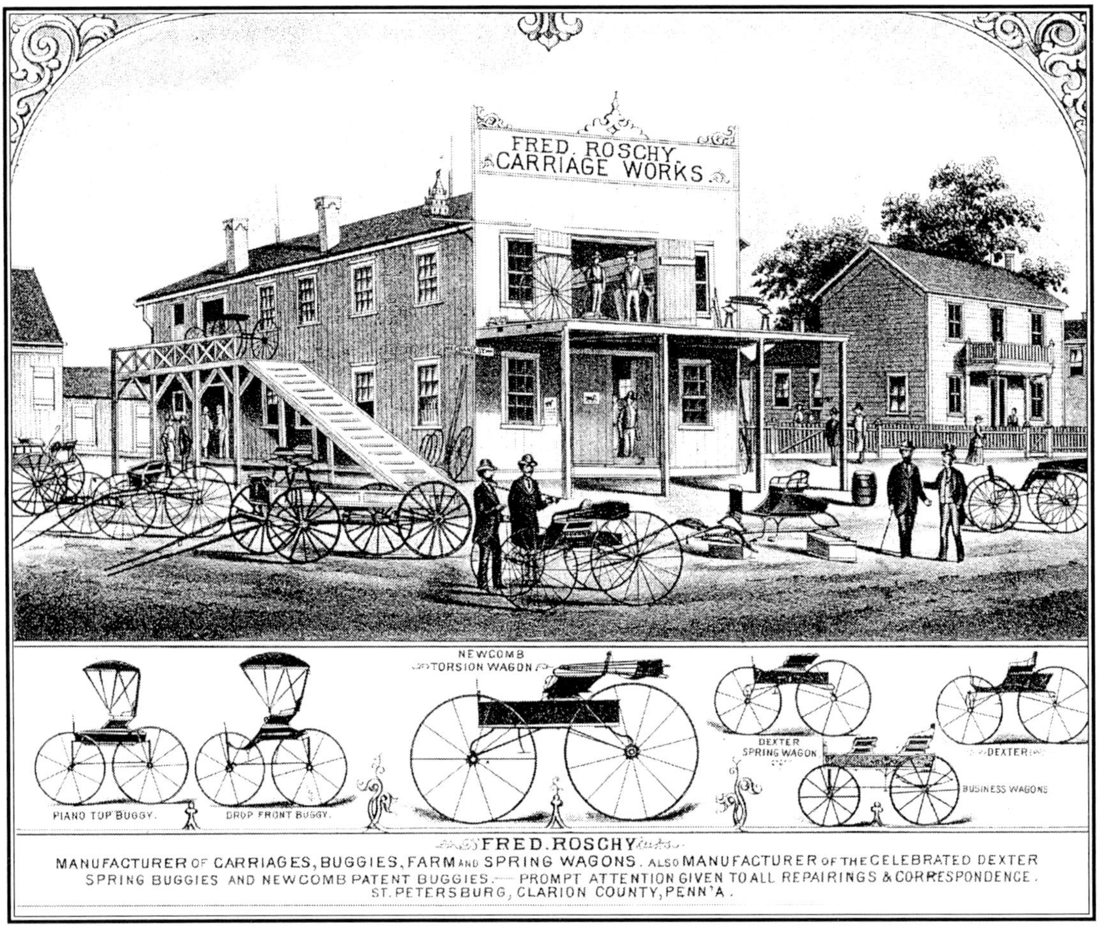


*Fred Roschy Carriage Works, St. Petersburg – Caldwell's Clarion County Atlas 1877*

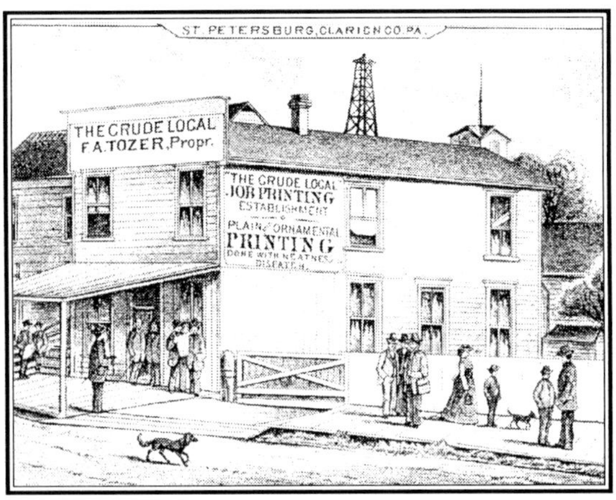


*The Crude Local office, St. Petersburg*
*Caldwell's Clarion County Atlas 1877*

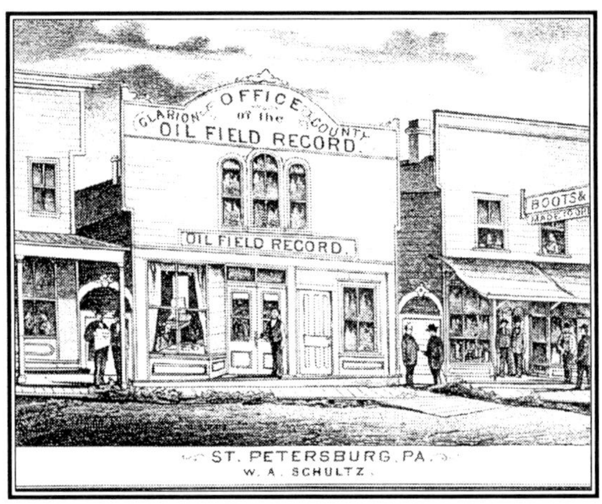


*Oil Field Record office, St. Petersburg*
*Caldwell's Clarion County Atlas 1877*

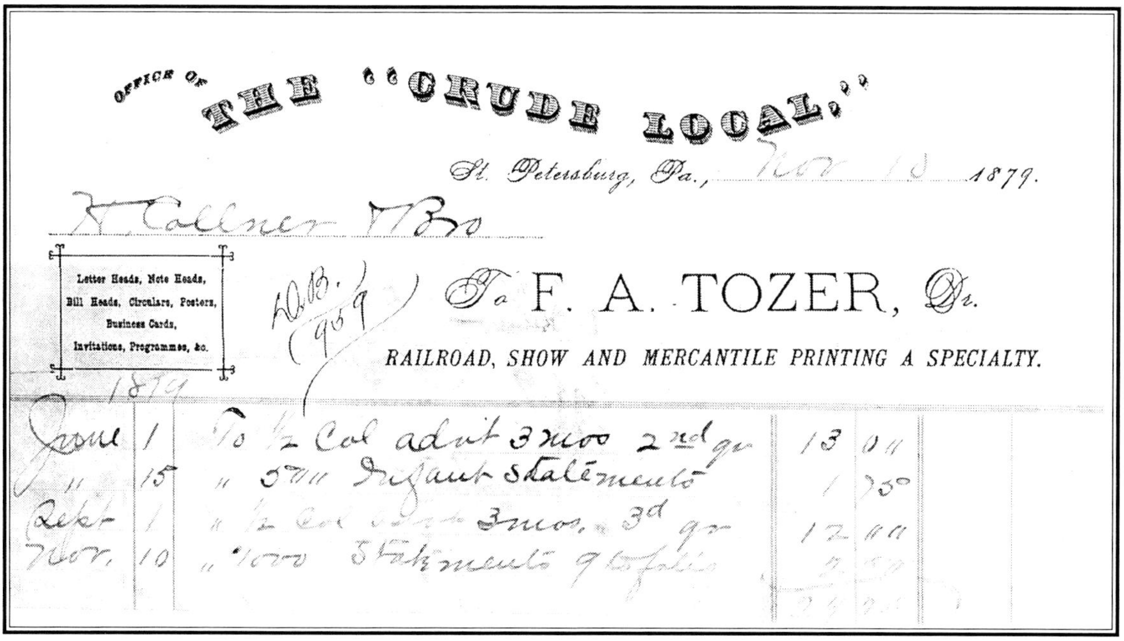

OFFICE OF THE "CRUDE LOCAL."

St. Petersburg, Pa., Nov 13 1879.

H. Collner & Bro

Letter Heads, Note Heads, Bill Heads, Circulars, Posters, Business Cards, Invitations, Programmes, &c.

To F. A. TOZER, Dr.

RAILROAD, SHOW AND MERCANTILE PRINTING A SPECIALTY.

| 1879 | | | | |
|---|---|---|---|---|
| June | 1 | To ½ Col advt 3 mos 2nd qr | 13 | 00 |
| " | 15 | " 500 Infant statements | 1 | 75 |
| Sept | 1 | " ½ Col advt 3 mos. 3d qr | 12 | 00 |
| Nov. | 10 | " 1000 Statements 9 to folio | [illegible] | [illegible] |

Receipt from The Crude Local, 1879

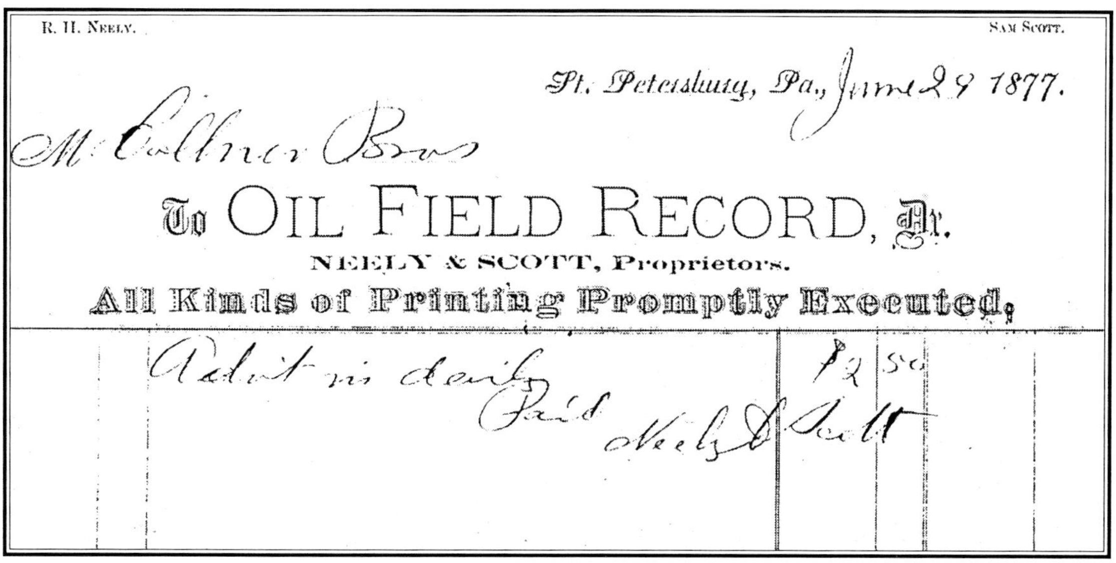

R. H. Neely. Sam Scott.

St. Petersburg, Pa., June 29 1877.

M. Collner Bros

To OIL FIELD RECORD, Dr.

NEELY & SCOTT, Proprietors.

All Kinds of Printing Promptly Executed.

| | | |
|---|---|---|
| Advt in daily | $2 | 50 |

Paid Neely & Scott

Receipt from the Oil Field Record, 1877

# PRICE LIST
—OF—
# TORPEDOES

From Sept. 10, 1884.

| SIZE. | Diameter Inches. | Length. Ft. Inches. | PRICE |
|---|---|---|---|
| 3 Quarts | 2½ | 2–11¼ | 14.50 |
| 4 " | 2½ | 3–11 | 15.75 |
| 6 " | 2½ | 5–10⅝ | 18.50 |
| 8 " | 2¾ | 6–5 ¾ | 21.00 |
| 10 " | 3⅛ | 6–3 ¼ | 23.75 |
| 12 " | 3⅛ | 7–6 ¼ | 26.50 |
| 16 " | 3⅛ | 10– ⅜ | 31.75 |
| 20 " | 3⅛ | 12–6 ⅝ | 37.00 |
| 24 " | 3½ | 12– | 42.50 |
| 30 " | 4 | 11– 5⅞ | 50.50 |

ALL WORK AT WELL OWNER'S RISK.

(OVER.)

*Price list of torpedoes, 1884*

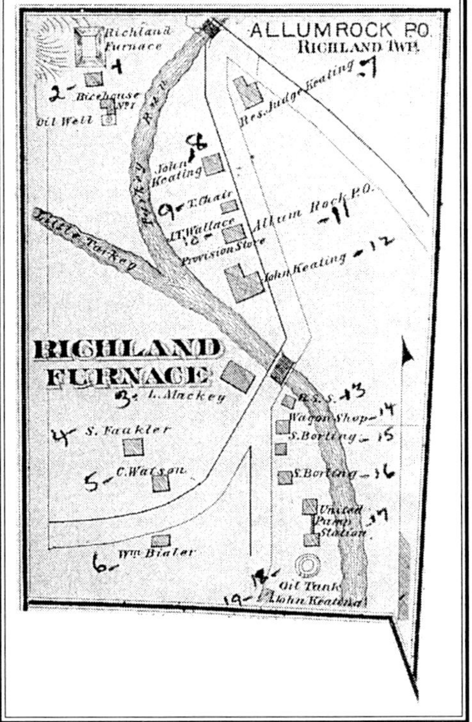


*Map of Richland Furnace*
*Caldwell's Clarion County Atlas 1877*

# TO PRODUCERS.

EDENBURGH, PA., .......................... 188

I am now prepared to offer you the advantages of the following prices in Nitro Glycerine **Torpedoes,** and will explode all Torpedoes **over 30 Quarts** in wells at **40 cents per pound.** All work at **Well owner's risk.**

**TERMS STRICTLY CASH.**

WM. L. JEFFORDS, Agent,
EDENBURGH, (KNOX P. O.,) CLARION CO., PA.

(OVER.)

*Notice of William L. Jeffords, Torpedo Agent*

*Home of M. L. Lockwood, Richmond – Caldwell's Clarion County Atlas 1877. Note page 288*

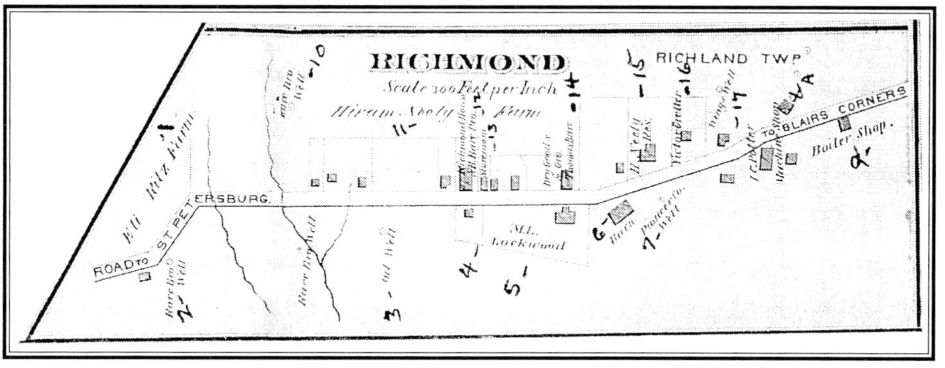


*Map of Richmond – Caldwell's Clarion County Atlas 1877*

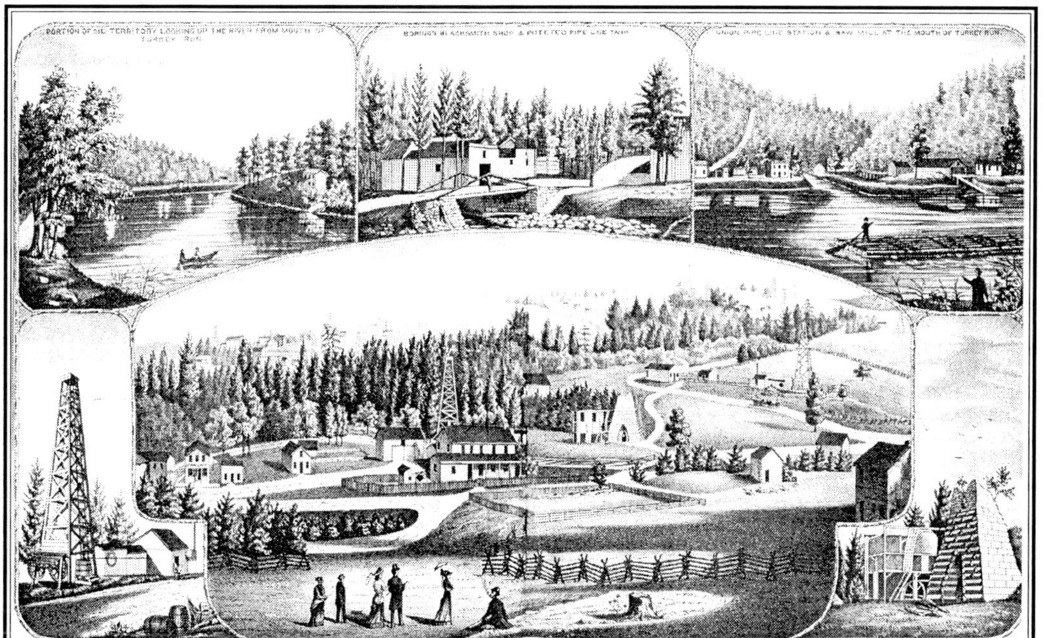

*Furnace Grounds and Lands of Judge Keating - Richland Furnace*
*Mouth of Turkey Run- Caldwell's Clarion County Atlas 1877*

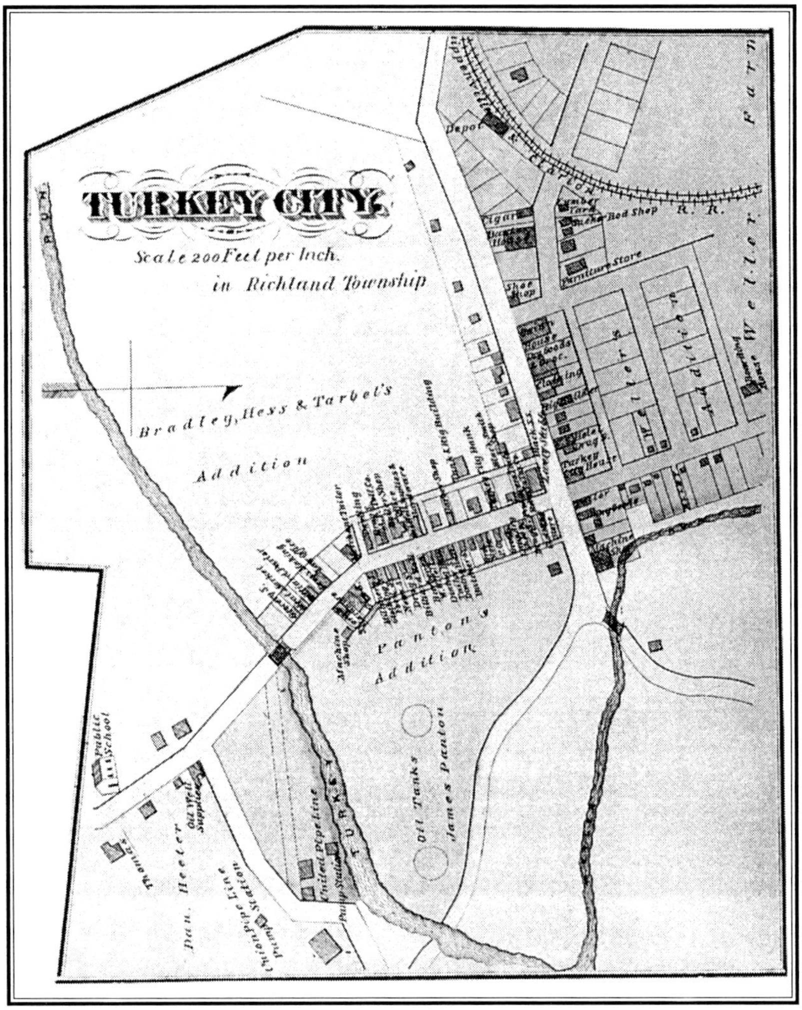


*Map of Turkey City – Caldwell's Clarion County Atlas 1877*

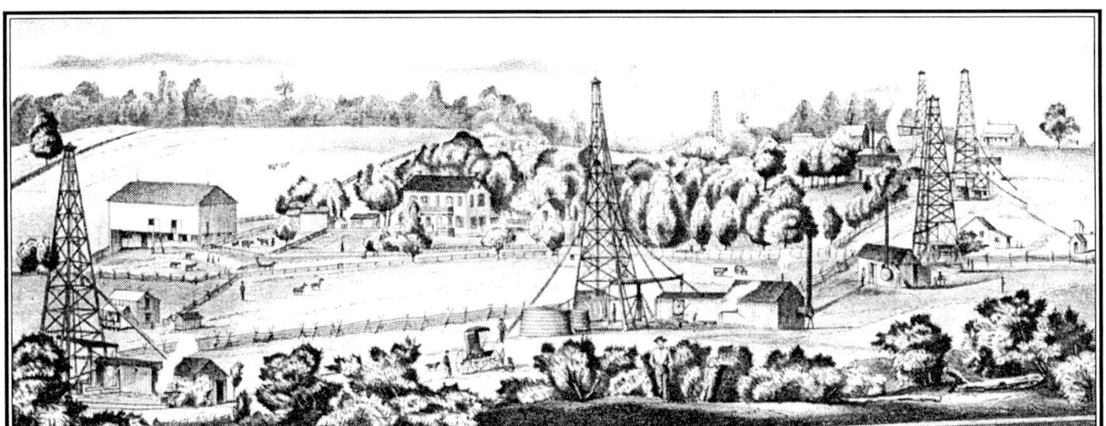

*Mishal Fillman & Son's Farm, Richland Township – Caldwell's Clarion County Atlas 1877*
*Note page 25 and 147 August 6, 1872. Three men were burned to death on this farm.*

*Berlin Oil Farm, Beaver Township – Caldwell's Clarion County Atlas 1877*

*Residence of Susannah Heeter, Monroeville – Caldwell's Clarion County Atlas 1877*

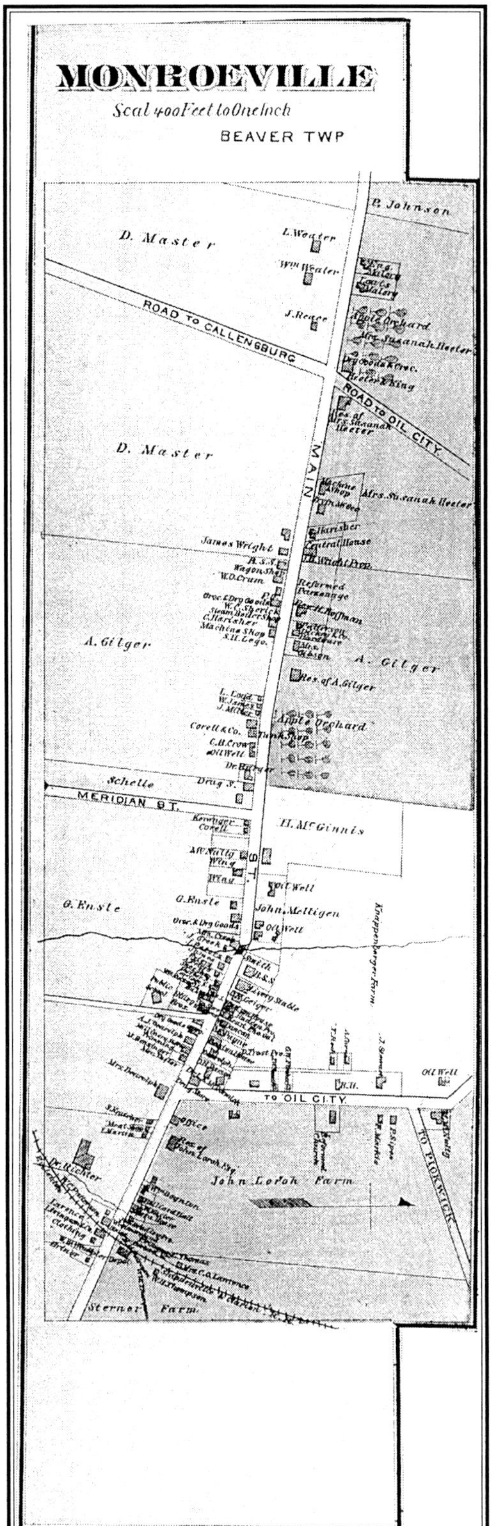


Map of Monroeville – Caldwell's Clarion County Atlas 1877

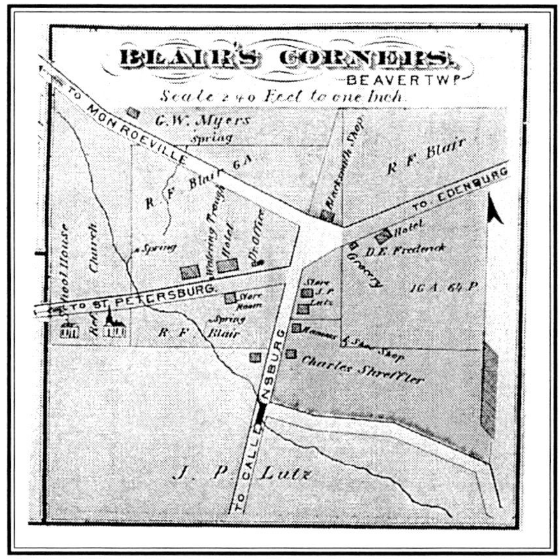


*Map of Blair's Corners – Caldwell's Clarion County Atlas 1877*

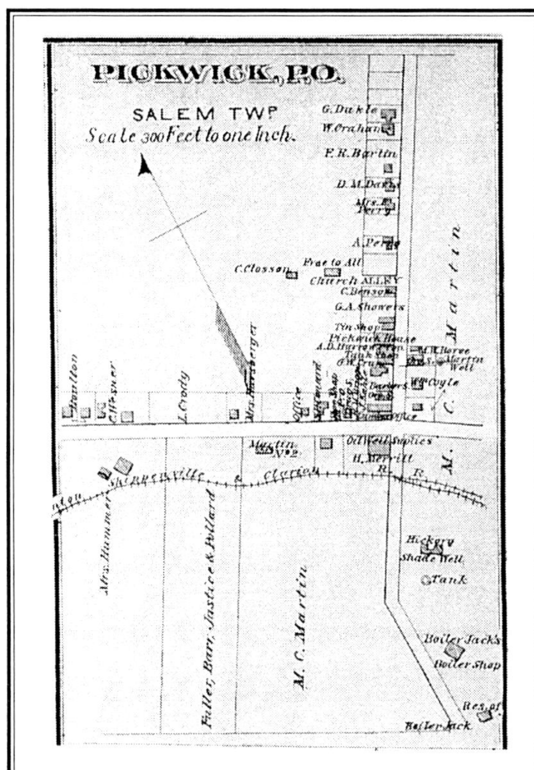


*Map of Pickwick – Caldwell's Clarion County Atlas 1877*

*Blair Residence and Hotel, Blair's Corners – Caldwell's Clarion County Atlas 1877*
*The Hotel was built during the Civil War with Robert Blair as proprietor. This hotel was torn down in the 1940's.*

*Knappenberger Oil Farm, Salem Township – Caldwell's Clarion County Atlas 1877*

*Delo Oil Farm, Salem Township – Caldwell's Clarion County Atlas 1877*

*Cropp Farm, Salem Township – Caldwell's Clarion County Atlas1877*

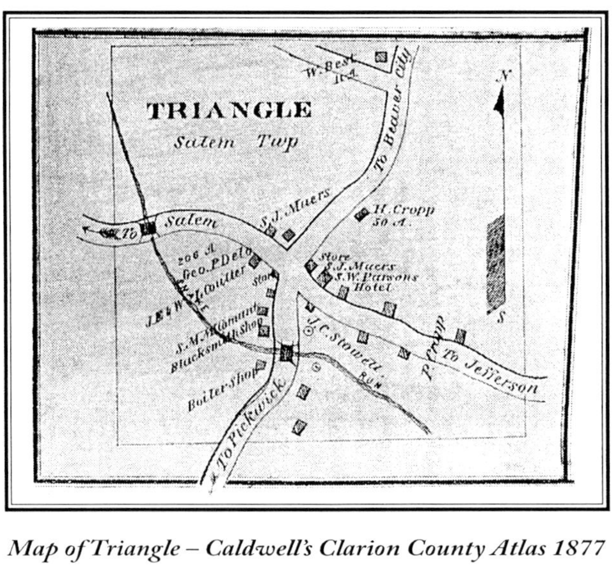


*Map of Triangle – Caldwell's Clarion County Atlas 1877*

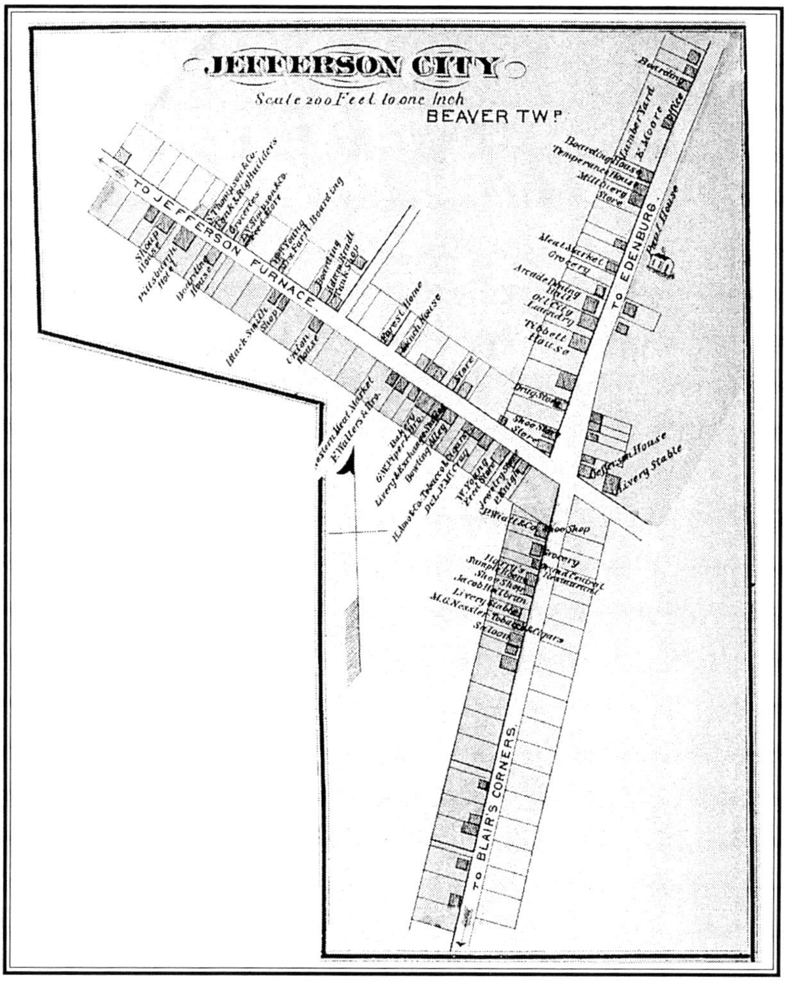


*Map of Jefferson City – Caldwell's Clarion County Atlas 1877*

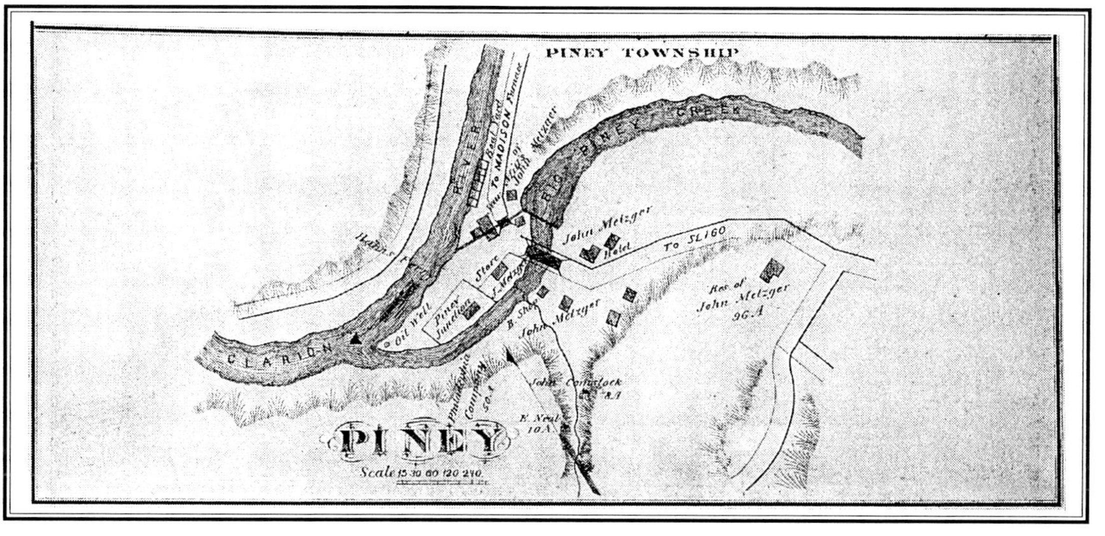


*Map of Piney – Caldwell's Clarion County Atlas 1877*

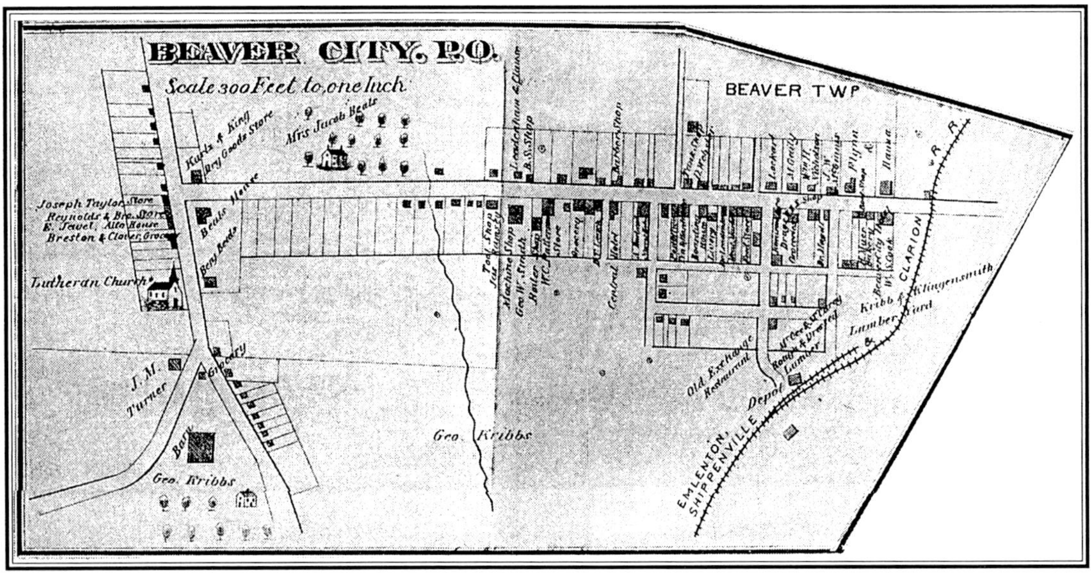


*Map of Beaver City – Caldwell's Clarion County Atlas 1877*

*George Kribbs Farm, Beaver City - Caldwell's Clarion County Atlas 1877*

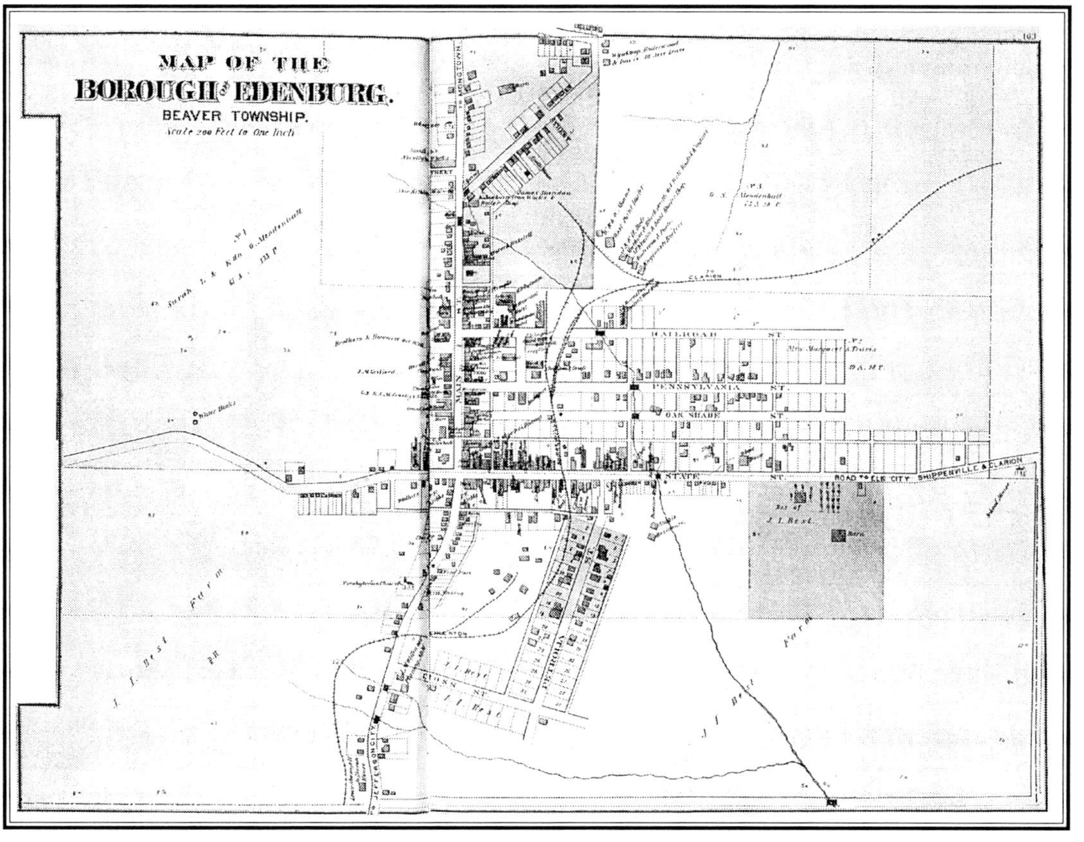


*Map of Edenburg (now Knox) – Caldwell's Clarion County Atlas 1877*

*Main Street, Edenburg (now Knox) – Caldwell's Clarion County Atlas 1877*

*Bull Wheel Factory, Edenburg – Caldwell's Clarion County Atlas 1877*

*Winger & Knight Oil Farm - near Edenburg, Beaver Township - Caldwell's Clarion County Atlas 1877*

*Residence of N. L. Moore – Superintendent of the Columbia Oil Co. Farm near Edenburg Caldwell's Clarion County Atlas 1877*

*Residence of G. A. Clark, Edenburg Caldwell's Clarion County Atlas 1877*

*A. R. Black Oil Farm - near Elk City - Caldwell's Clarion County Atlas 1877*

*A. R. Black Oil Farm - Beaver Township - near Elk City Caldwell's Clarion County Atlas 1877*

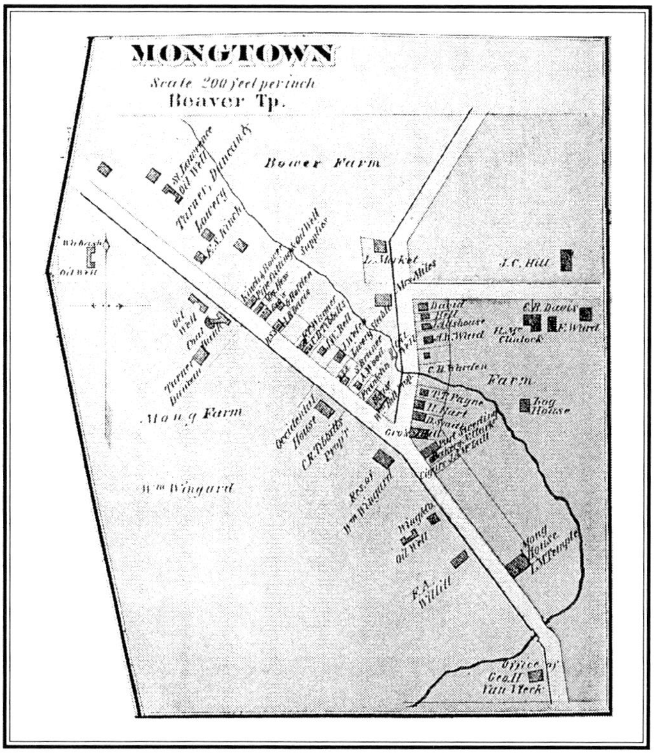


*Map of Mongtown – Caldwell's Clarion County Atlas 1877*

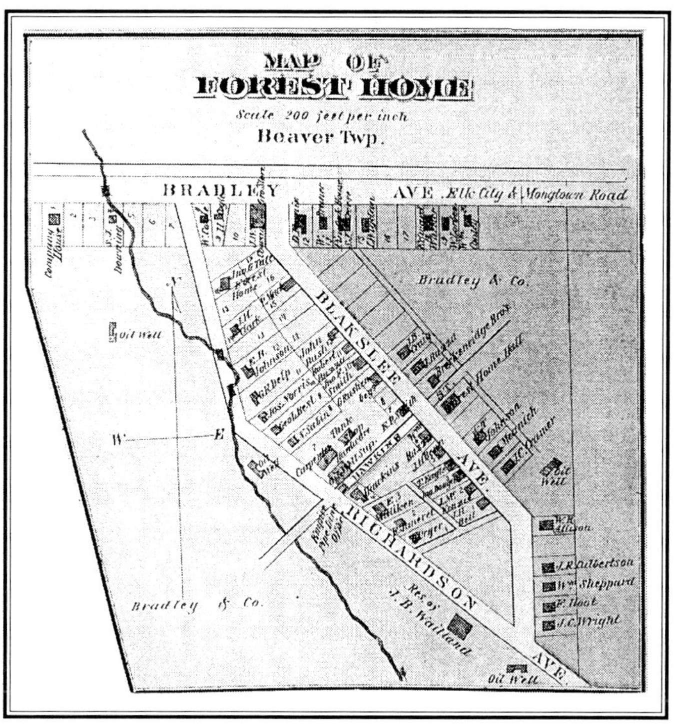


*Map of Forest Home – Caldwell's Clarion County Atlas 1877*

*Home Farm of Ross Haney – Beaver Township, Clarion County   Caldwell's Clarion County Atlas 1877*

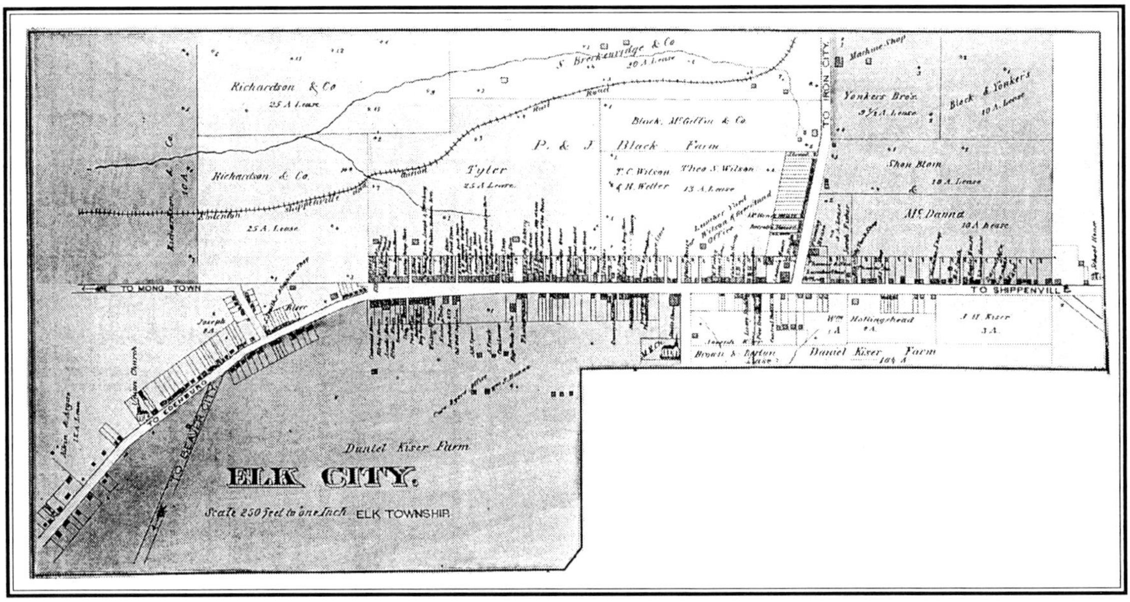


*Map of Elk City – Caldwell's Clarion County Atlas 1877*

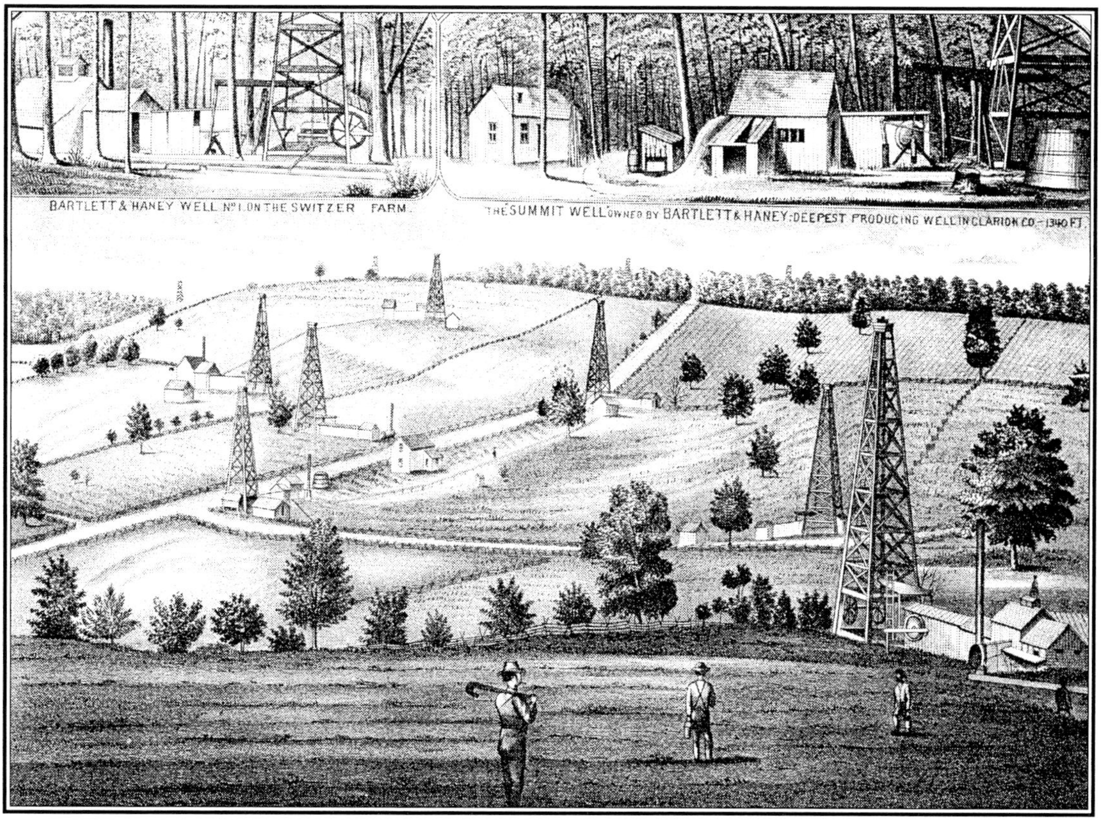


*Oil Farm of Bartlett & Haney - Beaver Township, Clarion County, Caldwell's Clarion County Atlas 1877*

## Oil News, July 24, 1872

It is again reported that they have struck a good sand rock in James' well at Blyson Run. The wells in Salem, Ashland and Beaver townships have not yet proved successful, though several of them are not completed. The development appears to extend up the general course of the Clarion River.

The lightning struck the derrick of the oil well on the John M. Kurtz farm, near Salem, about ten days ago and it burned down. The owners contemplate building a new rig and drilling the well deeper, hoping yet to get oil.

The center of the oil excitement now is Fairview, Butler County. So far as heard from there are no dry holes there.

The steady flow of oil operators to that part of the oil country embraced in portions of Clarion, Venango and Butler counties, of which Parker's Landing is the nearest center, has caused an almost total suspenision of extensive developments in that portion of the oil region surrounding and above Oil City.

*The Progress*, St. Petersburg, is out with its oil report No. 76 from which we glean as follows:

On the D. Ritts farm there are 37 wells, producing an aggregate of 657 barrels, and 13 wells drilling.

On the Keating farm there are 37 wells, producing 423 barrels and 18 wells drilling.

On the J. J. Ashbaugh farm there are 28 wells, producing 472 barrels.

The D. Ritts farm has decreased 106 barrels, although there have been 7 wells struck during the previous month, and on the Keating farm there has been an increase of 26 barrels.

## The Foxburg Fire, July 31, 1872

The fire at Foxburg on Saturday night was a very destructive one. It broke out in a building adjoining the Continental Hotel in Foxburg and so rapidly did the flames spread that before their progress was stayed every building, with the exception of the Continental, on the lower side next to the river below the railroad track, was destroyed. Very little property was saved from the burning buildings. The principal losers are as follows: Mr. Stewart, clothing store; Mr. Hadley, Olive Branch restaurant; Union Express office; Porterfield & McComb, hardware store; Western Union Telegraph office; Porapa & Wilson, grocery store; Weaver & Son, National dining rooms. The railroad platform was also destroyed. Total loss about $50,000; partially insured.

## Oil News, August 2, 1872

Mr. J. H. Heiner, of Kittanning, has succeeded in getting a 75 barrel well at the Stone House, near Lawrenceburg.

Sam M. Kier is again in luck. He has got a splendid well. It is known as the Birch well on the Charles Ritts farm, near St. Petersburg. He and M. J. M'Auley are the owners. It was struck on the 19th and has been yielding from 175 to 200 barrels daily.

On Saturday last a new well on the Lewis Collner farm, St. Petersburg, reached the third sand, with every indication of being a big one. It is owned by S. M. Kier, R. L. Brown and Simon Truby.

## Petersburg Oil Field, August 3, 1872

A well owned by Fuller, Hess, Tarbell, and others, on the David Shoup farm east of Petersburg, is in the sand with good show of oil.

Well owned by Painter, Haymaker, and others, on the David Shoup farm, in sand, oil flowing from the hole, and indications of a good well.

Milligan well, on David Shoup farm, down and pumping. For several days after being started it continued to yield 60 to 65 barrels per day.

No. 2 Pioneer on the Keating Farm, on the Clarion River a short distance below the mouth of Turkey Run, is down and pumping about 30 barrels per day. Owned by Turner and others.

Lydia well, on the Smith farm, south side of the Clarion River, below the mouth of Turkey Run, has been drilled some depth into the third sand and is flowing large quantitites of oil.

Germania well No. 2, on the David Shoup farm, east of Petersburg, is down and pumping about 30 barrels per day. Dower, Fertig and others are the owners.

A well on the west side of Turkey Run near the mouth, Keating farm, is in third sand, and for some days past has been flowing large quantities of oil. The drilling still progresses. When finished, a big well is expected.

A well on the Hiram Neely farm, near Turkey Run, commenced pumping on Wednesday of last week, and in 24 hours thereafter it was still yielding oil at the rate of 50 or 60 barrels per day.

A well on the D. Masters farm, northeast of Petersburg on the east side of Big Turkey Run, has been some depth into the third sand, and oil is now flowing from the hole. John McLaughlin, John Rishel and others are the owners.

*Bend of the Clarion River at Grass Flats, 1873*
*Courtesy Drake Well Museum*

The latest intelligence from the Petersburg district represents new wells as yielding oil as follows: Haymaker well, pumping 40 barrels. Fuller well 40. Cramer well 40. All on the D. Shoup farm.

A well owned by a Pittsburgh company on the Hiran Neely farm has just been struck, and is said to be pumping 45 barrels per day.

The Chambers well, "Little Turkey," was increased by a torpedo to 50 barrels per day last week.

They are starting a well on the Edinger farm, up the hollow east of the old Furnace well.

Gordon has a new well on the old Shoup farm, over the hill northeast from Hess's well, doing 50 barrels a day.

## Horrible Accident in St. Petersburg District
## Three Men Burned to Death
## August 6, 1872

It seems that five men were at a well on the Fillman farm, either as drillers or lookers-on, when the tools struck through into a crevice of the sand rock, and an immense volume of gas was liberated from its long imprisonment and rushed with great force up through the well and almost instantly communicated with the fire under the boiler and exploded with great force, enveloping the engine house and derrick. All this was the work of an instant, and the men working at the well had no time for flight, but were caught in the burning gas and three of them, Nelson Vensel, Michael Shoup, and Alexander McGee, were burned to death. Martin Frank was dangerously and John Malthorn slightly burned. All rest in the St. Petersburg cemetery.

## Oil News, August 6, 1872

Dr. Hammer is putting down a well on the John Martin farm, and one of the Godleith Exley farm. Each are now several hundred feet deep. He has a rig up for another well on the Jefferson Furnace land.

The first well on the Daniel Heeter farm on Big Turky Run, in Richland Township, which at first was rather small, has increased to a 40 barrel well.

A well on Birch lot, in Petersburg, owned by Jackson and others, is completed, and starts up at about 75 barrels per day. This yield has been kept up for several days and owners and others count on a big well. The parties who sunk the well purchased the ground, thus saving the royalty which is usually paid.

## Oil Intelligence, August 8, 1872

The Geo. Heeter well, Keating farm, is being cleaned out.

The Cramer well, D. Shoup, Sr. farm, is doing about 25 barrels.

The Painter & Shoup well, Shoup farm, is doing 17 barrels.

The McCool well, on the Logue farm, south side of the Clarion, is doing 25 barrels.

Tabor & Co. are going to start six new wells on the Logue farm on the south side of the Clarion.

The Milligan & Copnnelly well, on the D. Shoup, is still keeping up to its first production, 50 barrels.

Capt. Chalfant has commenced drilling on the Chalfant No. 2, about 100 yards from *The Progress* office.

Brown and Crawford have a new well on the Logue farm, which is down and doing 45 barrels. It is called the Bee Tree well.

The Campbell well, on the William Smith tract, south side of the Clarion, is about 8 feet in the sand and promises to be a large well.

Painter & Allshouse have the tools out of their well, No. 2, on the D. Ritts farm, and will probably strike the sand this week.

The Logue well No. 1 is a new strike. It is on the Logue farm, south of the Clarion River. There is every indication for a large well.

A very important field is being developed on the south side of the Clarion river on the Logue, Smith and Fox farms. On the A. Logue there are several wells just about down. Logue No. 1 will doubtless be a very good well.

Development will shortly commence on the Miles Logue, adjoining the A. Logue. Taylor & Co. are the operators, we believe. This farm was sold about one year ago for $12,000 on the supposition of its being "dead beat" territory.

There are two new rigs going up at the mouth of Turkey Run on the Keating farm. This will open up much more territory, if they prove successful, since there is no well within a considerable distance.

Logue well No. 1 on the Logue farm, south side of Clarion, is flowing a heavy stream, and promises to be a large well. The oil is thrown over the derrick. A Willoughby & Co. are the lucky owners. A year ago this was considered dead beat territory. – St. Petersburg, *The Progress*

## August 8, 1872

Charles Campbell, the man who was sentenced to our county jail for six months at the last court for stealing a set of jars at Petersburg, and who escaped a few weeks since, was killed in a fight at Louisville, Ky., a few days since. The principal witness against him was killed by lightning at Parker's a few weeks ago. – *Clarion Democrat*

## Youthful Depravity

A boy ten years of age was arrested by Constable Cribbs yesterday. He was taken to the lockup with the charge of being drunk and disorderly.

**If you feel dull, drowsy, debilitated, having frequent headache, mouth tastes badly, poor appetite and tongue coated, you are suffering from torpid liver, and nothing will cure you so speedily and permanently as Dr. Pierce's Golden Medical Discovery.**

**A health beverage, Tozar of the Westmoreland Hotel, Antwerp City, pours "Peruvian beer" to be drawn fresh from the fountain.**

**"A good smoke driveth away melancholy" saith Sir Walter Raleigh. If he had smoked Citizens Corps brand he would never have had the blues.**

## St. Petersburg News, August 10, 1872

The following will appear in *The Progress* of this morning:

The Risher well, on the Master farm, is doing 25 barrels. Immense quantities of second sand gas escapes, while there is very little third sand gas. For hundreds of yards around the well the trees are withered and dead, the gas effectually destroying every particle of vegetation with which it comes in contact. From present appearances, it is not at all improbable that the entire tract of woods, about 200 acres, will be destroyed.

The Heeter well, on the Dan Heeter farm, is one of those wells of which we read in fabled stories of Oil Creek. Months ago, several parties took the lease, and went to work as "wildcats." They were not over and above the needs of men, generally, in this world's possessions, but had pluck and energy. So day after day they worked, until their means were exhausted, and somewhat more. Even they commenced to despair when just before they touched the sand they had utterly "no show" at all. They had, from all appearances, a "dry hole." So much were they discouraged, and so well satisfied that the well would never amount to anything, that two brothers who own three-sixteenths in the well, came into St. Petersburg and offered the entire interest for $200, one half down, and the balance at the expiration of three months. They were uanble to dispose of it even at that price; and with heavy hearts they returned to the well. During the night, or next morning, they reached the sand. They had barely gone a foot in it when a tremendous flow of oil and gas occurred, rising to a height of 60 feet. This continued at intervals until she was tubed, and now 65 barrels will not touch her daily production. It is said that $200 will not buy three-sixteenths of the entire well. One of the happy owners called at *The Progress* office yesterday, saying that he guessed that he could afford to take a paper with an interest in an 85 barrel well. He was on his way back to Westmoreland county where, we will suppose, (if the natives are like some other people), that he will be looked up to as one of the rising men of the day, and his acquaintances, instead of calling out on the streets, "Hello there, you, Bradsby," will say, politely, "*Mr.* Bradsby, allow me to congratulate you on your success. I was confident that it could not be otherwise, my dear sir!" and so they'll all toady. All honor to Petroleum!

***Oil Wells in the Clarion County Field***
***Venango County Historical Society***

## Oil Intelligence, August 10, 1872

The prospect wells are doing about 35 barrels between the three of them. They are on the Collner.

James Hardison is starting a new well on the Foust farm, near the Evans No. 2. It is called the Crawford well.

The Palmer well on the Collner farm is down to 35 barrels. – *St. Petersburg The Progress*

## Oil News, August 14, 1872

A View of Brimstone – The Petroleum Centre *Record* says: Messrs. Hess and Tarbell, well known oil operators, are sinking a well on the Hale farm, some distance north of Petersburg. The well was started on the fourth of July last, and was named the Independence well. It is being drilled by Mr. Michael Hess and, we are informed by Mr. Hess, that the day before yesterday the drill penetrated a vein of genuine brimstone, at a depth of nearly 800 feet, and near the top of the third sand. A few feet above the brimstone vein, a vein of hard coal was struck. Since striking brimstone the drillers are somewhat of the opinion that the lower regions can't be far off, and they handle the tools carefully for fear of new and startling developments. New wonders and discoveries of other valuable minerals are continually being made while in search of oil.

## Oil News, August 16, 1872

The Coulter well No.2 on the Foust farm is down about 800 feet.

The Main Branch well, Capt. Spurr, was torpedoed last month while doing 6 barrels. The result has been that she is now reduced to a 3 barrel well.

The dispute in regard to the Porterfield & Treat No. 9 has not yet been settled. It will be remembered that Evans claims that it is on his "protection." It was expected that a definite settlement would have been reached last Saturday, and that the well would go down, but nothing has been at the work as yet. It would appear to make but little difference, since there are no wells in the immediate vicinity which are now producing any great quantity of oil.

Joseph Foust is starting a new well in his orchard, just above the Porterfield & Treat well.

The Harris & McLaughlin well No. 3, on the Collner farm, is in the sand some six feet but has not, as yet, any show at all.

Payne well No. 2, on the Collner farm, is doing fully 40 barrels. It is a new strike.

The Ralph well is doing about eight barrels. It is on the Foust farm.

The Nancy well is troubled with her cups, they not working right. She is now doing but ten barrels, but it claimed that were she in order she would produce 40 barrels.

The A. W. Robinson well on the Joel Fink tract is down about 925 feet.

The O. K. No. 1 well has been pumping by heads for some time back, owing to the scarcity of water. No. 2 cannot drill from the same cause.

*Oil Wells in the Clarion County Field*

Whittling No. 7 is down about 400 feet.

On last Monday morning the Hannah well No. 9, A. S. Palmer & Sons, was torpedoed, and her production increased to 40 barrels.

The Porterfield & McCombs rig, which was burned on the afternoon of the same day that Foxburg took fire, has been rebuilt, and they are about ready to resume pumping.

A. S. Palmer & Sons have three new rigs up at Grass Flats. They are all Hannahs, and two of them are already drilling. – *St. Petersburg The Progress*, Main Street.

## August 20, 1872

Producers to hold mass meeting at Parker's Landing at the Opera House
On Thursday August 22, 1872
We must unite and suspend all operations beginning Sept. 1, 1872 for a period of six months!

## Oil News, August 21, 1872

The Curran well on the M. Shoup farm will reach the sand this week. It is about 300 from the Mason well on Turkey Run, and 100 from the Conkle & Vick well.

Lithgow & Dougherty's wells are pounding away and will, from their situation, develop the Turkey Run territory pretty well. There are four of them.

The Jackson & Vensel rig on the Fillman farm, which was burned two weeks ago, is flowing some oil. They are trying to get hold of the tools, which were dropped at the time of the fire.

The Broad Top well on the Shoup farm, Turkey Run, is down about 400 feet. This well would in all probability have been down some time ago, had they not been very unfortunate with their tools. The well stood for somewhat over a month.

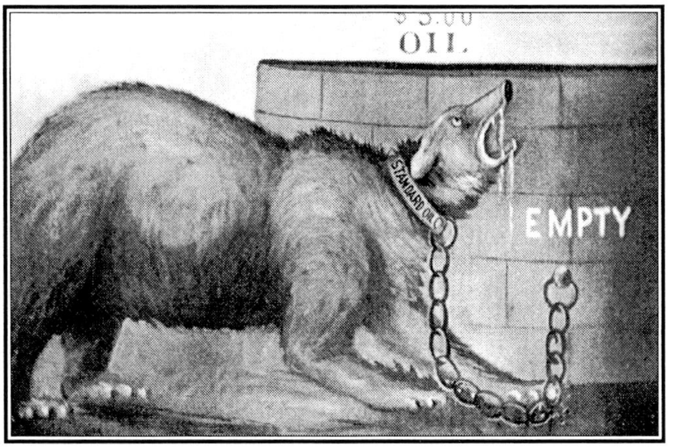


*This characature shows the relationship between the oil man and the Standard Oil Company. The oil men were holding out for $5 a barrel but the greedy wolf, representing Standard Oil, wouldn't pay, so everyone suffered.*

*Clarion County Oil Field Scene*
*Venango County Historical Society*

The Hulings well, on the Klingler farm, is in the sand and will be tubed in a day or two. The indications are favorable for a paying well. It is about 100 yards from the Vensel & Jackson well.

Smith & McMillan have a rig up on the William Eddinger farm, east side of Turkey Run, and just above Keating's furnace.

J. H. Jack has bought into the Salem well on the Heeter farm at the rate of $4,600 per sixteenth.

The Myers well on the M. Shoup farm is good for 45 barrels per day.

The Conkle well, owned by Conkle, Vick & Co., on the Shoup farm, is in the sand and has a most favorable show. It flows frequently and from the indications it will probably be a 25 or 30 barrel well.

Warren Oil Company well No. 4 on the D. Ritts farm is down 700 feet.

The Jackson well on the south side of Clarion River is doing 50 barrels.

Bradley & Chambers have a new rig up on the Knight farm on the west side of Turkey Run. It is about a quarter of a mile north of the Salem well on the Heeter farm.

The Panton well on the Heeter farm is down about 700 feet.

The Salem Oil Company has a new rig going up near the flowing well and will be ready for operations in about two weeks.

William Heeter and William Jack have a new rig up at Turkey City and will be ready to commence drilling in a week or two.

Robins has a well under way on the Neely farm.

Mason & Boyd are interested in the Isaac Neely farm and have a well down about 400 feet.

Andrews, Myers & Co. are operating on the William Neely farm and have their first well down about 600 feet.

A new rig is going up on the Master farm. It is owned by McLaughlin and others.

The Chambers well on the Knight farm is down about 160 feet. It is just east of the Salem well on the Heeter farm.

Paul Neely, Hunson & Co. have a new rig up on the William Neely farm and will commence drilling in a few days.

*Alfred Smiley – This photograph owned by Smiley, is in his book, "Oily and Otherwise, A Few Scraps," dated 1907. A. W. Smiley learned the pipe line business back when it began. The management for Van Syckle's pipe line company consisted of M. E. Van Syckle, superintendent, George E. Coutant, superintendent of teams, A. W. Smiley, assistant superintendent and general agent at Pithole in charge of the buying department, and Emmett Fleming, secretary. Mr. Smiley came to Foxburg in 1872 as superintendent of the Mutual Pipe Line Company.*

John Turner has a new rig up on the Keating farm. It is called the St. Lawrence.

Crowell & Hoffman are drilling a well on the Isaac Neely farm.

Richard and Alfred Neely have a well drilling on the Isaac Neely farm.

The Paul Neely well on the Isaac Neely farm is doing about 15 barrels.

The Scott well, just back of St. Petersburg, is doing eight barrels.

Hulings and Crary are starting a new rig on the Stubble farm on the east of Turkey Run.

The Spear well on the E. Davis farm, on the east side of Turkey Run, owned by H. Spear & Co., is down about 100 feet.

The Davis and Neely well on the E. Davis farm is in the second sand.

The Wing & Wilson well on the Fry farm is doing about eight barrels.

R. Neely & Co. have two wells under way on the Isaac Neely farm.

The Lady John well on the Isaac Neely farm, owned by H. Spear & Co., is down 400 feet. – *The Progress*

***Foxburg 1876***
***Courtesy Chris Fleming, Foxburg, Penna.***

# A.W. Smiley and His Friend John Wilkes Booth

The article below comes from information taken from a book written in 1907 by the late Alfred Wilson Smiley of Foxburg, Penna. Squire Smiley was a very important man at one time, having worked and followed the oil fields from Titusville to this area.

Mr. Smiley arrived in Titusville, Penna. in the early spring of 1860 and got a job putting down a well that was in sight of Colonel Drake's Well. Mr. Smiley was 17 years old at this time and became good friends with Col. Edwin Drake. In his book Mr. Smiley says this about Drake: "It was my pleasure to become somewhat personally acquainted with Col. Drake, and now, after a lapse of over forty-five years, my memory of the man is distinct to the effect that he was a jovial, kind-hearted, polite, polished gentleman."

*John Wilkes Booth, note silver cane*
*Courtesy Drake Well Museum*

Mr. Smiley arrived in Franklin, Penna. in October 1863 and worked there for several years at different jobs. During this time, Mr. Smiley again met with a very well-known yet infamous person, John Wilkes Booth. In his book Smiley writes, "It was during the summer of 1864 that I got acquainted with John Wilkes Booth who, on the evening of April 14, 1865, assassinated President Abraham Lincoln in Fords Theatre in Washington, D.C.

"Our first acquaintance was due to the fact that for a short time we occupied the same room at the United States Hotel. Mr. H.B. Smith, a clerk in one of the railroad offices, and myself had a double room at the hotel, that is, a room in which there were two beds. The hotels were all crowded in those days, and Smith and I, to accommodate the landlord, frequently doubled up to let in some good fellow for a day or two, when not able to get in elsewhere. On this occasion, the proprietor, Mr. H.W. Watkins had requested us to let in a couple of gentlemen for just a couple days. The newcomers to our room were John Wilkes Booth and Mr. J.H. Simonds. Booth and Simonds, instead of staying there only a couple days, remained for 10 days. They then secured rooms at Mrs. Webber's on Elk Street, but still took their meals at the United States Hotel.'

"Booth was very stylish in his dress, rather good looking, cold in manner towards strangers but, when acquainted, quite social and very polite. At that time I was rather an expert billiard player and Booth was also a good player, and many games we played, each earnestly trying to beat the other. So in this was I became acquainted with him, and have yet in my possession several little trifling articles which he personally presented to me.'

"The small silver-headed cane in the picture of Booth is yet in possession of the writer, having been given to me by Booth in the summer of 1864. Booth was a trained athlete and was very strong and quick. I was also pretty strong and active, being then 21 years old, and did not fear any common man in a rough and tumble. I recall an incident, however, when we both got the worst of the encounter. The occasion was at a dance at the Marshall House located at the Franklin end of the bridge over the Allegheny River.'

"On the night of the dance, there were lying near the bridge several steamboats and a number of lumber fleets, and the crowd assembled at the dance was rather a motley one. There were in attendance a number of deckhands from the steamboats and fresh water sailors from the lumber fleets. A fight started and in our desire to see how things were going, we crowded in quite close, too close, to where the melee was being pulled off. This attracted the attention of several of the fighters, and Booth and I both were dragged around the large room several times, regardless of how much they soiled our clothes, and threw us out, with orders not to return to the dance.'

"I have many times been asked the question, what in my judgment, prompted Booth to commit that horrible crime. My answer now and always has been that I do not know. I never heard him express himself either way; I never heard him talk a word of politics. I never heard him say a word about any of the leaders on either side of the Civil War, which was still in progress, or to ever make any reference whatever to either the North or South. He left Franklin in the Fall of 1864 and I never saw him again.'

"John Wilkes Booth was a one-thirteenth owner in the Homestead Well on the R. Hyner Farm near Pithole City. Booth paid $13,000 for his share in the early Fall of 1864, but after retaining it only a few months, sold to one of his partners and left Franklin, Penna. On the night Booth assassinated President Lincoln, according to a story, a visitor to the Homestead Well held a lighted cigar too close to the mouth of the oil pipe, setting fire to the well. Before the fire, the well pumped somewhere around 50 barrels a day, after the fire it increased its flow to over 250 barrels a day."

Squire Smiley died in 1927 at the age of 84 years and is buried in the Foxburg Cemetery, along with the many memories that are now forever lost. His wife Ellen died in 1932, at the age of 84 years. The silver cane is at the Drake Well Museum (The Author)

*Cyrus D. Angell*
*Courtesy Drake Well Museum*
*Refer to page 476*

## The Belt Theory

In the early days of the petroleum industry there was no scientific method by which deposits of oil could be located. Theories, however, were not lacking. The most valuable of these was that held by C.D. Angell. After drilling several wells below Franklin and making some observations, Angell concluded that petroleum existed in "belts" or courses. He adopted the theory that two "belts" existed, one running from Petroleum Centre to Scrubgrass and the other from St. Petersburg to Butler County. While the mass of producers discounted the theory, they were nevertheless influenced by Angell's observations, which had much to do with the opening up of new territory in Clarion, Armstrong, and Butler Counties.

Angell likewise established the existence of a belt about thirty miles long and from three to five miles wide, extending from Scrubgrass southward and slightly westward through Venango into Clarion, Butler, and Armstrong Counties, comprising the so-called lower or southern field. Wildcat operations had been under way in these localities for several years, but had met with scant success, although there were a few small cases of production which served as the points required for Angell's investigations and surveys. Curiously enough, Angell's activities in the lower counties were only moderately profitable, but many other prospectors who accepted his conclusions and followed his lines made big strikes. The producing areas of Emlenton, Foxburg, St. Petersburg, Parker's Landing, Petrolia, Karns City, Millerstown, Brady's Bend, Antwerp, Greece City, and a score of others opened in the lower field before 1875, were all on the Angell belt or on subsidiary belts leading from or running parallel to the main channel. Parker's Landing and Petrolia, in Armstrong and Butler Counties respectively, were probably the most prolific of these territories. Both reached their peak of development in the early 1870's, and for a considerable period each produced more than six thousand barrels of oil a day. At one time the wells of Parker's Landing and the near-by districts of Foxburg and St. Petersburg were yielding a little more than one-third of the entire Pennsylvania production. *(Early Days of Oil, Paul H. Giddens, 1948)*

## Rutherford B. Hayes, President of the United States 1877-1881

On October 30, 1875, after a trip through the Oil Country, the third term Governor of Ohio and his party stopped at the famous Fox House and took lunch. Governor Hayes was returning home to Fremont, Ohio after a pleasant tour of stomping throughout Pennsylvania with Governor Hartranft. The afternoon of October 29 was spent giving a speech in Franklin, Penna. and on to Oil City, Penna. for an evening lecture.

Rutherford B. Hayes was nominated for President by the National Republication Convention in 1876. President Hayes served his country from 1877 until 1881, when he was defeated by James Garfield who was later assassinated.

The Fox House was built in 1871 by the Fox Estate and burned to the ground on February 27, 1883. The following article is from the *Oil City Derrick*, February 28, 1883:

## Hotel Burned
## The Foxburg House at Foxburg in Ashes
## Narrow Escape of the Servant Girls
## Colonel Gregory a Heavy Loser.

At an early hour yesterday morning, Foxburg, the busy little town down the Allegheny was again and for the fourth or 5th time visited by a disastrous fire. The flames were first discovered in the rear part of the Foxburg House, the well known hostlery of which the genial Colonel W.S. Gregory is proprietor. The alarm was quickly given, and none too soon, as the dining room and kitchen girls, who sleep in that part of the building, had only time to escape in their night clothes as the flames and smoke took possession of their apartments. Had the discovery been made five minutes later, some of the girls would undoubtedly have been burned to death. By

*Fox House, Foxburg Penna.*
*Courtesy Chris Fleming, Foxburg Penna.*

this time the whole house was aroused and the boarders commenced removing their effects. The citizens responded quickly, and mainly by their efforts, the furniture and firearms of the hotel were removed, but in a damaged condition.

The fire department worked nobly and, the hotel being plastered, the fire did not make as rapid progress as is usual in oil town conflagrations.

By the greatest efforts of the firemen and citizens the flames were confined to the hotel, though the drug store adjoining on the south and the office of the Fox estate on the north, were in imminent danger, and the former was on fire several times.

The loss to Colonel Gregory is quite heavy, and as he was absent at the time, it is not known whether there was any insurance on the property or not.

The burning of this popular and well known hotel will be of serious inconvenience to the travelling public. For years past, the Valley trains have stopped here for dinner and supper and what provision will be made now for feeding hungry passengers has not been decided upon.

A subscription was at once started for the benefit of the maid servants who lost all their clothing, and over $100 was raised in less time than it takes to write it.

It is understood that the intention of the Fox Estate is to build immediately a handsome hotel on the same site which will be completed early in the summer.

## St. Petersburg, August 22, 1872

*The Progress*, in speaking of the direction that development in that district is taking, says: The development has been extending to the east rather than north, as heretofore, and the indications are that the chief point of operations during the coming fall will be on the Knight, Master, Isaac Neely and William Neely farms. Among the farms which have begun to develop with great rapidity is the E. Ritts farm, lying between Ashbaugh and Turkey Runs.

The D. Shoup, Sr. farm is also a great point of operations at present. During the last few weeks operators have been somewhat checked, owing to the low price of oil, which is at this time almost down to the cost of production. By the success of the Forest well, on the Master farm, and of the Devore well, beyond Monroeville, fully eight square miles of new territory has been tested. Operators still hold to the "belt" theory, oil seeking leases on the direct line, between the new wells. The coming month promises to show a decrease in the production. Operating will not be as heavy as during the summer, unless oil commands a higher price. The demand for leases is light, the holders asking from $50 to $200 bonus, and one-fourth royalty, even though the location be one, two or three miles beyond the actual development.

*The Fox Farm – The Fox family was from Philadelphia but were granted thousands of acres in and around Foxburg. A large stone mansion was built and the Fox Estate was in prime oil territory. Samuel M. Fox and Joel Fink of St. Petersburg were partners in the search for and discovery of petroleum beginning in the Fall of 1865. Mr. Fox died relatively young on Christmas day 1869 and is buried in Philadelphia. Joel Fink built a large home in Petersburg which is now the residence of the author, Gary McKinney. Joel Fink followed the oil and gas business, owned the Indiana, Pennsylvania Water Company, lived in Livermore, Pennsylvania, and died in 1907. He is buried in Indiana, Pennsylvania. Clarion County History 1887 A. J. Davis*

## A Mean Act

Special Dispatch to the *Herald*, Foxburg, August 22, 1872

Some malicious person or persons early this morning opened the stop cock to the tank attached to Sampson Well No. 1, and well No. 48, Fox farm, and before it was discovered one hundred and twenty-five barrels of oil ran out. The well is owned by Sampson and others of Emlenton, Penna.

## Oil News, August 22, 1872

Soliday well No. 3 on the E. Ritts farm, Ashbaugh Run, is down about 200 feet.

The Hennage & Satterfield well, on the E. Ritts farm, is just through the boulder.

Coulter & irwin have a well on the E. Ritts farm which has reached a depth of 800 feet.

Martin & Co. have a new rig under way on the D. Shoup, Jr. farm. It will be ready for operations in a week or two.

Well no. 2, Finley Oil Co., on the Finley tract, is down 750 feet.

Chambers well No. 2, on the Slicker farm, is down 940 feet.

The Hess well, on the Hess tract, has been drilled to a depth of 200 feet and is progressing favorably.

McKeever & Metzger have been some five or six weeks at their well on the Slicker farm, but have had many accidents. They are now down about 600 feet.

McCracken & Hughes well, on the Eddinger farm, have drilled to the depth of 1,000 feet, and will probably strike the sand at 1,150.

The Patterson & Dickey well, on the Eddinger farm, has reached the depth of 900 feet.

The Pioneer well, on the H. Neely farm, is down and is making a good show. She will probably prove a 25 barrel well.

Rice & Bros. have two wells going down on the H. Neely farm. No. 1 has reached its 800 foot.

The McGee well, on the H. Neely farm, is down about 900 feet. – *The Progress,* St. Petersburg, Penna., Main Street

***Clarion County Oil Field Scene***

## *Titusville Herald*, August 23, 1872
## Producers Convention
## Mass Meeting of Oil Men at Parker's Landing
## Harmony of Action
## Drilling to be Suspended for Six Months

Parker's Landing, August 22, 1872 – The call for the mass meeting for the producers and others interested in stopping the over-production of oil, at this place, today was very generally responded to by producers in this district and from the "Up River Districts," over 200 being present. The meeting was called to order in Spencer's Opera House at two p.m. The representatives of the St. Petersburg, third "district" were Marcus Hulings and Barney Foust.

The oil men signed a pledge that they do hereby bind ourselves to each other not to commence the drilling of any more wells for a period of six months from the first day of September next, or lease any ground for the purpose of operation and to use all honorable means to prevent others from boring.

The only way for oil prices to increase is to stop the production. A speaker, Capt. J. S. Bredin, noted that within a radius of ten miles, today there are over 227 wells going down. If all sign the pledge the day is not far distant when oil will be four or five dollars per barrel. All districts in the oil region must pledge and agree for our plan to work. Telegrams were received from throughout the oil regions in support of the pledge.

Mr. C. P. Hutchinson remarked, "We are poor in money, but rich in oil."

## Let's Back the Suspension, August 24, 1872

**The Suspension of Drilling Operations** – The result of the mass meeting of producers at Parker's Landing must be extremely gratifying to the people of the entire Oil Region, as it evinces a spirit of determination to control themselves at any sacrifice, which is the primary and initial step towards controlling others. Knowing their strength in this respect, it will not be at all surprising if this movement leads to the most strict and rigorous measures for regulating the production of oil. The agreement to stop drilling for six months will, it is hoped, be signed by the greater majority of landowners and leaseholders throughout the Region, if not by every individual who owns a foot of territory in the oil-producing districts. The effect of such prompt action will probably not be felt immediately, but if successfully carried out cannot fail within from 30 to 90 days to raise the price of oil to a figure which will increase the value of property and make every business man within the Region so much richer. The saving in one year with oil at $5 per barrel would, it is estimated, be equivalent to about $10,000,000, and with a certain knowledge of this fact, the only wonder is that some uniform concert of action has not been arrived at before.

*Loading oil in iron tank cars to be transported to refineries.*
*Venango County Historical Society Franklin Penna.*

## Oil News, August 30, 1872
## St. Petersburg – Three New Strikes
## Special to the *Herald* – St. Petersburg, August 29

Last evening, at five o'clock, the Wilson Irwin well, on the Eddinger farm, commenced flowing, and up to one p.m. today, when I visited it, had flowed 300 barrels. It will probably prove a good hundred barrel well.

The well on the same farm owned by Jas. Hughes and Ben McCracken reached the sand on Tuesday, and is now pumping about 80 barrels.

On the Neely farm the Pioneer Oil Company struck a well on Monday, which is doing 50 barrels. It is known as Pioneer No. 4.

## Oil News, August 30, 1872
## The Producers' Combination
## Probable Co-Operation on the Part of the St. Petersburg Operators
## Special to the *Herald* – St. Petersburg, Aug. 29 – 9:30 p.m.

In conversation with some of the leading operators here tonight, I learn that a hearty cooperation in the movement to advance the price of crude oil will doubtless be pledged on the part of the St. Petersburg operators.

From Antwerp City – The following were reported from Antwerp: M. Hulings, Joseph Gates, C. E. Johnson, A. A. Hulings, Wm. Robinson.

Mr. Karns offered the following resolution, which was carried: Resolved, That there is an entire unanimity of feeling in regard to not starting any more wells, and we call on all producers in Pennsylvania to join with us in good faith in stopping development.

## Mutual Pipe Line, Foxburg, August 31, 1872

Mr. A. W. Smiley has been appointed Superintendent of the entire Mutual Line. Mr. Smiley needs no introduction to the producers of the Region; in fact he may be called the Nestor of the pipe line business, since he was early connected with the first pipe that ever was laid in the Region, that from Pithole to Miller Farm, and himself pumped the first barrel of oil ever piped. Mr. Smiley's many friends will be pleased to learn of his promotion, which has been faithfully earned and fully deserved.

## Producers Meeting at St. Petersburg, September 5, 1872

The adjourned meeting of the St. Petersburg Producers this afternoon at the Aiken's Opera House was not very largely attended. Are plans to stop the drilling falling apart?

## An Oil Well on Fire, September 9, 1872
## An unknown man possibly burned
## Special Dispatch to the *Morning Herald*

The derrick and the engine house on the Edinger farm near Antwerp, in the St. Petersburg district and owned by J. W. Irwin and others form Petroleum Center, burned last night.

## Another Oil Failure, September 10, 1872

Well known oil operators McPherson and Brothers of Parker's Landing filed for bankruptcy. Also speaking of failures, there will be more if the price of oil continues to drop.

## Oil News, September 11, 1872

To show that the shut down is not working as of September 11, the following wells are continuing to pump.

The Hammer well on the G. Exley farm, Beaver township.

The well on the Montas farm.

The well on the Painter farm, Beaver.

The Hulings well on the George P. Delo farm.

McElhatten well at Tippecanoe.

Capt. Hassen of Oil City purchased 170 acre farm on Sweitzer Run at $100 per acre.

The Deloe well.

The Hamor well, Sweitzer Run east of Beaver Creek.
The Prospect wells of Foust Farm, St. Petersburg.
Eudora well on the Clarion River.
The Conkle well, Turkey Run.
The Faith well, Turkey Run.
The Fountain well.
The Kingfisher well on the Dittman farm near Antwerp.
The Coor #3 on D. Ritts, Antwerp.
The Northside well, Hardison Bros., on D. Ritts. – *The Progress*

Author's Note: At this time I have decided to not list all the well names, drillers, owners and farms. I feel I have made the point that oil wells were put down anywhere and everywhere, almost to a point of being out of control. The oil "belt" was being defined and there was so much oil being produced it was almost worthless.

The student of early oil in Northwestern Pennsylvania can further the study of individual wells and locations by studying the *Titusville Morning Herald* at the Public Library in Titusville, Pennsylvania and the Oil City Derrick at the Public Library in Oil City.

## St. Petersburg *Oil Field Record*, September 12, 1872

The following article is from the September 12, 1872 issue of the St. Petersburg *Oil Field Record.* It again shows the lawlessness in these early oil boom towns.

It has been the habit for some time past of a lot of drunken hoodlums to drive the road from St. Petersburg through Antwerp City and on to Turkey City and make night hideous with their revelry and profanity, until at last they are likely to come to grief. The gang has been followed by citizens and identified and they will all be arrested and punished as severely as the law will permit. It is a shame to allow such hoodlums to be at large in a civilized community. We understand that the thieves attempted to break into the residence of Samuel Goughler on last Thursday night, but did not succeed as the family was awakened and caused them to beat a hasty retreat. A load of buck shot awaits their return.

On Saturday night a crowd of about 100 people gathered on the eastern end of Main Street, Antwerp City. Two men from Titusville, Penna. instigated a free fight resulting in a riot. It was a disgraceful fight with several arrests made by Constable Cribbs. Results of the riot were many bloody mouths, broken noses, and black eyes. Cause, too much elixir!

## Explosion and Fire at Antwerp
## Narrow Escape from a Large Conflagration
## Special Dispatch to *the Herald* – September 24, 1872

At nine o'clock last night a boiler at the Colemen well, on the John Ashbaugh farm, owned by Robert Bryan and others, exploded and immediately the derrick and engine house were wrapped in flames. The Antwerp City Water Works, situated about 1,000 feet from the fire, furnished a line of hose and an abundant supply of water, which combined with the exertions of the crowd present, soon extinguished the flames. The engineer was severely, but probably not fatally, burned. It is supposed that he was asleep at the time of the explosion.

## Petersburg, September 25, 1872

The rig at Neely and Davis well on the E. Davis farm, took fire on Sunday night last and was wholly consumed. The cause of the fire, we believe, was the ignition of the gas, which issued forth in a large volume. We believe that no one was injured. – *Petersburg Progress*

## September 27, 1872

At this point in time the six months' shutdown has been reduced to a thirty day shutdown. Will this thirty day shutdown even be successful? The pledge now is to shut down from Sept. 25 until Oct. 25, 1872. Many producers think the shutdown is a good idea to make oil scarce at the refinery but if they shut down their own operation, it could mean bankruptcy.

## Oil News, September 28, 1872

Fox and Fink, No. 13 and 14 on the Fox farm, are producing about 30 barrels per day. Mr. Joel Fink is a well known Petersburg resident, with much success in oil producing – *The Progress*, St. Petersburg, Main Street.

## Production Stopped, October 1, 1872

Nearly every well between Tidioute and Fairview shut down. So far as known there have been but few acts of violence. Notices were left at several wells that if they did not suspend operations immediately, their rigs would be destroyed. Several engines in the Petersburg district have been torn to pieces, a number of holes plugged and some engine houses destroyed. The cry throughout the oil region is "shut her down."

Roberts Torpedo Co. has pledged no torpedoes during the 30 day shutdown.

## Petersburg, October 3, 1872

Oil four dollars today with more buyers than sellers.

## October 5, 1872

There is not a walking beam to be seen moving in this section except one
well in Antwerp City.
This is not neighborly; Shall it continue?

## October 11, 1872
## Destructive Fire at St. Petersburg, Oct. 8
## Thirteen Buildings Burned, Loss $8,650

About half past six o'clock this morning our citizens were alarmed by a cry of fire from the extreme eastern end of the town, and in less than half an hour not less than 1,000 people were on the ground, and all that could were actively engaged in rendering assistance. The chance of saving any of the houses nearby was slight, as the town has no facilities for fighting fire, not even water. A meeting has been called, and a hook and ladder company will be formed at once. – J. M. Place

## October 14, 1872

Officers Cribbs and Sipler apprehended J. C. Andrews who escaped from the Burgess's office at Antwerp City some time since. He was arrested for appropriating money belonging to his employer, Robert Bryan. Andrews was taken to Clarion yesterday by officer Sipler.

## October 18, 1872

A Burned Rig the Penalty of Pumping During the Suspension

This morning at about 3 o'clock, a well owned by S. N. Dunlap, on the Jamison farm near Foxburg, destroyed by fire.

## October 24, 1872

St. Petersburg – Another Fire, The New Water Works

The recurrence of fires will hasten the building of the water works.

## October 28, 1872

Turkey City or Gimlet Town will be the center for operations next spring 1873.

## Down River Items, November 7, 1872

During the six months' suspension, all the wells have resumed pumping or drilling but so far as known no new ones have been commenced by producers who signed the pledge to refrain from drilling new wells for six months.

At Foxburg, work is progresing on the new bridge.

The ferry at Parker's gave out last Saturday, cause breaking of the cable. Passengers are now taken across in small heavily loaded shiffs. The cable will be replaced and the bridge is about half completed.

## The Horse Disease, November 9, 1872

From *The Progress*, Nov. 6 – The dreadful horse disease has at length appeared among the horses in this vicinity. Robert Bryan has three which were attacked a day or two since, attended with the usual symptoms. H. Finlay, at East Foxburg, has five horses attacked by the disease. Owners cannot be too careful of the manner in which they treat their dumb servants. We published a letter from J. P. Vasey on the symptoms and treatment of the disease.

## Oil News, November 13, 1872

The Petersburg region is awake again since the suspension, and is the largest and best producing district in the entire oil region. Developments are extending rather too rapidly for the good of the oil market. – *Clarion Democrat*

## *The Progress*, November 25, 1872

Prospective Large increase in Production. The Development of Turkey Run.

It would appear from present indications that the extent of territory that will be operated next spring 1873 will be largely on the increase. The new development on the different farms east of Turkey Run have opened up such an extensive field that it is becoming of serious moment to the whole oil country whether or not it should be developed to its full extent.

## Torpedo, November 25, 1872

The Roberts Torpedo Co. of Titusville, cash required on delivery, lists its agents in the Petersburg, Antwerp, Turkey Run vicinity as Geo. Irwin and J. W. Vincen, office at St. Petersburg.

## A Chapter of Horrors, November 28, 1872

An unknown man murdered today in a Petersburg restaurant. A black man named Lucas became involved in an altercation, drew a revolver and killed the unknown man. Lucas then escaped.

## *St. Petersburg News*, December 2, 1872

The new St. Petersburg Water Works is almost completed.

The new Petersburg Hook and Ladder Company is anxiously looking for the arrival of its new truck. Now fires in St. Petersburg will be a thing of the past.

Two kerosene lamp explosions – A lamp exploded in the apartment of Capt. J. M. Place over the Post Office on Friday. The flames were promptly extinguished. On Saturday another explosion occurred in the drug store of Mr. A. O. Bliss. Mr. Bliss threw an overcoat over the flames and extinguished the fire.

A young man was found dead on top of a tank on the Galey tract near Grass Flats. He suffocated on the oil fumes and his head was over the open hatch hole of the tank. The unfortunate was a son of Mr. Jamison, seventeen years of age.

## The New Iron Bridge at Parker's to Open

Celebration of the Event, Ball and Banquet
December 6, 1872

*The bridge at Parker joining Parker City, Armstrong Co. with Perry Township, Clarion Co.*
*Venango County Historical Society*

## St. Petersburg, December 14, 1872

The Antwerp Pipe Line Company is overcowded with work, and has been compelled to run on Sunday several times.

St. Petersburg has still another of the necessiaries to a city. She has street cars, and comfortable ones, too, having a comfortable stove for heating, and closed in carefully and fitted regardless of cost. It is owned by the St. Petersburg and Foxburg Car Company and is mounted on "bobs" for the season.

## Fullerton, December 19, 1872

Burning of an oil car – On Wednesday evening between 6 and 7 o'clock an iron tank car full of petroleum caught fire on the siding at Fullerton along the Allegheny River between Foxburg and Emlenton. The oil was owned by the Antwerp Pipe Line Company, the car and oil were destroyed. A spark from the locomotive is supposed to have caused the fire. The car was one of a train but was detached before any of the others was ignited.

## St. Petersburg, December 20, 1872

Once a week a stage from Clarion brought the mail to Petersburg. Mr. Lewis Collner was the first postmaster with a salary of $13 per year. Mr. J. M. Place is now the growing city's well-known postmaster.

## The *Titusville Morning Herald*, January 10, 1873

The oil region oil news is collected by 46 different "News Agents." The men record daily oil and related events throughout the oil territory. From Titusville, through Petrolia, the *Morning Herald* is sold. At Foxburg, Whitcomb and Bone, News room; at St. Petersburg, J. M. Place, Post Office news room; and at Antwerp City, Crawford and Palm, news room.

## Accident at Foxburg, January 14, 1873

A special telegram to the *Herald* informs us that last evening about half past six, a man named Bacon, being in Lawrenceburg, was struck by the engine of the Parker accumulation train, a half mile south of Foxburg, thrown from the track and fatally injured. At the time of the accident he was deeply intoxicated. This adds one more to the list of rum's victims.

***Post Office in St. Petersburg with J. M. Place standing to left of the doorway. J. M. Place was the first postmaster appointed by the President in St. Petersburg. He was well-known throughout the oil region as a newspaperman. Venango County Historical Society***

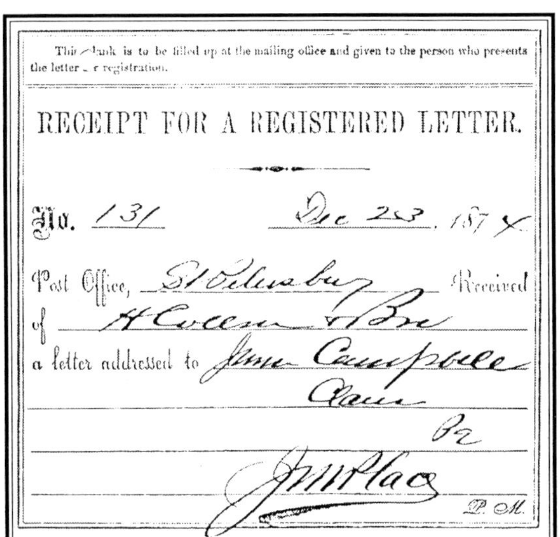

This blank is to be filled up at the mailing office and given to the person who presents the letter for registration.

RECEIPT FOR A REGISTERED LETTER.

No. 131 Dec 23, 1874

Post Office, St Petersburg Received of H Cullen & Bro

a letter addressed to Jim Campbell

Clarion

Pa

J M Place P. M.

***A St. Petersburg Post Office document from 1874.***

## In the Oil Region, January 14, 1873

The Post Office at Petersburg has risen in rank so that the President of the United States appoints the Postmaster. Mr. J. M. Place was appointed.

The "oil belt" is extended east toward Edenburg and south to Greece City. The oil business is not good, with producer against producer, to drill or not to drill, to pump or not to pump. The Producers meetings are leaning toward failure, what will the outcome be? Can refiners be trusted?

## January 17, 1873

The pipe line leading from Fairview and Petrolia to Brady's Bend is nearly completed. This will prove of great accommodation to producers, as oil will be delivered on cars at Brady's Bend.

## Bridge Destruction, January 20, 1873

The pier for the new bridge across the Allegheny River at Foxburg was struck by an imense body of ice last Friday. Not one stone is upon another. This will retard the construction until low water next summer. The Parker bridge was only slightly injured.

Coal now sells at Antwerp City from 12 to 14 cents per bushel. Coal is used to fire boilers at the wells, for heat in the home, and the tool dresser's forge.

Our stationary stores already show signs of the coming St. Valentine's Day.

## January 21, 1873

Oil $1.75 per barrel at the well

## January 25, 1872

The citizens of St. Petersburg, without distinction of party, have placed in nomination for burgess Mr. H. Derousse, formerly of Petroleum Centre.

## It's Time

**"Hang up the shovel and the hoe, Take down the fiddle and the bow."**

## From the *Emlenton Sun*, February 3, 1873

The new well on the Young farm is reported doing 50 barrels per day. This well tends to lead operators in the direction of Shippenville, as it is a little east of developments between there and Petersburg. The distance from St. Petersburg is about four miles. Depth of well is between 900 and 1,000 feet.

*The "Bay Fannie" well drilled by Elias Ritts and John V. Ritts, his son. The well was named after Elias's horse. St. Petersburg, Penna.*

## The River Pipe Lines, February 5, 1873

Tankage in "Pipe Line Men's" terms means the amount of oil the pipe line company can hold in their oil storage tank farms.

The Mutual Line's tankage is 60,000 bbls.

The Union Line 60,000

The Emlenton Line 20,000

The Antwerp Line 30,000

The Fairview Line 50,000

Total tankage 220,000 barrels.

Capt. E. L. Ray, newly appointed cashier at the Foxburg Savings Bank, comes here from the bank at Petroleum Center.

Two well known men from Turkey City had a personal encounter on a side street Saturday night at Antwerp City. Most fights have the same outcome but the punishment received by one of the participants is not often inflicted. His mustache was twenty inches from tip to tip, and in the fight one side of the adornment was torn off. A few vigorous jerks brought it out by the roots. The lips were severely torn, but will heal in time.

## February 17, 1873

Three hacks run daily between Turkey Run City and St. Petersburg.

Some of the enterprising citizens of Emlenton are contemplating the establishment of a bank at that place.

*Brady's Bend, 1876 – John Mather Photo*

## February 20, 1873

Ninety thousand barrels will soon be the capacity of the Antwerp Pipe Line Company tanks at Fullerton.

## An Oil Town Burned, February 25, 1873
## St. Petersburg Chastened by Fire – Great Destruction of Property – The Best Part of the Town in Ashes
## Detailed Statement of the Losses
## Full Particulars by a *Herald* Correspondent

A burning brand. Three or four hours of excitement, smoke and flame. St. Petersburg in ashes. That is the story in a few words.

It is little like a public holiday here today. There is very little rejoicing, although every one seems in good spirits. It takes more than a mere fire to discourage the people, and this disaster will be forgotten before summer sets in. But while others are parading, making speeches and performing other duties consequent upon the day we celebrate, the St. Petersburgers are searching the ruins for possible articles of value, looking for houses in which to place their families and wordly effects, figuring up insurance and preparing for rebuilding.

The great fire, for such it may be called, broke out a few minutes past six o'clock on the afternoon of Friday, the 21st inst. The alarm was soon given, and in a short time the utmost excitement and consternation prevailed. It originated near the very center of the business portion of the town, in a saloon owned by Fred Hepp, formerly of Titusville. J. H. Henry, Chalfant & Graff's bookkeeper, was passing at the time and noticing the smoke issuing from different points, sounded the alarm. He then attempted an entrance, but was unsuccessful, the locked doors resisting his progress. Assistance quickly arrived, however, and the doors were broken in, when the interior was found to be a mass of flame. It needed but one glance to show that the fire was beyond control, as the flames burst out in all directions as soon as the doors were forced in. The crowd augmented each monment till nearly every person in the village and vicinity were at the scene. In a few short moments the flames enveloped the entire structure, and, notwithstanding the utmost superhuman exertions of the citizens, soon grasped the adjoining buildings and decided the fate of the greater part of the business portion of the town.

Although previous small fires had warned the people of the necessity of water works, and other means of protection against impending disaster, no water was to be had. No means of protection were at hand, and the wooden buildings, crowded together on either side of the street, were almost literally at the mercy of the flames. The water works, for which provision was made some time since, were nearly completed, the connections only remaining to be finished, but as is usually the case, the disaster happened before they were ready for service. And to add to the danger of the moment came a brisk breeze that pushed the flames on and through the light wooden walls that, instead of opposing their progress, added to their volume.

After the saloon was fairly in a blaze Aiken's opera house ignited. This large frame structure must be familiar to the great majority of oil men. It sprung suddenly into life at Petroleum Centre when that borough was a "red hot town," and its sanguine residents predicted for it life-long existence and fame. The opera house did a thriving business for a long period and then, as the Sam Weeds and other farms "played out," and operators removed to more inviting fields, patronage fell off, its seats were removed, an improvised orchestra occupied the stage and the actors the auditorium, in other words the opera house was transformed into a concert saloon. Finally this business was found to be unremunerative, the structure was taken down and removed to St. Petersburg where it was re-erected with additions and improvements. As may be supposed, the dry pine boards, seats and galleries fell a willing prey to the "devouring element." The heat was intense, and although willing hands were waiting to tear away the flimsy structure, they were driven back and could only seek to prevent the fire extending.

From this point it spread rapidly along the street "scooping up" the post-office, the Adams House and other buildings on the same side. From the opera house, too, it leaped across the street to the Savings Bank which was soon in ashes. On that side of the street were the Barney Vensel residence, the St. Petersburg Hotel and L. Collner's residence, which were destroyed. Next to Collner's house was his store, which was saved, and checked the course of the fire at that point. The opposite side of the street was burning, and immediately on beyond the store, the flames again leaped the street, taking Collner's warehouse and Bradley & Whiting's office. Then they went on in their destructive path till *The Progress* office was reached and destroyed, and were then checked by the gap left by burning down Silberg's clothing store. On the Opera House side it extended to and destroyed S. Cook's residence and was stopped at Chalfant & Graff's office adjoining by water from a well in the rear of the premises.

(Reproduction of Newspaper Dealing With Disastrous Fire of 1873)

**Relief**

We the undersigned Relief Committee solicit subscription for the relief of the sufferers from the fire:

A. K. Wing,
Chairman.

Emlenton Pipe Company, one hundred dollars; Antwerp Pipe Co., one hundred; Blakslee Bros., fifty; Chalfant, Graff & Co., fifty; Pickwick Club, fifty; Capt J. H. Welsh, twenty-five; Frank Fertig, twenty-five. A. A. Dam:, Titusville, ten; A. K. & W. O. Wing, fifty; E. L. Bradley, twenty-five; Thompsou & Patterson, thirty; M. M. Burke, five; Frank Hogan, five; J. H. Elliot. five; H. Derousse fifty; C. W. Little five; J. M. Fink, five; C. Lehman, five, Brown & Raine, five.

This is the largest paper weich we can print on our press, as our large hand press and great quantities of type destroyed.

**Removals.**

Dr. Dorn cau be found at his residence.

R. W. DeHaven, has removed below. Bliss' Drug store, will open at the old staud shortly.

Post Office is at Wing's Store, Mail to the United States as usual.

Seymor Taylor can be found at there store at Antwerp.

S. Frankstein has re opened his store at Sherer & Cooperman's store.

J. H. Welsh has opened up opposite the Monitor House.

The PROGRESS has been removed, what is left of it, back of Lehman's Boot & Shoe store.

Chalfant & Graff can be found at the old stand.

THEFTS.—During the night, and just after the fire, thieves were at work carrying off goods. They passed through Antwerp, carrying arms full of spoils saved from the hell of fire that swept all else away. The hills were covered with stolen property conveyed away by those low born whelps who thought to enrich themselves at the expense of the sufferers.



# THE PROGRESS.

## THE FIRE FIEND.

*Over Forty Business Houses Burned!!*

LOSS $200,000!!

*Supposed To Be The work of an Incindiary*

THE FIRE OF 1873!!!

At 6½ o'clock last night a fire broke out in Fred Hepp's saloon, adjoining the Opera House. In an instant the entire building was in flames and shot like a demon of hell high into the air, enveloping the Opera House in a sheet of flame from end to end. The European House adjoining it took fire, the Adams House, Boston Clothing House, Hall's Restaurant, Seymour & Taylor's Hardware store, Mat Colwells Billiard room, Wm. Richards Barber shop I Michalisky's dry goods store, Wilson's dry goods store, Jamestown clothing store, Winsor's hardware store, Mr Ritts' dwelling, Jahringers jewelry store, Wilcox & Butt's furniture store occupied above as a residence, J. H. Welsh's wholesale liquor store, Post Office News Room, sweeping down the street carrying with it Smith Cooks new residence. On the opposite side of the street the fire commenced with Savings Bank, spreading east ward to Barney Vensel's office and residence thence to the store of R. W. DeHaven, the St. Petersburg hotel, then jumped across a wide street and caught Lewis Collner's residence then jumped two hundred feet to D. Herron's office, taking the entire row of buildings to the "Progress" office, and the next one beyond owned by H. Fulmer. By the noble efforts of the Fire Department, and the aid of hundreds of people who flocked from Antwerp, Turkey City, Richmond and Foxburg. In all forty business houses were destroyed, including three hotels; the Bank, three hardware stores, many groceries, &c., in fact, the ENTIRE business part of town.

Fred Hepp
J. B. Akin; Opera House
L. Goldschmidt
Adams House
C. A. Kibbe
Hall Bros
Seymour Taylor & Co
A. K. Wing
T. C. & G. W. Wilson
Jamestown Clothing Store

*The Progress newspaper detailing disastrous fire of 1873 in St. Petersburg.*

For more than three hours the fire raged with unabated violence; for more than three hours the citizens toiled with the energy of desperation before their efforts were effective, and then, at ten o'clock at night, they looked upon the blackened ruins of over 40 business houses, offices and dwellings. Many had lost their all, many were homeless, all determined to make the best of circumstances. Men, women and children were engaged in searching for household goods, which in the excitement of the time were tumbled into the street, fathers were anxiously in quest of lodgings for their families, here and there careful persons were kicking the fire off of charred pieces of timber and watching to prevent the fire breaking out a second time. All, after the previous intense excitement, were subdued and weary, with the exception of the usual number of thieves who take advantage of every such occasion to make brutes of themselves by getting drunk. By 12 o'clock, however, all valuables had been secured, the homeless had found accomodations for the night, and all was quiet.

It is a sad fact that a number of these pitiful wretches who gloat on the misfortunes of others, were present and followed their debased calling. A large amount of dry goods, hardware and household treasures were stolen and carried off by these vile sneak thieves. So bold were they that in one instance they attacked a young man who was guarding goods near Shaffer's livery stable, knocked him down, abused him in a shameful manner, and carried off all articles of value that they could secure. It is gratifying to know that one man was compelled to disgorge his ill-gotten gains and driven from the town. Perhaps it might have been better had they strung him up to the first convenient crossbeam, but then the St. Petersburg people are more particular as to the feeling of humanity than the little matter of treating such a man as he justly deserved.

That the fire started in Hepp's saloon is all that is definitely known in regard to its origin, but it is believed by many that it was the work of the proprietor. This suspicion obtained such firm hold that Hepp was arrested, and given a hearing before Esquire Reichart, this afternoon. Various witnesses were examined. The prosecutioin found that he had made some singular expressions that could be construed into threats before the fire and that he had shipped a quantity of goods to Titusville during the past week. His wife also left for Titusville in that time. This, in connection with the fact that he had failed to procure a license for the sale of liquor the coming year, and that he had a large insurance on this property, told strongly against him. Mr. Hepp testified that in consequence of the local election, his saloon had been closed during the day; in the evening he kindled a fire preparatory to "opening up," after the polls closed. After the fire was fairly started he went to a neighboring barber shop to get shaved, and while there the saloon took fire; he knew nothing more of the matter and of course, disclaimed anything in the incendiary line. The Justice did not deem the testimony sufficiently strong to hold him, and he was accordingly discharged. Opinion in regard to the matter is divided, there being many who believe Mr. Hepp's statement and that he could not be guilty of such a dastardly act.

A number of the citizens met at the Monitor House this morning for the purpose of organizing a relief committee to solicit contributions for the benefit of those in need. The following gentlemen were appointed such committee: Captain Welch, A. K. Wing, Eli Ritts, P. McCormick, Joel Fink and Charles Blakslee. And it may be said to the honor of those who subscribed, and to the oil men as a class, whom they represent, that in two hours time nearly $1,000 was raised. The oil man, wealthy or poor, in good luck or bad, is always ready to assist those poorer than he.

Among the subscriptions are the following: A. A. Dame, Titusville, $10; A. K. Wing, $50; Captain Welch, $25; E. C. Bradley, $25; the Pickwick Club, $50; Chalfant & Graff, $50; Frank Fertig, $25; Emlenton Pipe Company $100; Antwerp Pipe Company, $100; Blakslee Brothers, $50.

There are many in destitute circumstances and assistance from neighboring towns will be gladly received. The present extreme cold weather, the consequent necessary clothing and fuel; the present lack of employment, and financial stringency, make it difficult for the burned out of house and home to "keep the wolf from the door." Subscriptions sent to the committee will be gladly received and properly accounted for.

The borough election was in progress when the fire broke out, but adjourned immediately.

The Dramatic Society was rehearsing "Hasher and Slasher" when the alarm was given. They adjourned to another stage of action.

*The Progress* man, Barclay, issued an extra today. He will continue the publication of *The Progress* as soon as he can procure a new press.

Postmaster Place "opened up" in the New York bakery this morning and the mails now arrive and depart as usual. He will put up a new building as quickly as possible.

The present unsettled state of affairs will make it difficult for some time to come to give an accurate list of the losses. The following list is as near correct as can be procured, and will not vary materially from the truth:

J. H. Loomis, building, and Hepp's saloon fixtures, stock, etc., $2000. Hepp's insurance, $1,500.

Jerome Aiken's Opera House, dwelling, billiard room and restaurant, $10,000; insurance $4,000.

Shaffer's livery stable and one horse, $1,200.

G. W. King, Adams House, furniture, etc., partially destroyed, estimated $10,000; insurance $7,000.

Pacific and Atlantic Telegraph office, in Adams House, loss light.

Alden & Co., clothing, saved considerable stock, loss $10,000; insurance $6,000.

J. M. Place, drug store, loss $6,000; insurance $2,000.

J. A. Christy, drug store, loss $6,000; insurance $2,000.

A. K. Wing, building and tentant's goods, total loss $10,000; insurance $7,000.

Wilson & Co., drygoods store and dwelling upstairs, loss $3,000; fully insured.

Dunkle's livery stable, loss $500; no insurance.

G. W. Winsor, hardware, loss $8,000; insurance $7,000.

C. A. Kibby, clothing, stock partly saved, loss on building and stock $1,000; covered by insurance.

Hall's restaurant, loss $2,800; insured.

John Ritts home, etc., loss $2,000; insurance $1,500.

J. H. Zavinger, jewelry in Ritts house, loss $500; insurance $200.

L. Frankstein, clothing and dwelling, saved portion of stock, loss $2,500; insurance $700

Seymour & Taylor, building, and Wing & Blakslee, hardware stock; considerable stock saved, loss $8,000 to $10,000; insured.

Producers' Association Building, $500.

W. Sayles, household goods, $1,500.

Johnson's drug store, loss $3,000; insurance $2,500

Bradley & Whitling's building, loss $3,000.

L. Collner, dwelling, warehouse, goods, etc., loss $6,500; insurance $3,500

Robert Bryan, residence and offices, loss $1,800; insurance $1,200.

St. Petersburg Hotel, owned by Asa Balliet, occupants hotel, R. DeHavens, clothing store; Western Union Telegraph and Roberts' Torpedo Agency. Total loss estimated $16,000; insurance $9,000.

Barney Vensel, residence, Emlenton Pipe Line Office in same building, loss $6,000; insurance $1,500.

Fulkerson's restaurant and dwelling, loss $2,500.

Smith Cooke residence, furniture, etc., loss $5,000; insurance $4,000.

Chalfant & Graff, hardware, loss by fire, accident by derrick falling on buildings and thieves, from $3,000 to $4,000.

*The Progress* office, F. H. Barclay, proprietor. Loss $3,500; insurance $1,500.

Occidental hotel, Smith & Heimer, proprietors. Loss $1,000; insurance $1,500.

Henry Fulmer, clothing store, etc., $3,500; insurance $1,500.

Hart & Hogan's bowling alley. Loss $600; no insurance.

The St. Petersburg Savings Bank lost about $3,000, but have some insurance.

The above list does not include many small losses of amounts not yet ascertained. It is estimated that the aggregate loss will amount to $110,000 or $150,000, which is largely covered by insurance. The businessmen here appreciate the value of insurance, and nearly all will resume their old vocations as soon as buildings can be erected. Few of them are crippled, and the loss is comparatively light to what it would have been without insurance.

## The St. Petersburg Fire, February 26, 1873

In twelve minutes from the time the fire was discovered in Hepp's saloon that structure and Aiken's opera house were in ashes. Quick work that.

The water works were completed with the exception of a few connections, and the water was to have been turned on Friday. For some unknown reason there was a delay and a large amount of property was destroyed in consequence.

The Adams House will be rebuilt immediately. Teams were engaged yesterday hauling lumber on the ground.

For once mud was in demand. Large quantitites were thrown on the burning buildings and much property saved by the application.

So quick were some of the buildings burned that nothing at all could be saved. In one house, it is reported, a lady snatched her baby from the cradle and escaped, leaving all else to the flames.

There is a strong feeling against Hepp.

The burned district will soon be covered with new buildings of a better class generally than those destroyed.

A fire engine is wanted and will probably be purchased at an early day.

## Oil News, February 27, 1873

Blanchard & Siggins' well on M. Shoup Farm, Turkey Run, was completed on Monday last. It started up at 175 barrels per day, and still holds good for 150 barrels.

The wells throughout the old territory are falling off at a rapid pace.

## Antwerp Oil News, February 28, 1873

The Knight well, No. 1, upper Turkey Run, 40 barrels.

The Wilkinson well, Slicker farm, 35 barrels.

The Donaghy well, Isaac Neely farm, 60 barrels.

The Bradley and Chambers well No. 4 on Little Turkey Run, 50 barrels.

The W. P. Findley well, David Shoup farm, 15 feet in third sand.

The Emlenton Pipe Line completed a new pumping station at Turkey City today, having a discharge line from that point direct to Emlenton.

The Antwerp Pipe Company has put a large Cameron pump at their station at Antwerp.

The Blanchard and Siggins well on the M. Shoup farm, 150 barrels.

An unknown man had been found lying near Antwerp hollow, Thursday last. The man told officer Cribbs that he intended to commit suicide. To his surprise he was taken directly to the lock-up and charged with insanity and will be left there until his excitement cools down.

*Wells along the Clarion River at Grass Flats near St. Petersburg in the 1870s. Note the steepness of the river bank and photographer's wagon in the foreground. Courtesy Drake Well Museum*

# St. Petersburg
# In Slippers of Velvet He Walked in Shadow
# Aiken's Opera House – The Main Hotels – The Savings Banks
# *The Progress* Office – Incidents of the Fire – Pilfering
# On the Burnt District
# Beds and Bedding – The Borough Offices
# The Problem of the Day

The ancients in their classification of fire as a devouring element were indubitably correct. A glance at Main street today with its charred rubbish and blackened ruins, attest his voracious appetite. Warily appeared in our midst, as is his wont. In slippers of velvet he walked on March 1, 1873 shadow. Unheralded was he, but right nobly did he perform his hellish mission.

With the origin and scope of the late conflagration readers of *the Herald* are familiar.

Some particulars hitherto unmentioned, incidents, amusing and otherwise, with a glance at the borough under the resume of today, we feign would group together.

*The Collner Store, Main Street, St. Petersburg. Present site of St. Petersburg Post Office. The Collner Store, during the oil excitement, did $350,000 in business a year with daily sales between $1000 and $1500. Lewis Collner was born May 11, 1811 in Germany. He learned the shoemaker's trade and earned 40 cents a week. He dreamed of freedom and opportunity and borrowed thirty dollars from his brother and at the age of 26 headed for America. He landed in New York City with one dollar in his pocket. He arrived in Petersburg on September 10, 1838. About that time he started a store and was proprietor until 1871 when his sons took over. Mr. Collner was postmaster from 1864 until 1872. He died in 1894 and is interred in the St. Petersburg Cemetery.*

*NO SMOKING around oil wells! – Drake Well Museum*

Aiken's Opera House – This edifice, the second to fall, was a massive wooden structure more useful than ornamental. Three times had this building been removed and reconstructed. Twice had the gaping fire fiend raged around it, leaving it unscathed. The ground floor on the evening of the fire was occupied as a billiard hall and restaurant, while the upper portion constituted the hall of the Opera House, with an auditorium capable of seating eight hundred persons. As the flames lit up its sides and rapidly enveloped it, it gave from its location the assurance that the business portion of the borough was doomed. In a quarter of an hour from the alarm of fire, Aiken's store was a burning brand. As the flames arched upward, now driven eastward by the wind, now licking across the street, the spectacle was grand in the extreme. Even Nero might thereby have been provoked into cynical fiddling. The gentlemanly proprietor, Mr. Jerome Aiken, is well and favorably known to the Oil Region, and the sympathies of many friends are extended him in this, his reverse.

The main hotels, the Adams and St. Petersburg, fronting each other on Main street, were early victims of the fire. Their loss is really felt by our citizens, as they were, strictly speaking, the only houses that came under the statutory definitly of the word inns. Around the hearth of the Adams stove, Titusville and Upper Creek men were wont to rally, while the genial host and happy hostess rendered it home, indeed.

The Savings Bank followed in order, the office of the Emlenton Pipe Company, Its books and papers were rescued, and at the close of the succeeding day it was installed in a new location with capital stock unimpaired.

*The Progress* office and press was severely tried but not destroyed by fire, as the proprietors announce that they will issue as usual the paper on a smaller sheet.

Incidents of the Fire – Beds and shelter at midnight were of course a rarity, and many were the lodgings improvised for the night. A party of girls from the Adams House took refuge in a blacksmith shop and fitted up apartments adjoining the forge. The excitement attendant upon a fire occasions laughable freaks, while too often, as in Hamlet, it "makes cowards of us all."

A gentleman of literary taste ran frantically into his room, and overturning a pile of books upon his table, secured from the bottom a copy of "Gulliver's Travels" and departed, leaving Shakespeare and other standard works to their fate. A felonious individual, bent on making the most of the occasion, secured a sealed can beneath his coat and left for the outskirts of the borough to enjoy his feast, but to his chagrin, in removing the lid, the contents were found to be concentrated lye.

A nervous oil broker, rooming over a hardware store, was at the polls when the alarm was given. Rushing up street, he sought his room, kicked through the window, throwing therefrom a table lamp and small mirror, and following these effects himself. His papers rescued he entrusted to the care of a friend, who found on inspection that they consisted of three blotters, a love letter, some waste paper, and a schedule of his victories at auction pitch.

Pilfering was entered upon at midnight on an extensive scale. Clothing, cigars, books, furniture and property, both personal and mixed, rapidly changed owners and even the heavier commodities, such as provisions and hardware, were not exempt from felonious fingers. While petit larceny was too frequently perpetrated during the night, no wholesale bearing away of booty was entered upon as the *Courier* letter writer would have it. The baggage wagons and ponderous trains of plunder that passed through Antwerp moved only in the distorted imagination of the *Courier* correspondent.

Several buildings are in course of erection, and one at least is complete since it has a name and location. In a hemlock structure ten by twelve is compressed the office, counting and sales room of G. W. Winsor, while over its sign is the inscription, "The first hardware store opened after the fire!" The Adams House is to be rebuilt, at least temporarily, while the majority of those who suspended trade from "local causes" are preparing to follow suit. It is gratifying to learn in this connection that insurance companies are coming promptly to the front, and, through their agents, responding nobly to the demands upon them.

Winsor & Alden's Building, its upper portion, was devoted to offices and private rooms and were occupied by representatives of the various callings and professions. From its location, as well as genial character of the inmates, it had become one of the fixtures and favorite resorts of town. The group here formed a modern "kit-cat" club, where, humor was the order of the day and jest in rouge. The circle is, however, broken and its reorganization a matter of doubt.

Bed and bedding, at present writing, command a premium here. Table fare, with or without price, can readily be procured, but happy is he who has a comfortable couch on which to repose.

Your correspondent confesses his inability to longer cope with his surroundings. His nights are sleepless turnings, thinking meanwhile of better days, when a mattress was his, and feathers, not a satchel, lay beneath his head. The hospitality of our citizens cannot be too highly commended. At the residence of the Messrs. Blakslee and others family raids are rendered welcome and the latch-string is universally out.

The borough officers elect were installed on Wednesday morning. The inaugural ceremonies were, we learn, brief but impressive. The message of the Burgess was withheld for the present, but will be transmitted to the Council at an early day.

We cannot refrain here from felicitating our citizens, whether politically opposed to the successful candidates or no, on the result of the late election.

The new burgess, Harry Derrousse, is a gentleman of intelligence, of deserved popularity, a citizen alive to the interests of the borough and who, amid the cares of his business, consents to discharge the duties of an office which he will honor, but in no contingency can it honor him. The selection of Councilmen J. K. Elliott, J. H. Sutherland, H. Culver and William Blakslee, is a further guarantee that the interests and welfare of our people will be narrowly watched.

*Lewis Collner, first postmaster of St. Petersburg, businessman and oil producer.*

St. Petersburg today is suggestive of the forlorn hope. Whether it will rebuild and attain its former prosperity, or its business men seek other fields, is the unsolved problem of today. But whatever may be its future, the late fire will constitute an era in its history and many reminiscences painful and otherwise, will date from the evening of February 21st.

## Notes from Antwerp, March 3, 1873

Horse stolen– On the night of the 26th a valuable horse, the property of Mr. Joel Fink, of Petersburg, was stolen. No clue to the thief has been discovered.

The burnt district of St. Petersburg already shows marks of enterprise, being dotted over with substantial buildings and the debris rapidly ceasing, preparatory to the erection of other more commodious buildings than before covered the sites. The wary and energetic pioneer druggist, J. A. Christy, Esq., layed the foundation, completed and occupied his building in the short space of one day, and intends fitting up a fine building at an early day. On the ruins of the Petersburg House, Mr. Balliet will shortly build a fine structure, and again be ready to accommodate his hosts of friends and patrons, who will be glad to enjoy again the general hospitalities of this prince of landlords. The oldest inhabitant, L. Collner, Esq. also a sufferer by the fire, has reopened his store, and no doubt feels that when once he was lord of all he surveyed, he had been cruelly dealt with by those innovators who have surrounded him with buildings so inflammable as to endanger his fine property. The new Lodge of A. F. & A. M., in Petersburg, is progressing finely, having already over forty members. The I. O. of O. F. Lodge, in the same place, is filled to overflowing every meeting.

A well on Hummel, near Delo farm, was struck recently, producing one hundred barrels daily. Bradley & Chambers "Alve" well No. 1, formerly called No. 1 well, on Slicker farm, is doing sixty-five barrels.

## March 6, 1873
## A Rotten Egg Vender in Trouble

Edward Unicum is an egg vender who was arrested some time ago at the instance of one of our tradesmen on the charge of selling a barrel of rotten eggs, representing that they were fresh, and charging full price for the same. The prisoner was taken before Justice Clark yesterday morning, and after a full hearing of the case, the Justice bound him over to stand his trial at the next term of court on the serious charge of obtaining money under false pretenses.

## Letter From Antwerp, March 10, 1873

Antwerp City, Penna., March 7, 1873, To the Editor of the *Herald:*

This morning between 5 and 6 o'clock the engine house of the Armstrong well No. 4 in Antwerp was destroyed by fire. The fire originated in the roof about the smokestack. A day or two since a well owned by Watson & Bayne on their twelve acre tract adjoining Antwerp was completed and is producing one hundred barrels daily. A large tank of oil belonging to the same well caught fire and with no little trouble was extinguished in a short time thus arresting what might have been a great oil fire.

New strikes are continually being made in Turkey Run district all averaging fairly.

There are twelve wells on the Eli Logue farm, Perry township, now pumping, many of them of large yield. Several others are nearly down. No dry holes have yet occurred on this farm.

The Charles Pollock well farm adjoining the Logue farm, Fox heirs, is likely to prove good territory. Piper, Caldwell & Co. are now putting down a well on the Pollock farm, and the Spring will bring further operations.

## Letter from St. Petersburg, March 14, 1873

St. Petersburg, March 11, 1873, From Our Regular Correspondent – But a short sojourn here, since the great fire, is necessary to impress a stranger with the great recuperative powers of the borough. The street down through the heart of the burnt district and the grounds bordering, are literally cumbered with material for rebuilding. One hears in a hotel of an evening more talk about contracts for the erection of stores and hotels than about oil farms, leases and wells, or the price of oil, or the price per foot for drilling oil wells.

Commencing at the head of the burnt district, the first building projected is that of Postmaster Place, but above that is the store of Messrs. Wing, Wilber & Co, who were burned out a few doors below the Adams House but who, fortunately, were heavily insured. Between these, the Antwerp Pipe Company are erecting their office. Next in order comes the Adams House, Charles King, Proprietor. This has been replaced by a temporary one-and-a-half story structure, eighteen by twenty-four feet, which will be prefaced with a three-story building as soon as the question of insurance is adjusted. Next to the Adams House comes the drug store of Dr. F. J. Johnson, which is under contract. The foundation is now ready for the superstructure upon which work will commence during the present week. Messrs. A. K. & W. O. Wing have material ready for a store to occupy the site of the one which went back to its original elements, as soon as the insurance company determines whether they will rebuild or pay the seat. Adjoining, T. C. & G. W. Wilson have a two story store nearly ready for the reception of their goods. Next to this, Messrs. Cooperman & Co. have bought a lease and contracted for the erection of a dry goods store. The next lease is the property of Mr. Zahringer, a jeweler, who will soon build, and the next is occupied by J. A. Chustz's omnipresent "Pioneer Drug Store." A small shanty has already been erected and contains what was saved of the stock, but the carpenters are rushing up a more imposing structure which will be filled with the most desirable emetics and all kindred goods as soon as completed. On the opposite side of the street, Mr. Henry Fulmer has commenced to replace his hotel, and next to this is the material, on one hand, for *The Progress* office, and on the other for Hawkin's hotel and grocery store. The St. Petersburg House is represented by a two story shanty with a reception room and bar room on the first floor and a couple of sleeping rooms on the second. It is reported on the streets that a joint stock company will take the St. Petersburg House in hand and replace it with a building adequate to the wants of the place. Messrs. Wilcox & Butt, furniture dealers have purchased the store heretofore used by Messrs. Shear & Cooper and will fill it, as soon as they can get full possession, with a large stock of goods. These gentlemen own a large store house in Rouseville which they will remove to this place, to occupy the site of their old store. The St. Petersburg Savings Bank building is under way, as is also the Emlenton Pipe Line office. S. Frankenstein has purchased the lease next to the Pioneer drug store and has commenced to gather the material for a clothing store. The ground formerly occupied by Bennett's hardware store is to be occupied by Mr. N. Lehman with a dry goods store.

The people here generally express great satisfaction at the new arrangement of the Allegheny Valley River Railway time table, by which they can get the *Herald* as soon as the Oil City papers.

## David Jones

David was born on October 4, 1846 in St. Helena Parish, Louisiana on a sugar plantation owned by his parents. During the Civil War, the family plantation was burned, his father was killed, and his mother and sister died of pneumonia. His brother, George, joined the Confederate Army and was killed. David then joined the Confederate Army as a drummer boy.

*David Jones and wife.*

After the war, he worked in Chicago but was kicked in the chest by a horse. Doctors told him to go to the oil fields of northwestern Pennsylvania, where the crude oil fumes would heal his collapsed lung.

About 1874 David bought into a wagon shop with Mr. Burchfield which became Burchfield and Jones. David ran a very successful hardware, wagon shop, and blacksmith business in St. Petersburg. Jones served as burgess of St. Petersburg and loved to read and collect books. When the great fire of 1892 broke out, the Jones family had gone to a big fair at Parker City with many other townspeople. A great storm approached St. Petersburg and lightning struck the Opera House which started the great fire of '92. Fortunately for David, a crippled employee who had not gone to the fair saved the house and shop from destruction by dragging wet rugs and quilts to the roof of both structures.

When gas was piped throughout the town, the main valves and gauges were installed in his wagon shop so that he could adjust and insure the proper flow of gas to St. Petersburg.

David died in 1932 and is buried in the local cemetery. Davey Jones served the Confederacy in war but never had a veteran's flag and holder on his grave. In the year 2000, I had his name and military information registered at the Clarion County Veterans Affairs office, and now every year a flag is placed on his grave on Memorial Day.

## Letter From Antwerp City, April 3, 1873

To the Editor of the *Herald:*

Went from Antwerp to Turkey Run City. I found the latter place in a very flourishing condition. It comprises three hardware stores, kept vairously by Patterson and Dickey, R. O. Meldrum, and Winsor Bros., four groceries, several boarding houses and two "gin mills." Well on Heeter farm tubed today; indications good. Several new wells are in progress on Slicker, Heeter, Neely and Shoup farms. The Irwin well on the Samuel Eddinger farm struck in August and produced for some time 250 barrels, now doing 120. The Jackson and Vensel well where four men met with a fatal accident last summer, is pumping fifty barrels daily. Baxter well is called the "Boss" well on Turkey Run. The Western Union Telegraph Company are extending their line from Antwerp to Turkey Run City. The office will be in the Patterson and Dickey's hardware store and ready to transact business some time this week. Our sister town, Richmond, on the Neeley farm, unmentioned in your columns, is more than holding its own. Among the prominent business houses of the place we notice Mr. Shoup in the variety store and lumber yard. Mr. Barr's hotel and grocery store. Mr. Gilbrath's feed and provision store, besides many others. We found several copies of the *Herald* in this place, indicating progression and intelligence. Though they have no news room these luxuries and necessities are easily obtained from our popular news dealers of Antwerp. Business is fair to middling in Antwerp; mud three feet deep and reported rising rapidly. Gin factories almost ready to shut down for a longer period than thirty days.

## Letter from Antwerp City, April 12, 1873

To the Editor of the *Herald:*

We had a little racket in our usually peaceable town, in the neighborhood of the Western Union Telegraph office, Friday evening last. A citizen of the town became enraged at the conduct of two young men of this place, and suddenly threw one into the centre of the street, and put a nice "head" on the other. These men are small boys, and as the one emerged from his downy couch, he presented a sandy appearance. This was proper enough, as the lads were stealing his apples and peanuts, and must be taught that is wicked to steal. A subscription paper was circulated among the boys, and it was not long until a sum sufficient to purchase a good suit of clothes for the sandy looking "victim" was subscribed, and they went home happier if not wiser boys.

The weather for a few days past has been very warm and nice, but seems to have little effect upon the mud, as it is quite as muddy as ever. The Allegheny river was on a fearful high day before yesterday. At Fullerton (which place is about two miles from here), the river was within three feet of its height in 1865. J. T. Myers, accompanied by several others, left Oil City day before yesterday in a skiff at 10:03 o'clock a.m., being some twenty minutes after the departure of the passenger train on the Allegheny Valley railroad, and "hauling up" now and then to partake of "local option," arrived at Foxburg (the destination of Mr. Myers), but a few minutes behind the train. The rest of the party proceeded on their journey to Parker's Landing, where a final tie-up was made.

The Ancient Order of United Workmen are doing nobly, I understand. Summit Lodge No. 16 was organized in December and already has a membership of nearly sixty, and the candidates thus far have been of such a class as to admit of few rejections.

Took a stroll around Turkey City and vicinity last Sunday. Just as we were going up a steep hill the rain came down most furiously which made us *climb*, and also gave us cause to make a hasty retreat to the nearest house. After the shower, which seemed in the momentum to give a green and restored look

*Turkey City in 1874.*
*Courtesy Drake Well Museum*

to every object, we arrived at the city of Turkey, where we received the genial hospitalities of Sam Loughner. Although the quietness of a holy day prevailed, a mark of ambition was visible. Baxter well, on Turkey Run, is boss in name if not in production, although the Blanchard and Higgins well is the greatest producer. Our little friend Dr. Patterson has returned from Cleveland, looking as fresh and gay as you please. Also the genial face of Alf. Burnett, of the firm of Burnett & Billhart, again became upon us after a visit home. We welcome you back, boys.

Later. – A Mr. C., a prominent citizen has just returned from "a walk." It was scarcely necessary for him to explain how he became covered with mud up to the waist. More anon.

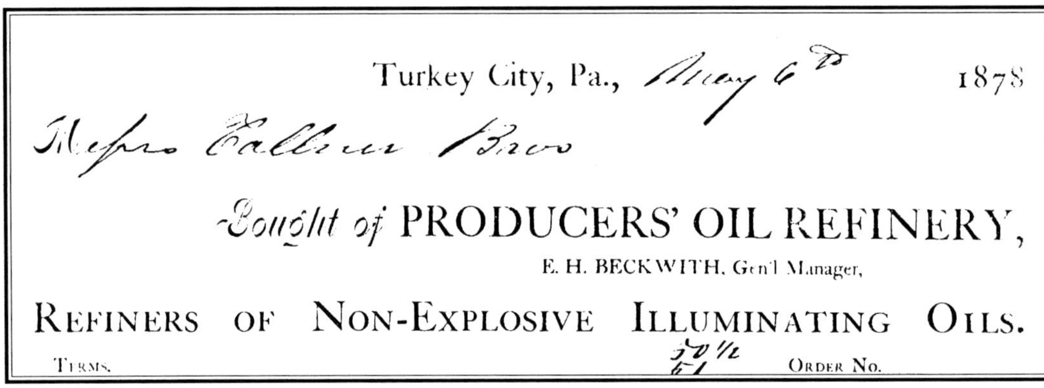
Turkey City, Pa., May 6th 1878
Bought of PRODUCERS' OIL REFINERY,
E. H. BECKWITH, Gen'l Manager,
REFINERS OF NON-EXPLOSIVE ILLUMINATING OILS.
TERMS. ORDER NO.

S. S. GRAHAM,
Dealer in
Books, Periodicals, Stationery, Wall Paper,
GENTLEMEN'S FURNISHING GOODS, NOTIONS, ETC.
Fine Tobacco and Cigars a Specialty.
Turkey City, Clarion Co., Pa., Jan 16

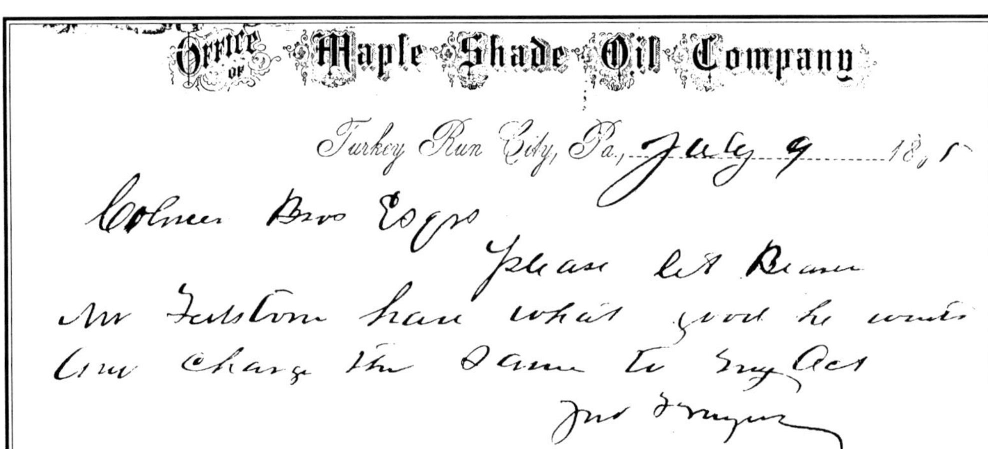
Office of Maple Shade Oil Company
Turkey Run City, Pa., July 9 18

*Oil era Billheads, Turkey Run City*

# The Fire Fiend, April 24, 1873
# *Morning Herald*, Titusville, Pennsylvania

A fire broke out this morning between three and four o'clock in the St. James Hotel on Main street. A lamp exploded in the parlor, and although the guests of the hotel promptly rushed to suppress the conflagration, the flames rapidly burnt through the front sides and roof of the building. The fire raged furiously consuming about half a block from Clark's grocery to Barney Kirley's sucker-rod factory. The handsome residences of Messrs. Blakslee, on the opposite of the street, were badly burned, blistered and scorched, and the window glass cracked and broken, but the efficient services of the fire department, and the utility of the water works, saved the lower end of the town from demolition. Loss about $25,000 with insurance of only about $2,000.

Special Correspondence of *the Herald*, St. Petersburg, April 23, 1873. – Fire, like a troubled conscience or evil genius, still haunts our borough. It is the prevalent belief that he dwells on the outskirts, and delights at stated periods to descend among us and ply his hellish avocation.

At an early hour this morning as the March winds, which have been prolonged into April, were howling around tenement corners and sleepers tucked the clothes about them preparatory to indulging in the final morning nap, the hideous cry of "fire" was heard from the watchman as he ran to and fro through Main street. His sonorous voice was a summons for a general rising and hasty attiring of our community, notwithstanding the fact that by past bitter experience they have become inured to such sensations.

The arising flames located the conflagration in the west or lower end of Main street, and were found to issue from a restaurant situate in a portion of the previously unburned district, and known as the St. James Hotel. This building was a somewhat commodious structure, built of wood of light material, an inviting mass of fuel for flame.

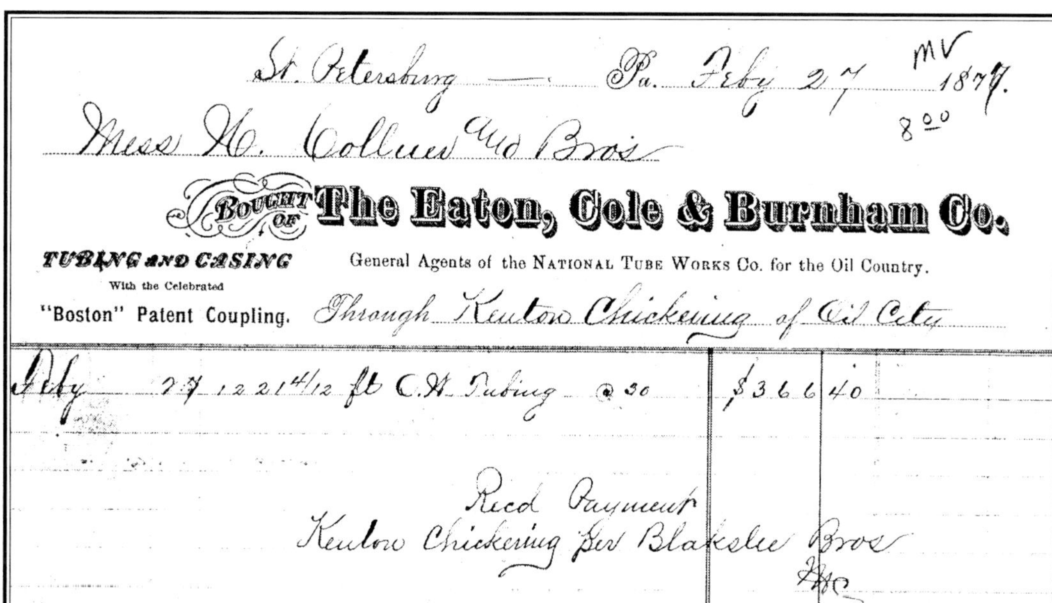
St. Petersburg — Pa. Feby 27 1877

Mess H. Collner & Bros

Bought of The Eaton, Cole & Burnham Co.

TUBING AND CASING With the Celebrated "Boston" Patent Coupling.

General Agents of the NATIONAL TUBE WORKS Co. for the Oil Country.

Through Kenton Chickering of Oil City

| | | |
|---|---|---|
| Feby 27 1221 4/12 ft C.H. Tubing @ 30 | $366 | 40 |

Recd Payment
Kenton Chickering per Blakslee Bros

*1877 receipt from The Eaton, Cole & Burnham Co. to Collner Brothers*

The origin of the fire is involved in that mystery which ever shrouds any evil or calamity originating in a beer saloon. John McWillan, the proprietor, is a gentleman whom to know is not "always to love," and is sometimes said to be far from reputable as a citizen. It is stated that a party, of which he was a member, had not retired at 3:30 this morning, and that the night had been devoted to a general debauch. McWillan alleges that while removing a kerosene lamp from the table, it unaccountably exploded and fired the carpet and furniture in an instant. Parties who arrived at the front of the building as the flames issued from the roof, found the host of the St. James lying upon the sidewalk divested of quite all his clothing, and writhing as if in agony, suffering evidently from the effects of his soda fountain, which rumor says, is provided with a whiskey jet.

The spread of the fire was in either direction from the hotel. The adjoining building on the east was the first to fall. The lower rooms were occupied by Messrs. Clark & Simons, as a grocery store, and the upper portion as the offices of Justices Richard and Booth. Very little of the contents of the building was saved. The flames in their advance westward consumed the restaurant of Charles Curley, feed store of J. Ishman, and sucker rod shop of Barney Curley, and other adjoining but unoccupied buildings. The loss sustained by parties as far as can be gleaned at present writing, are as follows:

Clark & Simons, grocers, loss $1,000; insured for $3,000. Charles Curly, loss $800, no insurance; Barney Curly, loss $1,200, no insurance; J. McWillan, loss $1,200, no insurance; M. M. Both, loss $200, no insurance; C. Reichart, loss $200, no insurance.

On the south side of the street stands the residences of the Messrs. Blakslee and their extensive machine shop, and the Orr House. At a period when the heat was most intense a strong north wind arose, sweeping the flames across the street, which for the time threatened to envelope them all. To the heroic efforts of the hose company and citizens their preservation must be attributed. The Blakslee residences present today a blackened front, which the enterprise of the owners and painters' ink will, however, soon eradicate.

The water works, in the hands of the hose company, who deported themselves as veterans, received last evening their first practical test. How gratifying to those who have encouraged and sustained their erection. It is patent that in their absence the flames could not have been checked, and the lower portion if not the entire borough would have been at the mercy of the devouring element. With such water facilities the disastrous fire of February could have been avoided. Away with the denunciatory criticism of the past. Away with any personal feeling that tends to diminish their efficiency. Men now see in them a palladium of safety, women repose more sweetly by night from a knowledge of their vested worth. The water works are a public benefactor, a conservator of the borough. Let them be sacredly guarded as such.

## Oil News, April 26, 1873

The Best well on the Delo farm will be down in a few days. A large well is expected.

The Harley and Crane well on the Best farm, is progressing satisfactorily and will be finished within the week.

The Johns well on the Best farm is completed and is pumping twenty barrels.

The Dick Owens well, on the Johnson farm, is in the stray sand. The drillers have been pounding away on a piece of steel and a pair of grabs for several days.

Another well on the Johnson farm has the tools in the hole, but we have not been able to learn the name.

*Another view of the trestle and furnace near Jefferson City, 1875. Venango County Historical Society*

William Lusk's well, on the Panton tract, is about down and, as he owns the entire interest, naturally feel considerable interest in its success. We hope that it will prove good, for he richly deserves success.

F. T. Davis has started a new well on the Jordan tract.

Many new wells are starting here to help in the general depreciation of oil.

The Polly Fry well, on the P. Fry farm, is doing about fifteen barrels.

The Lady Mand well, on the Fox farm, is doing twenty barrels.

Hugus, Rugh & Co's new well on the D. Heeter farm, is good for fifty barrels. This company has been very fortunate in their operations; well No. 1 struck last August, flowed eighty barrels for at least four months, and at present is doing forty barrels.

John T. Myers, drilling well on the M. Shoup farm, is now down about 400 feet.

M. Owens, on the Johnson farm, is in the sand, and the show is a good one.

Phillip Brothers tubed their well on the Fillman farm last Friday. She is doing forty barrels.

Hardison Brothers, on the Knappenberger farm, has commenced drilling.

Smith Cook's well, on the Hummel farm, is almost in the sand. Smith expects a good well.

Siggins well, on the Hummel farm, is in the sand. She is looked upon by operators as going to be a fine well.

The Prentice well, on the Master farm, is down, and is doing twenty-five barrels.

McCumber & Tabor well, on the Exley farm, is down 600 feet. – *Petersburg Progress*

## April 28, 1873
## Tramp Among the Wells

A correspondent of the *Erie Dispatch* writing from Foxburg, under date of April 22, says:

Yesterday being somewhat of a pleasant day, your correspondent concluded to take a short tramp among the wells along the Clarion river as far up as Grass Flats. For a distance of perhaps three miles, not a new rig can be seen among the hundreds of old ones. This strip of territory above mentioned was a few months since considered the very best of oil lands in the lower districts, though now scarcely one-sixth as much oil is produced from it as heretofore. Among the many good wells situated on this line, may be mentioned the following: "Cook" started up about a year since at 100 barrels, now doing from ten to fifteen barrels; "Hannah" started at thirty, now doing eight; "Mary R. Fox" started at ninety, rates thirty now; "Riddle" started up at 114, now doing between seven and eight barrels; "Phillips" started at 150, now doing forty; "Nellie" started at 105, now doing ten. I could mention many more in connection with these, but the above will suffice. These wells, however, I mention in preference, from having given accounts of them in previous letters.

*Producing well near Jefferson City, 1877.*

This territory along the river has held its own equally as well as the farmers lying back of it. Further to the northeast about Triangle and Turkey cities, Dogtown and Beaver City, new strikes are being continually made. These, however, are interspersed with an occasional dry hole, while also out on the "front" to the southwest of the belt they are getting some very fine wells at the "send off," but they don't keep the production up "worth a cent." It is not at all surprising to hear of a well which has started at 150 barrels down to 30 barrels in a couple of weeks' time. Dry holes are coming in fashion, as they number of late not a few.

You will notice from the above that operations are now being carried on entirely, or nearly so, at the extreme ends of the belt, and while new wells start up which have a tendency to increase the daily production, others in the middle of the belt are on the decline; enough so to make the amount of oil produced daily no greater than that of early winter.

There are no recent "strikes" to report from the lower end of the belt.

## Fire and Boiler Explosion, April 28, 1873

Special to the *Herald*, Antwerp City, Penna., April 27 – The Hulings & Craig well, on the Stubble farm, was burned last night about 8 o'clock. The well had just been torpedoed and pumped about two hours, when it commenced to flow large quantities of oil, some of which escaped through a ditch leading to the engine-house, when it ignited, and in a moment everything surrounding it was enveloped in flames. About twenty minutes after this the boiler exploded with terrific force. The dome of the boiler was thrown to a distance of 400 yards, and one of the flues penetrated the wall of a house some distance from the accident, finally lodging in the family bedroom. We have the pleasure to state, however, that no one was injured, although many narrow escapes are reported. The entire structure is a total loss, except the engine, which is but slightly damaged.

Another explosion was reported early this morning, but nothing definite is known concerning it at this time.

# The New Fire Test Bill, May 1, 1873

(Author's note: This bill saved a lot of heartache by keeping lamps from exploding in a person's hand. This saved lives and homes from burning down.)

The new petroleum fire test bill goes into effect today. The provisions of this bill will render it illegal for any person to sell oil for illuminating purposes, anywhere within the Commonwealth of Pennsylvania, the fire test of which shall be under 110 degrees Far., under a penalty of from $250 to $500 fine, or imprisonment, or both, at the discretion of the court. "Any person who sustains damage to his property or injury to his person by reason of a violation of any provisions of the act, the person guilty of said violation shall be liable to the person injured for all damages sustained thereby." The Governor will today appoint an inspector for each county, whose duties shall consist in a thorough inspection of all the oil manufactured or sold in their respective counties, with power to prosecute all parties violating the act.

*Clarion County oil well scene, 1879.*

The importance of the bill to the Petroleum trade cannot be overestimated. It is simply the commencement of a raid against the venders of unsafe kerosene. The interests of the producer and refiner, also of the entire home trade, have been seriously injured by this nefarious traffic. It has sown the seeds of distrust against our great staple as a safe illuminator in the minds of the public, which it may take some time to eradicate. It is to be hoped that all other States in the Union, for their own safety, will follow our example until this dangerous commercial and social ulcer has been torn out by the roots. *the Herald*

# New Strike, May 3, 1873

Special Dispatch to the *Herald*, Antwerp City, Penna., April 30 – A well on the lower Heeter farm, on Big Turkey, owned by parties unknown, was struck last evening when it commenced to flow largely and in twenty minutes blew the casing head off. Today it is reported as continuing to flow at the rate of two hundred barrels, although they are drilling.

There are two wells in the Martin & Exley farms, on Beaver, which are doing one hundred barrels each. This part of the country is but little developed, and presents a most inviting field to operators.

## Letter from Antwerp City, May 3, 1873

Antwerp City, April 28, 1873 – Editor *Morning Herald*:

Items are notoriously scarce here at present. There has been no unusual doings, not even a murder, for some days. The Bryan well, on J. J. Ashbaugh farm, just tubed is presenting every appearance of a good well. Several new wells are now in progress on Stubble, M. Shoup and Heeter farms and still further on the belt. New rigs appear in nearly every direction, making prospects very good in this district. Hulings and Craig well, on Stubble farm, that was burned on Saturday evening, has not been started up yet. The well had been pumped only about four weeks, producing about fifteen barrels daily, and from the appearance after torpedoing and when it was burned, a large producing well is expected as soon as it shall be put in running order again.

There has been nothing more ascertained in regard to that second explosion on Sunday morning, and we have come to the conclusion that some seriously frightened person by the former explosion imagined or dreamed he heard another. However, if anything more definite should become known, I will inform you. There was a large crowd present at the time of the explosion, and it was almost miraculous that more were not hurt or killed. The loss caused by the fire and explosion is estimated at $1,000.

Mason well, on M. Shoup farm, recently drilled deeper, is now doing full fifty barrels. From this point we had a splendid view of the city and the plains of *Turkey* still presenting a lively appearance, although some of the most important option houses are now the things that were. (I fancy it would be delightful to view these plains of *Turkey*, which are so brightly arrayed by those gorgeous gas-lights, from the summit of the surrounding hills in the evening.) Some of our noted liquor sellers are now at Clarion to answer the call of justice in regard to this matter.

The Hotel de Murphy seems to have fewer patrons than before. Our citizens were disturbed by a disgraceful row in this place early on Sunday morning, but as yet no arrests have been made. Where were our police, or have we none?

"Osiosco," correspondent of the *Derrick*, proposes to bid farewell to every feat of Antwerp, and visit the home of his childhood for a short time. Also "Billy" Britbart, our merry little jeweler, is absent on a visit to friends in Indiana county. "Oh Billy, my darling, we miss you; we miss your sweet winning smile," etc.

Friend W., of "Bonengelo," correspondent of the Emlenton *Messenger*, don't visit us as often as he did. He is always jolly and has news to tell, and we miss his genial face very much.

More anon, Biraseus.

## Engine House Burned, May 3, 1873

Special Dispatch to the *Herald*, Antwerp City, May 2 – The engine house of the Anderson well, on the D. Ritz farm, was burned this evening about 8:30. Great difficulty was experienced in saving the derrick.

## Fire on the Heeter Farm, May 15, 1873

Special Dispatch to the *Herald*, Antwerp City, Penna., May 14 – Heeter well No. 2 on the Heeter farm was burned this evening. It had just been cleaned out and commenced pumping. The cause of the fire was the escape of gas which ignited and the entire structure, together with 200 barrels of oil, was consumed.

*Imperial Refinery fire, Venango County, September 10, 1875. Frank Robbins Steroscopic View*

## In the Oil Region, May 15, 1873

Work has been commenced on the St. Petersburg Hotel. The Foxburg bridge has been recommenced and it will be ready for travel some time in July.

## May 17, 1873

Henry Carpenter was arrested early yesterday morning in the act of carrying a wooden pail with twelve pounds of nitroglycerine in it. Mr. Carpenter was arrested as he rested on the front porch of the meat market on Antwerp City's main street. It appears that Mr. Carpenter manufactures the explosive about two miles below the city along the Clarion River. He was on his way to a well owned by C. Vensel and others on the J. Ashbaugh farm. He was taken before Justice Patterson and fined $100.

## Oil News, May 19, 1873

A torpedo agent at St. Petersburg has been fined $60 for carrying nitro-glycerine through the streets.

## Clarion County, *The Progress, St. Petersburg*, May 28, 1873

A rig on the Fox farm was burned Friday—The Blakslee Bros. tried their new pump last week as a fire extinguisher. So great was the force with which the water was forced through the hose, that two men were unable to control the nozzle. The water works are most excellent and will throw water over any building in town, but the pump at the machine shop will double the distance with ease.

A Mrs. Chittenden living near Keating City fell into a fit yesterday morning while taking on many glasses of "Peruvian Beer" at the saloon in Antwerp. She was taken to Justice Patterson's office where she was revived and left for home.

*Clarion County Oil Field near Elk City.*
*Venango County Historical Society*

## Oil Field Record, June 1, 1873

James Walters fell from a rear window upstairs in Mrs. Mackey's Antwerp City Hotel, to the ground, a distance of 35 feet, Saturday night. He was picked up in a bruised condition and taken to Dr. Dorn's office, but his injuries were not considered dangerous. He was intoxicated at the time.

## St. Petersburg Circulating Library, June 24, 1873

Mr. J. M. Place, the wide-awake newsman, has established a "Circulating Library," greatly to the advantage of the people of St. Petersburg. What would the old settlers have thought of such an enterprise there a few years ago?

## St. Petersburg, June 27, 1873

Report of the Laying of the Cornerstone of the New Odd Fellows Hall Special to the *Morning Herald*, St. Petersburg, June 26

*View of Edenburg (Knox). Note the mud street.*
*Venango County Historical Society*

## Expanding New Territory, June 27, 1873

*The Progress* reports that last week a well was struck on the Bowers farm, about a mile northeast of Edenburg, far in advance of the extreme front, on the line drawn by John Turner, the owner of the well. Although dry holes were obtained on both sides, east and west, the Turner well proves to be a good seventy five barrel well, thus opening up an immense scope of fresh territory and helping to "bear" the market. This well was considered by all except the parties directly interested as being in dry territory, and the result has had a tendency to send old leases to the top of the market.

*The Progress* gives the following account of a narrow escape of a workman:

On Saturday last, T. Griffin, a boiler maker from St. Petersburg entered the fire-box of a boiler at Turkey City to make some repairs. The engineer had allowed the fire to go out, and as he supposed, had shut off the gas entirely. After being within some little time, the workman noticed that raw gas was circulating through the tubes, and started to go out, but before he could accomplish it an explosion took place, burning him terribly. It is unknown by what means the gas became ignited, but it is supposed that some of the coals remained. The injured man is doing well.

## July 1, 1873

The stable on Spring Street near the "waterworks" at Antwerp City has a very offensive odor.

The heavy rains of the past forty eight hours has greatly mitigated and washed away offensive surface debris in the streets and open lots.

## July 1, 1873, Beaver City

Willie Byers, a little child of but a few summers, was attacked yesterday morning on third street by a savage dog who might have killed him but for the presence of persons in the vicinity who drove the ferocious animal away.

## The Expanding New Territory, July 14, 1873

The *Clarion Democrat* says that the oil well on the Bower farm about a mile northeast of Edenburg, in Beaver Township, owned by Duncan, Turner and Lowery is yielding sixty barrels per day. The pipe lines have been laid to that point and one line, we are informed, is immediately to be extended up to the well at "Berlin Hotel" on the "Pike," west of Shippenville, where five hundred barrels of oil is in tank. Another well is being put down near Berlin's, also one on the Graham, one on the Gates, one on the Kribbs, one on the Armstrong and another on the Bower farm, and many others in the same neighborhood are preparing to drill. It is now a fixed fact that the great Parker and Petersburg "oil belt" extends up into Ashland and Elk townships and as soon as oil becomes a better price, lively times may be expected in that quarter.

## Oil News, July 14, 1873

The oil prospects in the lower district are even less encouraging than in this region, the upper district, because of the greater average cost of production and transportation.

The market is dull and the present conditions are not bright.

A down and out oil man estimates the average production is a little over 25,000 barrels a day, and that there are 100,000 persons directly and indirectly dependent upon oil production, the average income for each of whom is but fifty cents a day. Of course this estimate includes women, children and non-producers. We hope he soon feels better!

## Oil News, July 17, 1873

Antwerp Pipe Company received 70,000 barrels of oil from the "field" in the month of June, and the shipments by rail for the same month were greater than for any preceding month since it has been in operation.

The Chalfant well, in the rear of Chalfant, Graff & Co.'s hardware store, was torpedoed a few days since, and is now doing fifty barrels. This is one of the few wells which has kept up the production well, and has already produced some $14,000 worth of oil, even at the miserably low figures which have ruled since it was struck.

In the Hess well, which is being drilled near the Chambers wells on the Slicker farm, a streak of lubricating oil equal to the Franklin oil was struck at a depth of 306 feet. The well was tubed and pumped for a few days and produced a barrel or two per day, which would not pay, and drilling was accordingly resumed again with hopes of better results when the third sand is reached.

In a well drilled by Mr. Lowe, on the Hiram Neely farm, a large vein of gas was struck on Friday last at a depth of 300 feet. As this is hardly half the depth, gas is usually found, the drillers were taken completely by surprise, and could not believe that the roar, which resembled the sound of a boiler blowing off steam,

*M. E. Hess – Hess drilled the first well on the David Shoup farm, one-half mile from the famous Huling's Antwerp No. 1. Mr. Hess started as a driller in 1864 and operated for oil on Cherry Run, Pithole, Petroleum Center, Shamburg, and McClintockville. Michael Hess in 1873 put down the first well on the Shoup farm and the Daniel Knight farm. In 1874, Hess drilled the first well on the Moon farm in Ashland Township and the first successful well on the R. J. Dahle farm in Elk Township in 1875. He is also credited with one of the first wells on the J. I. Best farm at Edenburg. Hess was a true "wildcatter." In 1876 he again put down the first well on the Camp Ridge and David Whitehill farms. He and E. C. Bradly, Esq., put down the first wells on the Egypt farm in Beaver Township. When Michael Hess came "wildcatting" in Clarion County, he brought with him $60,000 profit from the Venango fields but testing so extensively for oil and in consequence of endorsements, became financially embarrassed. In other words, he was "dead broke." Mr. Hess later reorganized and enjoyed a good life as burgess in Edenburg. He died in 1901 and is interred in Knox Cemetery. History Clarion County 1887*

was produced by the rush of gas until it came up and filled the derrick. The forge being located inside the derrick, fire was communicated almost immediately. "Billy Walker," the driller on duty at the time, like another Joseph, escaping, leaving his coat behind, which contained ten or twelve dollars, was consumed. The fire was put out but the rig was considerably damaged, and drilling will be resumed again in a few days.

## In the Oil Region, July 17, 1873

Jas. Fowler, Esq., living near Foxburg, when crossing the new bridge, made a mis-step and fell upon the rocks some thirty feet below. It is feared that he received fatal injuries by the fall.

*The Progress* says that on Monday last the lightning played strange freaks, and almost added a few fresh corpses to the old grave yard, inasmuch as it struck the house of John DeRolph, splintering the roof, opening the side and destroying a great pile of dishes which were in the sink at the time. Mr. DeRolph was sitting about twelve feet distant at the time, with a small child on his knee. The shock was very violent and as soon as he recovered sufficiently, he went out and discovered that the building was on fire; with some difficulty he succeeded in extinguishing it. This was a most miraculous escape and we understnd that it is not likely that Mr. D. would invite the visitor to spend the night with him. During the prevalence of the same storm, and at nearly the same instant, the lightning rod on Collner's store was struck, but the conductors being good no damage was done, except melting the steel points off the top. The workmen in the Methodist church were given a very severe touch of the headache by the playing of the electric fluid over their tools and work benches.

## July 22, 1873
## Accident with a Lawn Mower

Little Bertie Simmons, a boy seven years of age, the son of Mr. Wyram Simmons, met with a very singular accident yesterday, by which he lost both his thumbs. It appears that he and some other children were playing with a lawn mower and by some means or other he fell with his hands on the "cutter" while in motion. The thumb of the right hand was completely cut off, as was the left thumb. Dr. Dorn was summoned and performed the amputation and repair.

**Five hundred choice watermelons just received at Champman's Food Store, Main St., Antwerp City.**

*Wells along Turkey Run at Turkey Run City, 1874. Courtesy Drake Well Museum*

## Oil Tank Struck by Lightning, August 13, 1873

Special to the *Herald*, Antwerp City, Aug. 12 – The Antwerp Pipe Company's oil station near Keating's furnace, during the storm of today was struck by lightning burning a tank of oil. Loss estimated at about $1000. The building and machinery, owing to the promptness and activity of the employes of the line, were saved.

## Letter From Antwerp, August 19, 1873
## Improvements – Old Oil Wells – Local Gossip
## Oil News – Turkey City

From our Occasional Correspondent – Antwerp City August 15, 1873 – You must really think your correspondent guilty of great intelligence in delaying so long to furnish a few items of events. Since I last wrote of Antwerp, many changes have occurred. This once thriving and well-to-do place, like the late of numerous other oil towns, has met with a turn of the tide and the most remarkable and thrilling incidents of her time have been recorded. I would not have you infer, however, that our little town is by any means dead or played out, for such is not the case. And were it not for the notoriously low price of oil, I should anticipate quite lively times here yet, although some of our large business firms have deserted us and fled with the excitement of that mighty element, oil. The old wells here and in this vicinity are holding out nobly. Huling's well, No. 1, the first well struck at Antwerp, and which produced when struck and some time subsequently, over one hundred barrels daily, is now doing fifteen barrels. Several other wells that were thought almost worthless have been increased to twenty and thirty barrels. I rode out to Triangle City a few days since. I had heard so much about that section of the country that I became anxious to visit it, and indeed was greatly edified with the grand scenery of numerous groves and stately mansions of ancient style of architecture. Triangle presented the appearance of a fast-growing place, there being already a hardware store and a first-class hotel, called the Triangle City Hotel, where I stopped to break my strong-growing fast. If you please, I will return to the first town en-route, Keating's Furnace, of which I will speak briefly, as I do not intend to intrude on your valuable space too much or abuse my privilege.

This place is located about one and a half miles from Antwerp, and in a very pleasant and picturesque valley, where the Antwerp Pipe Company, which is managed by men of such good propriety and business talents, selected the site for their new pump station; and indeed the structure adds only to the beauty of this magnificent place. It is a first-class station in every respect, and "gets away" with more oil than any other in this district.

"Pulling" has become so common that I am reluctant to do anything of the kind, nor indeed do I consider what I shall say but just and proper. Great credit is due D. D. Simons, J. Huston, T. Dowrey, B. G. Johnson, E. G. Wright and others, who, with seemingly inspired clients, were successful in saving the Antwerp Pipe Station from a tank of burning oil only about six feet distant.

Turkey City whose plains are so grand, is the next place we struck. Everything seemed brisk as usual here. This town has already been spoken of through your columns. We merely passed through Dogtown, but were nevertheless able to judge that, from appearance the citizens are of an industrious and enterprising class. There are numerous wells in this section, which average fairly. The next flourishing town we arrived at was Pickwick, where we stopped and called upon our genial friend, Lee Taylor, who has charge of Seymour, Taylor & Co.'s hardware store. This natural genius fully represents the town right in his store with a pick, to which is a ball of candle wicking attached, thus illustrating the city of Pickwick. There are a number of stores and boarding houses here, and things look generally lively. Beyond Triangle City, the most notable place in that section is Beaver, the location of the Antwerp and Vandergraft and Foreman Pipe Lines.

## Letter From Antwerp City, August 30, 1873
## Fires – New and Old – Large and Small

Birches' tank shop, at Richmond, was discovered on fire about 4 o'clock yesterday morning, but it was under such headway that no effort to save it was made. It is generally supposed to have been the work of an incendiary. The loss is estimated at about $1,000.

Another fire occurred at the boiler house of the Pioneer Oil Company's well, No. 4, in the same place and on the same morning as that of the above. The company was experimenting with oil for fuel, and by some miscalculation, the smokestack was heated too hot for the roof boards in its immediate vicinity, which resulted as aforesaid. A timely discovery by the engineer prevented any serious damage. A well belonging to the same company, which has averaged twenty barrels daily for over a year, was lately torpedoed and its production increased to fifty barrels.

A well on the Eli Ritz farm, which was recently completed, is flowing at the rate of sixty barrels per day. Another well on the same farm will be completed within a week. This farm adjoins Antwerp on the east and has proven excellent territory.

The Martena well, on Martena farm, Beaver district, was tubed today and is doing fully 200 barrels. The well is owned by J. Mehrten.

## From Antwerp City, September 3, 1873
## Church Dedication at St. Petersburg – New Hotel

The dedication of the Methodist Church of St. Petersburg took place yesterday (Sunday) morning. The weather was fine and the mild breezes were rendered sweet and holy by the occasion. At the usual service time the streets were thronged with people on their way to the church, which was also crowded to overflowing. The services were conducted by Rev. I. C. Pershing, D. D., President of the Pittsburgh College, assisted by Rev. D. Leslie, Presiding Elder of Franklin District, and others. It is not at all complementary to say that Mr. Pershing is a most fluent and effective speaker. Indeed the whole audience seemed absorbed in his discourse. This church, which is a grand structure and an honor to any community, was commenced about the 1st of June last. The contractors were McCamant & Harley, who have a large business in the same capacity in this section, and render general satisfaction. The audience room is 35 x 70 feet, and capable of comfortably seating 400. The cost of the church was something over five thousand dollars, and up to Sunday morning's services there was back on the debt over $3,000. By the effort and influential sermon of Mr. Pershing, $4,500 was raised which was about $1,200 more than the incurred debt. As a friend said, the preacher first touched the heart and then the pocket. The citizens of that enterprising town are indeed praiseworthy for their efforts and kindness, and the Lord will bless them. The church is located near a beautiful grove, and looks like a palace in a retired position.

The new hotel in St. Petersburg, called the St. Petersburg Hotel, erected on the same grounds of the former hotel, by Mr. R. Dutt, was opened yesterday. Now, gentlemen, good accommodations are to be had and at reasonable rates.

## FIRE!
## The Destruction of Antwerp
## *Oil City Derrick* of September 3, 1873
## Large Fire at Antwerp!
## All the Business Part of the Town Destroyed!
## Twenty-Eight Buildings Burned – Loss $28,000 Insurance $6,000 – Suspected Incendiarism
## Special to the Derrick by W. U. Line

Antwerp City, Penna., September 3 – This morning between 3 and 4 o'clock fire was discovered issuing from the basement of a building owned and occupied by J. C. Patterson as a grocery store. When the flames were discovered they were beyond control and those living in the house had scarcely time to save themselves. The alarm was given all along the street, but owing to the early hour, the people were slow to get out. A number of persons had narrow escapes from the fire.

The flames communicated to the opposite side of the street, and it was found utterly impossible to check them until the fire had reached a point on either side, where there was a little open space. At this juncture the timely arrival of the Bonner Hose Company from St. Petersburg saved the lower portion of the town. Messengers had been sent for the company, and they were on the ground within an hour after the fire broke out, so rapidly had they moved. They rendered most valuable service.

*A typical oil country fire. This is what the Antwerp City fire would have looked like, 1873.*

The entire loss will be $28,000 on which there is an insurance of almost $6,000. The number of buildings destroyed was 40.

The fire is supposed to have been the work of an incendiary. A horse was stolen from a stable nearby and other circumstances tend to confirm the suspicion. An attempt was also made to fire Richmond, but it was discovered in time to prevent its execution.

## The Antwerp Fire
## Business Portion of the Town Destroyed
## First Discovery and Alarm
## Rapid Progress of the Flames – Little property saved
## The Burned District List of Losses and Insurance
## From our Special Reporter

Again has a serious conflagration visited the Oil Regions, and the greatest part of an important village has been wiped out as the result. At four o'clock on Wednesday morning a man named Bowen, who occupied a room in the rear of Hatch's Cigar Store, was suddenly awakened with a feeling of suffocation. The apartment was full of smoke and hastily springing from his bed, the frightened man discovered flames bursting through the walls on the adjoining premises, known as the "New York Grocery." Instantly giving the alarm the citizens were quickly aroused from their slumbers, though not before the flames had gained such headway as to render it impossible to save the building where they originated. The grocery was owned by Squire Patterson, who used the upper story as a dwelling. He and his wife barely escaped with their lives, the Squire not having time to draw on his boots and Mrs. Patterson was still in a morning gown.

The wind was blowing northeast and in a few minutes the fire reached Hatch's Cigar Store, and likewise caught Mrs. Mackey's Hotel on the opposite side of the street. Here the progress of the flames might have been stayed without further serious loss, but unhappily the wind steered around to the west, driving the blaze along both sides of the street at a furious rate. The unsubstantial wooden structures burned like tinder, in many instances according no opportunity to save movable property. There were no fire engines, water was scarce and the inhabitants could do little to impose the ravages of the devouring element. One house after another succumbed beneath the fiery embrace, till almost the entire business portion of the town was destroyed.

An effort to stay the flames by cutting down "Curly" Green's building proved futile, despite the best exertions of the citizens, aided by many others from St. Petersburg and neighboring areas. The flames raged until they reached vacant plots on both sides of the street, where they were finally subdued at about eight o'clock, leaving numerous families homeless and only a mass of smoking ruins to indicate what had so recently been a flourishing community.

The following information lists part of the burned district, with the names and location of the sufferers, will enable the reader to form a correct idea of the extent of the conflagration:

| North Side | South Side |
| --- | --- |
| Powell Office | |
| Petroleum Centre House | Hatch Cigars |
| Hart Machine Shop | Patterson Groceries |
| Alley | Dwelling |
| Mackey Hotel | Crawford & Palm |
| Vacant Lot | Bryan Hardware |
| | Stowell Boarding House |
| | Westmoreland Bakery |
| | Black, Vacant |
| Dwelling | Hugus Boarding House |
| Adams Dwelling | O'Neill Restaurant |
| Vacant | Smith Meat Shop |
| | Dwelling |
| Haymaker Dwelling | McKeever, Vacant |
| Vacant | Maple Shade, Vacant |
| | Green Saloon |
| Hulings Dwelling | Williams Barber Shop |
| Vacant | Mayo Billiards |
| | Hayden Hotel |

It will be perceived that the above includes business houses of all descriptions, hotels, and residences. There is no question that the fire was the work of an incendiary, possibly some poor wretch who sought to be revenged upon Squire Patterson for having meted out strict justice in his particular case. It is supposed the fire was started in the basement, to which it was easy to gain access at any time. When Mr. Patterson closed up for the night no light was left burning, nor had any fire been kindled after the noon meal. The finest and largest building in the place was Mrs. Mackey's hotel, the burning of which entails heavy loss upon a highly deserving lady. From the fact that business had diminished greatly during the last year, in consequence of the great falling off in the production in the surrounding country, it is not likely any considerable portion of this town will be rebuilt. Indeed Antwerp has seen its best days, and not a few of the former residents months ago removed to other localities.

*Dr. J. B. Donaldson, St. Petersburg.*

The scene yesterday was exceedingly dismal. The rear portions of nearly all the buildings on the south side of the street had been set on posts, owing to the slope of the ground and those charred supports of the destroyed erections were generally all that remained to tell the tale of devastation. Heaps of ruins were scattered over the burnt district, while here and there could be discovered a stray article of furniture, someone's boat treasure.

**J. B. DONALDSON & CO.,**

**St. Petersburg, Pa.,**

DEALERS IN

DRUGS, CHEMICALS,

FAMILY MEDICINES,

**PAINTS, OILS, DYE-STUFFS,**

**WINDOW GLASS,**

Fine Soaps, Sponges, Brushes, Perfumery,

Pure Wines and Liquors for Medicinal Use,

**Fine Segars and Tobacco.**

Physicians Prescriptions Accurately Compounded.

BOOKS AND STATIONERY.

B. L. Fahnestock & Co's Pure White Lead Constantly on Hnad.

*An ad of Dr. J. B. Donaldson, St. Petersburg.*
*Arters Oil Field Directory 1872*

# School House
# Antwerp City

The school house, built in 1871, which stood on the hill above the town and escaped the flames, was moved down in the fall of 1882, and is the only building left to mark the spot where the town once stood. The school house was moved to a site on the lower side of the road, where Theodore Booher's home is now.

I have in my possession a metal piece of a school desk from the school. It was found on the site while raking the yard.

# George Washington, Jr.

After the fire, a person styling himself as "George Washington, Jr." remained at Antwerp providing tongue-in-cheek reports of the fire's aftermath and how Antwerp was going to rise from the ashes. In the October 18, 1873 issue of *The Derrick*, he provided the following list of what the stricken businessmen and women were doing.

***Antwerp City School – This picture of the Antwerp City School is the only photo found so far of a building in the oil town. What pictures that did not burn in the fire have just vanished over the past 130 years. The following article is from the Oil Field Record, a St. Petersburg newspaper dated November 4, 1872: "Antwerp City has recently had the misfortune to experience several cases of diphtheria. We have at present two cases of diphtheria in town, neither of which are considered very dangerous. In order to protect our students, the school building is undergoing the cleaning out process and in addition is receiving a couple coats of whitewash throughout. Our school is in a very flourishing condition just now, and we hope for a continuance of the same. The learning process is much more gratifying to both parents and teachers when there is no dissatisfaction in the school room.***

First Ward:

Telegraph office – discontinued

Antwerp Pipe Line Co. office – moved to Turkey City

Ann Mackey (Antwerp City Hotel) – gone to Scrubgrass

L. J. Lily, boots and shoes – gone to Turkey City

Bethune, store – gone to St. Petersburg

Schwartz, grocer – going to stay at Antwerp

Simon Hugus – gone to St. Petersburg

J. C. Patterson – gone to Foxburg

Second Ward:

Crawford & Palm, newsroom – going to stay at Antwerp (They left Antwerp for Edenburg a month later.)

Hulings, hotel – only one left in Antwerp

Simonton, torpedoes – gone to Turkey City

The correspondent, himself, left Antwerp about two weeks later.

On June 4, 1875 the following appeared in the *Oil City Derrick*:

The school building yet survives. One well is drilling on the J. J. Ashbaugh farm.

Hulings old well, the first put down in the third district, is pumped by heads and still producing several barrels of the greasy fluid each day, though hundreds of its latter contemporaries have long since gone where the woodblue twineth.

Five buildings constitute the sole remains of Antwerp. Only two years ago the place was among the liveliest in the Oil Regions, threatening to become a serious rival to St. Petersburg. But the wells declined with unexpected rapidity, a destructive fire hastened the final catastrophe, and scarcely enough of the once flourishing town is now left to form a sickly hamlet. Scores of blackened posts, which once supported balloon superstructures in which business of every description was carried on, stand like grim sentinels to indicate the former magnitude of the deserted village. The change is indeed painful and suggestive, nor does Pithole itself present a picture of desertion so overwhelming and complete. There is not a store of any kind, hotel accommodations expired with the big fire, the telegraph office has long been a thing of the past, sidewalks are unknown, visitors are rarely seen, and the fated town of Antwerp exists alone in the memory of fellows who came around in quest of fortunes and made their exit dead broke.

*General view of a train wreck on Antwerp curved trestle, Antwerp City area.*

*General view of a train wreck on Antwerp curved trestle, Antwerp City area.*

## The Lower District, September 4, 1873

There is now no more than one engine sold in the Lower District, compared to ten three months ago. There is also an absolute death of orders for drilling tools, excepting for furnishing wells underway.

The credit system in the "Lower River Districts" has now nearly ceased, the prominent firms selling only for cash or its equivalent. If this is rigidly adhered to it will greatly tend to place the producing business upon a legitimate basis.

## Destructive Fire at Antwerp – Forty Buildings Burned September 4, 1873

A fire occurred at Antwerp yesterday morning which destroyed nearly all the business portion of the town. Something like forty buildings, including three hotels, were burned. The fire originated in the basement of the grocery of Thompson & Patterson, and spread with such rapidity that the family of Mr. P., occupying the second story of the building, had barely time to escape. They had neither time to dress or secure a change of clothing. The loss to this firm is $3,000 on which they had an insurance of $2,000. From this building the flames spread to those adjoining, until almost every business house in the village was destroyed. Among those whose losses are the heaviest, is Mr. Robert Bryan, dealer in hardware, who puts the amount at $10,000, on which he has an insurance of only $2,400, other policies which he held on his property having expired but a few weeks since, and which he had neglected to have renewed. Crawford & Palm, news dealers, lost everything, as also did Simon Taylor & Co., manufacturers and dealers in sucker rods. In fact all the sufferers are heavy losers, as there was but very little insurance in the village, the various companies having declined to take risks there.

Since writing the above we have received the following:

## Addition Particulars
## Special Dispatch to the *Herald*
## Antwerp City, September 3

The disastrous fire of this morning, which occurred about 4 o'clock, began in the grocery store of J. C. Patterson. It was so silent in its march that several persons narrowly escaped with their lives. The craving flames crossed a chasm to the dwelling house of M. J. Rugh on the same side of the street, which was immediately enveloped together with Crawford & Palm's newsroom, Robert Bryan's hardware house, A. R. Black's billiard hall, Westmoreland bakery, West-moreland House, Frank O'Neil's ice cream parlor, Smith & Co.'s meat market, dwelling house of Mr. Lessig, a large unoccupied building, Chapman's feed store, a building formerly occupied as a clothing store, D. L. Green's saloon and barber shop, May's billiard room and Heator's saloon. The devouring element in the meantime passed to the north side of the street, whence it originated, to the Antwerp City Hotel, owned by Mrs. Ann Mackey, which was consumed, J. D. Hearts' machine shop, dwelling of E. Rhodes, Adams' dwelling, and a laundry were also destroyed, besides numerous other buildings. A. A. Bowen barely had time to reach the outside when the building tumbled. He at once awoke Squire Patterson who, with his lady, reached the open air in time to see their couch drop to the cellar. The place so busy one day since is now covered with gloom. Parties are working among the debris. Many families are homeless and seeking shelter. The incendiary did his work surely and can now gloat over the work of ruin with devilish complacency. The losses are as follows: D. Powell's office, $100; J. D. Host, machine shop, $400; Ann Mackey, $4,000, insurance $2,000; Truck's hotel bar, $500; E. Rhode,

*This lithograph represents the destructive fire at Antwerp City, September 3, 1873. A reporter for the Crude Local, a Petersburg newspaper, reported that "with the wink of an eye, everything went Hellward." – Harper's Weekly*

$350; Adams, $200; Haymaker's dwelling $300; Gradner's dwelling $125; Seymour & Taylor's sucker rod shop, loss $700; J. W. Harris' cigar and notion store, loss $1,200; J. C. Patterson, New York grocery, loss $3,000, insured $2,000; Crawford & Palm new rooms, tobacco and cigars, $2,000; Robert Bryan, hardware, loss $9,000, insurance $2,000; J. C. Stowell, boarding house, loss $1,200; Westmoreland bakery, loss $500; A. R. Black, billiard room, loss $400; S. Hugus, Westmoreland House, loss $1,500; Frank O'Neil, ice cream saloon, loss $500; S. P. Lessig, dwelling, loss $1,000, insurance $500; Smith & Co., butcher shop, $300; John Williams, barber, loss $100, insurance $150; May's Billiard Hall, $250; D. L. Green, restaurant, $150; P. Huston's Hotel, loss $300; A. A. Bowen, personal property, $300; L. C. Pierce, personal property $120. The foregoing losses foot up about $28,000

The following lines were first printed in the *Oil City Times* by publisher John McLaurin in the 1870s. *Antwerp City Revisited* was originally written about the ghost town of Pithole in Venango County, Pennsylvania. It has since been revised using Antwerp City as the poem's ghost town subject.

## Antwerp City Revisited

Not a sound was heard, not a shrill whistles scream,
As our footsteps through Antwerp City we hurried;
Not a well was discharging an unctuous
stream
Where the hopes of the oilmen lay buried.
We walk'd the dead city till late into the night-
Weeds growing where wheels once were turning-
While seeking to find by the struggling moonlight
Some sympton of gas dimly burning.
No useless regret should encumber man's breast,
Though dry-holes and Antwerp City may bound him;
So we lay like a warrior taking his rest,
Each with his big overcoat 'round him.
Few and short were the prayers we said,
We spoke not a sentence of sorrow,
But steadfast gazed on the place that was dead
And bitterly long'd for the morrow!
We thought, as we lay on our primitive bed,
An old sand-pump reel for a pillow,
How friends, foes and strangers were heartily bled
And ruin swept on like a billow.
Lightly we slept, for we dreamt of the scamp,
And in fancy began to upbraid him,
Who swindled us out of our very last stamp
In the grave we could gladly have laid him!
We rose half an hour in advance of the sun,
But little refreshed for retiring!
And, feeling as stiff as "a son of a gun"
Set off on a hunt for some firing.
Slowly and sadly our hard-tack went down,
Then we wrote a brief sketch of our story,
And struck a bee line St. Petersburg's fair town,
Leaving Antwerp City alone in its glory!

## Incendiary Attempt at Richmond, September 4, 1873

A pile of combustible material was placed by some malicious party or parties near Shoup & Graham's store in Richmond last night. In the morning it was discovered to have been lighted, but fortunately only smouldered away, thus preventing what might have resulted serious.

***Pumping oil from wells to tank. Note the many criss-crossing pipe lines.***

## September 4, 1873

A well known as the Modoc, on the Shoup farm near Antwerp City, struck the sand yesterday and when the drill penetrated to the depth of six feet, the well commenced flowing at the rate of one hundred barrels per day. The sand found is of the best quality. The well is owned by Heeter & Brooks.

## News From the Front, September 9, 1873 St. Petersburg and Vicinity

Mr. J. G. Black has opened a new hotel on the ground where the old St. Petersburg House stood. It is elegantly furnished and is kept in first-class style and largely patronized.

A new brass band has been started here called the Petersburg Cornet Band, and is highly creditable to the town.

The chief topic in the town just now is the decision of the Clarion Court in the case of Derouse versus Vensel, for mayhem. It was positively sworn that Vensel bit Derouse's ear off, but the Grand Jury failed to find a bill. This probably will not be the termination of the case.

The Bonner Hose and Petersburg Hook and Ladder Companies, James Elliot, foreman, will turn out on the 10th inst. To attend the grand parade and picnic at Parkers. The boys anticipate a good time.

Two fine halls are being built here, one as a Masonic Hall and the other for the Odd Fellows, both to be located on Main street. The Odd Fellows Hall will be dedicated on October 15th and the Masonic Hall on the 26th inst.

The production of the Petersburg District all told is approximately, including the Beaver District, 5,000 barrels per day.

The average product of the wells here does not amount to ten barrels each. The largest is 200 barrels.

Two new iron tanks are going up at Fullerton, both 22,000 barrels capacity.

There are five hardware establishments here all doing a thriving business as follows: Chalfant, Graff & Co., R. Bryan, D. Goettcha, Seymour, Taylor & Co., Blakslee Bros. And G. W. Winsor.

*Large capacity iron oil tanks, similar to those at Fullerton.*

There are three drug stores here doing a good business.

Dave Duggan, once the favorite and popular night clerk at Tom Goodwin's Hotel in Titusville, is here engaged in pumping wells.

The Pickwick Club, of Petersburg, were indicted for riot at the last term of the Court of Quarter Sessions. The Court quashed the indictment last Friday on the ground that the prosecution was malicious and the charge without foundation.

A new Methodist Church has been built here and was dedicated last Sunday. The structure is sixty feet long by forty feet wide. The bell weighs 1,200 pounds. The total cost was $6,000 and the church is now entirely out of debt.

*Foxburg's main street, 1872. AVRR depot station.*

The German Lutheran Church, in process of construction, is now under cover. It will be dedicated about October 1st.

A grand banquet was given in the Church Grove of St. Petersburg last Saturday afternoon by the Ancient Order of United Workmen. Able speeches were delivered and a general good time was had. Music was furnished by the Orchestra Band of that place. After the feast the members of the Order, led by the band, marched to their respective halls and gathering dispersed. In the evening, the Bonner Hose and the Hook and Ladder Companies were out in full costume parading.

# Antwerp City, September 12, 1873
# The Effects of the Fire – Police Force Organized – Personal Notes – A Marriage – Picnic Season, &c., &c.

Since the excitement of the fire things have been rather tranquil, that is compared with former days. It is doubtful whether the town will recover from the calamity which has befallen her. The City Council, however, are seriously agitating the question of rebuilding the burnt district, and from the proceedings we are ready to expect almost anything. The Council spoken of has assembled twice, Thursday 4th, and on Monday evening last. At the adjournment of the last meeting this (Wednesday) evening was set for another conflux, but probably on account of a great many members leaving to enjoy the various occasions of today, little business will be transacted.

This Council was organized after the fire, and has done some very important work. It has now under consideration the propriety of a constitution for the territory of Antwerp, which will be read at next meeting. A police force has been organized and several others were elected for various offices in the town. The place has been divided into two wards, First and Second, respectively. Under the provisional government a committeeman from each ward has been appointed, who assume all the responsibility of aldermen of a city. The Council of this city does business after its own propriety.

Mrs. Mackey, formerly proprietress of the Antwerp City Hotel, which was destroyed by the late fire, is now keeping the Antwerp Pipe House in the first ward. The machine shop of J. D. Heart has been rebuilt, which is the only construction since the fire.

A report is current that a new well on Bower farm, a little northeast of Edenburg, owned by Lee & Balliet, is flowing a stream of full size of the casing.

A picnic was held in Chestnut Grove, on the Heeter farm near Turkey Run, yesterday by the Chestnut Ridge Sunday School and a wonderful event was enjoyed by all.

(Author's note: What they wrote about Antwerp not recovering reminds me of all the boiler explosions, etc., when the reporter would write that it is doubtful if the driller will survive the accident that has befallen him. Antwerp City would not survive the accident that had befallen her.)

*1872 Foxburg Savings Bank Draft.*

# News From The Front, September 23, 1873
# Foxburg, Pennsylvania

The grading on St. Petersburg street to the bridge is nearly finished, and a wooden culvert has been constructed to carry off the water.

The new school house was opened last Monday for pupils, under the able direction of Professor Gosdban and Miss Dees. It has been largely attended during the week.

The Fire Department of Foxburg are getting up a ball to come off October 9th. It promises to be quite a select and enjoyable affair and will be held in the Fox Hall.

McCray & Manderville, coal dealers, are doing a heavy business. They supply their customers from Pittsburgh and the mines along the Valley Road. They also deal largely in anthracite.

Porterfield & McCombs, dealers in hardware and dry goods, are also doing an extensive business in their line.

The Foxburg Savings Bank is doing a very large business, particularly for producers, pipe line companies and shippers. They occupy a fine brick building on the corner of Petersburg avenue and Railroad street.

The Mutual Pipe Company have a large and commodious building opposite the depot for their offices. Mr. Charles P. Hatch, General Manager of the Union & Mutual and Butler Pipe Companies, makes this his headquarters. Mr. A. W. Smiley is the Superintendent of the Mutual, and Mr. C. J. Eastwick has the general supervision of the books and accounts of the company.

Messrs. Joseph Bushnell and J. A. Bostwick have opened an office here. Mr. Joseph Seep attends to the affairs of Mr. Bostwick and Mr. Joe Battles to those of Mr. Bushnell. These two firms are about the heaviest purchasers of oil in this section.

C. C. Bone & Co. keep a variety store and news depot here, and they say that the call for the Titusville Herald is daily increasing.

The Berger Family opened the new Fox Hall last Wednesday evening to a large and appreciative audience.

The traffic over the bridge is daily increasing.

A lamp was tipped over in Jas. T. Fowler's furniture store last Monday evening, which set fire to the premises. The fire was extinguished with great difficulty by pails of water after destroying several chairs, tables, etc.

The health of the community is excellent, and no sickness of any kind prevails.

The Ancient Order of United Workmen have a very flourishing lodge here. They number about forty members.

Mr. C. C. Bone is the only law officer in the town, and he acts in the capacity of attorney and Notary Public.

Rowdyism is altogether unknown in this vicinity.

Thompson & Wood have a large and well stocked drug store here, doing a good business. Mr. Wood is a chemist and druggist of long experience and thoroughly understands his business.

Mr. George Day, who has been in the grocery business here for several years, has made enough money and retired.

The brick toll house at the east side of the bridge is covered in, and the lathing and plastering will be commenced at once.

## St. Petersburg

There are three livery stables in town doing a brisk business.

The German Reformed Church will be completed in about two weeks.

A new turnpike road is to be built between Petersburg and Emlenton; $2,500 has already been subscribed for the purpose.

The St. Petersburg Driving Park is nearly completed. It is situated at the west end of the town, and will, when finished, compare favorably with the best half mile track in this section.

There are several fine horses here, among which may be mentioned the teams of Mr. Joel Fink and Mr. Smith Cook's, two enterprising oil operators.

Blakslee & Brother's machine shop is doing a large business in drilling tools and general well repairing.

Mr. Barclay, of the *The Progress*, has had a very fine sign presented to him by the citizens.

There are two public schools here. They are attended by about 300 scholars.

The regular police force of this town consists of one stalwart individual named Michael Johnson in blue coat and gilt buttons. He looks as if he had belonged to the Broadway squad, and is a match for any ten ordinary men. He reports the town very quiet at present, the last disturbance being about two weeks ago when he arrested two drunks and disorderlies.

The population of this town is increasing and notwithstanding the low price of oil, business is quite brisk.

There are about twenty new buildings going up, located principally on the central streets.

The Fire Department is thoroughly organized and comprises sixty active members.

The Petersburg House has been open about two weeks and under the management of Mr. J. G. Black, late of the National Hotel, Franklin, has become deservedly popular and is crowded with guests.

The Methodist Church is filled to overflowing every Sunday. The Rev. J. F. Stokes is the pastor.

The road between Foxburg and St. Petersburg is kept in excellent condition and no better highway exists in the Lower River Districts.

Captain Byrom and O. C. Phillips started from Titusville in a buggy at 7 o'clock Friday morning, and passed through St. Petersburg on their way to the front at 12:30 p.m., beating the early morning train five minutes.

The Odd Fellows and Masonic halls are nearly completed.

The Berger Family performed at the Opera House last Monday and Tuesday before crowded audiences.

The resident population here is about 3,000.

The members of the Dime Society hold Sociables here every Wednesday evening at the houses of the subscribers.

## Reconnoitre of St. Petersburg Oil Fields, September 25, 1873

Monday morning, September 15, at 7:35, I left Titusville on an exploring expedition in search of oil lands belonging to F. H. Gibbs, situated on the East Sandy Creek, immediately

*Triangle City House, downtown Triangle City. 1873*
*Courtesy Drake Well Museum*

across the Clarion county line in Venango. To start with I purchased a guide, the *Morning Herald,* which I perused en route to Oil City, and with my cigar and Herald my short trip was anything but monotonous.

Reached Oil City and found Valley train waiting our arrival. A few moments and we are off; nor did I care for an insurance ticket, having satisfied myself that the fertile imagination of the *Derrick* man had manufactured a scare-crow out of his own passions, and having read that the biggest crop of accidents this season were on other roads. Arrived at Foxburg in safety at 12:30 p.m., and did not hesitate to say "good bye" to the Valley Road for a few days. Three miles ride in an old Pleasantville stage, and we stopped in front of the "Boss" Hotel, named after the town, Petersburg, where I was sumptuously entertained by the host. After dining, Mr. Guffey, who represents the Gibbs & Sterrett Manufacturing Company at this point, kindly placed himself and rig at my disposal to escort me to my destination, Triangle City, and of course I did not refuse his offer. I am indebted to Mr. Guffey for a pleasant ride and can well understand after a two hours' talk why he is a general favorite among all classes. Reached Triangle City at 5 p.m., five miles from Petersburg, passing along the belt which runs east of north thirty degrees, we see the Irving well, Logan farm wells, Milligan well, Turkey City and Jackson well, and at Triangle we are greeted by a familiar voice, which I recognize as J. D. Kiley's, an old friend from New York State, who enjoys a high reputation all through that district. Mr. Kiley informed me at once that I was in his custody and from that moment until I left him my every want was responded to. The next day Mr. Kiley and the undersigned started with our compass at an angle of thirty degree east of north, and when five miles on our route we notice on the right of the belt the Edenburg well, pumping 130 barrels. This encourages us in our advance, and we push forward through swamps, brush-heaps, and every old thing indicative of good oil territory.

As we progress the scene varies from the educated oil operators to the genuine "Buckwheats," who stare at a white man as a rare curiosity. They look at

*Ferry crossing the Allegheny River at Scrubgrass.*
*Venango County Historical Society*

a man as he passes from every window and corner, when invisible from these points they climb to the tops of houses, trees or any eminence to catch one more glimpse of the rarity before they return to their buckwheat fields to resume their cradling. We reach East Sandy and the county line at 9 p.m. and found that the property we were searching for some 400 acres, lay one-quarter of a mile west of our line. Whether the belt extends to this point is a question to be decided. We ran the line twelve miles from Triangle City and on this line are located the large wells from Parkers to Triangle, a distance of twelve miles. The territory in this section is lasting and the sand-rock very thick. We were well cared for by one of the natives, a German who thought we must be "very awfully tired" after such a walk. It is needless to say the old man lost money on our hash, which tasted "very awfully good."

All the land on the line is leased up to this point and a well will be commenced this month, one-quarter of a mile from Sandy, on the line. We returned to Triangle the next day, and in the evening we were highly entertained at the residence of Mr. Lavens of the firm.

## Letter From Antwerp, September 26, 1873

Our city is decidedly sluggish, probably on account of the low price of oil. Business of almost every kind, comparatively, has pretty nearly deserted us. The travel between St. Petersburg and the numerous little towns out front keeps a continual stir on the street, however, which in a degree gives things a lively aspect. But two groceries, one boot and shoe store, two first-class boarding houses, one news room, the officers of the Antwerp Pipe Company, Mr. Hulings, and the Western Union Telegraph Company, which, together with a number of residences, constitute our little village. Since the fire no despicable characters have infested our town; this is seldom true of all oil towns. A more pleasant and retired place to live in that Antwerp cannot be desired. This town is located between St. Petersburg and Turkey City. Mr. Hulings offers lots at extremely low rates which affords a good opportunity for homes for the approaching winter. A well owned by Dennis & Conley Bros, on the Delo farm, was tubed a few days since and is now pumping fifty barrels.

C. Blackman & Co. are finishing a well at the mouth of Deer Creek on the Clarion river. The above place is about fourteen miles due east of here. A four-barrel well was struck there some twelve years ago, but the business at that period was not properly carried on, and it is thought to be valuable territory which, if in course of development, it should prove such, would open a large field for operation.

The Lee & Balliet well on the Bower farm is doing not over 100 barrels. So says Mr. B., a part owner of said property.

The Berger Family exhibited at St. Petersburg on Monday and Tuesday evenings of this week. The house was copiously filled both evenings. Indeed those from the front were sufficient to overflow the hall. A first-class troupe always receives a liberal support at that place.

## Antwerp, October 14, 1873

Alike the unsteadiness of the present weather is the prosperity and progression of general business of the oil country, always affording something, whether good or bad, to comment upon. Men of all circles are complaining bitterly of the "tight times." Although it may seem strange, but not entirely so when we consider the "bears" and "giant monopoly" of the country whose influence sways business to its potent will, I say it may seem strange that every person has feelings that there is some unjust, base cause on the part of a few parties who are in reality sole agents of this "Giant Monopoly," which causes such a depression in business affairs at the present day. But this small job (though extensive in its effect) will, sooner or later, be uprooted, and "after the famine there shall be a feast," and our feast is coming, after these "bears" and monopolies shall have been exterminated from our midst. With the autumn there comes a cloud of gloomy aspect spreading over our contemplated plans and blighting our prospects for success. Why, within the past year there has been a frightful screen interwoven between the business of this country and prosperity, beyond which lies a secret, of what kind of character I will leave the answer to your selves.

Within the last week two more families have taken leave of Antwerp, besides two houses have been torn down. There are no manner of hopes for the maintenance of such places as this when oil and business is so downcast. The Antwerp Pipe Company intends to move their headquarters from Antwerp to Turkey Run City some time this week, which excites a general desire of "pulling up of stakes" among what few families who yet remain. As soon as the Antwerp Pipe Company discontinue their contribution to the Western Union Telegraph office here, there will be another removal, which will undoubtedly take place this week. One of the two remaining groceries will be moved to St. Petersburg shortly, and Mr. Lilly, the prominent boot and shoe man, who has tried hard to keep the business of the place together, talks some of moving his business to Petersburg.

Mrs. Mackey, the genial and hospitable old lady who formerly kept the Antwerp City Hotel, and who at present is furnishing the board for the Antwerp Pipe men, will start for Scrubgrass, whence she came, to again try her luck in the old town, after having lost her fine hotel and furniture here. May she have success, for she is one of peace and good will to every person. 'Squire Patterson has a very good business thus far, but will go with the rest of us. He will make Foxburg his place of business by next week, and I may say to the citizens of that place to sober up, for a man of justice and good rules will be among them. Our heartfelt wishes go with him and all the rest. Within one week Antwerp will be pretty well cleared out and only the marks and ruins of business will be left to tell the story of by-gone days. [Author's note: It looks like Antwerp City died today.]

A very disorderly rough stepped into a house in town last Saturday and grossly insulted the tenants, who were two ladies, whereupon one of them administered a blow with a poker, I guess, which floored the chap and somewhat sobered him too; at least he became sensible of the fact that he had no more business there then, and "took a little walk," but soon returned again with an armful of stones and showered them against the door. His bump of caution then warned him to fall back till the enemy should not be on the reconnoitre, when, about 12 o'clock, midnight, he returned the third time and broke in the door with rocks, since which time he has been quiet, probably on account of the execution of law.

## News From the Front, November 3, 1873
## Foxburg

This town is fast playing out.

The Fox House was closed to the public at noon, on Saturday. The cause is said to be a lack of customers. Transient visitors now pass through here to the more commodious and convenient hotels of St. Petersburg.

The Union Pipe Line headquarters had been moved to Parkers Landing, and many of the principal residents have also left.

The town narrowly escaped a fire last Thursday night. A train of tank cars was being filled at the pipe line station, and a spark from locomotive No. 44 ignited the oil in one of the tanks. The engine immediately hauled the car away from the others, the man-hole was closed and the fire finally extinguished.

Owing to the closing of the Fox House, a bachelors' hall has been started by thirteen of the young men doing business here. They have rented a house and engaged the services of a first-class cook. Their club room will be fitted up with taste and every comfort and convenience will be supplied. They expect to live less expensively and more independently in this way than is possible at a hotel.

Five Indians were employed on the river below here last week in making up rafts.

Mr. W. H. Fithian, of the Union Pipe Company, moved to Parkers on Friday.

Thompson & Wood's drug store is largely patronized and seems to be a general headquarters for business men in the evening.

## St. Petersburg

The new turnpike road between here and Edenburg is nearly completed. The distance is about nine miles.

Cronan, the billiardist, made a big sensation here lately. He went to Chicago to play a match game.

The road between Foxburg and St. Petersburg is in good condition, and it is quite a luxury to travel over it after experiencing the deep ruts and capacious mud holes which obstruct the roads in Butler county.

The building season here is nearly over. A few residences are, however, still going up.

The fire companies of the town are thoroughly organized and ready at a moments notice for every emergency.

Mr. George Sargent, formerly of Petroleum Centre, has fitted up a first-class restaurant here on the European plan.

There are only 100 drinking saloons in full blast in St. Petersburg at present. The proprietors complain that business is not quite so brisk as usual.

The driving park is now finished. It cost about $7,000 and is unquestionably a credit to the town.

The Turkey Valley House was recently purchased by Mr. John C. Gaughly. It will be run in future as a first-class hotel.

The Argument Court will be held at Clarion next Monday. Some important cases from St. Petersburg will be heard.

Blakslee Bros., proprietor of the large machine shops here, are still doing a very extensive business both in manufacturing and repairing.

Mr. W. W. Johnston, attorney-at-law, has moved his office to the building formerly occupied by L. Blufawitch. He is doing a large professional business and is considered one of the leading lawyers in the county.

There has been a subscription raised in town for the erection of a brick German Reformed Church.

The Rev. J. Stocker, formerly of Rouseville, has settled here as pastor of the new Methodist Episcopal Church.

The German Lutheran Church is nearly completed. It is a very fine architectural structure and will be ready for dedication next month.

The St. Petersburg House is the principal hotel here just now. J. C. Black & Sons, proprietors, keep it in first-class style.

The Courier House Minstrels played here on Friday night to a large and appreciative audience. They were heartily encored throughout the performance.

The constitution of the new Masonic lodge here last Wednesday was a very fine affair. A banquet was held in the evening at the St. Petersburg House, which was largely attended.

The production of the Clarion or Petersburg district, seems to hold its own at present better than almost any other portion of the Region. Although there has been an average decline among the old wells, there have been several very good strikes in the neighborhood of Dogtown, which has temporarily made up for the loss. The production of the district, which embraces the entire St. Petersburg side of the river to the Front and as far north as Emlenton, is estimated at 5,000 barrels per day. Development here as elsewhere has almost entirely ceased, and there are only a few more drilling wells to be heard from. The probable falling off in the winter months will unquestionably reduce the production very considerably.

## Stop Drilling!, November 18, 1873
## Who Will Lead Off?
## Three Dollars a Barrel or Nothing

We presented to our readers yesterday "a plain, unvarnished tale," of the deplorable condition to which the producing interest was at last reduced. We clearly showed how the confirmation of the recent discovery of the fourth sand had put an entirely new face on the state of things in this region, distressing as they were before. Whilst every one has been expecting

***$3 oil – good times for the producer***
***Venango County Historical Society***

and hoping for a decline in production at this season of the year, this new fourth sand monster is suddenly let loose upon the region, overpowering the strong and weak alike, upsetting all previous calculations, and causing at last the hopeless victims to cry out "what shall we do to be saved?" Stop drilling!

## Ruin or Salvation, November 19, 1873<br>The Producers Last and Only Chance<br>Petroleum Three Dollars or Nothing<br>Drilling Must Stop!

The condition of affairs in the Oil Regions has become so serious as to excite the gravest apprehension. Banks, railroads, municipal corporations, owners of refineries, machine shops, barrel factories, sympathizing with the chaotic and ruinous state of oil development, look with painful anxiety and dread at the immediate future before them. The mercantile interest also shares the general depression, and the skilled mechanic and common laborer, when he hears of the financial embarrassments which have pursued and overtaken the most substantial citizens and the most long-established institutions may well stand aghast at the prospect.

What has brought the Region to this deplorable condition which borders on complete paralysis? The answer is easy and simple. The market reports of the *Herald* have indicated from day to day the low point to which we were hastening. As the mercury falls to zero in mid-winter, so we have passed below the freezing point and are hastening with alarming precipitancy to zero.

For the past year we have all been hoping against hope, catching like drowning men at the most attenuated straws, predicting, wishing and thinking, declaring and contending that "the

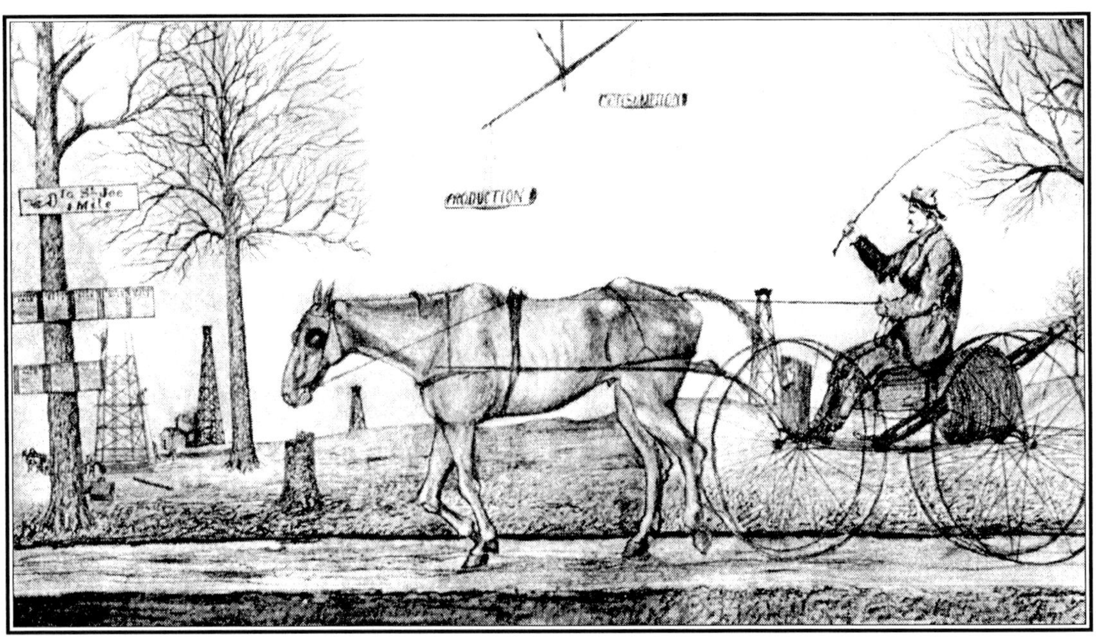


***70 cent oil – bad times for the producer***
***Venango County Historical Society***

bottom" had been reached, and as week after week, month after month, season after season passed away, bringing only disappointment and grim disaster, we have still pursued the phantom and hugged the delusion, and where are we today? Oil wavering between the miserable pittance, 37 cents per barrel at the wells, and the almost equally contemptible trifle of 50 cents.

We say, then, the drill must be stopped. This alone will prevent our ruin and ensure our salvation! It is the producers' last and only chance! Let every drill stop where it is till the first of April. The third or fourth sand rock is the best tankage for oil at less than three dollars a barrel. Nature will surely guard it for the short period of four months as safely as she has guarded it since the dawn of creation. Whilst in her keeping there can be no loss by fire, no shrinkage, no evaporation, no charge for storage, and there will be no risk in holding Nature's certificates.

## Progress of the Movement, November 20, 1873
## Three Dollars a Barrel or Nothing
## The Ball in Motion
## A Hearty Response

We yesterday received the following highly important dispatch from Parkers Landing:

To the Editor of the *Herald* – A public meeting will be held in the store of Shirley & Speer, Parker City, at 2 p.m., Friday, 22d inst., to take concerted action as far as possible, to avert the impending ruin that threatens the oil producers. All are earnestly called upon to attend. – Signed "Many Producers"

Every day that drilling is continued the Region is growing poorer and poorer. Thousands of people will be made penniless, and the mass of the producers made bankrupt, and for what

*Oil Exchange, Main Street, Parker City.*
*Courtesy Drake Well Museum*

purpose! Simply that half a dozen "shorts" may profit by the general misfortune of their fellow citizens. The South Improvement Company, which roused our people as one man, never threatened more disasters of the Region than the present suicidal mania for drilling for ten cent oil! There never was a more insane waste of wealth, even by the enginery of war, than the producer is daily assisting and consenting in contrary to all the analogies and examples of other producing and manufacturing interests in the length and breadth of the land, where suspension of operations is "the order of the day" until the markets of the world justify a resumption. Now or never! Three dollars oil or nothing!

## Stop Drilling!
## Ruin or Prosperity! Three Dollars a Barrel or Nothing
## The Parker's City Meeting Today
## The Grave and Vital Questions of the Hour
## November 21, 1873

The meeting to be held at Parker's Landing at 2 p.m. today will be one of the greatest interest and importance. It involves the prosperity of the whole Region. It comes home not only to the Producer, but to all other business classes. The world's real demand for our product probably does not exceed 30,000 to 33,000 barrels daily. This enormous product was taken off our hands because it was so cheap. It was not a consumptive but a speculative demand that called for it. Now the speculative demand has also ceased. Crude oil, as everybody knows, is worth little or nothing commercially speaking, till it is refined, but now the refiners can find no market for their product. Refined was sold yesterday in New York for 13 cents, and even at that price it goes begging. The European markets are glutted and the home market is in the same condition. Refiners are discharging their men and shutting down their works, their futures being nearly worked up, and the home market being overstocked and unreliable and prices continually on the decline.

***Burial site of Marcus Hulings and family, Grove Hill, Oil City***

A year ago, when we had a surplus production, we thought when we got top dollar, oil and evil would cure itself, but what are we to think now of the madness of the producer in continuing to drill with oil at fifty cents per barrel at the wells! If the entire oil business was conducted by an association on the same principles on which it is today, in contempt and violation of the law of supply and demand, the association would be bankrupt in thirty days. As the case stands, however, the producing business is carried on by men who come and go, are here today and gone tomorrow.

## The Effort To Stop Drilling The Fourth Sand Men Refuse November 22, 1873

One more effort has been made to prevent, if possible, the impending bankruptcy and ruin of the Oil Region by the threatened flood of fourth sand oil. At the urgent appeal and solicitation of loading operators we have for a few days past laid before the public the condition of affairs, and urged the only remedy which seemed calculated to meet the emergency, namely, the immediate stoppage of all drilling wells.

For this purpose a meeting was called at Parker's Landing yesterday, the proceedings of which will be found in another column; but those who had been looked upon as leaders of the movement found the opposition among the fourth sand operators to be insurmountable, and therefore concluded that the objects of the meeting could not be accomplished, and that it was all in vain to make inflammatory speeches and pass sounding resolutions.

*For Sale Cheap*
*Courtesy Drake Well Museum*

The fourth sand operators claim that they can afford to produce twenty-five cent oil, and that it would be foolish for them to stop drilling when they may be within a few feet of a 1,000 barrel well. The fourth sand men seem to have very little sympathy with the third sand men, and if there is an overruling Providence in the Oil Region, it is only a question of time when a race of fifth sand operators shall arise to drown out or use up the unmerciful fourth sand men.

Seriously speaking, although the Parker's meeting adjourned to meet again at Petrolia, it is perfectly plain that without the cooperation of the fourth sand operators, it is useless to make any further efforts in this direction. We, therefore, candidly advise the friends and promoters of this last movement to drop the subject at once, and forever. In sounding the alarm at the urgent and repeated solicitation of Butler county operators, we have discharged our duty, and whatever may be the result there, six or twelve months hence, no blame can be laid at our door.

## Hold Off on the Furs, November 29, 1873

We know of a dozen wives who read the "oil market" and "the news" from the front every morning in *the Herald* the first thing, as their husbands have promised them that "set of furs" when oil goes up to two dollars. Wretched sophistry of deceiving man! Well he knows that winter will be past, and the furs laid aside, before oil will touch that happy figure.

*The Old Engine House*

*The Old Engine*

*For Sale Cheep*

*These photographs show the results of unproductive wells and cheap oil. This is the story of an oil producer/driller who came upon hard times. Those who struggled to make ends meet or lost in their bid for wealth far outnumber those who became rich. Their tales are not as glamorous and are rarely told.*

*Some of the "For Sale Cheap" photos have been credited to John Mather. Obviously posed, these pictures create the dejection, the negotiations and the sale—sealed with a handshake. Courtesy Drake Well Museum*

## Attempted Suicide, November 29, 1873

St. Petersburg, Nov. 28 – About 10:30 this morning, Mr. J. W. Simpson, news and variety dealer, attempted suicide by shooting. The ball took effect in his forehead, and passing through and lodging in the back of his head. The attending physician thinks there is little chance of recovery. Financial embarrassment is supposed to have caused the rash act. *The Oil Field Record*

## Tough Times

Two weeks after the burning of Antwerp City, the country suffered through the panic of 1873 and plunged into a depression.

## The Garden of "Eden"

Mr. J. G. Mendenhall, in the year of our Lord 1830, traveled through the wilderness in search of a place to settle down and construct a home for his new bride. He and his wife purchased and cleared twenty acres of forest, where in the 1870's the oil boom town of Edenburg grew. The name "Eden" was chosen because of the beauty of the area and Mrs. Mendenhall compared it to the Bibles description of the Garden of Eden. As the small village grew, "burg" was added and thus the name Edenburg. The Mendenhall family bought two other farms near Edenburg in order to expand their productivity in the farming business. He named one farm Egypt and the other Jerusa-

*Main Street, Edenburg, Clarion County, 1877. The house on the left is the J. G. Mendenhall Mansion. All was lost in the terrible fire of October 13, 1878. Venango County Historical Society, Franklin, Pennsylvania. John J. McLaurin's Personal Notebook*

lem. When oil was discovered both farms proved excellent oil territory and brought the family much wealth. In the early days of Eden, mail was carried on horseback from Emlenton Penna. to Shippenville. When a Post office was opened, Mr. Mendenhall was appointed the first postmaster.

## The Emlenton and Clarion Railroad

As new wells were drilled and the oil belt was being pushed onto the Elk City fields another railroad was being surveyed. In January, 1877, the first railroad constructed was the narrow gage called the Emlenton and Clarion Railroad and was hailed with much "Joy and Gladness" prior to its entry into Edenburg. Mr. R. W. Smith of the Lehigh Valley was the first ticket agent and general manager of the railroad. Mr. Smith said, "We here in Edenburg will feel awful big as a town, a great deal larger than St. Petersburg or Clarion, and about fifteen-sixteenths as big as Pittsburg."

*Main Street, Edenburg, between 1876 and the fire of January 13, 1877. Looking North on Main Street from the corner of Penn. The large building at the left was the Bennett House. Courtesy of Barbara L. Hajel*

## Another Railroad to Edenburg

On December 2, 1876, a meeting was held at the Fox House, in Foxburg, Penna., by the friends of the proposed narrow gage railroad from Foxburg to Edenburg. Many enterprising gentlemen gathered to bring together the necessary capital required to build the line. A correspondent of the *Edenburg*

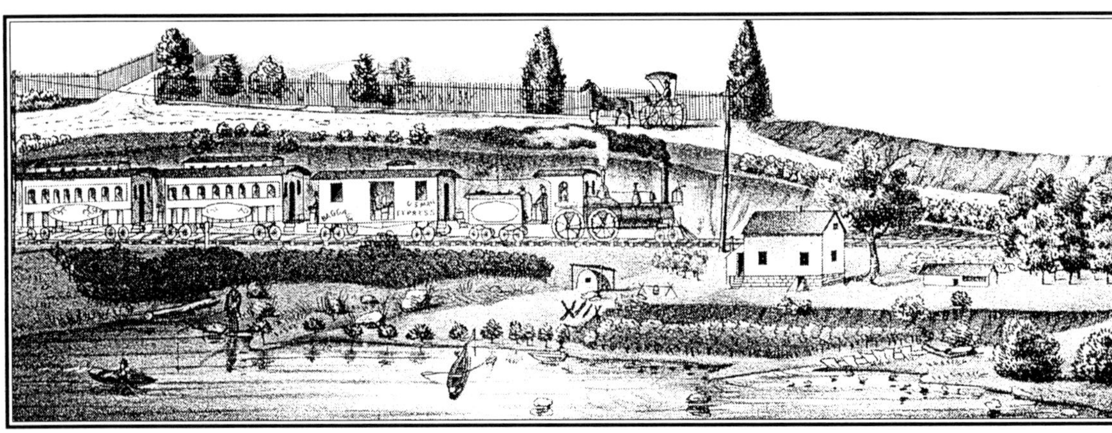

*Foxburg, St. Petersburg, and Clarion Rail Road on its way to Edenburg*

*The Bennett House*
*Courtesy of Barbara L. Hajel*

*Daily Herald* furnished the following list of persons present at the meeting and who are interested in the enterprise: Barney Vensel, James M. Guffey, J. B. McElwaine, J. Ashbaugh, Charles Vensel, C. King, A. Myers, J. Akin, J. V. Ritts, G. W. Butt, Charles Blakeslee, Captain Chalfant, A. W. Smiley, W. C. Mobley, Superintendent of the Parker and Karns City Railroad, Wm. Fox, Squire Fullerton, Capt. McCray, J. Mandeville and S. G. Bayne.

The Foxburg, St. Petersburg and Clarion Railroad was formed with William Fox, president; J. M. Guffey, vice president; W. S. Watson, secretary and J. V. Ritts, treasurer. This new line, along with the Emlenton and Clarion R. R. carried all types of freight, machinery and passengers throughout the Clarion County oil fields and tied the oil towns together from Foxburg to Elk City. These two rail

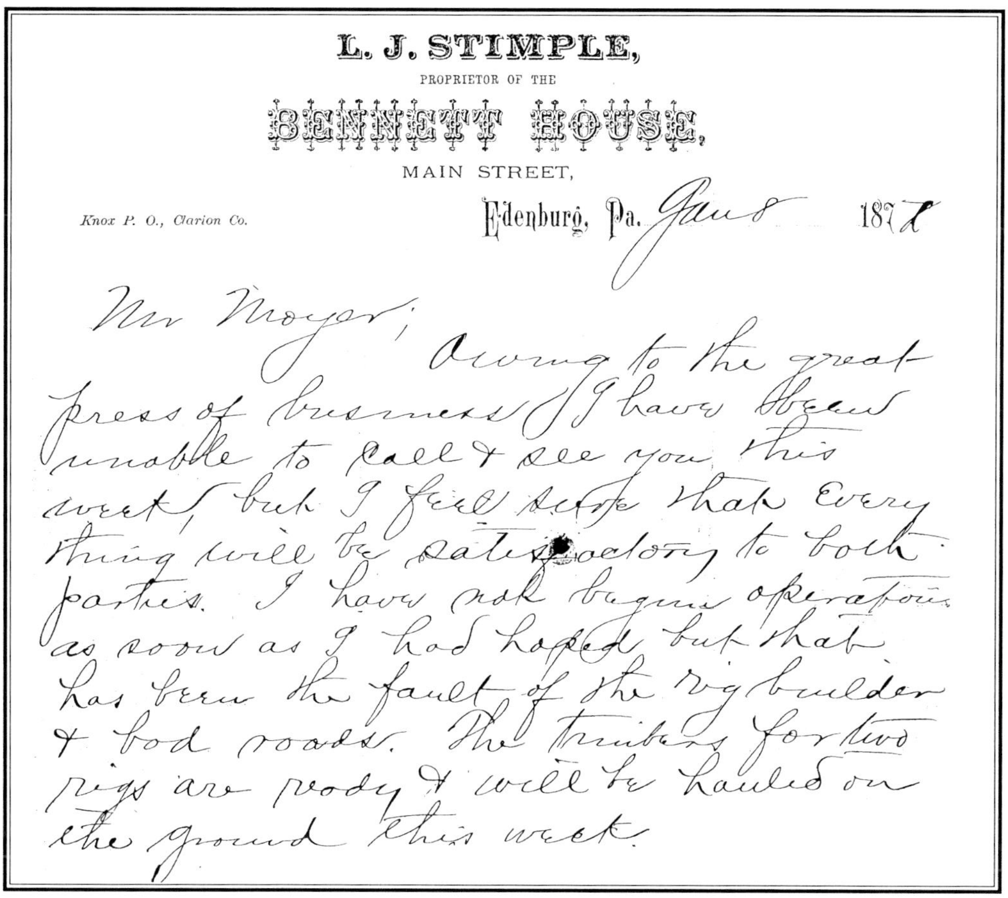

L. J. STIMPLE,
PROPRIETOR OF THE
BENNETT HOUSE,
MAIN STREET,

*Knox P. O., Clarion Co.* Edenburg, Pa. Jan 8 187[illegible]

Mr Moyer;
Owing to the great press of business I have been unable to call & see you this week, but I feel sure that Every thing will be satisfactory to both parties. I have not begun operations as soon as I had hoped, but that has been the fault of the rig builder & bad roads. The timbers for two rigs are ready & will be hauled on the ground this week.

*This note to Mr. Moyer from C. W. Van Flick. "Not begun operations; rig not up and the roads bad." Courtesy of Barbara L. Hajel*

*Main Street 1876-1877 Edenburg*

*Edenburg, 1877. Photograph shows State Street looking East toward Elk City and Shippenville. Old Catholic church on the left at top of hill with the Best residence to the right. Stereoscopic view, Frank Robbins, Photographer, Oil City, Penna. Clarion County Historical Society*

lines came together and connected at Jefferson Station then continued on as one line to Edenburg. Soon the cry was "On to Clarion!"

***1877 Clarion County Atlas***

## The Boom

Edenburg remained a quiet small crossroads village until the "Wildcatters" pushed the search for oil on ahead, following the "oil belt". Edenburg was right in the middle of the oil. Bearing sands and the boom was about to begin. The oil boom began and the population grew to more than two thousand by 1877, with more people pouring in each day. At the height of the excitement there were seventeen hotels and boarding houses, law offices, two banking establishments, two iron works, real estate offices, a bull wheel factory, countless drinking houses, boiler shops, and wagon factories. As the older fields slowed down and production started to fail, people moved on. Speculators, contractors, drillers, tradesmen, laborers and workers of all kinds came to Edenburg. Southwest of Edenburg, Beaver City sprang from a farmers field and the oil flowed. Many productive wells were drilled and soon Beaver City had two doctors, lawyers, a post office, five hotels, two lumber yards, machine shops, livery stables, many stores and shops and last but not least a church. Shady City was located east of Edenburg and the children of Edenburg went to school there until a school was finally built in Edenburg in 1877. Mongtown was the site of the famous "St. Lawrence" well and at Forest Home the Antwerp Pipeline Pump Station was located. Edenburg was in the autumn of 1876 organized and chartered as the Borough of Edenburg. A town council was organized, a mayor elected and a police force formed. Again "Riff Raff" drifted in and caused considerable problems in the new Borough. A report on fines paid to the Justice of the Peace for one month in 1878 shows fines paid in the amount of one thousand eight hundred dollars.

***1877 Clarion County Atlas***

## Newspapers

Edenburg was like many other oil towns in the Upper and Lower fields. Newspaper men were attracted to the excitement of the boom. Edenburg has been fortunate to have had many ambitious newspapermen with more then ten papers being published since 1876. Most were short-lived with the first being the *Edenburg Daily Herald.* It was published by J. M. Gifford and is considered Clarion County's first daily newspaper. The first issue hit the frozen streets on December 26, 1876, and proclaimed in bold print that its motto was "Independent in Politics, Neutral in Nothing." The daily circulation for the *Herald* was large and the paper was read throughout the Clarion County belt. Poor Mr. Gifford had the misfortune of having his press room and shop torn down during almost every fire that ravaged Edenburg. This was done to hopefully stop the further spread of the flames. It seems poor Mr. Gifford's shop was in the middle of every fire and the act of tearing down his press room numerous times was successful in stopping the fires but disastrous to his printing business. Poor old Mr. Gifford finally died at the early age of 33 on August 9, 1880, of consumption.

## Edenburg Fires

The Borough of Edenburg has been ravaged by an incredible number of disastrous fires. Between the years 1877 and 1880 the town was burned down six times. Each time the Borough was rebuilt and the citizens carried on. One or two fires were discouraging, but six was almost enough to cause the townspeople to pack up their bags and call it quits.

The first fire happened on January 13, 1877, involving two of the most important streets in Edenburg, Main Street and State Street. These streets were comprised of buildings built close to each other and considered the business district of the oil town. The flames were first discovered at eight o'clock in the evening and started in a gambling den on State Street. The townspeople set up a bucket brigade but the flames were soon out of control and destroyed the entire north side of State Street and moved north on Main Street for several blocks. The loss was set at $50,000 with twenty-two buildings destroyed. Rebuilding began on the morning of January 14, 1877, on still hot coals.

## Corbitt Boarding House

The horrible cry of "Fire" again rang out on June 19, 1878. The fire started in the Corbitt Boarding House on Pennsylvania Avenue, a residential sector of Edenburg. The bucket brigades were again called upon to fight the consuming flames. The winds were blowing from the northeast and the citizens' efforts were useless. When the fire was finally brought under control and the gusty winds had ceased, over fifteen buildings and residences with a loss of $20,000 were lost. After the fire, Miss Carrie Corbitt, daughter of the proprietress of the Corbitt House was the person suspected. Due to a lack of evidence and Miss Corbitt's unstable state of mind, charges were dropped.

*In the 1870's when Knox was known as Edenburg, the fire department was known as the Chandler Ladder Company, which at that time was commanded by Anson Beatty. The equipment they had was a hand-drawn, four wheel ladder cart containing ladders, leather fire buckets, axes and pike pole and was built by Mr. McElhatten, a local Edenburg resident. Later when a water system was developed in the town, a hand drawn hose cart was added.*

*Edenburg Fire, October 13, 1878. View looking down State Street toward Elk City. The Kribbs store is on the right. Old man Gear and his son are standing on the boardwalk at the left. The new Catholic Church is in the background. This great fire destroyed over 170 buildings and was the worst single fire Oildom had ever suffered. (Courtesy of Barbara L. Hajel, Edenburg Pennsylvania)*

## Arson Suspected

In October of 1878, it was finally discovered that the fires were being set deliberately in hopes of again burning the town down. Three or four fires had been set at different locations throughout town but all were discovered in their early stages and extinguished before they were out of control. Now in order to protect the town from these unknown fire bugs, a volunteer watch of one hundred forty men was organized to patrol the streets and protect the Borough. Six man shifts were formed to patrol the streets and stop the fires. Fire carts were purchased, a water works was built and a volunteer fire company was formed. Everyone thought that now the Borough was safe.

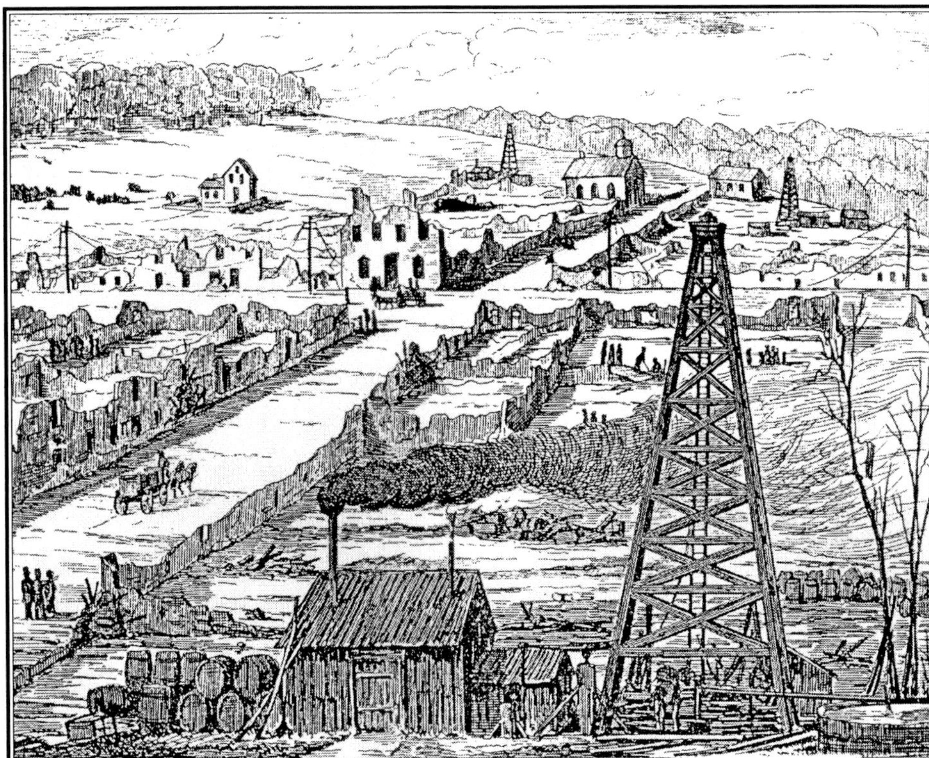

"I OFTEN BUSIED MY PENCIL MAKING SKETCHES OF THE RUINS"

*S. G. Bayne's sketch of Edenburg's great fire of October 13, 1878. View of State Street.*

## Oildom's Most Destructive Fire

Despite this valiant attempt to stop the fires and protect the town, the third and most destructive fire occurred on October 13, 1878. The fire and flames were discovered in the rear of Wilbur's Livery Stable on Main Street. When the fire company arrived they found the building engulfed in flames, and it was discovered the fire had been started with crude oil. The firemen still felt the flames could be brought under control. The fire hoses were quickly hooked up and to their horror, all the hoses had been cut with a knife and no hose could carry the precious water. The arsonist who had set the fire had done his job well. He not only started the fire where it would do the most damage, but he also sabotaged the means of controlling the fire.

*Edenburg Fire, October 13, 1878*
*Courtesy of Barbara L Hajel*

*Edenburg Fire – View of the great Edenburg fire, October 13, 1878. Photographed by Harry Corell, Jefferson City, Penna. This photograph was taken from the top of the Kribbs building looking North along Main Street towards Mongtown. Over 30 acres of buildings burned to the ground. This stereoscopic view was for sale at Little's Jewelry Houses, Edenburg, St. Petersburg, and Petrolia. Courtesy of Barbara L. Hajel, Edenburg, Penna.*

The fire was now completely out of control, flames leaped from shop to shop, home to home, and all the citizens could do was watch in horror. Flames swept over State Street and northward onto Main, ending with the total destruction of Railroad and Ohio Streets. When the smoke cleared, it was discovered that over one hundred seventy buildings were gone with a loss of $400,000, making this fire in Edenburg, Clarion County, the most devastating fire ever known to Oildom.

## Six Months Later

Six months later, fire once again raised its ugly head. On April 19, 1879, someone had poured crude oil on the floor of a vacant building that had been spared in the previous fire. The flames roared down Main Street and for the fourth time, Edenburg was on fire. Over ten buildings were destroyed with a loss at $15,000. No one was caught and again the town began to rebuild.

## Fire destroys the United States Hotel

On May 22, 1880, the townspeople were attending a circus just outside the Borough limits. As the fire grew higher and higher in the vacated United States Hotel, the cry of "Fire" was heard again. The flames were soon visible through the canvas of the tent. People panicked and the circus tent was torn to pieces in the rush. This fire of May 22, 1880, swept away seventy more buildings which included banks, offices, hotels, stores and the Post Office. As oil was starting to play out, many people gave up in despair and left.

## Fire Number Six

The sixth fire broke out on August 23, 1880, in the Livery Stable on Main Street. Again it was started with crude oil and quickly spread to other structures nearby. Englets Bakery was quickly torn down. Salt and wet blankets were used to cover the roofs of endangered rooftops nearby. The firemen did a noble job and these measures helped minimized the destruction.

*8,000 barrels per day near Bear Creek*

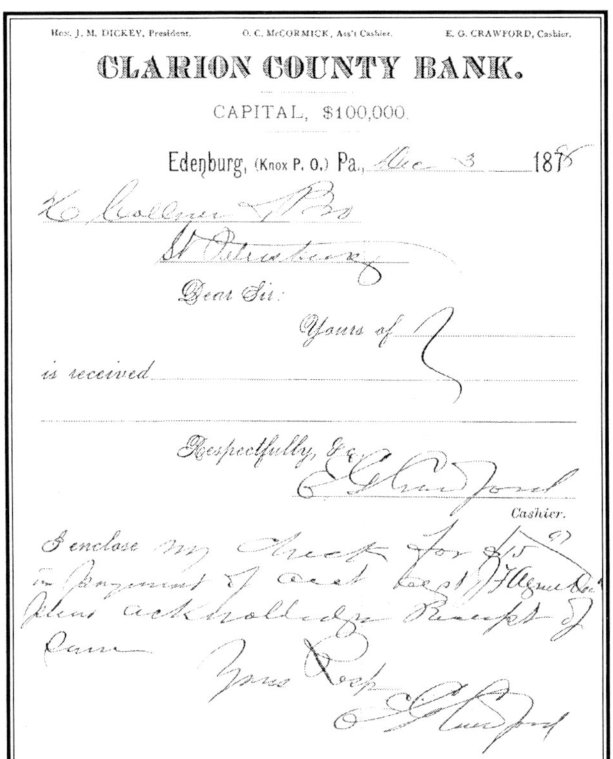

Hon. J. M. DICKEY, President. O. C. McCORMICK, Ass't Cashier. E. G. CRAWFORD, Cashier.

CLARION COUNTY BANK.

CAPITAL, $100,000.

Edenburg, (Knox P. O.) Pa., Dec 3 187[illegible]

[illegible]

Dear Sir:

Yours of

is received

Respectfully, &c.

E G Crawford

Cashier.

I enclose my check for $15 in payment of acct [illegible] Please acknowledge receipt of same

Yours Resp

E G Crawford

*Thanks to Barbara L. Hajel*

This sixth and final fire most certainly dampened the spirits of many people, who lost hope of rebuilding Edenburg. The "Oil Boom" was slowly dying down in the oil territories surrounding the town and newer, richer fields were being discovered elsewhere. Many disillusioned and dispirited families left Edenburg for good. Thankfully many strong individuals remained and Edenburg rebuilt, grew and prospered. Today Knox, as it is called, has weathered many storms and is a small community with many fine citizens.

## The Clarion County Bank

In 1877 the Clarion County Bank was started on central Main Street by Messrs. J. M. Dickey, Eccles and E. G. Crawford. This bank was a reliable institution and did a fine business. One incident worthy of mention occurred on October 13, 1878. At this time, with the oil excitement, there were very many new and suspicious persons in Edenburg. As the terrible fire raced down Main Street toward the bank, Mr. Orie J. Blakely, then assistant cashier, put all the money and papers of the bank into a sack. He was accompanied by only one other person. These two men carried the sacks full of money, gold, and papers through the crowd to Mr. Blakely's home on the Bower farm, and thus saved the bank. *Compendium of Edenburg.*

# Advertising Business Directory
## *(1877 Clarion County Atlas)*
## Edenburg

Allison, W. B., Pumper
Austin, John, Attorney-at-Law, State Street
Austin, Herron & Co., Merchants
Atwell, R. J., Justice of the Peace
Almis, G. W., Laborer
Arnce, Robert, Laborer
Austin, D. W., Clerk
Abbott, S. D., Agent
Anderson, A. P., Carpenter
Allison & Co., Hardware Dealers
Alden, A. A.
Abbott & Denslow
Abbott, Geo, Hotel Keeper
Amo, Hugh
Berlin, J. M., Dealer in Dry Goods, Ladies' and Gents' Furnishing Goods, Millinery Goods, and also a variety of Hats, Caps, Boots and Shoes, all of which we offer to the public at greatly reduced prices. Main Street.
Brothers, J. M., of Caning & Brothers, Dealers in Real Estate and Oil Interest
Brown, M., Clothing Merchant
Bush, John, Tool Dresser
Brown, N. W., Tank Builder. Shop and Residence on State Street
Beardsley & Herron, E. C. Beardsley & D. S. Herron, complete stock of hardware and oil well supplies
Bryan, Robert, Dealer in all kinds hardware, tin, sheet iron, tubing, casing, ropes, big irons, belting, stoves, sash, doors, &c., &c. All sizes of casing, tubing and gas pipe cut by steam power. State Street, next door to Edenburg House
Botsford, Frank, Contractor for drilling oil wells
Best, John I., Farmer
Barter, J. S., Laborer
Beatty, J. H., Laborer
Burshfield, J. S., Laborer
Brown, L. C., Laborer
Bell, Charles, Merchant
Brown, Jas. W., Carpenter
Beers, H. L., Agent
Barton, Will T., Laborer
Beatty, G. W., Laborer
Burk, Oran, Laborer
Burk, Frank, Laborer
Brown, W. A., Printer
Burkholder, Joseph, Blacksmith
Bunnell, G. I., Jeweler
Beatty, Wm., Carpenter
Burns, Daniel, Machinist
Burns, Foust & Co.
Bradford, Wm. H., Liveryman
Bole, J. & F. H., Machinist
Bunnell, Benjamin, Oil Operator
Brown, Charles, Rig Builder
Bates, Wm., Feed Store
Burnsley, Samuel, Laborer
Best, Daniel, Laborer
Bolan, Charles, Laborer
Burchfield, J. S., Laborer
Brown, C. E.
Berger Bros.
Brumley & Cunningham
Burgess, H. R., Druggist
Bombach, Jacob, Laborer
Bash, John, Laborer
Babcock, M. H., Carpenter
Boyd, John, Laborer
Boles, Thomas J., Musician
Brown, Frost & Co.
Byson, C., Laborer
Bailway, E. & S.
Bush, Barney, Laborer
Brucker, J., Merchant
Borland, A. J., Laborer
Broocks, W. C., Laborer
Bolan, John, Laborer
Beatty, W. H., Laborer
Boyd, C. W., Laborer
Bates, Wm. M., Merchant

Bell, J. W., Laborer
Bell, David, Laborer
Brown, W. S., Merchant
Borland, H. A., Clerk
Crawford, E. G., Cashier Clarion County Bank, Edenburg, Penna. Capital $100,000. Does a general banking business
Cantherdy, C. F., Builder and Contractor. All orders attended to promptly and on short notice. P. O. Box 701.
Cary, M., Oil Producer
Caning, P., Real Estate Agent
Crawford, T. J., Postmaster
Cook, Robert, Pumper
Cook, I. W., Livery, Feed and Sale Stable, on Main Street
Carpenter, S. L., Gauger
Crawford, A. C.
Crawford, J. A., Molder
Cauthard, Charles, Carpenter
Cedwick, Emanuel, Shoemaker
Coast, W. M., Contractor
Campbell, M. L., Laborer

*View of the Great Edenburg Fire, October 13, 1878, photographed by Harry Corell, Jefferson City, Penna. This photograph shows the vault of the Clarion County Bank and that of a pipe line office. Main Street about opposite side of the street from Penn Avenue looking northwest. Courtesy of Barbara L. Hajel, Edenburg, Penna.*

Connelley, Geo. W., Banker
Cricks, C. C., Laborer
Coney, D. C., Laborer
Crawford, M. A.
Cribbs & Berlin
Clark, G. A., Laborer
Carroll, F. J., Laborer
Chambers, W.
Chain & Fink
Clearwater, Wm., Laborer
Campbell, Robert, Boarding House
Cribbs, James, Carpenter
Chambers, Wesley
Coldren, Abe, Oil Producr
Clark, D. G., Blacksmith
Campbell, H., Laborer
Castell & Kinnear
Castle, Henry, Blacksmith
Cristy, John
Carl, J. F., Laborer
Canthard, Charles
Curtis & McCutcheon, Teamsters
Cage, Charles
Carothers, Jas. A., Contractor
Campbell, E., Agent
Crusan & Hill
Chalfant & Graff
Corcoran, Pat., Laborer
Cole, James, Carpenter
Carrel & Kinnear, Blacksmiths and Machinists; Manufacturers of oil well tools. Repairing done to order on short notice and at reasonable rates. State Street
Carrel, Henry, Blacksmith and Machinist
Dean, J. P., Cabinet Maker
Dyer, A. N., Blacksmith
Denning, Harvey, Engineer
Douglas, W. A., Lumber Dealer
Drew, James, Hotel Keeper
Denslow, F., Hotel Keeper
Dickson, J. J., Carpenter
Dana, L. W.
Defiance & Wry
Davis, W. C. B., Oil Producer
Donahue, G. R., Laborer
Duning, O.

Davis, A. R., Blacksmith
Dewoody, G. S., Laborer
Davis, E. E., Laborer
Downey, James, Laborer
Dutton, W. B., Liveryman
Dana, Lawrence, Stage Driver
Dougherty, Wm., Laborer
Drummond, David, Monitor House
Effendi, H. B., Physician and Surgeon
Englert, F., Proprietor of Star German Baker; Manufacturer of bread, cakes, pies, &c. Confectionery, groceries, fruits, nuts, crackers and canned fruits are kept constantly on hand. State Street
Eichbaum, G. R., Laborer
Elston, J. S., Laborer
Elliott, John
Finch, W. R., Photographer
Fink, A., Proprietor of the Arcade Dining Hall. Meals at all hours. State Street
Fisher, Peter
Farthing, J. N., Oil Operator
Fleming, P. A., Laborer
Fulton, G. W. & Co.
Frost, Andrew, Laborer
Fretz, J. S., Laborer
Flynn & Hamm, Merchants
Forker, J. L., Carpenter
Forker, Wm., Carpenter
Frankenstein, S., Tailor
Fertig, Charles, Laborer
Forker, Perry
Gealy, J. T., Attorney-at-Law
George, J. M., Oil Operator
Gibbs, C. L. & Guffey, J. M., Agents for Sterrett Manufacturing Co.
Gibbs, Sam, Shoe Store
Gibson, J. B., Merchant
Graham, Edward, Blacksmith Helper
Greene, W. H., Contractor
Gehn, H. C., Pumper
Gehn, N. D., Pumper
Gilchrist, James
Gehr, Harvey, Laborer
Galsher, John
Goodyear, Geo., Manufacturer of and Dealer in boots and shoes. Repairing done on short notice. Shop on State Street, next door to Bryan's Hardware Store
Gifford, Frank, Oil Operator
Gifford, J. M., Editor of Edenburg Herald
Gourley & Kunkle
Groce, John
Gibbs, W. W., Tinner
Grimm, Charles
Goheen, H. N., Laborer
Glass, John, Laborer
Gosnell, J. R., Laborer
Galsher, John, Laborer
Grand, C. N. H., Laborer
Gross & Happe
Gehr, Nelson D., Laborer
Gibbs, W. W. & Co., Hardware
Hazelton, W. L., Driller
Higgins, A. P., Operator
Heasly, C., Teamster
Hochschwender, G. F., Restaurant, State Street
Hyde, J. C., Agent and Operator
Holden, G. A., Contractor of Oil Wells
Hirsch & Co., L.
Hardenburg, H. A., Proprietor of livery, sale and feed stables; carriages, buggies, &c. for hire. Particular attention given to boarding horses. Located on State Street, near railroad crossing
Higgins, L. L., Hotel
Hamilton, W. F., Carpenter
Hoyt, W., Laborer
Henderson, L. D., Agent
Hicks, Frank, Teamster
Herren, D. S.
Hill, W. K., Laborer
Hataling, John, Agent
Hamm, G. H., Clerk
Herche, Wm. M.
Hope & Wheeler
Hill, S. R., Laborer
Hollace, Daniel, Laborer
Hosack, J. S., Laborer
Hardenberg & Dutton

Hall & Langfitt
Hill, John
Hart, H. E., Carpenter
Hamilton, Richard, Laborer
Hays & Harvy
Hotaling & Howard
Herpet, H. H.
Hendren, M. C., Book Agent
Higgins, D. A., Blacksmith
Hess, E. F., Agent
Hess, M. E., Oil producer
Hidden, E. M.
Hughes, Richard
Jack, A. H., Shoemaker
Johnson, J. S., Laborer
Johnson, Henry, Sport
Johnson & Burkholder
Johnson, L. M.
John, Wm., Shoemaker
Johnson, E., Blacksmith
Johnson, John, Dealer in Real Estate
Kern, E. D., Architect, Builder and Operator
Kirkpatrick, W. M., Dealer in all kinds of Groceries
Kernick, E. M., Pastor of M. E. Church
Kinnear, P. V., Blacksmith
Kribbs, P. F., Dealer in dry goods, groceries, clothing, notions, hats, caps, boots, shoes, drugs, patent medicines, hardware, etc
Kooken, D. W.
Kilgore, Thomas, Shoemaker
Kehr, Geider & Co.
Kearney, E. & Thomas
Keller, C. D., Clerk
Kribbs, P. F., Blacksmith
King, Frank
Kinsel & Worth
Kern, E. L., Carpenter
Keeland, J. M., Saloon
Knight & Winger
Kinch, E. S., Merchant
Kerr, James, Landlord
King, Francis, Blacksmith. All kinds of working tools made to order. Blacksmithing in all its branches attended to. Repairing a specialty
Kelley, John, Carpenter
Love, Michae, Oil Operator
Long, J. J., Contractor and Builder. Plans and specifications furnished on application. Address, Freeport, Armstrong Co., Penna.
Lamb, S. F., Boarding House
Lyman, John, Carpenter
Luther, J. H.
Lewell, Livingston & Co.
Lupher, L. D.
Lavid, W. M.
Lovjoy, I. P., Agent
Lovjoy & Aikin
Lieble, Anthony, Clerk
Lambert, Lewis, Carpenter
Lusher & Mays
Lohr, S. G., Carpenter
Long, J. M., Grocer
Litz, Jefferson, Physician and Surgeon. Office and residence, State Street
Long, E. H., Hotel Keeper
Leedom & Patterson, Oil Operators
Moore, Bro. & Co., Dentists at St. Petersburg, Edenburg and Clarion. Preservation of the natural teeth a specialty
Millspaugh, D. T., Agent Pipe Line, Edenburg, Clarion County, Penna.
Mendenhall, G. S., Dealer in pure fresh drugs, patent medicines and ladies' toilet articles. Prescriptions carefully compounded. Opposite City Hotel
Miller, D., Oil Dealer
Markle, D. J., Real Estate Broker and dealer in oil wells and leases
Martin, Theo., Oil Producer
Magee, A. H., House and Sign Painter. All orders promptly filled at low rates
Moore, O. G., Dentist
Morrison, M. G., Lumber Dealer
Miller, J. H., Oil Producer
McGee, S. L., Attorney-at-Law
Mattren, Joel, Teamster
Mason, A. L., Coal Operator
Morrison, R. A., Rig Builder and General Contractor
McGuire, J. G., Machinist

Myers, J. C.
Miler, C. H., Blacksmith
McFadden, M., Oil Producer
Mendenhall, Mrs. J. G., Lady
Marr, W. C.
Moulton, R. M., Clerk
Moran, P.
Murray, Wm., Carpenter
Monroe, R. & W., Machinists
McQuewan & Douglas, Lumber Dealers
Monroe, Wm., Carpenter
Movil & Co., Machinists
Morrison, J. C., Laborer
McLaughlin & Klingensmith
Morgan, J. J., Laborer
McNaughton, G. W., Clerk
McMillen, D. M., Carpenter
McNamara, Michael, Restaurant
Moulton, B. M.
Morgan, R. J.
Miller, C. H., Blacksmith
McGuin, J. G., Laborer
Mullen, W., Laborer
Millspaugh, D. T., Clerk
McBride, C. E., Carpenter
Martin, Wm., Laborer
McPherren, J. A., Laborer
Morrison, J. H., Carpenter
McPherren, H. R., Insurance Agent
McKenna & Kent, Machinists
McLaughlin, Thos., Carpenter
McMannis, Richardson & Co.
Morrison, S. P., Agent
Mays, M. S., Laborer
Murphy, John, Laborer
Marwood, H., Merchant
Mendenhall, W. L., Carpenter
McCoy, Charles, Carpenter
McNutt, C. H.
Mendenhall, Hevis
McQuistion, J. P., Merchant
Martin, Thos., Oil Producer
Miller, Alex., Laborer
McCormick, M. M.
Mays, W. S., Boarding House
McCalley, John, Laborer
Marr, Wm., Agent
Munhall, Wm., Oil Producer
Mimm, John, Laborer
McCaslin, J.
McGren Bros.
McGee & Clain, Exhange Hotel
Nolan, Wm., Laborer
Neath, John, Laborer
Nessler, M. G., Laborer
Newkirk, C. H., Oil Producer
Noble, A. H.
Nebing & Welsch
O'Neil, Frank, Carpenter
Oles, L. K., Carpenter
Oaks, David, Carpenter
O'Haif, Wm., Laborer
O'Dell, J. P., Laborer
Orr, Robert, Landlord
Orr & Bryan, Landlords
Otterstatter, John, Bookkeeper
Oppenheinner, B., Merchant
Opennheimer, A., Dealer in Dry Goods
Orr Bros.
Place, J. M., Books, Stationery and News Dealer at St. Petersburg, Edenburg and Emlenton
Peterson, E. T., Cabinet Maker
Porterfield, L. C., W. U. Telegraph Operator
Powell, W. W., Physician
Proudfit, M. B., Laborer
Preston, James, Laborer
Page, A. K.
Pollock, W. H., Baker
Price, Charles, Painter
Potter, G. W., Oil Producer
Palm, J. A., Teacher
Palm, Thos. W., Laborer
Pulley, Henry, Carpenter
Phillips, M. F., Laborer and Pumper
Perrington, M. D., Druggist
Phillips, W. B., Clerk
Place, M. J., Clerk
Painter, J. B., Oil Producer
Platt, Wm., Laborer
Phillips, John W., Druggist
Pachen, Warrant, Tank Builder

Pierson & Dean
Peters, John C., Laborer
Powers, Wm. J., Laborer
Producers Torpedo Co., Headquarters at Franklin, Penna., F. L. Seeley, Secretary. Branch Offices: Edenburg, Elk City, Bradford and Petrolia, Penna., E. E. Swett, General Manager, Knox, P. O., Penna.
Phillips, E. H., Pumper
Putnam, H. D., Proprietor of Livery, sale and feed stable Carriages, buggies and horses for hire. Particular attention paid to boarding horses. Stable opposite depot and rear of Bennett House
Quigley, J. H., Laborer
Quinn, Geo.
Richardson, R. R., Oil Producer
Rutherford, Thos. B., Oil Producer
Rittenhouse, A., Surveyor, Gauger and Real Estate Agent
Rollens, Edward, Stonemason
Rath, John, Laborer
Reidinger, Jacob, Barber
Ross & Lay, Ross House
Ross, W. S., Hotel Keeper
Reiff, D., Baker
Reynolds, Samuel
Ryan, Thomas, Laborer
Russell, F. C., Carpenter
Reiff, J., Baker
Robinson, A. J.
Racy, Charles W., Clerk
Richardson, John
Richardson, Thos., Blacksmith
Reese, S. Contractor, Builder and Oil Producer
Selby, W. A. & J. F., Attorneys-at-Law
Sheley, C. L., Restaurant and Boarding House
Shaw, Charles E., Livery, Sale and Feed Stable
Spear, L., Carriage Shop. Wagons and carriages made to order. Repairing done neatly.
Smith, T. C., Contractor and Builder. Plans and specifications furnished on application.
Scanlan, James, Clerk at Monitor House
Siverling, A. B., Clerk at Fulton Market
Servey, D. J., Carpenter and Joiner
Springer, John, Elk City
Springer, J. S., Feed Store
Smith, W. J., Laborer
Shaw, J. P., Carpenter
Stevens, E. A., Carpenter
Shafer, Henry, Carpenter
Shultz, John, Laborer
Smith, Lafayette, Laborer
Still, Joseph, Carpenter
Smith, Jos. S., Gentleman
Swarts, W. L.
Smith, A. J., Machinist
Sherman, J. M., Carpenter
Sweeny, T. J.
Shannon, Sylvester
Stetler, H. L., Hotel
Scott, R., Clerk
Sweet, E. E., Agent
Shuster, F., Printer
Stuart, R., Laborer
Sweeny, J. D., Laborer
Smith, G. W., Laborer
Smith, Geo., Laborer
Smith, Daniel, Laborer
Swift, Charles, Laborer
Stroup, Edward, Laborer
Swift, J. J., Laborer
Stants, E. M., Laborer
Sheever, Wm., Laborer
Scott, F. G., Contable
Shargo, Peter, Laborer
Strubb, Henry, Carpenter
Sloan, A. A., Carpenter
Shell, Valentine, Laborer
Sackett, E., Clerk
Scanlan, James, Clerk
Siverling, A. B., Clerk
Spark, L. B., Baker
Signor, T. A., Carpenter
Shannon, S. A., Laborer
Stobell E., Laborer
Selby Bros.
Slagle, Valentine & Co.
Smith & Signor

Simons, B., Tobacoonist
**[The author likes this title.]**
Stuart, A. A., Merchant
Shan, C. E., Carpenter
Selby, W. A., Lawyer
Smith, J. M., Laborer
Smith, John, Machinist
Sheridan, James, Blacksmith
Showles, J. A., Saloon Keeper
Smith, S. C.
Steiner, Samuel, Merchant
Shaner, J. L., Jeweler
Swett, M. L., Agent
Swartz, Wm. Q., Laborer
Speer, Lute, Blacksmith
Stout, J. A.
Scott, J. C., Carpenter
Shanafelt, Louis, Landlord
Travers, T., Merchant
Templeton, James, Rig Building, Repairing, &c.
Thomas, G. H., Stage Machinist
Thomas, Alfred, Clerk
Turner, John
Tibbit, C. B., Laborer
Turner & Kuhn
Turner, F. M., Landlord
Travis, L. R., Oil Producer
Tucker, G. H., Surveyor and Notary Public
Thompson, Jos., Laborer
Temple, James, Tailor
Timlin, E. A., Grocer
Turner, W. B., Clerk
Tibbitts, C. H., Hotel Keeper
Thompson, J. Hill, Printer
Underwood, G. S., Oil Producer
Urey, H. F., Druggist
Van Vleck, G. H., Oil Producer
Vender & Son, Mrs. J., Dealers in Dry Goods, groceries, queensware, notions, &c, Uc. Main Street, next to Petroleum Hotel
Vedder, Jacob, Merchant
Volger and Harvey, Laborers
Vanhorn, W. B., Blacksmith
Vanhorn, I. T., Laborer
Vogt, Otto J., Clerk
Whitehill R., Carriage Shop. All kinds of vehicles made to order and repairing done neatly. Main St.
Wright, Wm., Coal Dealer
Winter, F. C., Grocery and Feed Store
Wynkoop, D. J.
Warnock, R. A., Proprietor of Warnock House. Persons will find in this house all the com forts of home. Good stabling connected with the house. State St.
Wickman, J. P., Oil Operator
Wildman, F. J., Wildman's Opera House, Edenburg, ten miles from Emlenton on the E. & S. R. R. The best and most completely furnished Opera House in the Oil Region. Situated on the ground floor; seating capacity, 800; large and convenient dressing rooms; house, 40 x 112; stage 35 x 40; procenium opening, 22 feet wide, 16 feet high; elegant practical scenery on flats, and all the modern improvements; star combinations and first-class attractions.
F. J. Wildman, sole proprietor. Address: Lock Box No. 10, Edenburg, Clarion County, Penn'a.
Wilson, F. O., Atlantic Pipe Operator
Willoughby, A. C., Oil Operator
Wilson, R. H., Carpenter
Winslow, C., Hotel Keeper
Williams, S. H., Laborer
Walker, Joseph, Laborer
Walster, Wm., Butcher
White, Frank, Laborer
Wiles, A. W., Carpenter
Wilson, James, Machinist
Whitmer, Wm., Lumber Dealer
Walton, John, Superintendent
Wood, Alex., Laborer
Winslow, J. S., Laborer
West, J. H., Shoemaker
Wiffong & Gerner
Wolfe, J. L.
Wilson, E. S.
Wallace & Harris
Wolfe, J. D., Hardware Dealer
Wion, James, Laborer

Wink, Henry, Laborer
Wyant, Paul, Clerk
Walston Bros.
Winter, F. C., Laborer
Wyncoop, D. J., Oil Producer
Wyncoop, Underwood & Davis, Oil Producers
Worthrope, E. W., Physician
Wagner, F. C., Laborer
Wildman, J. J., Artist
Warmingham, John, Jeweler
Wilber, J. H., Lumberman
Ward, A. E., Laborer
Willett, F. A., Laborer
Wilson, W. M., Saloon Keeper
Williams, G. A.
Wilson, W. N.
Wheelock, C. A., Butcher
Wenner, Jacob, Laborer
Walsh & Comey
Williams, J. M., Carpenter
Whitehill, Robert, Wagon Maker
Williams, Michael, Blacksmith
Young, John
Zuver & Brumby
Zuver, J. E., Clerk
Zuver, T. S., Lawyer

*Edenburg, October 14, 1878 – This stereoscopic view was taken the day after the fire. This view is looking down Main Street and was taken by photographer Harry Corell of Jefferson City. The clock in front of P. F. Kribbs reads 12:20.*
*Thanks to Barbara L. Hajel*
*Edenburg, Pennsylvania*

*No. 106 drilling for oil. Stereoscopic views of the Pennsylvania Oil Region, Frank Robbins. This photograph came from John J. McLaurin's personal notebook, Venango County Historical Society, Franklin, Pennsylvania*

## Emlenton Shippenville and Clarion Railroad

This photograph shows the trestle near Jefferson Furnace, a narrow gauge which began to operate in 1877. A few months later a narrow gauge was built from Foxburg to Edenburg. This connected with the Emlenton, Shippenville and Clarion Railroad at Jefferson Station. Later the track from Emlenton to Jefferson Station was abandoned. This railroad changed hands to the Pittsburgh and Western and finally the Baltimore and Ohio. Today only the winding path through the countryside exists.

***E. and S. Railroad trestle at Jefferson.***
***Photograph courtesy of the Knox Centennial Booklet, 1876-1976.***

## The Triangle City Field

John H. Lee and Frederick C. Plumer, both from Franklin, were partners in the Clarion oil fields. The firm of Lee and Plumer put down a test well on the Hummel farm near Edenburg in 1872. This successful operation opened and expanded the outer limits of the ever-growing Clarion belt. This gave rise to the "oil boom" towns of Pickwick (named by Plumer after the famous Pickwick Club of St. Petersburg), and Triangle City (named after the three-cornered plot it was located on). The distance between the two is about one mile. Their wells on the Hummel farm netted handsome returns.

John Lee was in the oil business 41 years, having started in 1860. He died of tuberculosis in 1901 at Franklin. Fred Plumer died at Franklin in 1879 and in the newspaper it stated, "Fred was a thorough man of affairs, prompt, courteous, affable and popular. His long sickness was bore cheerfully and he faced the end, he died at thirty-one years without repining." Before his death he left the oil business and bought a hardware store at New Castle, Pennsylvania. "Peace to his ashes," John J. McLaurin. (*Sketches in Crude Oil*, 1896)

*John Lee – Mr. Lee got his start in the oil business in 1860 along Oil Creek. Photograph from John J. McLaurin's personal notebook, Venango County Historical Society, Franklin, Pennsylvania.*

*Emlenton and Shippenville Railroad trestle near Jefferson furnace. Today, the student of oil history would be hard pressed to find clues as to the location of these historic sites. Brush and trees hide these silent locations.*

*Fred Plumer – Their firm, Lee and Plumer, was a major operator in the Clarion fields. Venango County Historical Soicety, Franklin, Pennsylvania.*

# Salem Township
# Clarion County

Salem Township, Clarion County, on the northern side of the belt, was tested but did not give a good showing. In 1872, Marcus Hulings and Company drilled a test well on the Delo farm in the eastern part of the township. The "Delo" well came in at 40 barrels, which created an extensive but short-lived excitement. As the testing continued, a number of wells were struck on the Exley, Hummel and Knappenberger farms. The best territory was leased by Lee and Plumer and was around the vicinity of the Hummel farm. Two wells were drilled on the Troutner farm but were mismanaged and then abandoned. About this time Salem Township was considered not profitable and the center of drilling moved back into Beaver Township.

*Salem Township, Clarion County – 1877 Clarion County Atlas*

*Union Pipe Line 1876. Edenburg Pump Station housed the large steam engine and the pumps used to force oil through the lines.*

## Beaver Township
## Clarion County

Beaver Township lies in the central part of Clarion County. Its greatest length from east to west is about nine miles and its greatest width is about seven miles. In Beaver Township, the oil excitement ran high for about six years starting in 1872 and hitting its peak in 1877.

In 1872 the first wells to strike oil were on the George Berlin, George Texter and Elias Ritts farms. The oil found there is from the third sand, or green oil. The total number of wells drilled in Beaver Township is unknown, but estimates run from 800 to 1000 wells. Many wells were good producers and created much wealth while others were "dusters." James M. Guffy, of St. Petersburg, and J. T. Jones, Wesley Chambers, and other operators kept branching out until this field was producing 10,000 barrels a day. In order to move the oil to the railroads and then to the refinery, pump stations were built by The United Pipe Line Company. Pump stations were at Forest Home, Shamburg, Jefferson City and Blair's Corners. Their network of iron pipe lines was extensive.

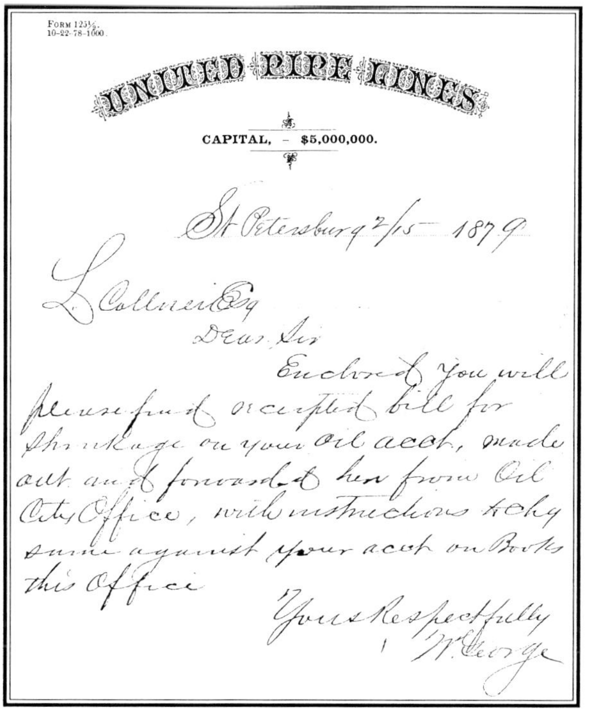

FORM 125½.
10-22-78-1000.

UNITED PIPE LINES

CAPITAL, - $5,000,000.

St Petersburg 2/15 1879

L Collner Esq

Dear Sir

Enclosed you will please find receipted bill for Shrinkage on your Oil acct, made out and forwarded here from Oil City Office, with instructions to chg same against your acct on Books this Office

Yours Respectfully
W. George

*United Pipe Line St. Petersburg Penna., Feb. 15, 1879*

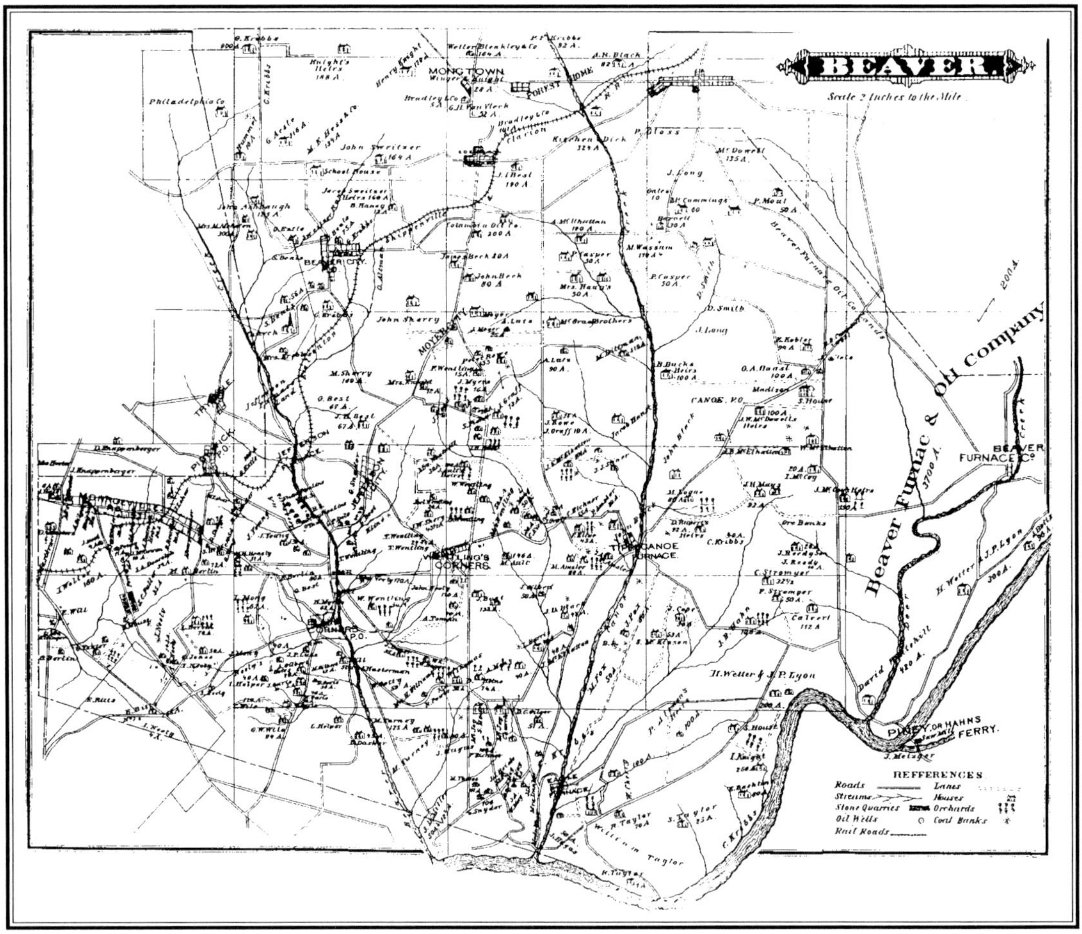


*Beaver Township, Clarion County*
*1877 Clarion Co. Atlas*

Other pioneer oilmen in the Beaver Township fields were Hahn, Kribbs, Frederick Prentice, H. Baliett, Fertig, Bayne, Plumer, Gray Brothers, Brundred, S. M. Kier, Turner, H. Wetter, Kelly, Jacob Black, George Berlin, and others. Major Henry Wetter, a man of honor and energy, was the largest operator in the Clarion district until swamped by the low price of oil. Death overtook him while struggling against heavy odds to recuperate his health and fortune. A terrible decline in the price of oil caused his failure in 1877 and compelled Merrick and Conley's Edenburg bank to close.

## A Wildcatter's Wildcatter
## John Loved the Chase

John H. Galey, 1840-1918, Pioneer in petroleum and metal mining. He discovered the Mid Continent Oil Field at both ends. In 1892 he completed the first commercial well at Neodesha, Kansas, the north end of the great oil province, and in 1901 he developed the great gusher at Spindletop, near Beaumont, Texas, discovering the south end. In 1880 he mined for silver in Arizona when the historic towns of Charleston, Paradise and Galeyville sprang into existence, the latter named for him. John with his father and family are listed in the 1870 census for Perry Township, Clarion County. His occupation is listed as "Oil Speculator" and his age is listed as 30. (Drake Well Museum, Titusville, Pennsylvania.)

*John H. Galey*
*Courtesy Drake Well Museum*

## "Mis-Sir" Buckwheat

A young lady from New York accompanied her father to the greasy, muddy oil fields of Clarion County. Her father was to meet John Galey to finalize a business deal. The carriage pulled up to the Triangle City hotel. The young lady had heard the stories of Mr. Galey's great oil deals, his bank account, and his wealthy lifestyle. Standing on the front porch

*Maple Shade Well in the Pleasantville Field, 1868*
*John H. Galey's first successfully drilled well. John is 3rd from left.*
*Courtesy Drake Well Museum.*

was a gorgeous chap. He sported a stunning suit, a huge gold chain, a diamond pin and polished boots. No expense was spared in his "get up."

The young girl excitedly exclaimed, "Papa! Papa! There is Mr. Galey!" Her father immediately looked up and recognized him as a neighboring bartender and spoiled gin slinger. "No, no Mary, that's not Mr. Galey."

Many of the "upper crust" from Titusville and other civilized communities looked down upon the old farmers of Clarion County as poverty stricken and slow. The "high class" outsiders called them "Buckwheats."

Just then, a half mile below the hotel, a man on horseback was seen approaching. The rider and his panting steed were covered in mud, the rider's clothes were tattered looking, and his hat was backwards on his head. The young lady, anxious to show how much she knew about the oil country, hazarded another guess. "Oh, Papa, Papa!" she said earnestly, "I'm sure that's a Buckwheat!" The father burst out in a hearty laugh as he greeted the unknown rider. "Mary, you are right," he said, "This is my friend and partner, Mr. Buckwheat Galey!" A good laugh followed the father's version of the day's incidents. How did you like that one, Mr. Galey?

## Pickwick Business Directory
## *1877 Clarion County Atlas*

Crum, G. W., Carpenter, Contractor and Tank Builder
Crody, Levi, Farmer
Campbell, G. W., Contractor and Oil Producer
Gallager, John, Shoemaker
Richard, J. L., Pumper
Setelle, I. H., Superintendent of Oil Business

*God Bless this Great Nation and all her Oilmen*

## Triangle Business Directory
### *1877 Clarion County Atlas*

Arnold, S., Carpenter and Builder
Bailey, John, Pumper
Coulter, J. E., Merchant
Coulter, W. S., Hardware and Oil Fixtures
Cronester, W., Teamster
Ginter, S., Drayman and Teamster
Harley, W. B., Merchant
Harley Brothers, Producers
Kelley, John, Driller
Muers, S. J., Merchant
McCamant, S. M., Manufacturer of wagons and buggies. Blacksmithing and repairing done to order. Horse shoeing a specialty.
Parsons, S. W., Proprietor of Triangle Hotel
Stole, J. H., Producer
Williams, Dock, Tool Dresser

## September 1, 1874

The market is weak and quiet, few sales reported. In the Clarion, Venango and Butler fields, oil is selling for between 95 cents to one dollar twelve cents per barrel.

## September 3, 1874

Sad news has reached the citizens of Edenburg. Tillman Jack, while at work on the "Jack well" near Turkey Run City was caught in the engine and crushed to death. Mr. Jack stumbled over a spool of rope and fell head first into the machinery, hurling him into eternity.

## Blair's Corners Directory
### *1877 Clarion County Atlas*

Blair, R. F., Proprietor of Blair's Hotel. This house is widely known and stands as a first-class house. Bills low. Good stabling also for horses.
Blair, M. D., Robert A., Physician and Surgeon. Office, Blair's Hotel.
Frederick, D. E., Proprietor of hotel. Accommodations good; satisfaction guaranteed.
George, A. H., Manufacturer of and Dealer in harness such as bridles, saddles, whips, coarse and fine harness. All work warranted.
Lutz, J. P., Merchant

## New Pipe Line
## Oil City, September 29, 1874

The new pipe line of the American Transfer Company from Turkey Run City to Oil City has been completed and successful, having but one pump station in route. James H. Snow was the superintendent in charge of construction. (Author's note: see page 297)

## The Angel of Death
## Edenburg June 17, 1875

Edenburg mourns the death of a well-known citizen. Young Heckerthorn was working on the J. D. Wolf oil well on Main Street. Work had been progressing at a rapid rate when about 3:00 the bull wheel burst. Mr. Heckerthorn, the driller, was killed instantly. A large crowd gathered to witness the horrible scene. The body was carried to his home and prepared for burial.

## President U. S. Grant
## August 17, 1875

*Parker City Daily*, Parker City. President Ulysses S. Grant and his party passed down the valley road yesterday en route for Pittsburg and from thence on to Washington. Returning from a trip through the oil region, many of the local citizens turned out. As the special train passed Foxburg and the station opposite Parker City, the people ran out waving their handkerchiefs and were likewise waved in return.

## The Gushing Damsel

"What are your favorite books?" the Gushing Damsel cried.
"Bank books and pocket books," the Oilman quickly replied.
(*Sketches in Crude Oil*, John McLaurin, 1896)

## Boiler Explosion
## August 22, 1875

The angel of death has again visited Edenburg. On August 22, 1875, Augustus Wilson and Alonzo Goss died in a boiler explosion. Mr. Wilson died instantly and Alonzo Goss died four hours later. In 1871, Alonzo Goss and Tip Johnston drilled the first well in the borough of St. Petersburg near the racetrack. Mr. Goss has worked the Petersburg, Edenburg fields for the past four years. The accident happened on the J. I. Best farm. The boiler came from Petroleum Centre. Much of the old equipment out of Venango County is very well worn and in need of replacement. It is believed that age played a large roll in the explosion. The community of Edenburg and the families mourn the loss of these two citizens. Augustus Wilson was laid to rest in the Union Cemetery at St. Petersburg, Penna.

## His Ruination

This small handout was printed in 1874 by the American Sunday School Union, 1122 Chestnut Street, Philadelphia. It was handed out in Sunday Schools throughout the Clarion oil region. The oil fields sometimes brought out the worst and many a young lad should have "taken heed":

## The Grocery Ruined Him

What little boy is that standing there in the grocery?

It is little Robert. His father was a poor, miserable drunkard. When he was a little boy, people liked him, and thought he would be good. He went to Sabbath-school, and his teacher thought much of him. As he saw many bad people drinking rum, his teacher was afraid he would learn to drink too. This little boy used to go to the grocery to buy things for his mother. His teacher told him that the bad men about the grocery would try to make him drink; and that, when his mother sent him there, he should just get the thing he wanted, and go right out, without stopping to speak to the men who were sitting there. He promised to do as he told him; and said he would not touch a drop if they should ask him a hundred times. But one day, after getting what he wanted, he stood sauntering about the store, to hear the men tell foolish stories. After a while, the men began to try to make him drink. He did not mean to drink at first, but he should have gone right out, and there would have been the end of it. But he still lingered; and it was not long before they persuaded him to drink. After he had taken one glass, he wanted more, and he kept on drinking till he became drunk.

After a while, the teacher learned what had happened, and went to see him. He burst into tears, and appeared very sorry.

But it was not a great while before the men at the grocery persuaded him to drink again. His teacher told him that he would certainly be ruined, if he did not stop drinking entirely, and keep away from the grocery. But he did not feel so sorry as he did at first. For the more people sin, the more hardened they grow. It was not long before he got drunk again. He began to be irregular at school. He grew worse and worse. He left the school altogether, and kept out of sight of his teacher. At the age of thirteen years he became a drunkard.

*The Grocery Ruined Him*

Boys! keep clear of all places where intoxicating liquors are made, or used, or sold.

**Happy the child, whose youngest years**
**Receive instruction well,**
**Who hates the sinner's path, and fears**
**The road that leads to hell.**

## Giblet City
## September 8, 1875

*Parker City Daily* – Turkey Run City has a most widely read newspaper. The "Gem" of the Petersburg field now has the *Gas Light Weekly* with a circulation of over 600 copies. We at the Daily wish the owners much luck.

## Pennsylvania Transportation Company

In order to move the over-abundance of petroleum from the Western Pennsylvania Oil Fields, Henry Harley of Titusville, Pennsylvania, and other investors from both Western Pennsylvania and the Eastern Seaboard, proposed a oil pipeline. The cost of transporting oil to the Seaboard by a pipe line would be covered by a 16 and 2/3 cents per barrel fee, while the lowest possible cost of movement by rail in full trains would be 40 cents.

In 1876 the Pennsylvania Transportation Company was formed with Henry Harley, Esq. elected President. The following letters containing pledges of support come from Western Pennsylvania Oil Pipe Line Companies. A pamphlet was printed and circulated explaining the: Extension of Pipe Lines to the Tide Water; The Financial Advantages; Opinions of Counsel; and Cost Estimates.

PIPE LINES PLEDGED TO CONTRIBUTE BUSINESS TO THE PENNSYLVANIA TRANSPORTATION COMPANY.

*Antwerp and Oil City Pipe Line.*

St. Petersburg, Pa., April 25, 1876.

HENRY HARLEY, ESQ.,

*President Pennsylvania Transportation Co.*

DEAR SIR:—The *Antwerp and Oil City Pipe Companies* are pleased to learn that you are making arrangements to construct a large pipe line to the seaboard. Our lines are now running from forty to forty-five thousand barrels per month. Our associates in business are also producing a considerable portion of this oil. We also have a large amount of Territory developed and undeveloped, lying directly on the oil belt, in Clarion County, and we pledge you our hearty co operation, giving your enterprise the preference over the transportation lines by rail. We will also unite with you all our business on a fair and equitable basis, but will require of you such protection as will be reasonable and just. Yours truly,

(Signed) H. M. HUGHES,

*Sec'y. and Supt. A. & O. C. P. Co's.*

*Relief Pipe Line.*

New York, April 25, 1876.

H. HARLEY, ESQ.,

DEAR SIR:—We have seen your project of a pipe line to the sea-board, and consider it practicable in every respect. We have no hesitancy in saying we should be pleased to see the line built, and will give you a hearty support in delivering you oil from our line at such times as we can. Feeling that it has advantages over railroad transportation, and also in sympathy with the producers of the oil regions generally. *We are with you*, and you may count on our support and co-operation. Yours truly,

(Signed) F. PRENTICE, *Pres't.*

S. A. WHEELER, *Vice-Pres't.*

*Relief Pipe Line Co.*

*Atlantic Pipe Line.*

New York, April 25, 1876.

MR. HENRY HARLEY,

*President Pennsylvania Transportation Co.*

DEAR SIR:—We understand that you are about to construct a pipe line from the Pennsylvania oil regions to the sea-board. We are engaged in producing oil in Clarion and Butler Counties, and in transporting oil by pipe line, are now handling about 2,500 bbls. per day. We have long since believed that a pipe line to the sea-board was not only practicable, but just the thing wanted, in fact the only proper method of transporting crude oil. We therefore heartily endorse your measure, and pledge you our aid and patronage. Respectfully yours,

(Signed) M. HULINGS, *Prest.*
J. T. JONES, *Manager.*

*Grant Pipe Line.*

Parker City, Pa., April 17, 1876.

We, the Officers of the *Grant Pipe Company*, take pleasure in endorsing the project of Mr. Harley, of laying a pipe to the sea-board, and will do everything in our power to aid and encourage its construction. We will be glad to become a feeder to it, and will at all times give it the preference over railroads at the same rates.

(Signed) D. S. DUNCAN, *Pres't.*
T. McCONNELL, JR., *Sup't.*

***This information comes from the Pennsylvania Historical and Museum Commission, Drake Well Museum, Titusville, Pennsylvania***

## Elk City, April 15, 1876

On Thursday last, a man by the name of Mason was arrested at Ben Hogan's Saloon and boarding house. Mason, who is from Triangle City, caused a terrible fight and shots were fired through a door. The shots were fired through the portal into a room of pleasure, just missing Hogan's manager and partner, French Kate. Ben Hogan and Kate have followed the oil fields from Pithole through Elk City. Mason was placed in the jail at Clarion. *Clarion Democrat*

## Hogan's Palace

Hogan's most widely known enterprise was his "Floating Palace" at Parker's Landing. A drawing of the "Palace" is on page 111 but a photograph on page 3 is believed to be a true photograph of Hogan's Palace. This converted barge was 140 feet long, 25 feet wide and was fitted up in a most elaborate and gorgeous manner. Ben Hogan's "Floating Palace" contained 40 state rooms, a stage with a full string and brass band, and a large ballroom for dancing. The bar on board contained the choicest beers, wines and liquors, and the best cigars were sold.

*Passing Time*
*This drawing comes from the book, "The Life and Adventures of Ben Hogan."*

Public balls were given three nights a week and the main theme was "Wine, Women and Song." The "Palace" was anchored in the middle of the Allegheny River between Clarion County and Armstrong County. This way Ben could avoid legal difficulties. Two boats were kept running from the shore to the "Palace," and on these came the Merchants and Oil Kings, Bankers and Brokers, Drillers and Thieves.

If an oil man spent one hundred dollars on champagne and then dropped another one hundred dollars on "Women and Song", nobody grumbled.

The only time local officials ever attempted to board the "Palace" was foiled when the boats mysteriously capsized and the lawmen were last seen floating rapidly with the current towards West Monterey.

Unfortunately, as we all know, "All Good Things Must End." The great "Floating Palace" ran against a snag near Pittsburg, Pennsylvania, and was sunk, thus ending its infamous career in the Oil Country.

Parker City at the time of the oil boom was "Red Hot". Its inhabitants became accustomed to such exclamations as, "I've lost my ear!", "I've lost my nose!", "I've lost my pocket-book!" The Last Statement was the Most Common of All!

Ben Hogan's wild career bounty jumping, his ring and gambling career. (1877, Bradford Penna. *Evening Star*)

## Step Right Up!
## August 1, 1876 Edenburg

C. M. Clawson was killed yesterday while drawing a torpedo from a well near Edenburg. The torpedo accidentally exploded and blew off his head.

No sooner is one torpedo agent blown up than another steps in to take his place. It is a dangerous business, but men will always risk their lives for money.

## Exley Farm
## September 4, 1876 Jefferson City

J. C. Clark, while pulling casing, was instantly killed by an accident at the McNulty Bros. well. The well is on the Exley farm near Turkey Run City. The body was taken to the family home and readied for burial.

## Elk City
## December 22, 1876

*Clarion Democrat* – Elk City is becoming one of the liveliest towns of the oil regions. Building is at a lively pace, with new businesses arriving each day.

## Bad Well

The good well is the humble bee
With honey on its wings;
The dry hole is the bumble bee
That buzzes loud and stings.
(*Sketches in Crude Oil*, John McLaurin)

## Jefferson City Directory
### *1877 Clarion County Atlas*

Bachtell, I
Yates, Miss Caroline, Proprietress of the Jefferson Hotel
Scott & Clark, Livery, Sale and Feed stables. First-class turn-outs at reasonable rates.
Coulter, W. L., General Dealer in hardware, tubing, casing and cordage and oil well supplies; stores at Triangle and Jefferson City
Texter, James K.
Gallagher, John, Boot and Shoemaker
Thompson, S., Tank Builder
Tibbett, C. H., Proprietor of Tibbett House
Yates, William, Lumberman
McCracken, G. H., Hardware
Rogers, I. C., Hardware
Wint, P.
Cristy, T.
Snyder, J. G., Farmer, 57 acres

***Oil Field Crew – Back row, left to right: John L. Cribbs and Billy Thomas. Front row: John Cooper, Dave Hess and John Hess. Note the cigars and the bottle. These guys were a tough bunch.***

## Broke
## Titusville, November 1, 1876

Well-known oilman, Jonathon Watson, fails. Watson, one of the early operators in the oil fields has had his ups and downs. It has been reported his liabilities exceed $100,000. (See page 10) *Clarion Democrat*, Clarion Penna.

## Elk City
## February 14, 1877

There will be a meeting tonight to form a temperance society. Visitors to this region have noted the absence of such an organization and, as one has said, "no place on earth is more in need of one." *Titusville Morning Herald*

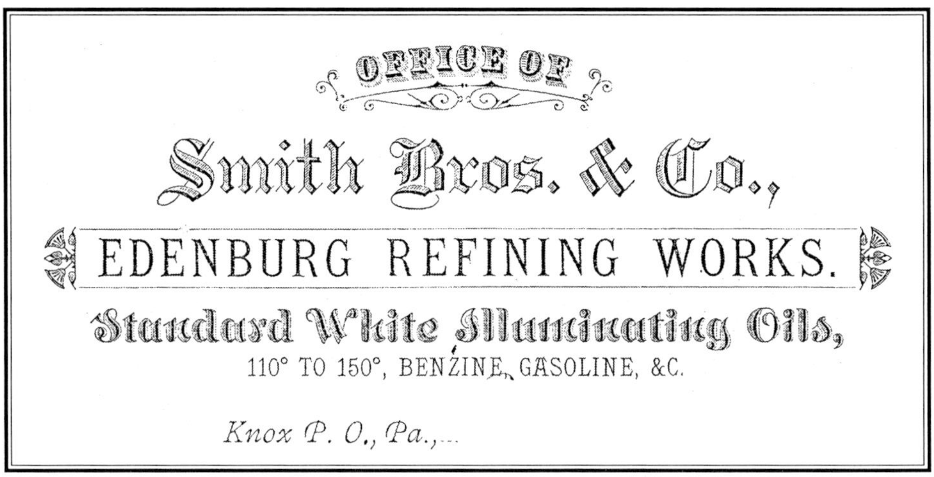
OFFICE OF
Smith Bros. & Co.,
EDENBURG REFINING WORKS.
Standard White Illuminating Oils,
110° TO 150°, BENZINE, GASOLINE, &C.
Knox P. O., Pa.,...

***Billhead for the Edenburg Refining Works. Courtesy of Barbara L. Hajel***

## Elk City
## Mills Spring Grove, Ice Cream Garden

The finest and choicest retreat during the heated term in the whole Oil Regions. The Gardens are located just outside of town, West End. A large platform has been erected for dancing purposes, and will be let to select and private parties. All the delicacies of the season can be purchased at the Gardens. Candies and Fruits in abundance.

## Home Sweet Home

A little girl knelt beside her bed in her humble home. The small driller's shanty was in the middle of the oil patch and was all the family had. As her mother stood by, the little girl prayed to have herself and all her relations cared for during the night. She then added, "And dear God, do try to take good care of yourself, for if anything should happen to you we should all go to pieces, Amen." Love, Belle Logue

## Man Killed in Beaver City February 23, 1877

On Tuesday evening as the nine o'clock train on the Emlenton and Shippenville railroad was nearing Beaver City, and after the conductor had announced the name of the station, saw a man lying across the track in the cut below the station. Before he could stop the train, the engine, tender and baggage car had passed over him, instantly killing him. No one present knew him and the mangled piece of humanity was put in the baggage car and taken to Edenburg. It was afterwards discovered to be the body of James Sullivan, a machinist from Beaver City. His employer had given him money to buy shoes, but it is supposed he spent the money on liquor. *Clarion Democrat*, Clarion Penna.

## Beaver City Business Directory
## *1877 Clarion County Atlas*

Acker, D. A., Boot and Shoemaker
Beals, Jacob, Farmer and Oil Producer
Brundred, B. F., Oil Producer, P.O., Oil City, Penna.
Christy, D. M.
Cuthbert, G., Engineer
Cook, W. S., Proprietor of Beaver City Hotel
Cross, J., Proprietor of billiard parlor. Fine brands of cigars and tobacco for sale.
Cage, R. E., Carpenter and Contractor
Fleming, J., Engineer
Gore, John, Blacksmith
Gray, J. A., Contractor and Builder
Haney, Ross, Farmer and Oil Producer, 250 acres.
Kribbs, W. G., Lumber Dealer and proprietor of lumber yards at Beaver City. Keeps constantly on hand a full stock of all kinds.
Kribbs, Geo., Oil Producer and Farmer, 200 acres
Kerstetter, T. J., Driller
Krouse, J. F., Engineer
Kurtz & King, Dealers in dry goods, groceries and notions
Lockard, ___, Proprietor of boarding house
Leach, D. F., Dry Goods Merchant
Mull, C., Carpenter and Contractor
Patton, J. V., Superintendent E. & S. R. R. Post Office, Emlenton, Penna.
Patterson, W. C., Hardware Dealer
Ramsey, J., Proprietor of blacksmith and machine shop. All kinds of tools made and repairing.
Scott, Jno., Driller
Sheffer, G. E., Tool Dresser and Driller
Sheffer, W. M., Proprietor of feed store. Cash paid for grain.
Schell, J. D., Merchant, dry goods, groceries and provisions constantly on hand.
Smith, G. W., Blacksmith and Machinist
Smith, G. C., Pumper and Operator

***Drilling by Horse Power***
***Venango County Historical Society***
***Franklin, Pennsylvania.***

## Atlantic Pipe Station
## Beaver Township, February 23, 1877

Jefferson City is our rival on the hill, but is today in something of a check. The Operators in this area sense the "death toll" may have sounded. The wells are showing a decline and today the Erquhart and Laven well came in as a complete failure. This well was on a 47 degree line from the Gates well, which opened up this new strike on the eastern belt and is not more than thirty rods from it. This causes serious concern to the numerous operators in the immediate vicinity that have been paying from one hundred to three hundred dollars per acre bonus and from one-sixth to one-fourth royalty. *The Edenburg Daily Spirit*

## "Wishee Washee"
## Edenburg, March 27, 1877

The Chinese population of Edenburg started a little war among themselves last week. Sam Wah Lee came and started a laundry which excited the wrath of two other Chinamen, who are also in the "washee" business there. They got Samuel into their establishment and after locking the door proceeded to "bouncee" their rival. Sammy got away and made information against his fellow countrymen for assault. They were arrested but not yet tried. We wish "Samee" good "luckee." *The Edenburg Daily Spirit*

## *Clarion Democrat*
## March 30, 1877

The recent decline in the price of oil and the unusual number of dry holes have had a tendency to check new operations. The Harrington well on the Long and McDowel farm south of Elk City is a dry hole. Many fear the worst and the Bradford field may be the next stop for many of our local operators. Joel Fink of St. Petersburg learned the oil business along Oil Creek back in the early 1860s. Mr. Fink as a young man floated rafts on the Clarion River to Pittsburg and here is where he learned the river business. Mr. Fink was a raftsman on Oil Creek before returning to Clarion County. Joel Fink and Samuel Fox, in the fall of 1865, began to operate for petroleum. Mr. Fink has been successful but the last well he drilled with Shamburg and Harrington has finished as a dry hole near Salem.

(In late 1879, Joel Fink removed his rigs and machinery to Indiana County, where he became a successful gas producer. He died at the age of 87 in 1907 at Livermore, Pennsylvania, and is buried in Oakmont Cemetery, Indiana, Penna.) "Peace to his ashes," John J. McLaurin. (*Sketches in Crude Oil*, 1896)

## "Pa Has Struck Ile"

I once was unknown by the happy and gay,
And the friends that I sought did all turn away;
Our dwelling was plain, and simple our fare,
And nothing inviting, of course, could be there.

But now, what a change! Our house is so grand,
Not one is so fine throughout the whole land,
And we can now live in the very best style,
And it's simply because my "Pa has struck ile."
'Tis strange what attention a fortune does bring,
At home or abroad how friends to one cling;
And now even strangers are courteous and bland,
To pay their addresses or take by the hand,
When before, on a walk if a neighbor I'd meet,
Cold was his look and quick his retreat,
But now in my carriage he greets with a smile
And it's simply because my "Pa has struck ile."

This poem was written after James Evans pushed down the first well in Franklin, Pennsylvania. That night his daughter, Miss Anna Evans, met her boyfriend at the door and exclaimed, "Pa has struck ile." The expression struck the town, making it a bigger hit than the well. – *Sketches in Crude Oil,* John McLaurin, 1896

*Gone is the old Dutch farmer's corn field. (Note the old split rail fence.) This oil field paid better than the corn. Photograph from John J. McLaurin's personal notebook, Venango County Historical Society, Franklin, Pennsylvania*

## The Sad Pumper
## July 2, 1878

George B. Ellsworth, a pumper, well known, commits suicide. He was employed by Strong and Fargo and was operating at a well near Pickwick, Clarion County, Pennsylvania. Mr. Ellsworth has been sad and not feeling well. He committed suicide and left his "earthly home" by drowning himself in a tank of oil. *The Edenburg Daily Spirit*

## Fire Bug
## Turkey Run City, July 15, 1878

Word has been received that terrible devastation has struck this once thriving hamlet of Turkey Run City. Once the "gem" of the Petersburg field with much growth and excitement, has now been struck a possible final blow. Business has moved on leaving a trail of devastation on its march to Bradford, Pennsylvania, and new fields. Turkey Run City has suffered an incendiary fire with a loss of $21,000 and most of the town is gone. The fire was first discovered at W. H. Daughenbaugh's furniture store on Rail Road Avenue, but the flames were beyond control. The alarm was sounded but due to the early hour and the lack of proper equipment, the fire took its terrible toll. A strong wind carried the flames from building to building until it was arrested by tearing down the machine shop of J. Needham, manufacturer of sucker rods. It is estimated that 37 buildings were destroyed and very few were covered by insurance.

The fire is supposed to be the work of an incendiary and these suspicions are supported by Postmaster S. S. Graham. Postmaster Graham witnessed two actors make a hasty retreat shortly before the fire alarm was sounded. The citizens and law enforcement officers of this area are on the search but little hope exists for their capture. Turkey Run City whose plains were so grand is now reduced to a mass of smoking ruins.

Signed E. M. Reardon, Cashier at the Turkey Run City Bank.

## Edwin Morris Reardon
## Head Teller, Turkey Run City Bank

Edwin Reardon was born in Shippenville, Penna., Clarion County, on April 4, 1850. His parents were William and Delilah Reardon. He was educated in the public schools and at the age of nineteen he took a position as bank clerk in the First National Bank of Sharon, Penna. In this position he remained for two years and then went on to the Sharon Savings Bank as teller. In 1873 he became teller fo the Lamberton Bank of Franklin, Penna. In 1877 Mr. Reardon took the position of cashier and manager of the Turkey Run City Bank, Clarion County, in the Pennsylvania oil

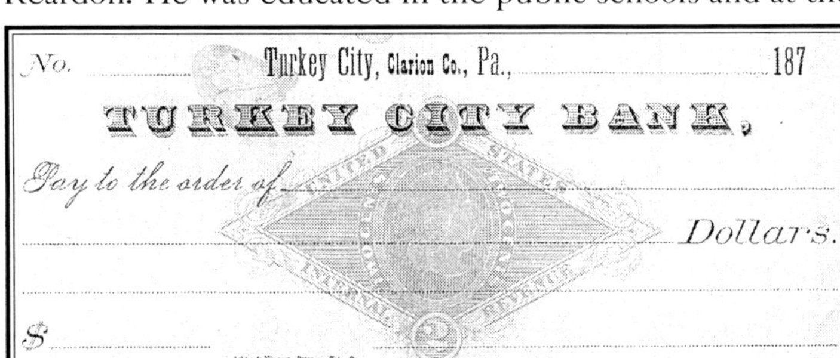
No. Turkey City, Clarion Co., Pa., 187

TURKEY CITY BANK,

Pay to the order of

Dollars.

$

*Turkey City Bank, Turkey City, Clarion County, Pennsylvania. Author's note: After searching 40 years, this check was finally delivered. Thanks to Homer Miller, Turkey City Historian, Bradford, Pennsylvania.*

district. In 1879 he moved to the Bradford, Penna. district and in 1883 he was attracted by the breadth of the West and went to Dallas, Texas.

Edwin Morris Reardon, a Clarion County boy, went on to found the American Exchange National Bank in Dallas, Texas, which at the time was the strongest bank in Texas.

In fact, financially, socially and in a civic sense, Mr. Reardon was a strong factor in the progress of Dallas, Texas. The day of Mr. Reardon's funeral all the banks in the city of Dallas were closed. This information shows the influence Edwin Reardon and many others in this history had on the "Building of America." Mr. Reardon died on January 15, 1924. *(Dallas Times Herald)*

## Turkey City Business Directory
## *1877 Clarion County Atlas*

Black, James, Proprietor of stage line
Buckholdt, C. E., Hardware
Buckholdt, A. P., Dealer in hardware and oil well supplies, paints, glass and oils, and general jobbers. Reasonable charges
Beatty, J., Gauger
Bouch, W. B., Proprietor of billiard hall
Burnap, E. A., Proprietor of machine shop
Baxter, C., Proprietor of hotel
Bithune, H. J., Oil Producer
Bolles, W. A., Oil Producer
Carlin, T., Oil Producer
Clearwater, Joseph M., Producer, Torpedo Agent
Conner, M., Contractor and Driller. All orders promptly filled.
Craft, J., Groceries
Clingensmith, J., Engineer
Duncan, R., Laborer
Diem, Joseph, Shoemaker
Duncan, G. S., Oil Producer
Dilks, J., Proprietor of boiler works, and all orders promptly filled to the entire satisfaction of customers.
Daughenbaugh, W. H., General Furniture Dealer. Looking glasses, picture frames, brackets, &c. Repairing neatly and promptly done. Rail Road Avenue, Turkey City, Penna.
Davis, J. L., Pumper
Easley, T. H., Salesman
Fisher, G., Barber
Freeman, W., Laborer
Freeman, G. W., Gauger Union Pipe Co.
Farguhar, Wm., Dealer in drugs, patent medicines, toilet articles, tobacco, cigars, &c. Main Street, Turkey City
Flynn & Hamm, Wholesale and Retail Merchants
Graffowitch, F., Merchant Tailor, next door to Turkey City Hotel. I desire to inform the public of my superior stock which I intend to make up in the latest styles of the tailoring art. Being a practical cutter, I guarantee a good fit. Clothes cleaned and repaired to order. Satisfaction guaranteed or no charge.
Gayetty, F. M., Proprietor of Gayetty House. First-class accommodations and good sample rooms. The table supplied with the best that the market affords.
Graham, S. S., Postmaster, Dealer in books, periodicals, stationery, gent's furnishing goods, toys, &c. Fine tobacco and cigars a specialty. Post office building, Main Street
Gwinn, D. M., Proprietor of the Gwinn House. The table supplied with the best the market affords.
Gwinn, Miss S., Millinery store and a full supply of millinery goods
George, H. M., Proprietor of drug store
Gardner, W. H., Proprietor of barber shop
Geer, E., Pumper
Hunter, R. M., Butcher
Hulseman, B. H., Oil Producer
Harrop, I., Pumper
Hathaway, M. B., Dealer in dry goods,

groceries, boots and shoes, hats and caps, clothing, chinaware, flour, feed and baled hay. Also, Proprietor of Turkey City Bakery. Lumber, lath and shingles, &c.
Heard, A. N., Dealer in groceries, provisions, notions, tinware and stoveware
Jobe, John R., Driller
Jones, S. M., Oil Producer
Lilly, L. J., Shoemaker
Lehman, J. M., Dealer in groceries and produce; also, all kinds of goods kept in a first-class store and a good supply of boots and shoes on hand at all times of the year.
Landis, A., Restaurant
Lively, T., Lively & Taylor, Turkey City, Clarion Co., Penna.
Lewis, T. E., Physician
Lively & Taylor, Central Carpenters. Tank and rig building done on short notice, and satisfaction guaranteed. Shop in
Turkey City, Clarion Co., Penna.
Lawson, S. C., Carpenter
Makin, N., Oil Producer
McElvaine, M. M., Hardware
Mooney, J. J., Boiler Maker
Myers, A. H., Photographer
McGee, S., Pumper
McKean, A. E., Mason
Mackey, Mrs. A., Proprietress of the United States Hotel. The table is supplied with the best that the market affords.
McCune, J., Driller
Maloy, P., Livery, feed and safe stable, opposite Turkey City Hotel. The public will receive good accommodations and at reasonable rates.
Mills, J., Lumber Dealer
McDonald, ___, Blacksmithing of all kinds done to order. Also, veterinary surgeon. I treat both horses and cattle. No cure no pay.
Needham, J., Sucker Rod Manufacturer
Nichols, C. B., Foreman Union Pipe Company
Noblit, W. G., Proprietor of blacksmith shop
O'Neil, M., Boiler Maker
O'Bryan, T., Driller
Parks, J. H., Druggist
Parks & Co., Druggists
Proctor, J. T.
Powers, C., Photographer
Pessell, C. W., Rig Builder
Panton & Bernie, Proprietor of machine shop. Repairing done on short notice and all orders promptly attended to.
Reardon, E. M., Cashier of bank
Robins, H. O., Oil Producer
Roberts, A., Groceries and provisions
Rowen, L. S., Groceries and provisions
Swan, J. C., M. D., Druggist
Smith, J., Shoemaker
Sanders, D. M., Oil Producer
Stephens, Emil, Hardware and tin shop. Tin, copper and sheet ironware manufacturer; dealer in stoves and hardware, woodenware and hollow-ware; also, smoke stacks made to order, boilers repaired, roofing, spouting and jobbing a specialty.
Stewart, W. J., Barber
Sencer, G. W., Union Pipe Company—Office Operator
Shepard & Books, Painters
Taylor, O. E., Attorney-At-Law
Taylor, A. M., General Carpenter. Tank and rig building on short notice and satisfaction guaranteed.
Wright, S. H. G., General Machinist. Repairing of every description executed with neatness and dispatch, and satisfaction guaranteed. Engine repairing specialty.
Wilby, R. G., Oil Producer
Walters, H., Laborer
Whitzel, J., Gauger
Wright, J. D., Engineer on Union Pipe Line
Weaver, G. W., M. D., Physician and Surgeon
Wright, S. H., Proprietor of machine shop

## Refinery Sold
## Monroeville, October 21, 1878

The oil refinery at Turkey Run City passes into the hands of John D. Rockefeller and the Standard Oil Company. This refinery has been very successful in its operations and is now being added to the long list of Rockefeller's buy outs. The Turkey Run Refinery will be closed and business transferred. E. H. Beckwith, General Manager, note page 184, Producer's Oil Refinery.

## Monroeville Business Directory
## *1877 Clarion County Atlas*

Burgar, M. D., W. F., Oil Producer
Bliley Bros., General Dealers in all kinds of hardware; also, tinware and woodenware. On Main Street, a short distance west of depot.
Bumpus, H. W., Oil Producer, Turkey City
Crum, O. J., Blacksmithing in all its branches; also, wagons and buggies built to order and on short notice
Carrobb, James, Oil Producer
Dickey, James, Oil City
Dearolph, A. J., Dealer in dry goods, groceries, clothing, boots, shoes, hats, caps, glass, nails and general merchandise
Ensle, G., Dealer in dry goods and groceries, hats, caps, boots, shoes, notions and everything usually kept in a first-class country store
Eddinger, F. F., Pumper
Forre, James B., Manufacturer of and repairing of buggies, wagons, sleds and sleighs, West End, Monroeville
Geiger, J. N., Livery, sale and feed stable; also, Dealer in flour, feed, hay, grain, apples, potatoes, &c
Hardison, W. L., Oil Producer
Hunter, D. A.
Hammond, W. J., Operator in Oil
Huffman, Rev. H., Pastor of German Reformed Church
Hendershot, M., Agent for Young People's Bible History Illustrated
Heeter & King, Dealers in dry goods, notions, hats, caps, boots, shoes, groceries and a general produce store, West End Monroeville
Heeter, S.
Krotzer, James, Dealers in groceries and produce. Also, carpenter, contractor and builder
Klingensmith, T. P., Oil Producer
Loroh, John, Justice of the Peace
Malloy, L. E., Oil Producer
Mortimer, Dr. J. L., Druggist and Dealer in patent medicines, paints, oils, window glass, etc. Prescriptions filled at all hours, Market Street
McGee, Irvin, Proprietor of machine shop, repairing of all kinds done on short notice
Master, C. D., Oil and Coal Producer, 21 acres
Master, D., Farmer and Oil Producer, 92 acres
Peters, A. J., Manufacturer of fine boots and shoes. Satisfaction guaranteed
Patterson, W. C., of the firm of Patterson, Dickey & Co., General Dealers in all kinds of hardware
Richards, Thomas, Boiler Maker
Sterner, Henry, Farmer and Oil Producer, 48 acres
Sterner, D., Farmer
Sherrick, W. C., Dealer in dry goods, notions, white goods, latest styles of ready-made clothing, hats, caps, boots and shoes.

Fine family groceries, finest teas, syrups, tobacco, cigars, foreign and domestic canned goods, flour, Queensware, wood and Willowware, &c

Trost, D., Proprietor of the Barton House, near the depot. Good accommodations

Tiffany & Son, J. N., Oil Producers

Vandersoal, S. W.

Weeter, Wm., Oil Producer

Wright, J. H., Proprietor of Monroeville Hotel and Boarding House

## Turkey Run Bank
## March 6, 1879

E. M. Reardon, cashier at the banking establishment here has received word that all banking will be closed on March 6, 1879. The bank building escaped the July 15, 1878 fire that left many of the town's businesses in ruin. The Turkey Run bank was associated with the banks in Foxburg and St. Petersburg but due to the sharp decline in the Clarion field, business has been closed. (see page 258) *The Edenburg Daily Spirit*

## Therefore
## Edenburg, March 12, 1879

At the producers meeting at Edenburg the operators and producers agreed that things could not get any worse. The Oilmen were asked to stop the "drill" immediately. It can only result in ruination to the industry.

It was resolved that, whereas! the shortest way to two dollar oil is through 25 cent oil.

Therefore! Be it resolved that we favor pushing the "drill" as "rapidly" and "diligently" as possible until the goal of 25 cent oil is reached.

It is sad we must hit rock bottom before we can turn our industry around! *The Edenburg Daily Spirit*

*Two Dollar Oil*

*Twenty-Five Cent Oil*

# The Edenburg Daily Spirit.

VOL.1. NO. 77. EDENBURG, PA., THURSDAY EVENING, NOVEMBER 6, 1879. TEN CENTS A WEEK

## THE SPIRIT.

THURSDAY EVENING, Nov. 6, 1879.

### SPIRIT RAPPINGS.

Edenburg can boast of nine physicians.

John Paul Logan lectured in Emlenton last night.

Chalfant will endeavor to close out the balance of the Beardsley & Herron stock to-day.

Swartz has taken back his meat shop, from Beatty, and now Beatty talks of starting up still another.

The Masons left for Franklin this morning. Mr. N. N. Booth will move into the house, on State street.

Mrs. Chris Robinson, of the Exley farm, is preparing to start for Colorado, where her husband is interested.

On Sabbath next, morning and evening, Rev. Fleming will preach on the subject of Future Punishment, in the Presbyterian Church.

Collner, the Democratic candidate for Sheriff, was elected by 512

(By Our Special Artist.)

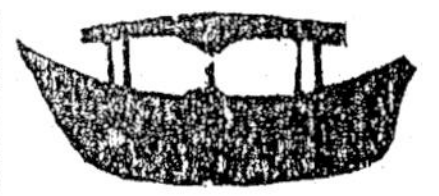

### A Salinacious Excursion.

At half past six o'clock last evening that desolate old sidewheeler, "Defeat," laden with a promiscuous crew of Democratic humanity, launched itself upon the waters of public ridicule, and is now plying the muddy Salt River. Simon Fraser is captain of the craft, and hopes to find the country in good condition up there, though the Republicans have inhabited it so long that there is but little doubt of its utter worthlessness. Charlie Wilton, first mate, will organize a Y. M. C. A. immediately on the barque's arrival in port, and with the assistance of Marion Clover, the coxswain, he expects to keep things lovely for a year, when they will be prepared to come down again. Pat Moran, the pilot, will "hug 'er close agin the bank" until he finds

### JOHN MUNHALL.

**He Dies, a Victim of the Caprices of Youthful Ruffians.**

Doubtless the sport-loving youths who, on Friday night last, took advantage of the so-called "All Hallowe'en" to commit depredations and unlawful acts, had little thought of the result that has followed their misdoings. They little thought, when rolling the ponderous express truck up on a car, that the result would be a human life lost, a home made desolate, and the anguish of many hearts cause the eye to redden with weeping. When they look now, that it is all over, those boys should make a solemn vow to let "halloweving," as they call it, alone forever afterward. No doubt they have done so ere this, and although the public does not know who the offending parties are, those parties know it themselves, and each and every heart should quake at the thought that they, individually and collectively, were the indirect murderers of poor John

## The Oil Market.

HIGHEST PRICE TO-DAY:
**$0 89,**

LOWEST PRICE, $0 88 3-4.

Closing Price, $0 89 3-8.

The following changes took place at the Edenburg purchasing office to-day:

| HOUR. | PRICE. |
|---|---|
| 10 20 | 89½ |

Refined in New York 7¼c. United shipments, 64,000 barrels. Money 7 per cent.

ONE YEAR AGO.

One year ago to-day crude was 83c. Refined in New York 9c.

## OPERA HOUSE,

EDENBURG,

**Friday Eve., Nov. 14th**

THE

'ENNIAL

## THE SPIRIT.

A Faithful Chronicler of Passing Events
EDENBURG, (KNOX. P. O.,) PA.

THURSDAY EVENING, Nov. 6 1879.

### TOWN TALK.

Who is your tailor? Marcus, to be sure. 73

‡‡Free oyster soup at the Newport House.

‡‡Finest Brackets in the county Bensinger's. 1

—Fresh home made Mincemeat at the Star Bakery. 65

Try the Schlitz Milwaukee Beer at the Newport House. 61

‡‡Best brands of tobaccos and cigars a specialty at the Postoffice. 33

‡‡Oysters, milk and plain stews, only 25 cents at the Newport House. 40

‡‡Sewing machine needles and oil at Bensinger's Brick Block Book Bazaar.

"Our Pets" are the best five cent cigar made. Warranted all Havana filler. At Bensinger's. 72

If you have the blues, get your Cigars and Tobacco at

THE PEOPLE'S DRUG STORE.

Another new lot of wall paper the latest fall and winter styles, just received at Bensinger's. 72

‡‡Sam Gibbs has a complete line of boots and shoes, and will not be undersold by any one. Call and see him. 57

Everybody knows it:

THE PEOPLE'S DRUG STORE

keeps the largest and best assortment of Drugs and Patent Medicines.

Marcus has opened in connection with his merchant tailoring establishment a cleaning and repairing shop. Clothing entrusted to his care will receive the best attention. 73

WANTED.—Every...

## NEW MILLINERY GOODS!

A Large Assortment of New Styles of *Ladies, Misses & Childrens* HATS AND BONNETS

Just Arrived

**From the Eastern Cities, at Miss Hettie A. Keatley's Fancy & *Millinery Store*.**

CORNER OF MAIN AND CHURCH Sts.

*E. Butterick's Patterns for sale.*

**DRESS-MAKING DONE IN THE MOST APPROVED STYLES.**

Prices to suit the times.
38w Edenburg, Pa., Sept. 22, 1879.

## BENSINGER'S,

SLATES, SLATES, SLATES, SLATES,

—*in the*—

Pain cannot stay where it is used. It is the cheapest medicine ever made. Five drops cover a surface as large as the hand. One dose cures common Sore Throat. One bottle has cured Bronchitis. Fifty cents' worth has cured an Old Standing Cough. It positively cures Catarrh, Asthma and Croup. Fifty cents' worth has cured Crick in the Back and the same quantity Lame Back of eight years' standing. It cures Swelled Neck and all other Tumors, Rheumatism, Neuralgia, Contraction of the Muscles, Stiff Joints, Spinal Difficulties, and Pain and Soreness in any part, no matter where it may be, nor from what cause it may arise, it always does you good. Twenty-five cents' worth has cured bad cases of Chronic and Blood Dysentery. One teaspoonful cures Colic in fifteen minutes. It will cure any case of Piles that is possible to cure. Six or eight applications are warranted to cure any case of Excoriated Nipples or Inflamed Breast. For Bruises, if applied often and bound up, there is never the slightest discoloration to the skin. It stops the pain of a Burn as soon as applied, and is a positive cure for Chilblains, Frosted Feet, Boils, Warts, Corns, and Wounds of every description on Man or Beast. Price, 50 cents and $1.00. Foster, Milburn & Co., Sole Prop'rs, Buffalo, N. Y. For sale in Edenburg by N. C. Andrews & Co. 73

## THE SPIRIT.

A sixteen column evening newspaper, published at

**EDENBURG**

(KNOX P. O.)

## RAILROADS.

*Emlenton, Shippenville & Clarion Railway.*

*Foxburg, St. Petersburg, & Clarion Railway.*

EASTWARD TRAINS.

| STATIONS. | 1. | 3. | 5. | 7. |
|---|---|---|---|---|
| | a. m. | a. m. | p. m | p. m. |
| Foxburg............De | 8 00 | 10 16 | 12 50 | 7 00 |
| East Foxburg | 8 08 | 10 24 | 1 00 | 7 0 |
| St. Petersburg...... | 8 15 | 10 31 | 1 20 | 7 5 |
| *Ashbaugh ......... | 8 13 | 10 34 | 1 23 | 7 [illegible] |
| Richmond............ | 8 23 | 10 38 | 1 28 | 7 [illegible] |
| *Keatings........... | 8 25 | 10 40 | 1 30 | 7 [illegible] |
| Turkey City......... | 8 28 | 10 44 | 1 35 | 7 [illegible] |
| Fullerton........... | 8 35 | 10 47 | 1 38 | 7 [illegible] |
| *Ritts Farm......... | 8 37 | 10 53 | 1 43 | 8 [illegible] |
| Blairs Corners...... | 8 43 | 11 00 | 1 50 | 7 50 |
| *Deloes............. | 8 46 | 11 03 | 1 52 | 7 55 |
| Jefferson City...... | 8 59 | 11 05 | 1 56 | 8 00 |
| Emlenton............ | 8 00 | ...... | 12 55 | ...... |
| *Kings.............. | 8 08 | ...... | 1 04 | ...... |
| *Agnews............. | 8 13 | ...... | 1 10 | ...... |
| *Butter Town........ | 8 18 | ...... | 1 15 | ...... |
| *McMichaels......... | 8 24 | ...... | 1 21 | ...... |
| Turkey City......... | 8 36 | ...... | 1 32 | ...... |
| Monroeville......... | 8 45 | ...... | 1 40 | ...... |
| Pickwick............ | 8 50 | ...... | 1 47 | ...... |
| Jefferson........... | 8 58 | ...... | 55 | ...... |
| Beaver City......... | 8 54 | 11 13 | 2 03 | 8 08 |
| Edenburg............ | 9 00 | 11 19 | 2 10 | 8 15 |
| Elk City............ | 9 11 | ...... | 2 20 | 8 23 |
| *Dale City.......... | 9 15 | ...... | 2 24 | 8 27 |
| Shippenville........ | 8 23 | ...... | 2 32 | 8 36 |
| *Clacks............. | 9 27 | ...... | 2 40 | 8 44 |
| *Patrick Farm....... | 9 22 | ...... | 2 45 | 8 5[illegible] |
| *Pike............... | 9 35 | ...... | 2 40 | 8 5[illegible] |
| *Bridge............. | 9 45 | ...... | 2 59 | 9 [illegible] |
| Clarion........Ar. | 10 00 | ...... | 3 12 | 9 [illegible] |
| | a. m. | a. m. | p. m. | p. m. |

WESTWARD TRAINS.

| STATIONS. | 2. | 4. | 6. |
|---|---|---|---|
| | a. m. | a. m. | a. m. |
| Clarion..........De | 7 40 | 10 10 | ...... |
| Bridge.............. | 7 65 | 10 22 | ...... |
| Pike................ | 8 06 | 10 33 | ...... |
| Patrick Farm........ | 8 10 | 10 35 | ...... |
| Blacks.............. | 8 15 | 10 41 | ...... |
| Shippenville........ | 8 22 | 19 58 | ...... |
| Dale City........... | 8 30 | 10 47 | ...... |
| Elk City............ | 8 34 | 19 55 | ...... |
| Edenburgh........... | 8 44 | 10 59 | ...... |
| Beaver City......... | 8 44 | 11 07 | ...... |

*Courtesy of Barbara L. Hajel, Edenburg Penna.*

## Teamsters

*The Teamster*
***Pleasures of Teaming in Oildom (Sketches by S. G. Bayne)***

Every piece of furniture, and even the coal for engines and the engines themselves, had to be hauled in to the well by teams, through the mud that was—

**Wholly unclassable**
**Almost impassable**
**Scarcely jackassable.**

It was the common custom for many teamsters to carry a keg of beer along, and when their wagon wheels sank to the hub, they wet their outsize vocal cords between cursing the poor struggling horses. Teamsters had an awesome vocabulary, "blasphemous, brimstone-tongued," and they were the meanest fighters around. In saloon brawls, a teamster often bit off an opponent's nose, a portion of the upper lip, or a chunk of ear. And he chewed it like tobacco. – *Early Daze In Oil,* Earnest C. Miller, 1974

## Fishing for Lost Tools

*Fishing With Pole Tools in a Well 1,500 Feet Deep*
*Venango County Historical Society*

In drilling oil wells, the heavy tools are apt to get fast so they can not be moved by the cable to which they are attached. The sand made by the drilling clogs the tools, fastening them firmly. These tools are made in five sections which are screwed together with right-handed threads and weigh combined nearly 2,000 pounds, and are about 70 feet long. The operation of getting them out of the well is called "fishing," which requires skill and experience. The pile of pole tools in the right of the picture are made of 3 1/2 inch iron, measuring 20 feet in length, each weighing 500 pounds and screw together with left-handed threads. In a well 1,500 feet deep, the pole tools will weigh nearly 20 tons. As the ordinary derrick would not hold up such a weight, a pair of heavy beams called "shears" are placed inside of it. To the apex of these shears is attached a five-pullied tackle, (even this purchase is not enough). The end of the rope is led around the "bull" wheel shaft, and power from the engine conveyed to the wheel, thus giving great additional power. When tools are fast, the rope is cut off close to their top by an ingenious knife, and the pole tools are let down, armed at their lower end with a "die" which has a left-handed thread cut in it. This die fits over the end of the fastened tools, tapers inward slightly, and when turned to the left cuts a thread on the tool. The upper end of the poles are square and project some distance from the well, and on this the large wheel fits (shown in the right of the picture). A rope is wound four times around the rim and led to the bull wheel shaft. The bull wheel is connected with the engine and turned slowly, exerting a tremendous strain on the poles. As the strain is always to the left, the die gradually cuts a thread in the fastened tools. The die being tapered, its resistance increases until one of the fastened tools unscrews. It is then removed and the operation repeated until the lower piece is reached. This must be loosened by cable tools (the iron tool leaning against the wheel in the right, is a "spud"), and is made to spud, or "spear," around the tool and loosens the sand which imbeds it. When this is done, the piece is easily raised by a "grab" made for the purpose.

This information comes from the back of the photograph, "Views of the Penna. Oil Region," published by Frank Robbins, Oil City, Penna., View No. 107. Messrs. Patterson, Mitchel and Co., of Oil City, are experts in the business of recovering lost tools. Their experienced agent, Mr. James Robinson, of St. Petersburg, Penna., is represented in the picture.

## Elk Township
## Clarion County

In 1875, E. C. Bradley and Co. drilled a well on the Johnson farm northeast of Elk City, and about the same time another well was put down on the Joseph Kiser farm. Both wells produced about fifty barrels a day and caused much excitement in that area. The price of petroleum was up at that time and thus encouraged extensive preparations for operating.

Most of the drilling was done in this territory between 1875 and 1877. E. C. Bradley and Co., Miller and J. Aikens, Patterson and Leedom, Turner and Wetter, J. M. Guffy, Frederick Prentice, H. Baliett, Hahn, Kribbs, F. Plumer, R. Richardson, Gray Bros., J. T. Jones, W. Chambers and Brundred were extensive operators in this territory. In all there have been over 1,000 wells drilled in Elk Township.

## Elk City

Elk City was the last of the great oil boom towns in the Clarion County belt. The Boom was almost Bust. Elk City was the most prosperous of Edenburg's neighbors. In 1877, with a growing population of well over two thousand, Elk City was as large as Edenburg. A boardwalk extended from Edenburg through Mongtown and Forest Home to Elk City, a distance of almost two miles. The board walk was used to keep your new boots out of the knee-deep mud and Sunday strolls by sweethearts and families. At Elk City's height of prosperity there were twenty hotels and boarding houses, two churches, a school, post office, three lumber yards, two machine shops, two photography shops, drinking establishments, stores and scores of homes. Today a drive through Main Street Elk City is like a drive through the woods; only a few houses are left to mark the former city.

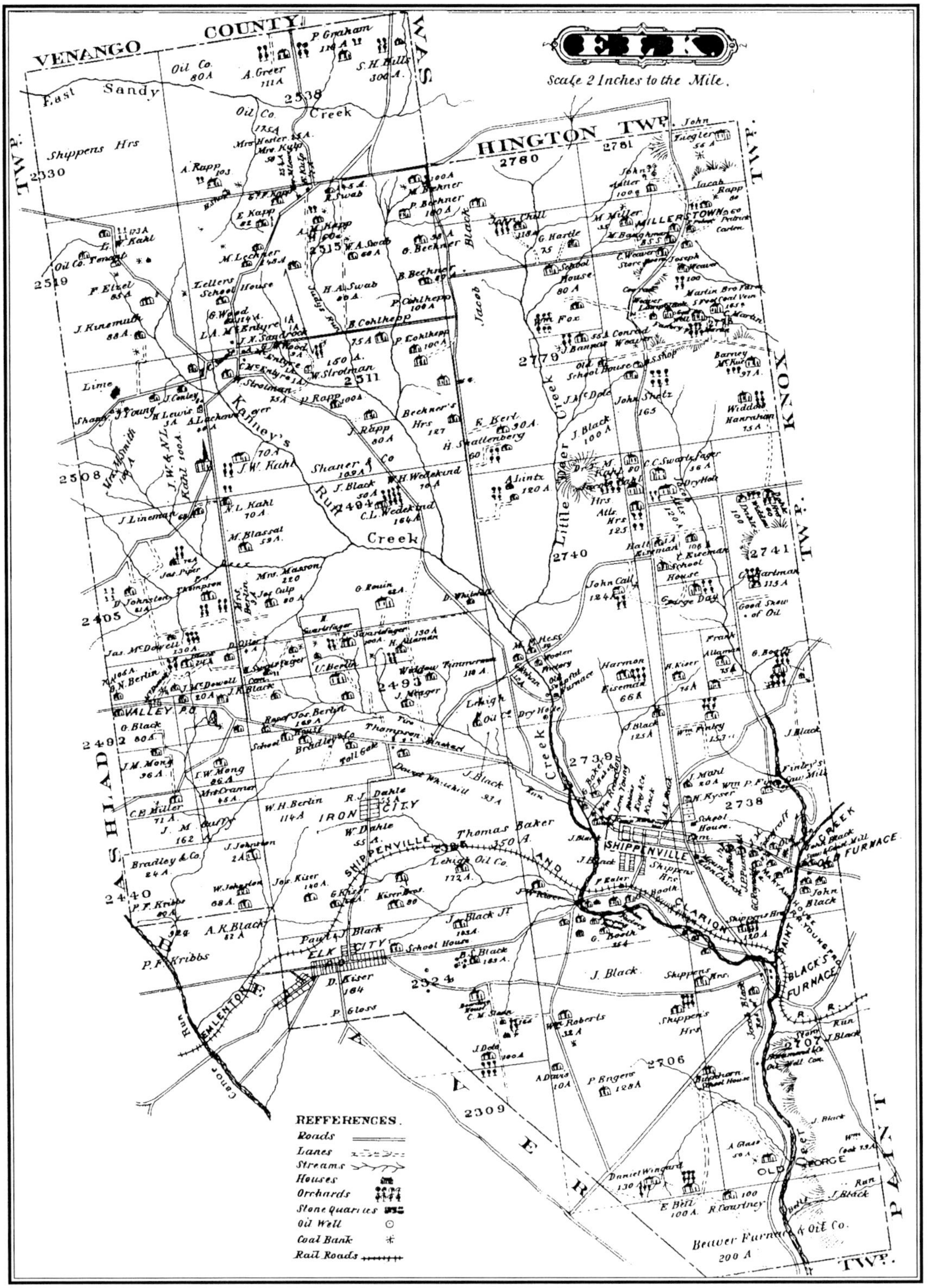


*Elk Township, Clarion County*
*1877 Clarion County Atlas*

*Elk City Main Street scene, 1877. J. R. Riddle and Co. Photographic View Artists. Stereoscopic and large views; oil scenery a specialty. Located at Fairview, Baldwin P.O., Butler Co., Penna. (Thanks to Barbara L. Hajel)*

*Elk City Station*
*This scene is a view of the Foxburg, St. Petersburg and Clarion Railroad station at Elk City. The station house is at the left side of the photo. This is a Frank Robbins stereoscopic view of the Pennsylvania Oil Region, Oil City, Penna. (Thanks to Barbara L. Hajel)*

Elk City was located on two farms owned by Damal Kiser and Paul Black. The lots in Elk City were surveyed and then leased to the buyer for a period of three years. Elk City burned "bright in the night" and then dwindled away with the excitement. The building commenced in 1875 with 1876 and 1877 being its peak year. The buildings were put up overnight, close together and in a very temporary manner. From the muddy streets, the buildings, shops and stores looked well kept and painted. The Elk City Post Office was established in 1875 and soon rose to the rank of Presidential appointment with a yearly salary of $2,000.

The usual criminal element took up residence there and in 1876 a police force was formed. Much of the town's boarding houses, hotels, businesses of ill-repute, and saloons were little affected by the insufficient force and lawlessness still reigned supreme. The infamous Ben Hogan and his associate, French Kate, set up shop in Elk City in 1876. He ran his saloon and related business until the "boom" went "bust" and then moved on.

Elk City had a moral and spiritual population. I.O.O.F. of Elk City, No. 948, was organized April 11, 1877 and meetings were held every Saturday evening until October 2, 1883. The "Mountain Lodge," No. 14 was organized July 27, 1877 and meetings were held until October 2, 1883. The Good Templar Lodge was formed on March 6, 1877 with 41 charter members. A public school was well attended on Fourth Street and had all the latest books and teaching materials with new patent furniture. Two churches were built in 1876, the Methodist and the U. P. Church which were full each Sunday. "Huff's" Bank on Main Street was first opened in Elk City and upon the decline was moved to Edenburg.

Other enterprising producers were Major W. T. Baum, C. D. Angell, Samuel S. Fertig, (brother of John Fertig, who drilled the famous Nobel Well in the Oil Creek valley), W. H. Clapp, Col. J. J. Carter of Titusville, Frank W. Andrews (who lost all in the Clarion fields, went to Bradford to start over and was killed after being thrown from his horse), and George H. Dimick, who later became President of the Bradford Commercial Bank.

## J. Watson and S. G. Bayne

The following information is taken from a long lost autobiography, "Derricks of Destiny." This book was penned in 1924 by S. G. Bayne, a real pioneer in the Clarion County oil fields. Mr. Bayne tells of his arrival in the Venango County oil fields and the "ups and downs" of the early days of the Petersburg boom. His story is told in a very unique and interesting way and illustrates his thoughts with comical characters. These stories were rediscovered, much to my pleasure, on a lonely shelf, long ago forgotten. Sam Bayne went from a "greenhorn" under the watchful eye of Jonathan Watson to a wealthy, well-respected banker and businessman. One of S. G. Bayne's favorite statements was, "The forget-me-nots of today are frequently found the has-beens of tomorrow." To all students of oil history, this book is a good read. Oh, by the way, welcome Mr. Bayne to your place in the history of Clarion oil, in Northwest Pennsylvania.

***Men Who Made Oil History***
***Samuel G. Bayne 1879***
***Samuel G. Bayne started in with Jonathan Watson in extensive operations in Clarion County, doing work all along the line from St. Petersburg to Edenburg.***
***Courtesy Drake Well Museum***

## *DERRICKS OF DESTINY*
## An Autobiography by
## S. G. Bayne
## 1924
## Brentano's Publishers, New York

### From Boom to Boom

Jonathan Watson was a typical petroleum pioneer. He was keen, alert, and had a way of snapping out his sentences. He had grown wealthy with the boom, and was reaching out for more power. After breakfast, he drove me out to his oil wells behind a team of high-steppers.

Acres and acres of what had once been farms stretched out before us and from them rose gaunt latticed derricks, black, greasy, hideous. In some places, these bristling scaffolds looked like forests of blasted trees. They were crowded into valleys, and others crowned the hills. On one tract of ten acres I counted no less than fifty-three derricks. The present generation knows about the oil booms in Southern California, where wells have been sunk in back lots and derricks rivaled the big sets of the Screenland, but in those days, the whole oil business was new and strange. Pioneers blazed the way, and they blazed it a bit too much at times, as many a charred mass of timbers over what had once been a gusher, showed. I plied Jonathan Watson with eager questions as we rode along the muddy roads.

"Bayne," he said, "if you'll stay with me, I'll make an oil man of you, you've got it in you."

"Where's the best hotel?" I asked.

"The Bush House."

"Lead me to it," I answered, "I'm yours for life."

From that moment I started right in to learn the oil industry in all its phases, and during the next two years succeeded in losing all the money I had brought with me. But before I again got on my financial feet I had learned the business from every angle, and had also retained that best of teachers, "Experience."

## The Old "Sugar House"

With my small capital, I decided then and there to become an operator on my own account. Technically, I was what oil men called a "greenhorn." I bought an oil well known as the Sugar House. [Put down in the fall of 1859 by William Bronsdall, Boone Meade and Henry R. Rouse, the Sugar House came in at 5 barrels in November 1859. It was tubed in December 1859 and began pumping 50 barrels per day in February of 1860.] It had been bored in classic ground, for it was the second well sunk in Pennsylvania. It stood only a few yards away from the "Father of All the Wells" none other than that which had been driven by "Colonel" Edwin L. Drake. I think that it was the sentiment of working on such a sacred spot that appealed most to

*Debut of the Author as a Driller*
*Sugarhouse Well, 1869*
*This well was put down very close to the original Drake Well. Samuel Bayne is shown sitting on a stump, wondering "Where's Toot?" (Sketch by S. G. Bayne)*

me. The achievement of the great pathfinder inflamed my imagination although even at that time he had ceased to be one of the leaders of Oildom.

"Sugar House" next to Drake's first well, was not a very valuable piece of property when, in 1869, I bought it from one "Pat" Harrington. Pat got only enough oil out of it to make a scanty living, as he was too lazy and shiftless to attempt anything more. I met him at the well and asked him innumerable questions. Next day, I bought the well from him for a small sum, as everything about it was obsolete. It had the old-time "Grasshopper" rigging, and the derrick was scarcely more than 15 ft. high. "Sugar House," however, had all the necessary factors, an asthmatic engine, a small boiler house, and even a wooden tank, as is shown in the sketch I made of it that day in May when I became an oil well owner.

## Anyone Seen "Toot"?

Pat pocketed the price of the property and agreed that he would instruct me fully in operation. "Now, then," he said, after a lengthy treatise, "if you do these things as I told you, and keep a stiff upper lip, I guess she'll hang together long enough to make you a small killing." With this cryptic statement he shuffled along in the direction of his favorite saloon in Titusville. He was known locally as "Toot" Harrington, and that was the last I ever saw of "Toot." I had supposed he would at least give me a fortnight's instruction in how to run the pump and the little steam engine, for I knew nothing about either.

After Harrington had disappeared in the woods I pushed my hands into my pockets and stood staring into space. I looked up from my dream, and seeing a young fellow chopping wood near by, I went over to him to see if he could help me. His name was Pete McArdle. I told him of my woes and unique situation. Pete was a decent sort of chap and said he had worked thereabouts "considerable", so we held an executive session and buckled down to the job before us.

The well was gassing quite a little, and Pete observed: "If we could collect that gas in a barrel and run it into the boiler you wouldn't have to chop any wood for fuel, and that would make at least a start."

No sooner said than done, so I gave him money and told him to run up to Titusville and get a solid, oak oil barrel and some tools, pipe and nails. He returned with the supplies in a few hours.

There was a great deal of lumber around the well, so we built a high platform and set up the barrel upon it. We next led the gas from the well into this barrel and from the barrel into the fire-box of the boiler. We lit this for a few minutes, and as it worked, our first step seemed highly successful, so we called off work for the day.

## Oh, No!! Bed Bugs

Pete had his own food in a cache, and he showed me the house of a Dutch family where I might obtain bed and board.

I was very tired, and after retiring fell asleep almost immediately. But the bed was infested with friendly enemies and I awoke in the morning with all the well-known hallmarks of an encounter. I never slept in the house again, but put up an amateur hammock outside. This was no hardship, as it was summer. I won't trust myself to say anything about the board.

*The Agent and His Victim* *The Victim Escapes*
*(Sketches by S. G. Bayne)*
*The fast-talking real estate agent from Titusville attempting to buy an old Clarion County "Buckwheat's" farm.*

Next day we began work on the well again, got everything in shape and slowly commenced pumping oil into the tank. All went well and we were feeling like the Croesus twins. I remarked to Pete, "Well, old man, we're not such fools as we look, are we, eh? This old well should give a good report of itself."

Then *whack! bang!* It did! The gas barrel exploded with a crack that might have been heard in Titusville. It rained splinters and shreds of iron hoop and barrel heads, but luckily we were in the shelter of the derrick; otherwise we might have experienced that "killing" Pat had promised me. We stood looking at each other, and finally broke out laughing to relieve our feelings. All our labor-saving devices had gone up to join Gabriel, and we realized we'd have to reorganize our operating policy. Pete chopped the wood and I took it to the engine shed in a good old-fashioned wheelbarrow.

The well would produce nearly twice as much oil if run night and day as it would when run twelve hours only. Oil was bringing more than $5 a barrel at the derrick, and we got in enough to net about five barrels daily, but we had to work hard. However, since I was little more than a lad, I was well satisfied with the results.

## Out of a Scrape

The oil region was a great place for changing and trading. The same chance that brought me into this scrape, as I called it, took me out again. I knew a champion wood-chopper; he had chopped seven cords of oak between sunrise and sunset on a bet of $500, and won it. I found this prodigy and presented the matter in this way: "Now, Wilson, you have saved some money. Here is an opportunity for you. You are a giant and can easily do the work that Pete and I are doing. Buy my well, and you can chop the wood for a day in less than an hour and run the well yourself, clearing thereby $25 a day."

*The Farmer Making Hay While the Oil Flows*
*Attempting to lease the old Dutch farmer's property.*
*(S. G. Bayne)*

He didn't utter a word in reply, but closed the bargain for the purchase of the well and prospered ever after.

I had stood on historic ground beside the first oil well ever worked. That well had introduced a great product, petroleum, which has spread from pole to pole and has done more for humanity, when everything is counted, than any commodity which has been produced since Adam was a boy. More than that, I had owned and operated the second well ever drilled, the "Sugar House," right within the shadow of the first.

I made mistakes and plenty of them, but I analyzed them and found that they were largely my own fault, although in a measure due to the unsystematic way in which business was often conducted in the petroleum regions. I made up my mind to stick until my last dollar was gone. All this time, Jonathan Watson seemed to have an eye on me, sometimes two eyes. His barn-like office in Titusville was a rallying place where oil fans within a radius of fifty miles congregated to make deals, absorb information and advice generally. Watson not only owned a large oil production himself, but also drilled wells for other operators. When a boom was on, and it usually was in those days, an amateur could learn more by sitting in Watson's office, and keeping his eyes and ears open, than by taking a college course. I quite appreciated this opportunity and soon became one of Watson's most insatiable scholars.

One day, to my surprise, he called me into his private office and offered to make me a member of the firm. This was a wonderful opportunity, which I lost no time in accepting. When the news of the partnership became known a great deal of hard feeling was engendered. There were a number of men who felt they had a prior claim on Watson's good will and should have been taken into partnership before a young Irishman, just over, and barely able to talk "United States," and a poor poker player, at that. However, the new firm of Watson & Bayne was an accomplished fact. It brought me into the vortex of action, and that quickly, as you will see.

## Early Days in the Oil Fields

Watson put me in charge of drilling operations in the field, making me leader in an absorbing and often uncertain game. My preceptor and partner who had told me that I had the making of an oil man in me said also that I had a capacity for hard work. The play was so swift, so filled with excitement, however, that it seemed more like sport than toil.

Even the selection of a location for a well was decidedly a gamble. The early prospectors had been guided by surface indications, such as a scum of oil in a spring or on the brooks and streams. Drake had put down his first well on the banks of Pine Creek. There grew up a belief that the underground flow of oil in some way followed the rivers on the surface, and that one should always drill in the valleys. This theory was knocked into a cocked hat when the wildcat speculators of the day started wells on the hilltops and saw them come in with black geysers. Derricks were put up in the woods, in the cow pastures, everywhere, and the chugging of the drill could be heard at nights where the owl hooted and the whippoorwill told his woe. Farmers were waylaid by promoters to sign leases which sometimes made them rich, and often left them poorer than Job's turkey, as they usually leased their lands on a royalty basis. Dry wells spelled disaster. Fortune shifted from getting a "duster" to striking a "spouter."

# That Odd Device

One of our most unusual, as well as heart-breaking experiences, was with the divining rod fakers. Petroleum is found, as we now know, in all parts of the world where there are sedimentary rocks. The scientific expert cannot always tell where the oil is, but usually he can inform his client where it is not. Of course, even though he may be able to predict with success that oil will be found, only actual drilling will demonstrate whether the liquid exists in paying quantities. In the early years of the oil fields, guessing, superstition, and especially that most palpable of fakes, the divining rod, flourished like the green bay tree.

***The Divining Rod Magician in Action***
***Could this divining rod magician be the same fellow that convinced Watson there was a big well in the middle of the Clarion River? (Sketch by S. G. Bayne)***

That odd device, the diviner's wand is a "Y" shaped switch of wood, usually hazel, which the seer seizes firmly in both hands and holds out before him in a horizontal position. So equipped he begins to walk slowly about the region under survey, and if the loose end of the rod is dragged down forcibly towards the ground by some unseen force, like the compass needle by the pole, he announces that oil is there. Of course, one must be a diviner himself in order to tell whether the force which draws the rod down is really external. In my own hands the hazel stick was always unresponsive; but many people believed in the diviners' claims of occult power, and in a region impregnated with oil it naturally followed that a majority of the wells which they located were good.

Mr. Watson became firmly convinced of the value of the hazel rod, so firmly, indeed, that two dry wells in succession failed to shake his faith. However, this bad luck did upset the diviners, and after a good deal of discussion they agreed to go over the whole region again at their own expense and rectify the error. The upshot of all this was that when they finally agreed upon the location of what they proclaimed a great new bonanza field, it turned out to be right in the middle of a good sized river! Abram James was a man of fine intellect and a hazel switch. In 1874 he proclaimed a good well in the middle of the Clarion River. He also was an investor and lost $6000. Excitement by this time was running high, and instead of being discouraged by this diagnosis, as the diviners no doubt hoped he would be, Mr. Watson declared that the well must be drilled. Others joined him, a great pool was made up, and it was decided to divert the course of the stream.

This undertaking was given to me for execution. I labored for months with a small army of workmen and finally accomplished it. Then the drilling rigs were set up in the dry stream course and with bated breath all waited for the outcome. Had the diviners happened to strike it lucky their fame would doubtless have been established for all time, but an all-wise Providence ruled otherwise, and every well came in dry! This disappointment nearly broke Mr. Watson's heart and it ruined many another man financially. It became the great fiasco of the time in the oil country. But it did accomplish one good thing. After that we were not troubled by any more diviners! The well was 1800 feet and dry.

We needed no magic or witchcraft to keep the pace in Oildom in those days, for the very air was surcharged with adventure.

## Killed in the Shooting

I shall never forget that afternoon I reached the Bush House, in Titusville, where I came face to face with a mournful procession which was bearing the bodies of seven men to their graves.

"What does this mean?" I asked a bystander."

"Killed in that shooting," was the reply."

What my informant meant was not that the unfortunate men had been victims of lawless gun play, but that they had met their doom in an explosion at a neighboring oil well, which had been "shot," that is, accelerated by nitroglycerine.

After the oil stratum was struck, it was customary to lower a charge of powerful explosive, which, on being fired, broke up the rock or sand and thus released an abundant flow. On account of the recklessness with which explosives were then handled this was, indeed, a hazardous business. The drillers in the early days took their lives in their hands. They were able fellows, taken all in all, but a rough lot, the original "rough necks," for this is the title given to them in the oil regions to this day. They had good hearts, all the same. The more I knew them, the better I liked them, and, eventually, as I had gained some skill in understanding them, they grew to like me.

Once a drilling rig had been set up, the hands worked in shifts night and day without let-up, until the well either came in or the owners were satisfied that they had guessed wrong. These shifts, or being "on tower," as they were called, were twelve hours long. As often ten rigs were all going at once within a small area, excitement was always in the air. The climax came when the oil bearing level had been reached and the long cylinders of glycerine were lowered. On the tops of these tubes was a fulminating cap or a cartridge, detonated by a "go devil." This device consisted of a heavy disk of iron, which was dropped into the well. It was fitted with curved wings, which increased its speed and gave it a rifled motion. The velocity of the "go devil," of course, was very high, and the nitroglycerine exploded with a terrific downward force. The effect was not manifest on the surface immediately, as often the released oil had to travel upward five hundred, or even a thousand, feet, before we saw it. It behooved all bystanders, none the less, not to linger after the "go devil" had bidden us good-bye.

The whole method of handling explosives has been improved since those days. Nitroglycerine, which is a very treacherous fluid, is now usually mixed with clay or sawdust, or some such inert substance. Although the speeding up of oil wells was generally done by professional shooters, accidents were of almost daily occurrence. I was not deterred by the sign of the sevenfold funeral which I had seen at the Bush House on the day of my arrival, and soon became as hardened a well shooter as the others of that most hazardous calling. Very often, though, I was brought up with a good sharp turn and confronted with the perils which were always close at hand.

I had been the Watson "outside man" only a few weeks when I had occasion to run up to a nitroglycerine factory. I had ordered an extra large quantity of the explosive for some wells beyond Titusville, and as I wanted it promptly I became a little impatient when there was a delay in delivery. The men who conducted that place were beyond a complaint of mine when I arrived. Where the factory had been, was a crater. One mistake too many had been made in the mixing room and the production crew had all been blown to pieces.

## One Hundred Quarts

As I covered a good deal of territory, I had a swift, little saddle horse which took me from well to well, though often I cut across lots on foot when the roads were deep in mud and the going was too heavy. [On February 23, 1876, I was walking home along the highway near St. Petersburg,] when I was overtaken by "Jim" Barnum, a professional well shooter, and at his invitation climbed up on his buckboard.

"I've got about a hundred quarts with me," he said. "I'll just drop them at the well and then leave you at your own door."

I did not feel like bothering him to go out of his way, so I thanked him and left him at the [bottom of the J. J. Ashbaugh hill near Petersburg. Jim traveled down a lane which led to the magazine where he stored the nitroglycerin. I cut across the fields to my own destination, but unfortunately within a few moments I heard a tremendous explosion.]

"There goes poor Barnum!" I exclaimed involuntarily, and walked back and found both him and his horses dead. [The largest piece of the luckless agent would not weigh two pounds and the next day one of his ears was found, a half mile away.]

## Run For Your Life

One of the wells at Petrolia had suddenly gone dry and I decided to give it the glycerine treatment. As I was lowering a heavy canister of the explosive into it, by a long copper wire, I heard an ominous roar below. I called my two helpers to run for their lives and I closely followed them. My ear, which had become well trained by that time, told me that the cylinder, although still unexploded, had started a commotion. It had just been enough to set up a flow of the oil and gas which had been accumulating in some vast and underground pocket.

A rumbling, black column, flung from below, rose out of the well, balancing on its crest the unexploded cylinder. It carried the tube seventy-two feet in air and hurled it against the crown pulley at the top of the derrick by which the giant drills were hoisted. Then came ruin! The huge iron wheel was blown to smithereens; the whole drilling rig crumpled; the machinery was wrecked; the adjacent oil tanks shattered, and thousands of gallons of oil let loose. Had we not taken to our heels just in time, the three of us would have joined that total loss.

## The Intoxicated Man

As I got off a train on one of my inspection trips about a month later, I noticed one hundred yards or so in front of me an intoxicated man, who was carrying a nitroglycerine can on a cane slung over his right shoulder. The can was sliding backwards on this support run through its handle. It would get to the crook at the top of the walking stick, and then slowly slip back again. Several times, the bearer kicked his foot against a stone, stumbled, righted himself and staggered on. I shouted a warning, but he was in no mood to listen.

On gaining the railroad track, he caught his foot. He reeled forward and fell in a heap. I saw a blinding flash; my ear drums nearly split with a frightful roar; and although I was at least fifty yards behind him then I was pitched backwards in a riot of stars. When I came to, I beheld a void in that part of the roadbed where I had seen the poor fellow. About it were twisted rails and splintered ties, but there was no trace of him as far as I could see. Then I noticed that there was a peculiar glistening coating on the smooth leaves of a black cherry tree which stood near the right of way. It was nothing, more nor less, than a varnish made of the remains of the luckless victim of his own folly. I finally found one of his heels, which I picked up and shipped in a cigar box to the man's family, with a letter telling the details of his untimely end. I still have the note of thanks from his wife, sent on receipt of this smallest of coffins.

The incident was reported at length in the press. I gave to newspaper men of my acquaintance all the details, as I felt such publicity would be helpful in arousing the authorities to the necessity of imposing greater restrictions on the handling and transportation of high explosives. Certainly, it was no more than right that none should be permitted to carry nitroglycerine unless he had a license, granted only on his furnishing credentials as to his reliability and sobriety.

## Wells On Fire

Losses from fire were colossal. Some of the wells burned for a long time before the proprietors could extinguish them and stop the waste. I remember one in Butler County which had been ablaze for months. Finally the owner offered $10,000 reward to anyone who could put it out. Victor Gretter took up the challenge. At great expense, he constructed a huge mattress, heavily weighted with lead. At each front corner he hitched many teams of horses which dragged it over the flaming hole where it was left to smother out the fire. The attempt was a complete success and both Gretter and operator were happy. (see page 26) A photograph still hangs of Gretter in the Masonic Lodge at St. Petersburg.

## Waste

What a wraith of waste hung over the young petroleum industry! To me many of the methods seemed amazingly thriftless. But the mining of petroleum had outpaced itself. Deeper drilling and the use of explosives had caused a production for which there was no really adequate means of caring. We graduated too quickly from the twenty-five barrel a day class. At first, the flood from below was pent up in reservoirs with dirt banks, where, of course, it deteriorated. Then an Iowa school teacher laid down his chalk and his penwiper and came East, where he built wooden tanks and made a fortune. Finally came the huge steel containers. These were better, but were erected more slowly. But storage and transportation of oil in such quantities as is now possible was not on our schedule.

I came to own a group of wells on the plank road to Pleasantville and my neighbors and I formed a small association for marketing our product in Titusville. We were obliged to have perfect oak barrels, tanks, stables, teams, blacksmith shops, feed stores, boarding houses and many other appurtenances. Of course, a staff of employees was needed to look after all these things. When our tanks were overflowing we filled the barrels and hauled them in wagons down the plank road to the Titusville refineries. I often led the procession as the marketeer.

Then commenced the bargaining with half a dozen men for the best price obtainable. Finally, selecting a single refiner, I would close with him without seeing a dollar in cash. Instead, I had to accept a four months' note, which, all too often would be defaulted upon at time of maturity and then we would lose an entire consignment. What with the often extortionate demands of the teamsters, who had established a monopoly, we found it hard sometimes to get out with a profit. However, oil was selling at about $5 a barrel, and even under such conditions I was making money, in various ventures. In fact, I had been able to branch out considerably, after my connection with Jonathan Watson had terminated.

In 1871, I had crossed the ocean, paid a brief visit to Ireland and made arrangements for the marriage of Miss Kelsey and myself.

While we were planning for our life of happiness together, we were in the hands of mighty economic forces. New factors had entered into the petroleum industry which were to struggle for the mastery. The signs of this great conflict were half revealed, though I was all unconscious of the impending crisis.

In Clarion County, with Mr. Watson, I had invested some money, obtained by selling wells at Pleasantville during the Petersburg boom on the Allegheny River. There came a tremendous flow of oil, and while this "Black Bonanza" was in full swing I naturally accumulated a considerable amount of money. I accordingly implored Miss Kelsey to cross the Atlantic, which she did, reaching New York late in September, 1872. [On October 2, 1872, I married Miss Kelsey and we went to our honeymooners' hotel at Niagara Falls.]

## Desperation

In gay mood, we arrived, at last at Petersburg. But we were hardly settled when Oildom was plunged into gloom and terror by the price of oil tumbling from five dollars to forty-nine cents a barrel, and the value of wells and their machinery going down to the bottom in the landslide. There was then no great Standard Oil Co. to support and carry us over a disaster, and we were in a deplorable condition. All this was caused by a flood of oil from new wells, for which there was no market, no tankage and no refineries. The business was not organized, and the producers were bewildered and did not know what to do. Some left the country selling everything to go home to their friends; some began farming, growing potatoes and other vegetables to live on. Many men couldn't raise even their fare, but the one railroad in the region took pity on them, and the conductors were instructed to give passes on brass fittings at so much a pound, which were resold to plumbers in Oil City. Those who had furs and jewelry sold them for what they would bring. I saw a prominent and wealthy well owner who had sold his topcoat, and when the weather got cold he was obliged to go about in his wife's dolman.

I didn't do a thing for three days but walk up and down trying to find a way to get the cash to make the credit come. No one had any money, or could borrow anything. All I had was wiped out, but in desperation I finally found an old farmer who would give me $800 for my holdings out of his private stocking. What a sacrifice to face! But if I agreed, I would be the only man in the district who could say he owned $800 and was entirely free of debt.

Half a million gone, and only $800 to show for it! In my dreams the furniture and walls were all marked with the talismanic legend $800, $800, $800. It was in point of fact an oleaginous Sedan, a greasy Waterloo. Many of the banks went up, and all of them were wobbling.

The largest bank was in Plumer, and it had been fitted in a most pretentious style. There it stood after the failure, wide open, with three cows and a calf using it as a stable. The owners, of course, took the money, but left the building just as it was. I purloined a paper-weight for a souvenir. This was a splendid bank and had done a large business. I rode ten miles on horseback to verify that I had heard about it, and it was worth the trouble. When I entered the bank the cows rose up and looked at me with an indignant and protesting air, just as the officials of a modern bank regard the examiner when he descends upon them without warning. Then the calf began to suck my fingers and butt me with its head in a playful spirit; but I had to be on my way, and left the place to whatever fate held in store for it.

Adjoining Plumer was that celebrated pool of oil known all over the world as "Pit-hole." When this great underground reservoir was broached and the wells broke loose, there sprang up a mushroom town. Thousands of fortune seekers rushed in, and stores, houses, dance halls, grew up as though by magic. Locations for wells had been sold for enormous sums. As the flow weakened and oil prices sagged wearily, fires became more and more frequent. Whenever a man was dunned for a bill or asked to settle his account, he looked at the creditor, sad-eyed and wan, and said, "Haven't you heard? I was burnt out at Pit-hole!" That gave him a stay and sometimes a general release, for oil men were proverbially free and generous about money matters.

I had been in difficult places many times, but this was my Trafalgar and Bull Run combined.

It was my theory, however, that this disaster was nothing more than the result of an embarrassment of riches. I believed that the old law of supply and demand would in time reassert itself, just as it has done so many times since in the petroleum industry. Petersburg, Titusville, Oil City were the precursors of Smackover and Los Angeles.

There was only one way out, and it led through the lands of Conservation and Distribution. We would, I felt, find the path slowly, yet surely.

So I rented a ten-acre field with a shanty on it for an office, and at once began buying up on credit adjoining wells that, although capable of producing handsomely, could then only be run at a loss. I hired teams and men and pulled out the casings and piled them all on the ten acres.

We moved to a three-room cottage and my brave young wife met every change with a smile. She who had been the center of an admiring circle in Belfast, now went singing about homely tasks. I have dined in most of the famous restaurants but I never enjoyed meals more than I did in those days in that cottage in Petersburg. In summer, we were embowered in rambler roses in our tiny home, and we kept in our garden many of the old-fashioned flowers brought from England. Our amusements were few, but we had our books, a carefully selected library with Master Shakespeare in the place of honor. Very often, Mrs. Bayne worked at her easel. A man in stress and storm never had a more peaceful nor a lovelier haven than it was my good fortune to have in those days when ruin came to us.

While we were working and watchfully waiting for the return of prosperity, we made as merry as we could.

## Fighting Wildcats

"If after the manner of men I have fought with hearts at Ephesus," St. Paul said, "what advantageth it me?" We who tried to make good in Oildom had fights on our hands continuously. They called us Argonauts, but much of our time was given, not to seeking the Golden Fleece, but to keeping from being fleeced or from having the wool pulled over our eyes. The bad men of the Wild West were lambs compared with the wildcats of the Wild East in the oil fields.

There had flocked into the petroleum regions adventurers and ne'er-do-wells, besides the thousands of men of ability and character who later were to become leaders of a new era. The riff-raff were ready and all too willing to jump each other's claims, to set fire to each other's oil, and to steal casings out of everybody's wells.

When the boom came to Petersburg and all Clarion County shrilled with the beating of the drills, the Wild East came into its own. We had pipe and tool thieves, compared with whom Wyoming cattle rustlers were plaster saints. They prowled around our leases, and came and went from one county to another, keeping generally about two or three steps ahead of the sheriff.

When the oil excitement broke loose on the banks of the Clarion River, oil supplies were then very high and hard to get, and if a contractor could furnish machinery, he was sure of a contract at his own figures.

## Keep Your Eyes Open

Two very enterprising contractors, whose names we will say were Charlie Clayton and Jim McGraw, conceived the brilliant idea of putting the Allegheny River between them in their operations. McGraw infested Butler County; Clayton was the curse of Clarion. Larceny was their liaison officer. They stole their supplies from local operators, then ferried their loot across the river at night, and each handed it over to the other party to use in his contracts. This worked to perfection for a long time, as the victim could never find a trace of the stolen plunder in his own neighborhood, and as a consequence, the more frequent the thefts, the more fast trotters the firm drove, and the more diamond horseshoes appeared on their ample shirt fronts.

***Our Trusty Detective Above the Cooperhead Line***
***The Clarion River hillsides are famous for their copperhead snakes. (Sketch by S. G. Bayne)***

I had good reason to believe that this precious pair was raiding my pipe pile, but I never could get actual proof. Finally, several other operators and myself employed a detective. Disguised as a laborer he went to a well which Clayton was drilling on the picturesque banks of the Clarion River, directly opposite the celebrated grove of rhododendra on the Logue farm. This patch had never been despoiled of its flowers, for Oildom regarded it as a thing of beauty which could be a joy forever to its occupants, as it contained the largest and most exclusive colony of copperhead snakes in the eastern United States. Our man stayed around the Clayton well for days. It proved to be a "duster," and as is usual when wells prove dry, tools were withdrawn. When the jars came up the detective scanned them closely for the names of the makers. At last he found a place where the strokes of the hammer had failed to obliterate the factory marks.

"Clayton," said he, "fine jars you have. What firm makes your tools?"

"Er, well, yer see," was the reply, "I don't know anything about them details. Jim McGraw furnishes the tools."

We then swore out a warrant and both partners were arrested, and they gave bail for their appearance before Squire Bostaph, out at his farm two miles from Petersburg.

The whole countryside was now in a fever heat of excitement. The two thieves were surrounded by a gang of bad men, and in Petersburg, at that time, a week without a murder or a shooting affray was counted as lost. I, therefore, prepared for action. Watson & Bayne were then drilling a dozen wells employing forty-eight men and when the day for the hearing arrived, I shut down all our wells and hired teams to take the drillers to the Squire's office.

Daniel O'Day, future Standard Oil magnate, then plain Dan, was busy a mile or so away introducing the first pipe line system practical in the oil country. I explained to him how necessary it was to have a few disinterested witnesses, and he let me have as many big fellows as he could spare. In order to make the joy of all complete, I borrowed enough revolvers at hardware stores in Foxburg and Parker's Landing to provide every man with one. This was an imposing bodyguard, and made a fine showing as we started for the field of battle.

## Squire Bostaph

When we reached the Squire's office, we found the Clayton-McGraw cohorts ensconced in the office, or lounging on the grass plot in front. There were nearly a hundred of this rough crowd, and all were armed with shot guns.

*Old 'Squire Bostaph raised his gun and stated, "Gentlemen, if there is going to be any shooting here today, I propose to do it myself." (Oil and Gas Man's Magazine, Butler, Pennsylvania 1908. Drake Well Library, Titusville, Pennsylvania)*

Squire Bostaph was a Dutchman, and, what with his short, white beard, his square face and his shrewd eyes, he might have been mistaken anywhere on earth for the late Paul Kruger, once head of the Boer republic in South Africa. As he entered his court or office shanty, the magistrate looked us all over sharply from under his eyebrows. He said he had forgotten something and excused himself. When he came back he brought a double barreled shotgun and laid it on the table in front of him.

"Order in the Court," announced the Squire. "Gentlemen, if there is going to be any shooting here today I propose to do it myself. Drop your weapons on the floor, and the first man who picks one up, I shoot on the spot!"

The Squire was a man of his word, as we all knew. We obeyed his order and gently put our guns down.

The defendants pleaded "Not Guilty" and declared that they could prove their innocence at the County Court House. They furnished the cash bail which the Squire had demanded of them, and then all were ready to leave.

## Look Out, Mr. Bayne!

*Samuel G. Bayne and his brave horse, "Buckskin," almost killed on their way home to St. Petersburg. (Oil and Gas Man's Magazine, Butler, Pennsylvania 1908. Drake Well Library, Titusville, Pennsylvania)*

I led the procession from the office down the dugout road to Keating's Furnace. I was driving a light buckboard drawn by my mare, Buckskin. The creature was remarkably intelligent, for she knew as much as an average man. That day she was strangely restless, and as we went along she glanced from side to side nervously. The road ran down hill at a steep angle, and at its left was a precipice, in the bottom of which were rocks and jagged stumps. At that point it was very narrow, in fact, it was much like a Rocky Mountain path, and there was room on it for only one vehicle at a time.

"Look out, Mr. Bayne, behind you," came the warning shouts of my men in the rear.

I saw, as I turned quickly, that Clayton and McGraw, driving one of their best teams to a heavy buckboard, were coming at me full tilt, with the evident intent of forcing me over the brink into the ravine. I waited until the last moment, and then, just as they were about to crash into my rig, I pulled its wheels up on the right side. The mare sensed the situation, for there she stood in her strange position, stock still. The men, surprised at this maneuver, could not get control of their team in time to prevent their horses and backboard shooting over the top of the steep embankment. Both animals were instantly killed, and the buckboard was smashed to smithereens, yet the two men escaped with only bad cuts and bruises.

Months passed, and still the case was not reached. The partners loudly proclaimed that they were going to get me out of the way before the trial. I did not feel like obliging them, so I kept in the middle of the road and traveled as much as I could by daylight.

One evening, when I was on my way back from a business trip to Oil City, I took the precaution to go through the train as it neared Foxburg, which was the nearest station to Petersburg. I saw Clayton and McGraw sitting in the back seat of the last car. Evidently they had followed me from Oil City and I rather suspected it was their intention to attack me on the three mile stretch of lonely road between Foxburg station and home. So I had a word with the conductor, which resulted in my dropping off the train as it slowed down for a hardly perceptible moment at Fullerton, two miles away. The train did not really stop, but I landed easily on the platform. The telegraph operator lent me his revolver. Feeling rather better prepared to meet wildcats, I then took a path through the woods, bound for home.

## No Thanks!

Petersburg was in those days a wide open and an all night town. Such places had the original White Way, for they were lighted up with the huge jets of flaming natural gas, which roared from the earthbound reservoirs. As the drillers, on their towers worked all hours of the twenty-four, and usually changed shifts at midnight, none of the public places ever closed. At three o'clock that morning, I walked into the crowded hotel and saw Clayton and McGraw.

"Boys," I said, going up to them, "won't you have a drink after your long walk?"

They stared at me.

"No, thanks," Clayton replied, "the drinks are all on us!"

This was only a truce, however, for I felt it was my duty to make a public example of those two defendants, and to do what I could to bring law and order into our section. When the trial came off, to relieve the county prosecutor as much as possible, I employed James Campbell as special counsel to represent the other producers and myself. The accused men retained an eminent, criminal lawyer, William Corbett. The case was a celebrated one in a way, and Judge Trunkey, late Chief Justice of the Supreme Court of Pennsylvania, presided.

I was much interested in the jury, for, although this trial was held in an oil country, not a man understood anything about the technical details of petroleum mining. Even the use of so common a term as "jars" meant nothing to the jurors.

"Your Honor," exclaimed the foreman, jumping to his feet, "this man, Bayne, has not told this jury what kind of fruit was in those jars!"

The jury did at last get all the facts, and the law, eventually, found Clayton guilty, but McGraw was acquitted.

Clayton was to have been sentenced that evening. Before court reconvened, his wife and nine children, ranging in age from a babe in arms to fourteen years, were brought to my hotel by the prisoner's most able lawyer. The sobbing family said they would certainly all starve if the head of it went to the penitentiary for a long term, and they asked me to appeal to the court for leniency. I did so, and as a result the light sentence of only a year was imposed. After it was all over an enthusiastic crowd tried to carry me to the hotel on their shoulders but I escaped with disheveled clothing.

But Clayton arranged for his own commutation. His wife, by the end of the week, smuggled saws and ropes to him in a basket of fruit, and he did the rest. Being a skilled woodsman and of strong physique, he got through the forests to the Ohio River, and eventually reached New Orleans, I am told.

He had arranged to meet his partner at night at a certain secluded point in Turkey Run, Clarion County, to have a settlement of their contracting business. With the aid of a crude oil derrick lamp and a large pine stump for a table, they adjusted their affairs. It was a strenuous settlement, as Clayton afterwards explained that each stuck his bowie knife into the stump beside him as evidence of good faith.

Nothing was heard of either of the men for some time after that, but one moonlight night, when

*An original St. Petersburg Police Badge.*

I was returning home on horseback through a belt of woods on the Heeter farm, a horseman rode silently up behind me. It was Charlie Clayton. I thought my time had come.

"Don't be alarmed, Bayne, I'm the last man in the world to hurt you. But I want to ask you a question. I know you can have me arrested for breaking jail, but will you do it, or will you give me a chance to start life again and behave myself?"

*Bayne's Midnight Ride to Clarion*
*My Horse Buckskin (Sketch by S. G. Bayne)*

I gave him my word, that as far as I was concerned, he should have his chance, and we shook hands on it. The best part of this story is that he did make good, for he afterwards won a fortune in the oil business in a legitimate way and became a quiet and respected citizen.

The oil regions, which went through all the phases of lawlessness common to rushes for gold and treasure, gradually took on the semblance of obeying statutes made and provided.

## Bayne's Midnight Ride

I remember one occasion on which I went to the help of an Irish friend of mine, Charlie Watt, who, without any provocation on his part, had been attacked by Chief Wilton of the Petersburg fire department. There was a Fireman's Carnival being held, and the dispositions of many of the visitors had a three alarm quality. Wilton knocked Watt down and put his foot on his neck to prevent his rising. I reached down and assisted Watt to his feet, but in so doing my elbow struck Wilton, who then transferred the attack to me. The battle went on with varying fortunes until I finally wore him down and put him on his back.

Wilton was a very popular chief and his men could not brook this public defeat of their leader. So they commenced to shove me toward an oil well supply store where I did my trading. The place was deserted by the clerks, who were all out enjoying the holiday. Rushing in, I immediately swung round and locked the door on the inside, ran to the exit at the rear and made my escape. I never stopped until I reached home, about half a mile distant. Soon I looked out of the window and saw the whole mass of excited men racing toward the house full tilt, each grasping an ax handled taken from the store, all anxious to show their chief what loyal henchmen they were, and to punish me for daring to dispute with him.

I left by the lee side of my house where I could not be seen, and under its cover gained the woods near by, still unobserved by the angry, clamoring mob, and gained the summit of a high, steep bluff running down to the Clarion River. Making my way down the incline, sometimes tumbling, sometimes tripped by the brush, I finally reached the shop of Abe Soliday, who built my oil tanks. I told Abe of my scrape, and he quickly hitched up his team. We drove down along the Clarion River to where it joined the Allegheny, and thence to Parker's Landing Ferry. From there we drove to Petrolia in Butler County, where I gave myself up to a Justice of the Peace, who fined me five dollars and discharged me. I was free once more, but was obliged to stay there until the excitement at home had subsided.

On my return, Wilton had begun suit against me for heavy damages for what he termed as assault, and also complained of me to the Criminal Courts. After a month or so, he gave out a story that he was sorry he had sued me and that he intended to drop the cases. Being very busy, I was only too glad to believe it. On a Sunday night, however, I received a message from my lawyer to the effect that if I did not get to Clarion by ten o'clock next morning my bail would be forfeited.

I jumped from bed, and saddled my old, reliable mare, Buckskin, which had always been ready for every emergency. It was snowing heavily at the time, and by a wrong turn I found myself in a freight yard between two moving trains, going in opposite directions. The noise was terrific. Buckskin stood through it all like a horse of bronze. Had she flinched both of us would have been killed.

By daylight, I reached the little hamlet of Shippenville, where after Buckskin had been freshened up with a hot bran mash and two quarts of oats, she plowed for seven miles more through the mud. I flung the reins over her head as we reached the Court House green. Glancing back, I saw her sprawling on the turf and rolling over and over again, as if in glee over our success. I dashed up the steps and entered the courtroom just as the clerk for the third and last time, was calling the case of "Wilton versus Bayne." My counsel, Judge Wilson,...escorted me up to the presiding justice. The case did not last long, for all proceedings against me were dismissed. I received many congratulations, not only on my vindication, but on what the newspapers of the day referred to as "Bayne's Midnight Ride to Clarion."

In those days at Petersburg, where tragedy so often followed on the heels of comedy, and the law was slowly asserting itself, we had to struggle hard to hold our own. Happily, the social conditions became better organized. I was able soon to devote more attention to activities of unusual interest which had always fascinated me.

*Charles W. Pratt*
*The firm of Bayne, Wilson, and Pratt were large operators. Mr. Pratt found out the hard way just how deep the mud was on Main street, St. Petersburg in the spring of 1874 - Courtesy Drake Well Museum*

## Charles W. Pratt

Charles W. Pratt, who has had a varied career in the oil business, was born in Allegheny City, August 27, 1847.

He was taken into partnership by Bayne & Wilson, the firm becoming that one which has become so well known wherever oil is produced, Bayne, Wilson & Pratt. Pratt's connection with the oil business dates back almost to the inception of the industry, and in it he has had the usual share of reverses which can be summed up in "hard luck stories."

Mr. Pratt landed in Foxburg, Clarion County, where he remained some time, finally selling what interests he had acquired, and in the spring of 1874 located in St. Petersburg. In the vicinity of Edenburg he operated on the Widow Black farm, the Gates and Jerusalem farms. This field occupied his attention until 1879, when he followed the migrating hosts of grease hunters to Bradford, remaining there but a short time when he went to Leadville, Colorado, to take charge of and take up mining claims for Bayne, Melvin & Fuller.

There is a legend in St. Petersburg as to how Bayne and Pratt got acquainted, which is as follows: In the spring of 1874 the mud was eight feet deep in the main street of St. Petersburg, and there was only one crosswalk, that in front of the St. Petersburg hotel and not wide enough for two people to pass. Bayne and Pratt one day met in the middle of the crossing. Pratt said, "Stranger, I was here first. You will have to turn back or jump off." Bayne replied, "I will flip a penny with you to see who makes the jump." Pratt acquiesced in this arrangement. Bayne tossed the penny and Pratt lost. Pratt jumped overboard and went out of sight. Bayne got frightened and made a break for the hook and ladder house, got a pole 40 feet long, with a hook on the end of it, and fished around for about 20 minutes and finally got Pratt by the coat collar and yanked him to the sidewalk. Bayne then queried Pratt as to his name, and Pratt told him. "You are the gamest man I ever met," said Bayne, and with that remark he presented Pratt with a hundred dollar bill. This was the beginning of a friendship that has lasted 24 years and bids fair to continue indefinitely.

When the active operations began in the vicinity of Pittsburgh, Mr. Pratt removed to that place and took charge of the office of Bayne, Wilson & Pratt, from which orders were booked from the producers operating in the territory adjacent to that city. Mr. Pratt, with his family, makes his home in the East End.

## Samuel M. Kier

As early as 1847, S. M. Kier sold oil as a medicine and cure in Pittsburgh. He is given credit as being the first to distill petroleum into an illuminent. He set up a one-barrel still on Seventh Avenue and became America's pioneer oil refiner. Mr. Kier was a producer in the Clarion field. S. M. Kier slumbers in Allegheny Cemetery. "Peace to his ashes," John J. McLaurin. (*Sketches in Crude Oil*, 1896)

*Men Who Made Oil History*
*Samuel M. Kier*
*Mr. Kier is given credit as being the first to distill crude oil into an illuminant and also was a heavy investor in the Clarion oil regions.*
*Courtesy Drake Well Museum*

## The Second Sand
## Is This Story Really True?

There's a story about a stranger who was slogging along Main Street and kicked what seemed to be a man's hat lying in the middle of the mud. From the depths a voice came angrily, "Say, that's my head in that hat. Don't you kick it again."

The traveler peered down, horrified. "You'll be buried alive," he screeched.

"Never you mind, stranger. I've got a good mule under me and he's got to the "Second Sand Rock", and I will be back on dry land in a minute."

## Men Who Have Made Oil History
## Edwin Colton Bradley

E. C. Bradley was born December 21, 1833, at Crossingville near Meadville, Pennsylvania. Attracted by the Pennsylvania oil boom, he went to the Oil Creek district. He had a hard struggle to gain a foothold, purchasing oil from the producers, barrelling it and hauling it to cars for shipment to the refinery. Mr. Bradley eventually established profitable and highly successful operations throughout the oil region. E. C. Bradley had his ups and downs but always rebounded to become a well-known figure in the Clarion County oil fields.

*Edwin Colton Bradley*
*Courtesy Drake Well Museum*

In 1872, he in company with H F. Whiting, Marcus Hulings, the Blakslee Brothers and R. Richardson, organized and built in Clarion County, the "Antwerp and Oil City Pipe Line," one of the first oil transportation lines ever constructed. In 1877 this line was taken over by the Standard Oil Company with seven other lines and merged into what was known as the United Pipe Line Company. Mr. Bradley's Edenburg oil office is on page 139.

The panic of 1878, when oil went from four dollars per barrel down to forty cents, caused E. C. Bradley's failure due to his kindness in helping his friends by endorsing their notes. In 1879 he moved his family to Bradford for a new start.

Many of the most prominent oil men in the history of the oil industry were his friends and business associates. At one time John Steele, better known as "Coal Oil Johnny," drove team for him along the oil creek, hauling oil. His career in the oil business made him a familiar figure in the old Pennsylvania oil towns of Rouseville, Oil City, St. Petersburg, Edenburg, Elk City and Bradford.

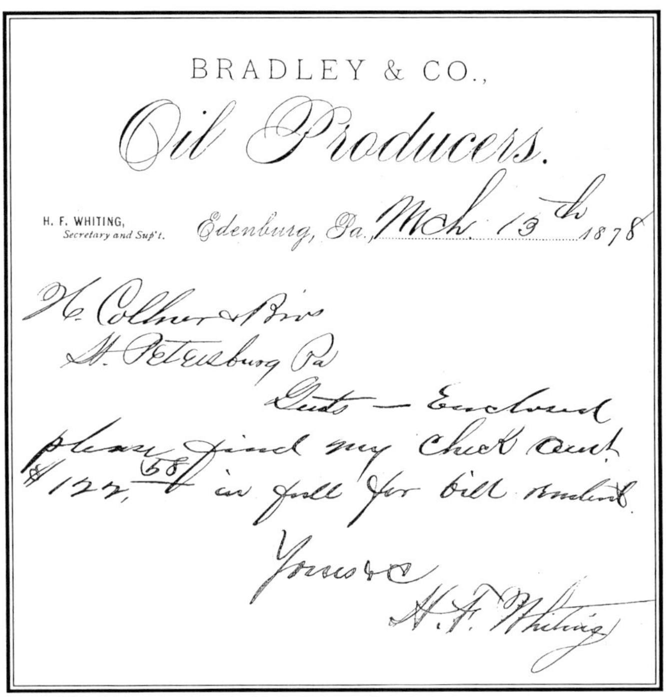

BRADLEY & CO.,

Oil Producers.

H. F. WHITING,
Secretary and Sup't.

Edenburg, Pa. Mch 13th 1878

H. Collner & Bros
St. Petersburg Pa
Gents — Enclosed please find my Check amt $122.68 in full for bill rendered.
Yours &c
H. F. Whiting

***Billhead for E. C. Bradley and Company***

Mr. Bradley owned and operated the Empire Gas Company at Wellsville, New York, a very successful natural gas company. In 1900 he was handicapped exceedingly by the loss of his eyesight but his indomitable will power overcame all obstacles and he remained in active control of his business up to six months before his death. Edwin Colton Bradley died April 22, 1913 at Wellsville and his death was greatly mourned. To you, Mr. Bradley, thanks. "Peace to his ashes," John J. McLaurin. (*Sketches in Crude Oil*, 1896) See page 475.

## Jerome B. Aiken

*Men Who Made Oil History*
***Jerome B. Aiken***
***J. B. Aiken was a hotelman and an oilman. In 1872, he moved the Opera House from Petroleum Centre to St. Petersburg and continued to follow the new fields. Courtesy Drake Well Museum***

Mr. Aiken was born March 18, 1843 on a farm at Johnsonville, New York. He is one of the oldest producers in the oil regions, having first made his appearance upon the scene of the great petroleum industry in 1860 and in the early days of historic Oil Creek. Mr. Aiken made his first investment in the oil business on the Eveleth & Bissell tract, just across the creek from the old Drake well, located on Watson's flats.

J. B. Aiken worked the fields with much success. He finally gravitated to Titusville where he made headquarters until 1870 when he drifted with the tide to Petroleum Centre. He made good business deals in Petroleum Centre and one of his purchases was the Opera House.

## Petroleum Centre's Opera House

The following information comes from the *Petroleum Centre Daily*, April 11, 1872: Jerome B. Aiken has taken down his large concert hall located on Washington Street and is about to remove it to Petersburg. A boat load sailed majestically down the placid bosom of Oil Creek this forenoon. We hope the good ship may avoid the many rocks and breakers to be found on the shores of the creek and the Allegheny River and arrive safely at its port of entry.

**May 5, 1872:** We are informed by cable dispatch that the good ship carrying the opera house is caught on a sandbar a short distance this side of Oil City in latitude 30, longitude 29 1/4, and practically wrecked. All hands were saved. Fate appears to attend the voyages of the fast sailing ship, Arkansas, this being the second time she was suffered wrecking.

**May 13, 1872:** The boat containing Jerome Aiken's opera house which ran aground near Rouseville recently has been unloaded and transported the balance of the distance to Petersburg by rail.

**June 14, 1872:** Our townsman, Mr. Jerome B. Aiken, has got his new opera house in St. Petersburg completed and an opening ball was given last night. Next week a theatrical company commences an engagement in the hall. Jerome is a good fellow and we wish him success in his new enterprise.

The opera house burned down in the fire of February 25, 1873 (page 171). Jerome Aiken's opera house, dwelling, billiard room and restaurant, $10,000 insurance and $4,000 insurance.

Mr. Aiken became a permanent fixture to this ancient oil town and one of its most progressive, enterprising and influential citizens. It was here that Mr. Aiken first embarked in the hotel business and bought the St. Petersburg Hotel where for ten years he conducted one of the best taverns to be found in the oil regions, while at the same time he was largely engaged in the production of petroleum. After sharing in the ups and downs for a decade in St. Petersburg, he set compass for Richburg, New York.

In Richburg, he weathered the storms for six years and after meeting his first absolute Waterloo in oildom and losing all he had acquired, he bid adieu to Richburg and set sail for the oil field of Washington, Penna., arriving there in 1886. After carefully looking the situation over he engaged exclusively in the production of oil, and from his first investment until the present time success has eminently crowned his efforts.

## Honorable M. L. Lockwood

"In 1865, I came to the oil regions of Pennsylvania and engaged in the production of oil. In 1871 I was married to Lydia H. Tompkins, of New York, and settled at Richmond, Clarion County, Penna." (Page 131) They lost a little girl in 1875 and her small memorial is in the Union Cemetery in St. Petersburg.

*Men Who Made Oil History*
*Hon. M. L. Lockwood*
*Marcus Lockwood started his Clarion County oil days in Richmond near Antwerp City.*
*Courtesy Drake Well Museum*

Marcus L. Lockwood was a well-known operator and prominent figure in the Pennsylvania oil fields. Mr. Lockwood was involved in the Clarion County oil territory from St. Petersburg through Elk City. His home was located in Richmond, one-half mile from the oil town of Antwerp City. M. L. Lockwood was on his own since the age of fourteen. While in the Oil Creek region, he became one of the oil scouts of that period. He operated with success in those regions before coming to the Parker's Landing and Petersburg territories.

Hon. M. L. Lockwood became active in politics and in movements for the betterment of the oil producers and the public in general. Mr. Lockwood often said, "A cat in hell without claws is less helpless than the oil operator." He was sent to the Pennsylvania State Legislature and a part of his work there was the championing of the free pipe bill. In national legislation, probably more credit is due Mr. Lockwood than to any other man for the work he did, with the cooperation of the late Senator Regan of Texas in the framing and final passing of the act which created the Interstate Commerce Commission.

In later years Mr. Lockwood was identified with oil developments in Oklahoma and Texas. The Hon. M. L. Lockwood died January 1, 1917, at his home in Tulsa, Oklahoma. To you, Mr. Lockwood, "Peace to his ashes," John J. McLaurin. (*Sketches in Crude Oil*, 1896)

## John H. Galey

John H. Galey was born in Clarion County in 1840. Entering the business early in its history, Mr. Galey was throughout his career a bold and enterprising operator, forging to the front mainly through his own perseverance and ambitions, coupled with a tenacity of purpose which never let a failure deter him from continuing explorations. Many of the greater fields and pools developed since the inception of the oil business owe their existence and life to the ambitions and determined character of this man. Some of the largest producers of oil the country has ever known have been the product of his keen foresight in pushing operations in what might look to another eye like unproductive territory.

Associated with Mr. Galey in his oil operations at Parker's Landing was William Hartley, of Bedford, Penna., and among the famous wells they opened up and which tended to stimulate the early development of the Butler, Armstrong and Clarion fields in 1869 and 1870 were the "Island King" and "Island Queen" wells on Stump Creek Island in the Allegheny River, and the "Annie Lee" well on the Fowler farm at Foxburg. (see pages 7 and 8)

In the spring of 1870, John Galey leased a tract of land belonging to Judge Keating, situated up the Clarion River at a place called Grass Flats. The well was a success, producing 60 to 70 barrels per day. This caused excitement, land was leased, and a forest of derricks grew. John Galey was a shrewd oil operator and one of his favorite remarks was, "That fellow's so cold blooded, he sweats ice water."

The following information comes from The Pennsylvania Historical Museum Commission, Drake Well Museum, Titusville, Penna. A special thanks to Susan Beates, Curator, May 2004.

*John H. Galey*
*An old Clarion County oilman*
*Old John was the "Wildcatter's Wildcatter."*
*Courtesy Drake Well Museum*

## A Short History of the Beginning Of the Mid Continent Oil Field

(Reproduced as originally written by Thomas M. Galey, August 1923)

This section of 4 1/2 casing was taken from old Norman No. 1 well located on the East side of the City of Neodesha, Kansas and located on the West bank of the Verdigris River, about 200' North of the bridge. It was donated by The Prairie Oil & Gas Co., a joint going to the Petroleum Exposition in Tulsa and a section to the Drake Memorial Museum at Titusville.

This well was the first commercial oil producer in the Mid Continent Field and its discovery lead to the formation of The Standard Oil Co. of Kansas and The Prairie Oil & Gas Co. It was discovered in 1892 and shot on October 4, 1893 from 805' to 817' with thirty quarts of nitro glycerine made by George Perry and brought by him in a spring wagon from Webb City, Mo. The initial production was about 12 barrels.

The casing used was 434' of 5 5/8 and 667' of 4 1/2". It was purchased from The Oil Well Supply Co. in Pittsburgh in 1891 and was first shipped to Osawatomie, Kansas together with a string of drilling tolls by Major W. M. Mills who was prospecting for gas. It is of wrough iron with a three inch weld and of fine quality. It was probably made by The Reading Iron Works at Reading, Penna. The well was abandoned by its owner, The Prairie Oil & Gas Co. in the year 1917 when water broke into the lower joints.

The first well in the State of Kansas was drilled in 1860 near Paola Kansas with unimportant results. The first gas well was drilled just East of Paola by John Werner, July 25, 1882. The first oil well was drilled West of the town in 1888, depth about 360'. A twenty five barrel refinery was built in the early nineties and some journal oil lubricants were sold to The Missouri Pacific Railroad, but the success of the venture was of no importance. The company continued a few years and went bankrupt and after five years more was taken over by The Standard Oil Co. of Kansas and dismantled. A barren well was drilled 2000' at Paola in these early days, also another at Howard Kansas and another at Atoka, Oklahoma. There were a dozen or more shallow gas wells at Paola and several small oil wells. Gas developments spread to Cherryvale, Osawatomie, Coffeyville and Independence.

Major W. M. Mills, who still lives at Osawatomie, had had indifferent success in his gas operations at Paola and elsewhere when he came to visit friends in Neodesha. Citizens of that town had made an effort to get together to test for a gas supply but without doing anything. Mr. Mills proposed to drill a well himself but was unable to get the 5000 acres he demanded until Dr. Thomas Blakeslee and other business men gave their aid and secured the leases on a 10¢ rental basis and including all the minerals.

Major Mills secured the services of McBride & Bloom of Independence, Kansas who had been successful drillers for gas and water and the well was contracted at $2.50 per foot. The forty foot standard derrick was erected on the four acres of the T. J. Norman land and in a few weeks operations ceased, the report being circulated that a small amount of gas had been found. Barton Blakeslee, then a boy, was warned away from the well. He suspected something unusual and hid in the bushes and at the time of drilling-in saw the operators covering up signs of oil. He was much excited but kept silence upon his fathers advice. Several days later much curiosity was aroused by a tool box in the freight station consigned to Ottawa for it contained two kegs of liquid.

Major Mills then completed another well a mile West during the winter and secured a producer an as the discovery was pretty much an object of gossip, he frankly admitted it. The Major was unable to satisfactorily pay the contractors and after trying in vain to interest Eastern capital, finally brought out John H. Galey of the firm of Guffey & Galey of Pittsburgh, Penna. This was early in 1893. Mr. Galey promptly took Major Mills into partnership and he shot the first well and put it to pumping. Active operations were begun on a great scale and after a few months Guffey & Galey bought out the interest of Major Mills for $4400.00 and he returned to Osawatomie to develop his gas business at that place, Paola and Spring Hill.

During 1893-4 Guffey & Galey leased over a million and a half acres extending from the Indian Territory line in a North East direction to the Missouri State Line. In 1893 fifteen wells were drilled and from that time to 1895 when Guffey & Galey sold out to The Forest Oil Co. sixty eight wells were drilled in the immediate vicinity of Neodesha and forty or more wild cat wells were drilled throughout their huge acreage. Some of these wells opened important pools now in old age. The oil developments by others began to spread here and there and pushed well down into Oklahoma under very discouraging conditions.

The earliest movement in Oklahoma occurred at Chelsea in 1886. Edward Byrd found there enough evidence of oil to interest several Eastern men who incorporated a company under the laws of Kansas and in 1891 drilled eleven wells, depth 325'. No realization could be made on the investment and finally the property changed hands, becoming the famous "Big Lease" of The Cherokee Oil & Gas Co. In 1896 the Curtiss law became operative which upset the plans of all Oklahoma operators. It was in 1893 that Michael Cudahy, the meat packer, obtained a blanket lease on 200,000 acres in the vicinity of the present city of Bartlesville. The Curtiss law cut him down to 640 acres with one well. Such were the difficulties in those days.

Hence, like the Drake well, the existence of oil was known in the Mid Continent for several years previous to the Norman discovery well. It was early in the century, preceding the Drake well, that a big well was discovered near the Cumberland River in Southern Kentucky during operations for salt.

The activities of Guffey & Galey convinced the Standard that the time for their entry had come. It is surprising the large number of wells which were drilled in the pioneering effort before the Norman well was discovered. By the close of 1895 Guffey & Galey found themselves with over 100 wells, 150,000 barrels of oil in storage, a huge acreage and inability to sell a barrel because of adverse freight rates which were controlled in New York City. They had nearly $400,000.00 invested and were heartily sick of it. Going to New York the matter was laid before John D. Archbold and Henry M. Flagler. Their Forest Oil Co. took over the property on October 30, 1895 for $150,000.00 cash to the great relief of Guffey & Galey whose pioneering effort had cost them heavy loss.

The Forest Oil Co. promptly put fifteen drilling rigs to work, taking over most of the men of the old organization, and prosecuted operations with vigor so that by July 1897 they had production in seven counties with 224 wells including 109 dry holes.

The Prairie Oil & Gas Co. was then organized to take over these Western holdings of The Forest Oil Co. which it did on Jan. 1, 1901. Included in this was a large acreage which had been leased by W. L. Mellon of Pittsburgh, then a young man and now Chairman of the Board of the Gulf Oil Corporation. Operations up to that time were of a decidely experimental character. Independent operators had succeeded in developing only small quantities of oil in remote sections, but they desired the Prairie to buy it. To accommodate a few of the heavier producers, tank cars were provided and storage tanks erected at Neodesha where a small pipe line system had been inaugurated. In 1903 operations around Chanute, Humboldt and in Montgomery County had grown to sufficient magnitude to induce the Prairie to extend its local lines to connect these fields and on August 1, 1903 they began the systematic registering and publication of runs and shipments.

To care for this oil, in 1895 the Standard Oil Co. organized The Standard Oil Co. of Kansas as its Western refining division and 1897 built a small refinery at Neodesha having a capacity of 500 barrels per day. In 1901 four more stills were added and increases have continued ever since that time.

Realizing early that the inauguration of a broad system of pipe lines and storage facilities would immediately stimulate development of the Mid Continent field, the Prairie promptly began an elaborate extension of all its facilities including the erection of 123 tanks at Neodesha, about $35,000,000.00 being invested in a few short years. Meantime the industry grew rapidly but the Prairie was able to keep pace with developments which ranged far South into Oklahoma. The biggest event was when connection was made with the Eastern Pipe Line System and Oklahoma oil became a factor in the competition with Eastern oil. In 1903 the great rush came when Eastern operators were attracted Westward by the discovery of the sensational greatest gusher at Beaumont Texas which was instituted by Captain Lucas and drilled and completed by Guffey & Galey. This completion lead to the formation of The Gulf Oil Corporation by the Mellon interests when they bought out the interest of John H. Galey for about $700,000.00. As usual, this discovery was preceded by the discovery of shallow production at Corsicana Texas which was instituted by citizens of Corsicana and taken over and developed by Guffey & Galey for a half interest and was the first commercial production in Texas. These wells flowed their production for years from a shallow depth.

Eastern oil men flocked into the Mid Continent and the Prairie soon found its facilities swamped with oil and the producers with low crude prices. This condition of excess supply or undersupply have continued to this day, history repeating itself with unfailing regularity.

The history of the early development of the Mid Continent field is a history of the Prairie Oil & Gas Co. and it is hoped that soon that history will be written while some of the principals are still living.

The major epoch making discoveries of oil in the United States are as follows:

1859, at Titusville, Penna., Col. Edwin Drake
1866, at Venture Co., Cal., A. J. Salisbury
1892, at Neodesha Kas., Maj. W. M. Mills

The only living discoverer is Major W. M. Mills of Osawatomie, Kansas who is one of the last of the old pioneers. He was born in Clarion Co., Penna. in 1851. The great decline in the price of Eastern oil in 1884 caused him to come west seeking new fields and he had a hard struggle but won out.

Col. J. M. Guffey still lives at an advanced age in Pittsburgh Penna. where he still maintains an office and has a fine home on Fifth Ave. In the early days when politics was big factor in business, he was a power in the Democratic party and was known as a "boss" until he finally locked horns with Bryan and both dropped out of the arena. He was gifted with financial ability and his alliance with John H. Galey under the firm name of Guffey & Galey made a strong team.

John H. Galey was born in Clarion Co. Penna. in 1840 and died in Joplin Mo. April 12, 1918 at the age of 77. He entered the oil business in the early development at Parkers Landing and by his boldness, hard work and pioneering effort he later made the firm of Guffey & Galey known wherever oil was produced. He located and drilled the first high grade oil gusher, the famous Mathews well at McDonald, Penna. This well made a worlds record of production, having been exceeded by very few high grade wells since. This discovery was preceded by years of strenuous effort in Pennsylvania in a buckboard with a team of horses. His greatest discovery was the famous Lucas Gusher at Beaumont which well made the record output of any well in the world to that date. He was a pioneer in the fields of West Virginia, Eastern and Western Ohio, Illinois, Kansas, Oklahoma, California and Texas. He was one of the first men to observe the anticlinal theory of oil and gas accumulation when he observed, in the Murraysville Gas Field, that where the Pittsburgh Coal lay high the gas volume was greatest.

This observation lead to his revealing to I. C. White of the West Virginia Geological Survey, his discovery who mapped for Guffey & Galey the axis of the major folds in West Virginia in an attempt to develop a large gas supply for the Westinghouse Co. at the instance of George Westinghouse. Dr. White later wrote the first practical description of the anticlinal theory of oil and gas accumulation.

Among odd wells he has drilled might be mentioned the first in the Colinga field in California. He drilled one near Washington D. C. he was reluctant to mention. A promoter salted a well near that city and when the bailer brought up oil, Guffey & Galey took over the property and drilled another hole. They found out too late that the discovery well had been salted. He once drilled a deep well a few miles North of Salt Lake City Utah which showed a little gas. He drilled a number of shallow holes at Seven Lakes New Mexico which showed a little free oil on fresh water. He was too far along in years to become a factor in the Louisana development but for years he maintained that some day big production would be found in Southern Arkansas, such was his instinct. Like many pioneers he died without large estate but his name will carry on with that of Theodore Barnsdall and many others who did the big business of the early pioneers. Refer to pages 463 and 464.

Thomas M. Galey
Independence, Kansas
August 4, 1923

## James M. Guffy

*Men Who Made Oil History*
*J. M. Guffy*
*In 1876, J. M. Guffy was a citizen and council member of St. Petersburg. In 1877, he was Vice-President of the St. Petersburg-Turkey City railroad. In 1878, he was the secretary of the St. Petersburg Driving Park. J. M. Guffy went on to become truly a leader in the world of natural gas and crude oil discovery and distribution.*
*Courtesy Drake Well Museum*

John Galey and J. M. Guffy always did things in a big way and developed the world famous McDonald field, in 1891, where they leased for oil a large number of the most prolific farms ever operated in the whole history of the trade. It was on one of these tracts that the famous Mathew well, which had a record for oil production much in excess of any ever drilled in America, was located. This well had the distinction of producing at the rate of 850 barrels per hour when it was nearly 60 days old.

John Galey was born and raised in Clarion County across the river from Parker's Landing. He formed a partnership with James M. Guffy of St. Petersburg, who was a successful operator and ex-Pithole town clerk. James Guffy was not only a nationally known oil prospector, but a good politician and money raiser. He was the chairman of the National Democratic Party during President Garfield's term. James M. Guffy often said, "Never take any man's dust, in oil or politics."

Mr. J. M. Guffy was instrumental in developing the gas fields of Western Pennsylvania, Ohio, Kentucky, Tennessee, Texas, and Indiana. His companies supplied gas to Pittsburgh, Johnstown, Wheeling and Indianapolis.

The biggest and best was yet to come. On January 10, 1901, Galey and Guffy, two old Clarion County oil men, hit the "Jackpot." While drilling a well near Beaumont, Texas called "Spindletop," the ground began to shake. Suddenly the well began to spout mud and the roar was deafening. Workmen watched in terror as four tons of heavy pipe shot up out of the hole, over the top of the derrick, and crashed dangerously around them. Oil erupted from the hole and shot 175 feet in the air. The gusher "Spindletop" was called by the Texas Dispatch "the greatest oil strike in the history of the oil business." This great well was gushing forth over 100,000 barrels of oil per day and opened up the Gulf Coast for major production. Mr. Guffy lived in a handsome home in East Liberty, Pittsburgh, and had a devoted wife and four children.

Enough oil was produced at "Spindletop" to give start to many major oil companies, one being the J. M. Guffy Oil Company, which evolved into the great Gulf Oil Corporation.

Mr. Galey was very busy after his brush with the law. (page 8). We, the students of early oil, thank you, Mr. Galey and Mr. Guffy. Job well done, boys!

*Men Who Made Oil History*
*P. F. Kribbs*
***Mr. Kribbs was a business man, community leader, and oil operator in the Edenburg area. In 1876, Mr. Kribbs was elected the borough's first burgess.***
*Courtesy Drake Well Museum*

## P. F. Kribbs

The father of P. F. Kribbs was Capt. George Kribbs. George lived on the family farm near Edenburg which later became the site of Beaver City, famous oil boom town. (Page 138) When George was 12 years old, in 1817, Christian Kribbs with his family moved into the wilderness and cleared the ancient forest. This farm was located about two miles northwest of the future town of Edenburg. About 1830 George, the father, married Susan Sheffer and commenced on the family farm where he raised a family of four boys, one of whom was P. F. Kribbs, and four daughters.

In 1838 the firm of Plumonds and Creary built Jefferson furnace, and Captain George Kribbs took an active interest in the business. In 1846, Captain G. Kribbs, D. B. Curll, Daniel Delo and Lindsey Pritner built Eagle furnace. In 1848, Captain Kribbs and J. B. Reynolds, of Callensburg, later of Oil City, bought out the above firm and the furnace was managed by Captain Kribbs. The same year Captain Kribbs moved his family to the furnace and Philip F. was placed in the store of J. B. Patrick as bookkeeper, later becoming J. J. Brodhead's bookkeeper. Subsequently, P. F. became bookkeeper and assistant manager of the furnace. In 1856 he married a daughter of Daniel Knight. In 1860 Captain George Kribbs moved back on his Beaver City farm, and P. F. Kribbs settled up the affairs of the furnace and in 1861 moved to Edenburg, Clarion County, and commenced the mercantile business with Porterfield, Titzell & Co. Later he bought out his partners and did business in his own name. In 1873 Edenburg became the center of an extensive oil field and at that time the only business place was P. F. Kribbs's store. He also was the owner of several farms, which he leased to F. Prentice, of New York City. He later on became quite an oil operator himself and at

***11:40 on the St. Peterburg, Turkey City Railroad.***

*1877 Clarion County Atlas*
*P. F. Kribbs Block, Edenburg Penna.*

one time was the owner of 120 oil wells. In 1876 Edenburg became a borough and he was the first burgess.

Mr. Kribbs was born on June 2, 1832, and passed away on June 18, 1907. He is interred in the cemetery at Edenburg. "Peace to his ashes," John J. McLaurin. (*Sketches in Crude Oil*, 1896)

## Myron Matson

*Men Who Made Oil History*
*Myron Matson*
*The firm of Matson and Cole drilled a well in 1874 known as the Paul Black farm well and opened up the Elk City field.*
*Courtesy Drake Well Museum*

Myron Matson was born on a farm near the village of Rushford, in Allegany County, New York, on September 13, 1850. In the year 1870, at the age of 20 years, Mr. Matson concluded to leave the old homestead and try his chances in the oil country and, with a pick and shovel, commenced his oil country career. In the fall of 1874 he, in company with Thomas Cole, purchased a string of drillers' tools and went to contracting under the firm name of Matson & Cole. The first well drilled under contract by the new firm was the one known as the Paul Black farm well near Elk City, Clarion County, which was completed in 1874. This well was a good one and it was the means of opening the Elk City field. Mr. Matson was in partnership with Mr. Cole about one year when he purchased the latter's interest in the outfit and continued contracting himself for several years, making considerable money. It was in the spring of 1877 that

Mr. Matson made his first venture as an oil producer. He leased 12 acres of land from the Kline farm, situated near Wentling's Corners, Clarion County, about eight miles from Edenburg. This well proved to be a good one, starting off at the rate of 150 barrels per day. Oil at that time was $2.50 per barrel.

In 1879 Mr. Matson left Clarion County for the more promising Bradford field. He was estimated to be worth between $700,000 and $800,000.

Personally, Mr. Matson was a most genial companion. He had none of that stiffness about him which so often characterizes men of great riches. He was a large, heavily framed man, with penetrating black eyes, black hair and black mustache. His features were prominent and his voice stentorious. He wore a No. 7 3/8 hat and weighed 265 pounds. In his domestic life he was an affectionate husband and father.

*Men Who Made Oil History*
*Hon. John Fertig*
*He was mayor of Titusville, Vice-President of the St. Petersburg Savings Bank, and a large oil producer in Clarion County.*
*Courtesy Drake Well Museum*

**I've Struck Oil**
**The "Oily fever," don't you see?**
**Infects almost every "live Yankee"**
**From north, from south, from east they come**
**With hopes of finding Petroleum!**

## Honorable John Fertig

John Fertig was born March 17, 1837 on a farm near Fertig's Post Office, Venango County. He assisted his father with the farm work until he reached the age of eighteen. After he received his education, he taught in the district schools. In order to increase his salary, John planned to be a school administrator, but destiny had prepared other work for him.

Inspired by the success of Col. Edwin Drake, Fertig began his career in the oil business on the upper McElhenny farm in December 1859. With a hemlock tree for a spring pole and a set of "pole tools," they kicked down a well 200 feet deep with no success. John Fertig and his three partners faced the facts of drilling deeper or go elsewhere. Captain A. B. Funk, on the lower McElhenny farm, also drilled down a spring pole well 200 feet, where he stopped. But in the winter following, Captain Funk placed an engine and boiler, and in the spring of 1861 he drilled with steam power to a depth of 440 feet where he struck oil. This well was called the "Fountain Well No. 1" and was the first "gusher" or free flowing well in the world. The yield of this well was 1,000 barrels every 24 hours.

Then Mr. Fertig and his associates placed an engine and boiler at their well and drilled down 440 feet and hit the "Fountain Well No. 2." The yield from this well was 500 barrels a day for 19 months. This well was struck on July 4, 1861. Mr. Fertig was guiding the temper screw when he felt the drill drop into a crevice. At once the fire under the boiler was extinguished, but not a moment too soon. The ground shook and with a terrible roar the oil shot upward far above the top of the derrick. Thus began a remarkable life-long career in oil production.

Hon. John Fertig operated in nearly every producing field east of the Pacific slope. He, as principle operator, drilled over 3,000 oil wells. In 1873 Mr. Fertig was elected mayor of

Titusville and served three terms. In 1876 he was elected to the Pennsylvania State Senate. For 17 years the firm of Fertig and Hammond was a large operator. Mr. Fertig was vice president and a large investor in the St. Peterburg Savings Bank, and was closely identified with Titusville's progress and prosperity. The Hon. John Fertig was spending the winter in Los Angeles, California when he had a hemorrhage of the brain and died March 19, 1911. He is in an eternal slumber at the Titusville Cemetery. Thanks John and "Peace to his ashes," John J. McLaurin. (*Sketches in Crude Oil*, 1896)

*Men Who Made Oil History*
*Hon. John W. Hammond*
*He was President of the St. Petersburg Savings Bank, businessman, and partner in the firm of Fertig and Hammond, oil producers.*
*Courtesy Drake Well Museum*

## Hon. John W. Hammond

The Hon. John W. Hammond was a partner in the oil firm of Fertig and Hammond. The company of Fertig and Hammond drilled numerous wells on the Fox Estate at Foxburg in 1870 and 1871. This firm became very successful operators in both the upper and lower fields. Mr. Hammond was from Erie, Pennsylvania, and one of the founders of the banks in Foxburg and St. Petersburg. All business dealings were ended when the Hon. John W. Hammond died in 1880. To you, Mr. Hammond, "Peace to his ashes," John J. McLaurin. (*Sketches in Crude Oil*, 1896)

## James H. Snow

*Men Who Made Oil History*
*James H. Snow*
*Mutual Pipe Line, Foxburg 1870*
*American Transfer Company*
*Superintendent in charge of construction, 1874.*
*Courtesy Drake Well Museum*

James H. Snow, general superintendent of the trunk line system of the Standard Oil Company, was born in Harwick, Massachusetts, December 16, 1850. He came to the Oil Country in 1868, and found employment with his brother William, who had preceded him, with Abbott & Harley's Pipe Line, at the National wells. He continued in the service of Abbott & Harley until the beginning of 1870, when he engaged with Martin & Harms to go to Foxburg and construct the Mutual Pipe Line. When that concern, about April 1872, changed hands, passing by purchase into the possession of the Empire Transportation Company, he was transferred with it and remained with the new concern about one year. He then built the Emlenton Pipe Line, owned by Jabez A. Bostwick, from Emlenton to Turkey City, and later on extended the same to Oil City. This line was subsequently known as the American Transfer Company. At the time of the general consolidation of the pipe line interests, which took place about 1877, he was again transferred, this time to the Bradford field.

Mr. Snow stood in the front rank of pipe line men for 30 years. He thoroughly understood the business in all its

*St. Petersburg Savings Bank*
*Blakslee and Ritts girls driving "Bay Fanny". The Hon. John Fertig and the Hon. John Hammond were large investors. Many old farmers who had made money in the oil excitement lost much of their investments in this bank.*
*Courtesy Marion Ritts*

branches down to its minutest details. Numerous inventions and mechanical improvements of unquestioned value stand to his credit and attest his mechanical skill and ingenuity. Hydraulic engineering was his hobby and when one makes a hobby of his business, he is pretty certain to succeed. Mr. Snow was vice-president of the Snow Pump Works at Buffalo.

## The Cry Was "On To Bradford"

*Clarion Democrat*, March 8, 1879, Elk City: One hundred forty three rigs, crews and machinery are being pulled from Clarion County and moved to the Bradford fields. The price of oil is not worth the cost of drilling for it. The drilling continues. Oil per barrel is now at 75 cents. It is agreed upon that the Clarion County oil field has been fully drilled and explored.

*Bradford, Pennsylvania 1876*
*This photograph shows Bradford just as the oil boom took off.*
*Courtesy Venango County Historical Society*

## Bradford, Pennsylvania
## The Boom Is On!

Located in the Tuna Creek Valley, wildcatters had by late 1875 penetrated the oil bearing sands around the sleepy town of Bradford. Train loads of oilmen from every part of the lower country crowded the streets and overran the hotels, saloons and dance halls. By 1878 the population of Bradford had swelled to over 4,000. Over 1,000 letters a day were received at the post office, 500 messages were received and delivered by the telegraph office, and at times up to 250 loaded railroad cars were standing in the yard. *Early Days of Oil,* Paul H. Giddens, 1947

# Emperor Dom Pedro at Foxburg
# Dom Pedro d'Alcantara
# An Incident – By C. C. Cochrane

It happened away back in the 70's, when Parker was the hub of oildom, and the picturesque Allegheny Valley from Oil City to Bradys Bend was a hive of industry, with its shrieking pulleys, its bull wheels, and the resounding anvils, beaten by the brawney tool-dresser, and often aided by his no less brawny mate, the driller, made the night as well as the day hideous with their din of conflicting noises.

At that early day but little use was made of natural gas, except to light up at night the surrounding country with a glare, which made the oil derricks look like gigantic ghosts of some past age. This part of this beautiful valley was at that time the cynosure of all eyes. The oil business was a budding industry that attracted attention of the entire world and brought visitors from every point of the compass. Travelers from England; from France; from Germany; from Japan and far-off China added to their itinerary a trip over the Allegheny Valley railroad from Titusville to Pittsburg, through the heart of the Pennsylvania oil regions, where, to the language of a foreigner, there were oil springs from which pure oil flowed like water.

Dom Pedro, the then Emperor of Brazil, was making a tour of the United States and of course his program included a trip to the oil regions. Mr. Thos. M. King was superintendent of the Allegheny Valley railroad at that time and had charge of the "special" which carried the Emperor and his suite.

The firm of Jesse Smith & Son had three pumping wells near Foxburg, and superintendent King wired the senior member of the firm that an arrangement had been made by which they would stop at a certain well of his, which was producing at that time about 20 barrels per day of the "Amber" oil, to show the Emperor the process of producing oil from the sand. This was before the Seaboard Pipe Lines existed and all the oil was transported from the region by rail in tank cars.

*St. Petersburg, 1896*
*Courtesy Marion Ritts*

*"Milking" the well head. Dr. C. R. Moore at a well near Antwerp Hollow. Note the wooden barrel to put oil in. The old Doctor was burned on June 30, 1914, so badly at a well on the Martin farm, that he died less than 24 hours later.*

I was then inspector of oil and happened to be at Foxburg on that day, and on receipt of the message started, in company with fifty others, ladies as well as gentlemen, for the "certain well," located about half a mile below Foxburg along the railroad track, where I pre-empted a reserve seat on a large oak stump to see the show. The well was immediately shut down to gather a head of oil and when the imperial party drove in sight was started to pumping and was doing its nicest, throwing the rich amber fluid into the tank in a two-inch stream in the presence of the great Emperor.

Those who had the good fortune to see Dom Pedro at that time will remember him as a very large, imposing and dignified looking man, with flowing white beard. He was dressed in some kind of uniform with a fatigue cap, which he doffed gallantly in the presence of ladies. Immediately on descending from the train he was taken familiarly by the arm by Mr. Jesse Smith and conducted to the top of the tank where he looked down into the foaming fluid silently for some moments and then as silently turned away without a word or remark. I could not help wondering what were the thought of each of the strange assorted pair, "Bons Camarades"—the great Emperor arm in arm with the unassuming Pennsylvania oil

*Shoveling coal into a boiler to make steam to run an engine at a well on the J. J. Ashbaugh Farm.*

producer—an incident that could not have occurred at that time in any other country on earth except the United States.

The imperial party quickly boarded the train, from the platform of which the train moved away, with a wave of his cap, Dom Pedro bade adieu to the assembled crowd, as well as to the Pennsylvania oil region, never to return again.

*Lewie Quirk oiling the engine at a well on the J. J. Ashbaugh Farm.*

*A "brave young lady" at a well on the J. J. Ashbaugh Farm.*

*Arthur "Boney" Edinger at a well on the J. J. Ashbaugh Farm.*

*Well on the J. J. Ashbaugh Farm.*

Soon after the return to Brazil, Dom Pedro abdicated his throne and went into exile, while his empire joined the grand galaxy of Republics.

## Loguetown

The Eli Logue farm near Grass Flats, was the location of the short lived, unstable oil boom town called Loguetown. Shabbily built buildings consisting of a school, post office, and a few homes sprang up. The farm contained 320 acres and has a beautiful view of the Clarion River. (Birds eye view in Clarion County Atlas, 1877) See page 124

A Story is still told that the Logues were so rich, that instead of washing dishes they just bought new ones. (The Author)

*Drilling a well.*
*Courtesy Venango County Historical Society*

This poem was published in Oil Well Reminiscences, issued by Oil Men's Association of Butler County, 1907. This poem brings the oil boom era all together with a few plain and simple words:

## The Dead, Unburied Cities
## By C. F. Allen

To the cities of golden visions and hemlock boards; favored more in cheerful drinks than paint; transient, but wonderful; famous for many days, and buried through all the years to come; one who knew your glory, shared your shelter, and (sometimes) drank your beer, lays on your cheerless graves the token of kind and reverent memory.

The snow upon the derrick floor
Lies white and cold and still;
The ghosts of drillers gone before
Crouch down upon the hill,
And Magic lifts her wand no more
Along the frozen rill.

As on Sahara's darkening sands
The tawney tents arise,
Yet all the waste deserted stands
Beneath the morning skies.
We saw the work of nimble hands
Fade out before our eyes.

The hemlock flung its fragrant shade
Above the pumper's door;
The muddy caravans of trade
Toiled painfully before;
And while the chopper swung his axe
The trader raised his store.

A railway shot its line of light
Among the slaughtered trees;
The derricks burned their lamps by night
Like boats on crowded seas;
And golden fountains shot in sight
On such lone lands as these.

A city built in sixty days
Along a single street,
Devoid of plaster—while it pays
Holds all the rest effete;
There is a glamor with the craze
Makes such delusions sweet.

And who recalls with wistful eyes
Its fortune's ebb and flow,
Will see the vanished city rise
As in the years ago.
And through its unforgotten skies
Hear elfin trumpets blow.

The derricks grow like fairy frames,
The golden fountains gleam;
A thousand half-forgotten names
Ring out as in a dream;
And every far-drawn vista flames
Through snowy clouds of steam.

Oh, city! Fade and melt away,
As all thy visions fled;
Through all the magic bugles play
Thou art forever dead;
The rabbits haunt thy brambled way,
The owl sits overhead.

The derricks lift against the sky
Deserted, black and lone;
And sad Oblivion, hovering nigh,
Looks down upon her own;
And low thy hemlock castels lie
As Pharaoh's halls of stone.

# EMLENTON

Emlenton is located in the southern part of Richland Township, Venango County, Pennsylvania, and is one of the oldest towns in this part of the state. It was a lumbering and trade center years before the discovery of petroleum. The area surrounding Emlenton has had several small wells, but none have been large enough to create a general excitement. Emlenton's location lies just outside the oil belt that created the boom towns of St. Petersburg, Edenburg, and Parker City. Its rapid growth dates from the oil developments of the lower field and Emlenton's oil shipping capability. It had a population of two to three thousand with churches, elegant public school buildings, hotels, stores, and a weekly newspaper. Emlenton is located eight miles above Foxburg on the Allegheny River, and on the line of the Allegheny Valley Railroad. It was an oil shipping point of some importance.

## Richland Township, Venango County

Richland Township is bountifully supplied with iron ore, limestone, sandstone, coal and petroleum. Iron ore and limestone can be found in almost any part of the township in abundance. Only a few years ago, the furnace men did a flourishing business, but recently petroleum has absorbed the attention of the people.

*RITCHEY FARM*

*The Thomas Ritchey farm, Richland Township, Venango County, located one and one-half miles from Emlenton, Pennsylvania. One of Thomas Ritchey's sons was killed by a boiler explosion at the No. 5 well in this lithograph. ("History of Venango County, Pennsylvania," 1879)*

The furnaces are now in ruins, and there is not a pound of iron manufactured in the township. The first oil well that was put down in this part of the county was in 1860 by Elias Widle and Co., on or near the mouth of Ritchey Run, but the excitement in other places drew the attention of operators from this section, so that not very much has been done until recently. In 1877, operations were commenced on the Ritchey and Wentling farms. Sometime previous to this, there had been several wells put down above Emlenton on the river near and on the Russell farm. On the Ritchey farm there are now ten wells producing 110 barrels per day. Six new wells are being put down on the farm, and also a number of others in the vicinity. The Wentling farm has but three wells, which are producing about 21 barrels per day. The largest well on the Ritchey farm when first struck produced 30 barrels per day. Enough developments have been made to show that Richland Township is well supplied with the oily fluid. Several years ago there was a test well started on Mill or Kearns Creek, but owing to the inexperience of the operators, they became discouraged upon reaching the second sand, and abandoned the well. This well was located near Giering's Mills. Mr. Giering sold some of his land for $1,000 per acre to a Mr. Pennebacker, but the excitement in other places drew his attention to another field, and nothing has been done to really test this territory. ("History of Venango County, Pennsylvania," 1883, Waterman, Watkins & Co., Chicago, Ill.)

## THE DERRICK
## June 9, 1874
## An Impending Danger to Oil Towns

A correspondent writes us that the pleasant little village of Emlenton narrowly escaped becoming the victim of an oil fire Sunday evening. The danger arose from the location of an oil tank on the hill above the town. There was nothing to prevent the burning oil from pouring down upon the residents and through the streets of the village. It would be a torrent of fire as destructive and more terrible in its details than the Mill River flood. Luckily, the torrent of flame was turned aside and it swept past the village leaving it uninjured and threw itself into the river.

The danger which threatened the inhabitants of Emlenton is hanging over nearly every village of the oil regions. Most of these towns are nestled in the valleys among the hills. The hillsides are used on which to locate large tanks of oil containing from ten to twenty thousand barrels of petroleum creating a small avalanche if let loose. Too often these tanks are placed in the very streets of the villages and in close proximity to the houses. It requires but a single flash of lightning to cause this amount of crude oil to burst its boundaries and precipitate its fiery contents down upon the village which stands in its very path. The consequences of this would be terrible and there is nothing to keep this from happening day or night during the summer time when thunder and storms prevail.

There is hardly a town in the region but where these tanks may be seen on the higher ground above it. In many instances, tanks stand on a bluff overhanging the village. These dangers are also being daily added to. A few weeks ago a thirty-thousand barrel tank was built upon the high ground above Siverlyville where the ground slopes directly towards the village. Should a stroke of lightning set this on fire during the night hours, there would in all probability be loss of life, and loss of property would be inevitable. Tanks of oil placed in such locations threaten human life and it seems to be as much a criminal offense as that of the town commissioners at Mill River in accepting the walls of a dam which was based on no better foundations than those of crumbling sand and dirt.

Emlenton is not the only town that has been threatened in this way. Last April, a large tank situated on a hill above Modoc City, Butler County, was struck by lightning. The particulars of this fire did not reach the newspapers, owing to the exciting scenes from Millerstown and other quarters which at that time limited the columns. The burning oil rushed down the hill towards one end of the town. Several buildings were destroyed and in one a family nearly escaped death. Rushing out of one door, their way was cut off by the burning streams and they had little time to escape by another passage before the building was on fire. Thankfully at that fire nothing but property was injured but this practice must be ended and the lesson must be heeded.

Our towns have escaped remarkably well thus far. We have no sickening details of a terrible catastrophe. But the dam at Mill River stood nine years before it broke away. Chicago had been a city years before her great tragedy and we cannot tell the hour when some sad accident may occur, and one of our pleasant villages be blotted out of existence.

Let the tanks of oil be located away from our villages and beyond all possibility of danger. It is not necessary that tanks should be placed in such close proximity to the towns, as suitable locations may be found elsewhere where no danger will be threatened to life or property and in so doing the world will be spared another terrible tragedy.

JAMES BENNETT,

Manufacturer of and Wholesale and Retail Dealer in

Tin, Copper and Sheet Iron ware,

PITTSBURGH AND BUFFALO HEATING AND COOKING STOVES,

Fire Fronts, Grates, Iron Mantles,

Hollow, Japanned and Pressed Ware,

Pittsburgh Plows, Mowing Machines,

HORSE RAKES, TORNADO CHURNS,

PATENT CEMENT CHIMNEY

QUEENSWARE, OHIO STONEWARE,

AND WOODEN PUMPS.

Main Street, near Depot, Emlenton, Venango Co., PA.,

*Branch Store at St. Petersburg, Clarion County, Penn'a*

Smoke Stacks and Oil Tanks made to Order.

# EMLENTON
## Oil City Derrick, June 9, 1874

The town had a narrow escape yesterday evening from being destroyed by fire. During the thunder shower, lightning struck one of the large oil tanks located inside the borough. The tank contained some 12,000 barrels of oil. The gas and oil burned rapidly for about thirty minutes. The oil was turned into the river, and between the wind and rain the fire was extinguished, thereby saving the Rev. Smith's house and the greater part of the business portion of the town. The oil was being pumped into the tank from St. Petersburg during the fire, there being no telegraph in the place to stop the operation.

This narrow escape should have the tendency to awaken the citizens to a knowledge of the danger which the location of these large tanks directly in the village occasion, and to have them removed beyond the borough limits.

*The covered wooden bridge over the Allegheny River at Emlenton was opened in October, 1856, and was washed away in the Great Flood on February 5, 1883.*

# EMLENTON
## The Derrick, June 13, 1874

The march of improvement has penetrated Emlenton quite extensively of late. Among recent improvements is a superb brick school house, two stories in height and fitted up in the best style. James Bennett is completing a substantial brick block on Main Street designed for an opera house, a first class store, the Emlenton Bank, two or three shops and several offices. D. D. Moriarty has taken possession of his new post office, which is situated in a fine two story building and supplied with all the modern conveniences.

The Presbyterians have begun building a brick church located on the site of the old foundry and to cost over $8,000. Crawford and Son have established their foundry in a new premises built with special reference to this branch of trade. A new machine shop is being built, employing a number of skilled workmen in an operation near the railway depot. A steam grist mill is projected and will be pushed ahead with the utmost dispatch.

The required amount of stock has been subscribed and no time will be lost in pushing the undertaking to a speedy close. A new addition to the town has been surveyed into building lots above the hill on which various residences will be put up this season. In fact, Emlenton seems to have suddenly received a liberal infusion of progressive spirit of the age, with gratifying results already accomplished and yet greater in the near future. (J. J. McLaurin, correspondent)

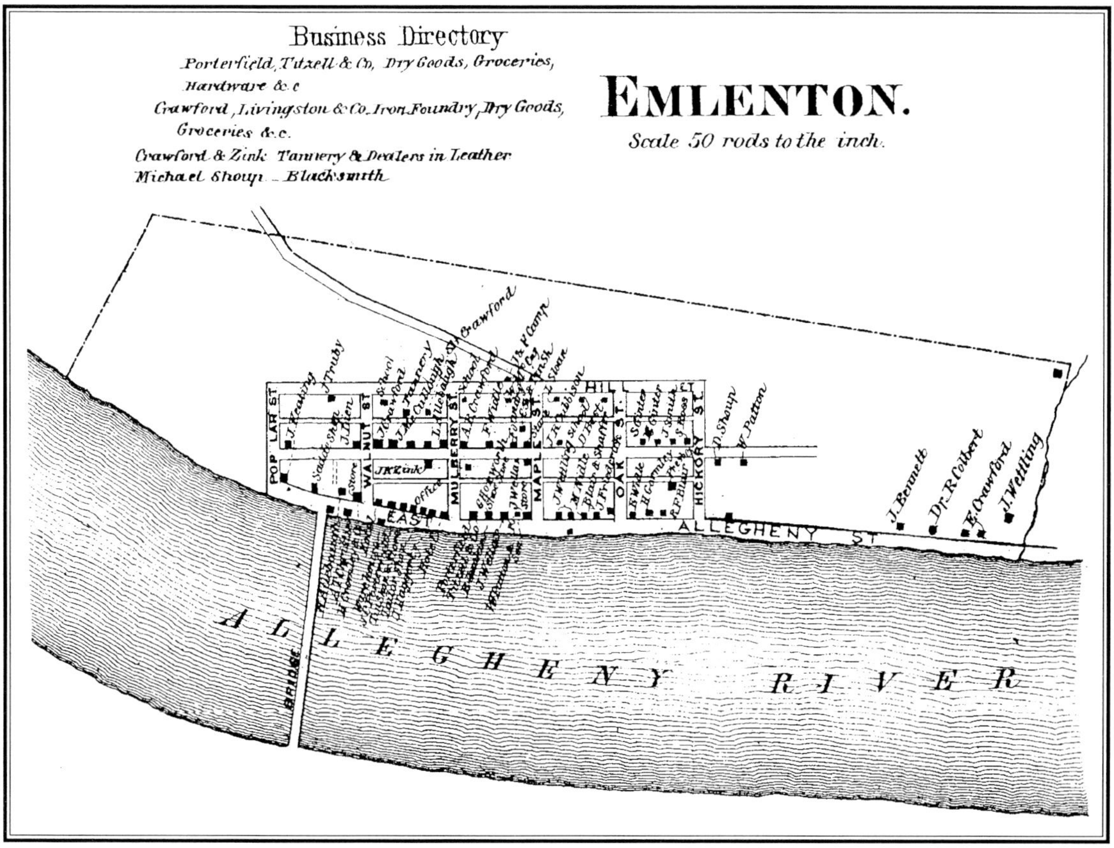


*Venango County Historical Society, Franklin, Pennsylvania*

## THE GREAT EMLENTON BRIDGE DISASTER

In the early days, crossing a stream or a river was always a dangerous and challenging feat. Horses drowned, people drowned, and goods were lost to the clear, swift flowing waters of northwestern Pennsylvania. Emlenton, Pennsylvania, was a small settlement on the Allegheny River created by the need for building stone at Pittsburgh. Around 1820, Andrew McCaslin came up the Allegheny River in a flat boat searching for stone to meet the needs of a thriving and expanding young city some seventy miles down river called Fort Pitt, or by this time the metropolis of Pittsburgh, Pennsylvania. McCaslin built a log house along the river at the future site of Emlenton, which also served as a store and tavern. Also at this spot McCaslin established a ferry across the Allegheny River and this small settlement became a drawing point for river men and settlers alike.

## IRON FURNACES

By the year 1838, some 20 iron furnaces were in operation within a sixteen mile radius of Emlenton. Rich veins of iron ore and bituminous coal were discovered in the surrounding hills and countryside. Supplies for these iron furnaces and the goods needed for the growing population were brought up river from Pittsburgh by boat.

Soon the small community had a steam sawmill, a tannery, a distillery, and a brewery. Log houses, scattered farms, and lumbering operations were here and there in the Scrubgrass and Richland areas. A first-class hotel was built called the Crawford House and by 1844 two large warehouses were erected along the waterfront. The surrounding areas were still quite wild but slowly farms and new settlers were moving into this beautiful area.

## A COVERED BRIDGE

Clearly the small community had outgrown ferry boat transportation and a meeting of residents was held May 2, 1854, at the Crawford House, and a bridge company was organized. Emlenton received its name from the owners of the land, Joseph M. Fox and his wife Hannah Emlen Fox, thus Emlenton. To build a bridge, a committee was formed with Samuel Fox elected president and a board of directors was named including John Keating, then postmaster, Joseph Weller, R. S. Porterfield, Henry Kohlmeyer, Samuel Anderson, and J. J. McGinnis. Later, J. F.

C. B. SEYMOUR. GEO. SHEFFIELD.

**C. B. SEYMOUR & CO.,**

*Wholesale and Retail Dealers in*

**HARDWARE**

ROPE, CASING, TUBING,

AND OIL WELL SUPPLIES,

PLEASANTVILLE, PA.

**C. B. SEYMOUR & CO., FAGUNDUS, PA.**

C. B. SEYMOUR & CO., TRIUMPH, PA.

SEYMOUR, TAYLOR & CO., St. Petersburgh, Pa.

PETROLIA FOUNDRY AND MACHINE WORKS.

**R. H. LECKY,**

MANUFACTURER OF

**Engines, Drilling Tools, Rig Irons,**

GRATE BARS, HORSE POWERS, THRESHING MACHINES,

**Plows, Stoves, Sled Soles, Coal Cars,**

Wheels and Axles, Brass Castings,

And all Kinds of Repairing Done,

EMLENTON, VENANGO COUNTY, PENN'A.

**M. NIDLE,**

MANUFACTURER OF AND DEALER IN

**Furniture, Mattresses, Mirrors, &c.,**

ALSO AGENT FOR

TUCKER'S PATENT SPRING BED,

WORK WARRANTED.

*River Street, - - EMLENTON, PA.*

J. C. SCOTT,

DEALER IN

**Rough and Dressed Lumber,**

Mouldings, Doors, Sash and Blinds,

Office and Yard Corner Main and Oak Sts., Emlenton, Pa.

*Bills of Lumber Sawed to Order.*

**J. J. GOSSER,**

DEALER IN

**Drugs, Medicines, Paints, Oils and Dye Stuffs,**

KEROSENE OIL AND LAMPS,

Perfumery and Fancy Articles, Pure Wines and Brandies for Medicinal Use.

Main Street, Emlenton, Pa.

*Physicians Prescriptions Compounded at all Hours.*

**CHARLES BABST,**

**DRUGGIST,**

AND DEALER IN

PAINTS, OILS, GLASS, VARNISHES,

TOBACCO, CIGARS, WINES, LIQUORS, PATENT MEDICINES, &c.

Corner Main and Maple Streets,

EMLENTON, PA.

Winsch was elected secretary and Joseph Weller elected treasurer. Sixteen thousand dollars was raised by selling stock at $20 a share and Daniel McCain, a bridge builder by profession, came to town with his family and the project was started. Work on the bridge progressed well and finally in October, 1856, a little over two years, the beautiful structure was completed.

## THE WINTER OF 1883

The winter of 1883 was colder and more harsh than the previous few, and considered the worst since 1865. The Allegheny River had a huge ice jam in the turn at Parker City and West Monterey. The river was at flood stage and the channel was running full of huge iceberg-like pieces of ice. The following true story is about a bridge, icy waters, a brave young lad, and many frightful hours. You, the reader, will long remember the following story—a story that will become one of the town's most cherished legends.

*J. J. Gosser, druggist and dealer in pure drugs, chemicals, paints, oils, dye stuffs, toilet articles and family medicines. Physicians' prescriptions carefully compounded. A large line of pure wines and brandies for medicinal use. Main Street, Emlenton, Pa. ("History of Venango County, Pennsylvania," 1879)*

The covered bridge at Emlenton was a toll bridge, and Tom Taylor, who had lost a leg in the Civil War, was toll taker.

When a heavy snowfall made sleighing the best means of transportation, farmers dumped wagonloads of snow on the floor of the bridge. The covered bridge served the countryside for twenty-seven years. The end of the structure was dramatic. On Sunday, February 5, 1883, flood waters had brought the river very close to the floor of the bridge. Great cakes of ice, coming down the swift river, were crashing against the arches of the bridge. All that Sunday, townspeople and folk from the countryside were there to watch the fearful and yet fascinating spectacle of a river gone wild. Venturous men and boys were out on the bridge enjoying the dangerous thrills of the vibrations felt as ice blocks, 18 to 24 inches thick, rammed against the structure.

Suddenly the bridge shivered. Men and boys raced to the nearest exit as the force of the blows on the bridge plunged it into the turbulent, ice-jammed waters below.

Frank Allebach and his brother, Harry, had spent most of the day watching the river, and in late afternoon had gone home for dinner. Scarcely had they sat down to the table when they heard a terrific roar and crashing sound, followed by shouts and cries from the crowd. Dashing outdoors, they saw a section of the bridge, upright, going down the river. As the ice-battered bridge turned about, they saw a boy standing in the open end, waving his arms.

## WHO WAS THE BOY?

Young Bingham Ross had walked that Sunday afternoon from his home to visit his uncle, Joe Ross, who lived in a house at the top of the Emlenton hill. Young Ross had joined several other boys who were watching the river, and along with other men and boys ventured out on the bridge to enjoy the thrill of danger. Just a moment before the fatal crash, a horse and buggy was driven by Asbury Middleton swiftly and safely across the bridge.

The crash that tore the bridge from its foundations caused everyone to make a dash for the nearest opening and his companions raced toward the Emlenton side. They made it, but Bing didn't. A terrific roar, a sickening drop, and he found himself on a section of the bridge that was tilting and turning in the rough waters. After a moment of stunned surprise, the lad made his way cautiously to an open end of the bridge. Incredible as the situation seemed, he felt for the moment secure. Water was playing over the floor of the bridge but it seemed evident that the bridge would keep afloat. Bing's appearance at the open end of the bridge brought forth shouts of horror and dismay.

Porterfield & McCombs,

Wholesale and Retail Dealers in

Dry Goods, Groceries,

BOOTS, SHOES,

HATS AND CAPS,

Ladies' and Gents' Furnishing Goods,

HARDWARE

CASING, TUBING,

AND

Oil Well Supplies,

EMLENTON, PA.

Branch Store at Foxburg, Clarion Co.

## GET HIM OFF!

"There's a boy in it."

"Some body get him off."

But no one knew what to do. Dilly Cubbison, a teenage girl, stood in the mud at the water's edge, near the tollhouse with several friends. She recalls the shouts of the crowd, the helpless panic. No one in the crowd thought of the bridge. It was the boy.

Someone got word to the telegraph operator at the railroad station a block away. He telegraphed to Foxburg and to Parker to be prepared to effect a rescue if the structure lasted until it reached those bridges. Coligan, a section boss on the railroad, and his section hand got on a hand car and called for volunteers. Hopping aboard were A. R. Newton, a marble worker; H. J. Crawford, organizer of Columbia Gas, then a lad of seventeen; and John Marshall, described as a husky he-man, always ready to lend a hand to someone in distress.

As the structure traveled slowly down river, pushed by the rushing waters, the handcar, pumped by four of the men, traveled slowly down the tracks, keeping abreast of the bridge.

The men shouted words of encouragement to the lad, reminding him that a rescue would be effected at the Foxburg bridge.

It was only half a bridge that carried the boy, the Emlenton half. The other section had smashed to smithereens when it hit the water and quickly sank.

Willis Shoup, then only a youngster, happened to be standing with Andrew Jackson on his farm several miles below Emlenton. Jackson's grounds were flooded and his kitchen had several feet of water in it. As Willis and Jackson stood on a high spot of ground and looked up river, they saw an amazing sight, a section of bridge sailing upright down the river. At that moment they saw the handcar with a half dozen men on it. The men were shouting to the boy on the bridge, "Climb out on top, get on the roof." As the bridge passed the Jackson farm, Willis says the boy was standing and waving his arms and sobbing, "Come and get me off somebody." Shoup and Jackson were helpless.

Word had reached Foxburg too late to arrange for a rescue. Horrified onlookers on the bridge at Foxburg watched the structure approach, battered and turning in the turbulent waters, and saw it smash against the bridge foundation. A cry of dismay arose as the thing again broke in half and a part of it disappeared beneath the angry waters. Miraculously, the boy was on the surviving bit of wreckage.

"Climb out on top," shouted the men on the hand car. Just then the wreckage turned completely over. Finding a hole in the roof, or was it the floor, above him, Bing grabbed at a

*Emlenton, Pennsylvania. This iron bridge replaced the old covered bridge.*

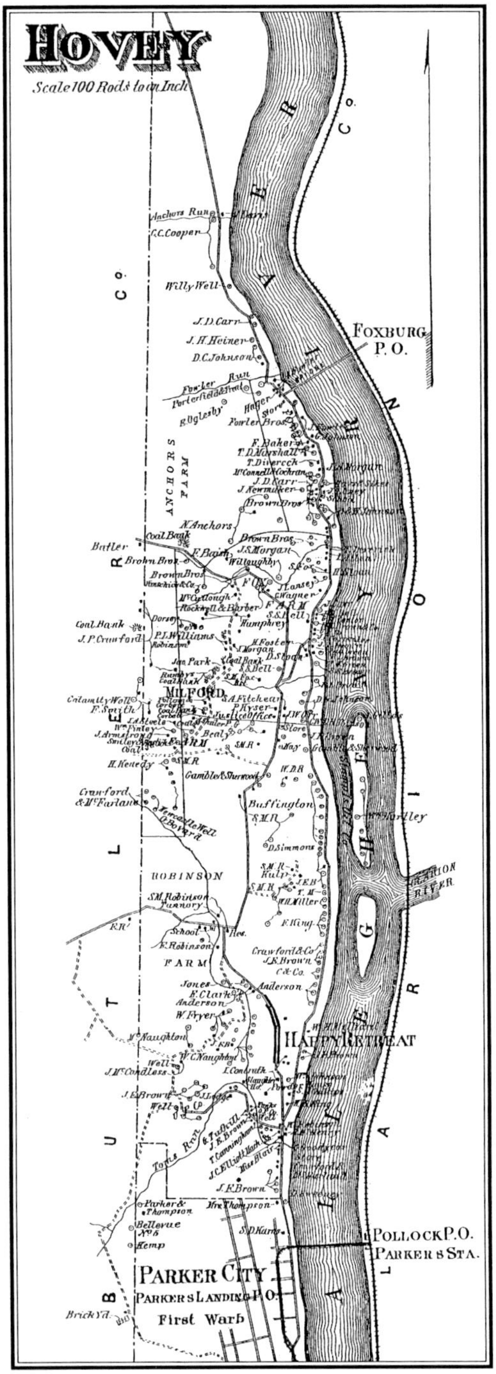


*Hovey Township, Armstrong County, Pennsylvania, 1876. This map shows oil territory between Foxburg and Parker City along the Allegheny River. Shown are the islands in the Allegheny River, the mouth of the Clarion River, the oil rich Milford farm, and a small village called Happy Retreat. ("History of Armstrong County, Pennsylvania, Illustrated," 1883)*

rafter, pulled himself up and managed to get through to the top. It was none too soon. The bit of wreckage was slowly sinking.

It was almost dark now, down by the river. Along the Parker bridge on the Emlenton side, lanterns had been strung every two feet. Under each lantern hung a knotted rope, its end dangling in the water. Lantern light flickered eerily across tense faces. The rescuing squad huddled against a sharp wind, yelling to each other above the crumbling of the ice, straining to keep their eyes focused on the small dark object in its approach.

All afternoon the boy had clung to the broken span. Sometimes it rode along like a boat, then buffeted by the swirling ice, it would dip and sway dangerously. Once, only an hour before, the span had turned completely over in the water, but when it righted, there was the boy, still hanging on.

There had been no way to reach him through the boiling flood. Men had gone up and down the railroad in a hand car watching him, calling to him. At first he had called back, but he no longer seemed to hear nor attempted to answer. The growing darkness and the cold wind that had sprung up at sunset increased his danger. He must be rescued now or never.

"If the span isn't crushed by ice," the anxious watchers said to each other, "If the boy can hold on a little bit longer." "If we can reach him before he is swept under the bridge." It was his last chance.

In spite of the onward rush of the flood waters, the wreckage seemed to approach the Parker bridge very slowly, pushed this way and that, bobbing up and down erratically. When Bing's rescue happened, it happened with terrible speed.

We have this from Bing himself as told to his good friend Orrie Hallack, an oil worker at Perryville on the hill opposite Parker.

"I got up on top, but I don't remember how. That was after the bridge smashed up against the pier of the Foxburg bridge. The thing was sinking. Water came up to my knees, then to my waist. When I came in sight of the Parker bridge, the current had shifted the wreckage to the Parker side. I knew it was now or never. Either they got me off or I'm done for."

As the structure approached the Parker bridge, the boy was aware of lights and people. The water was now to his armpits, but suddenly he was there at the bridge. Reaching his arms high, he grabbed at the iron railing. Aided, no doubt, by would-be rescuers, he was pulled over the rail to safety. Bing was pulled to safety halfway between the abutment and the first pier on the Parker side of the bridge.

It is inevitable that fiction would grow up around such an amazing epic. Hearsay has it that the boy's hair turned white during his perilous journey downstream. It is said that his hands were so tightly gripped on the rescue rope that someone had to produce a pocket knife to cut the rope away. He then turned, broke away from the rescuers, and ran, pushing his way through the throngs on the shore, and spurning offers to drive him home in a wagon, ran for dear life, not stopping until he had reached his home, some four miles away.

ARTERS'

Emlenton Borough

DIRECTORY FOR 1872.

A

Allebach Jacob, carpenter
Allebach Henry, watchmaker
Allebach Levi, jeweler, cor Main & Mulberry sts
Allegheny Valley Brewery, Sebastin Kreis, prop, Main st, h Water st
Armstrong S, carpenter, bds Valley House
Arner Leslie, tinner's apprentice, bds Moran House

B

Babst Chas, druggist, cor Main & Maple sts, h hill
Baish Jos. J, shoe shop, Main st, h same
Barnard L A, clerk
Barnard S. P, clerk, bds Alleghany st
Barnard J. A, clerk, bds Main st
Beck Wm, teamster, Exchange Hotel
Bennett Jas. house furnishing goods, stoves, tinware, &c, Main st, h Allegheny st
Black Amos, agent A V R R, bds Crawford House
Black Edward, assistant agent A V R R bds Crawford House
Bonner J. A, clerk, Crawford House
Boney John, justice of the peace
Boyce Rob't, tinner
Brown Dr. A G, of Hamilton & Brown, o & h Allsgheny st

C

Camp J. H, shoemaker
Camp B. F, carpenter
Carman Wm, blacksmith shop, o Main st, bds Exchange Hotel
Chew Thos, blacksmith, bds Water st
Corry Miss R. R, manager W U Tel Co, bds Crawford House
Crawford House, J. M. Sloan prop, Main st and A V R R
Crawford Ebenezer, carpenter
Crawford G. G, laborer
Crawford Sam'l. H. brickmaker
Crawford David, moulder
Crrwford J. A, carpenter
Crawford H. A, carpenter
Crawford A. B, carpenter
Crawford S. W, carpenter
Crawford W. D, carpenter
Culbertson Jno. R, carpenter, h Main st

D

Dam Wm, cabinet maker, h Allegheny st
Deem John. Z, confectionery, Main st, h on hill
Dolan Thos, tool dresser, bds Crawford House

E

Eichbaum Geo. R. machinist, h Main st
Emlenton Livery, Moran & Morgan, Allegheny st
Exchange Hotel, Jas. S. Hagerty, propr, cor Water & Allegheny sts

F

Fleming E L, clerk. bds Allegheny st
Forker W. A, clerk at Valley House
Fox J. B, minister
Frazier Jas, carpenter, bds Valley House
Freese Harry, tinner, bds Main st
Fulmer Jack, tinner. h Main st
Fulton C. S, moulder, h Stratonville

G

Gates G. W, dry goods, clothing &c, Allegheny st, h same

His sister Minnie (Mrs. Rhodabarger), of East Brady, vouches for the fact that he arrived home breathless, soaked and caked with frozen water, and not until he burst into the kitchen door and blurted the whole amazing story to his astonished family, did they have any idea that a mishap had befallen him.

Word spread quickly around the neighborhood, says Mrs. Rhodabarger. Neighbors came in bringing food, eager to hear for themselves the fantastic story. No one slept that night, she said. Until darkness faded and light was streaming in at the windows, family and friends feasted and talked.

Bing was none the worse for his dreadful experience. The next morning, with lunch pail in hand, he set off across the fields with his sister, Minnie, to the Kensington Schoolhouse. We can well imagine he was the hero of the day.



Gearing R. S, jeweler, Allegheny st, h same
Gearing H. L, of Lawall & Gearing, h Allegheny st
Ginter Henry, laborer
Goper J. J, druggist
Gorman Jas. T, house & sign painter, Allegheny st
Gormley Henry, merchant
Gosser J. J, druggist, Main st, h same

H

Hagerty Jas. S, propr Exchange Hotel
Hagerty Miles, clerk
Hagerty Geo, laborer
Halderman J. A, laborer
Hamilton & Brown, physicians, Allegheny
Hamilton Dr. B. F, of Hamilton & Brown, Allegheny st, h Shippenville Road
Hamond Saml, blacksmith helper, bds Water st
Harold M, oil operator, bds Exchange Hotel
Heasley J. H, salesman, h Allegheny st
Henesy Thos, foreman gravel train, bds Exchange Hotel
Hess Daniel, harness maker, bds Allegheny
Howard Daniel, blacksmith helper, bds Crawford House
Howell —, tinner, bds West Emlenton
Hoyt J. W, agt Howe machine, Allegheny st, bds same
Hunt A. W, cabinet maker, bds Allegheny st

K

Keffer F, tinner, bds Exchange Hotel
Kerr Mrs K, dress maker, Allegheny st
Kries S, teamster
Kreis A, painter
Kreis Miss Annie, millinery, Allegheny st
Kuhns J. K, carpenter

L

Lane R, plasterer, bds Exchange Hotel
Lawall & Gearing, furniture, Main st
Lawall Stephen, of Lawall & Gearing, h Main street
Leach L. I, moulder, h Water st
Lecky R. H, engine & machine shop Main st
Livingston Mrs E. L, millinery, Main st
Lovell & Redic, dry goods, boots, shoes &c Main street
Lovell Jno, of Lovell & Redic, h Shippenville Road
Lowall Stephen, cabinet maker
Lowery F. G, clerk
Lyman Wm, clerk Exchange Hotel
Lytle J. D, plasterer, bds Exchange Hotel

M

Mahaffy Chas, foreman machine shop, bds Valley House
Marshall Robt, carpenter, bds Valley House
Marshall D, carpenter, bds Valley House
McCabe Jas. tinner, bds Exchange Hotel
McCombs Jno, of Porterfield & McCombs h Allegheny st
McCoy Miles, machinist, h Main st
McCulla Wm, land dealer, bds Exchange Hotel
Middleton Columbus, butcher
Miller Benj, driller, bds Crawford House
Moran House, T. J. Moran prop, Allegheny
Moran & Co, dry goods & notions, Allegheny
Moran T. J, of Moran & Co, h Allegheny st
Moran P. H. tinner, bds Moran House
Morgan P. of Moran & Morgan, h Allegheny
Moriarty D D, post-master, cor Main and Maple, h cor Hill and Maple sts

N

Nidle Michael, furniture Allegheny st, h do
Nidle Chas. P, cabinet maker, bds Allegheny

O

Obeal Jno, cabinet maker, bds Water st

P

Palmer V. P, oil oper, bds Crawford House
Perry W, surveyor, bds Exchange Hotel
PERRY THOS. M, oil oper, w Emlenton
Petroleum Livery, Wm. Carman prop, Main
Porter W. L, editor & prop Rising Sun, h Richland tp
PORTERFIELD & M'COMBS, wholesale & retail dealers in dry goods, boots & shoes, &c, Allegheny st

While attending school, Bing worked at farming and carpentering. He also helped Cooney Weaver at his oil wells around the old May Walker farm.

Later he became a pumper in the McDonald oil fields. He lived for nine years after this episode. In the winter of 1892, Bing came home from the oil fields ill and died of typhoid fever. In a short time, his brother Piney died of typhoid and later that winter his sister Eva died of the fever. They are buried in the Matildaville Cemetery, near West Freedom, Pennsylvania.

Information on Bing's famous bridge ride came from an article written by Rachael P. Newton of Emlenton and reporter for the Emlenton News dated 1956. Also, thanks to Margaret Webb Sanders's "The Year of the Mintie May," unlocated but mentioned in Rachael's source and special thanks to the Venango County Historical Society, Franklin, Pennsylvania, 2006.



PORTEBFIELD J. C, of Porterfield & McCombs, h Allegheny st
Porterfield Chas, laborer
Porterfield R. W, laborer
Porterfield R. S, laborer
Potter G. W, harness maker, h Water
PRING J. D, harness shop, Allegheny st, h do
Penny E. B. carpenter, bds Ex Hotel

R

Reed Adave machinist, bds w Emlenton
Reeves E. N, plasterer, bds Exchange Hotel
Redic O. C, of Lovell & Redic, h Emlenton
Reighner Conrad, laborer
Reighner Jermiah, grocer
Ross J. H, shoe shop cor Water and Allegheny sts, h Allegheny st
Rupert W. C, shoemaker, bds Allegheny st
Rughley Alonzo, laborer, bds Valley House

S

Sampson & Stiles, physicians & surgeons office Main st near Crawford House
Sampson Dr. F. W, office Main st, h on Hill
Sands F. F. clerk, house Main st
SCOTT J. C, dealer in rough and dressed lumber, sash doors, &c, office Main st, bds Exchange Hotel
Shannon O. E, lawyer
Sheakley Isaac, blacksmith
Shoemaker J. W, carpenter
Shoup Franklin, carpenter
Sloan J. M, hotel keeper
Sloan H. R, tailor, h Main st
Sloan Harvey A, watchmaker, bds College st
Sloan Jerome, hostler, bds Valley House
Sloan Uriah, carpenter
Sloan Leslie, carpenter
Smith Joseph G, carpenter
Stephenson Franklin, laborer
Stiles Dr. Edward, jr. o Main st, h Main st

T

Taylor Thomas, toll gate keeper
The Rising Sun, cor Main and Maple sts, W L Porter editor and proprietor
Treat M C, operator
Truby Jacob, sr, teamster
Truby Jacob, jr, teamster

V

Vallentine Harry, driller, bds Crawford H
Valley House, David Vencel, pr, Walnut st
Vencel David, proprietor Valley House
Virtue A, minister

W

Weaver P. J, harness shop, cor Water & Allegheny sts, h do
Weller Joseph, dry goods, cor Maple & Allegheny sts h do
Weller Wm. E, laborer
Wellerey Emanuel, laborer
Western Union Telegr'h Co, Miss R E Corry manager, o A V R R depot
Whiting E, shoemaker, h Emlenton
Wick J. M, laborer, h Hill st
Widle Emanuel, carpenter
Widle Elias, saw mill
Widle Eugene, laborer
Widle Wesley J, laborer
Willson Augustus, carpenter
Wise Levi B, clerk

Z

Zeck Isaac, engineer, A .V R R, bds Crawford H
Zink Wallace, clerk, bds Water cor Alleg'y
Zink J. K, laborer

# THE GREAT BUTLER COUNTY

# 1869 - 1880 OIL REGION

From 1870 through 1875, Clarion, Armstrong and Butler Counties were known as the "Lower Region" and completely overshadowed all other oil fields. The "oil towns" of Parker City, Petrolia, Karns City and Millerstown were the major population centers of the Butler oil territory.

The Oil City Derrick newspaper reported that according to reliable reports, these busy centers had the "vim and vigor of Oil City, Rouseville, Petroleum Centre and Pithole, all done up in a single package."

Students of oil history, hold tight to your stove pipe hat! (The Author)

# A HISTORY OF BUTLER COUNTY OIL PARKER DESIGNATED QUEEN CITY DRAKE WELL TO BUTLER COUNTY

"The smell of oil from Drake Well reached Butler County with dispatch. In January of 1860, within six months of the successful completion of the Drake experiment, residents of Butler organized the Butler Oil Company. This first of many Butler County oil companies was born in secrecy lest its members be subjected to the ridicule of less adventurous members of the community. Nothing is known of its membership or its operations.

"Despite the apparent failure of the Butler Oil Company, John Negley and others organized the Butler Pioneer Oil Company in February of 1861. As its sole operation, the company leased a lot from Mr. Negley near the present site of the brewery building and began drilling. When no oil appeared at a depth of eight thousand feet, drilling was abandoned. The failure of the Butler Pioneer Oil Company to find oil on the brewery lot did not materially discourage those who had faith in Butler County oil.

"The next year, 1862, saw the organization of the Enterprise Oil Company at Prospect. Its organizers, including Jacob Zeigler, Mrs. Anne Bredin and others, drilled several wells at Harris Ford on Slippery Rock creek. These wells were drilled in wet and never produced oil. Undiscouraged by past failures, Jacob Zeigler, the Campbells, Bergs and McJunkins, together with other prominent citizens of the community, formed the Butler County Oil company in 1864. Although its operations were doomed to eventual failure, this company had under lease some of the richest oil territory in the county. It had a large portion of the land lying between Martinsburg (now Bruin) and Chicora under lease.

"Numbered among the farms covered by Butler County Oil company leases were the Gibson, the Fletcher, the A. L. Campbell and the McClymonds farms. In the productive years ahead a fortune in oil was to be extracted from each of these farms. In all, the company drilled five wells, unaware of the fortune within their grasp the owners abandoned their operations without finding oil. Had drilling been continued oil doubtless would have eventually been found. Discouraged by their failure to tap the underlying petroleum, the leases were abandoned and the company disbanded."

## Wildcatting Began

"While no substantial wells had yet been brought in, sufficiently hopeful signs of the existence of oil underlying Butler County had been found to encourage further exploration. Wildcatting was in full swing all over the county by the end of 1865. Wells at Millerstown (Chicora), Harlansburg, along Slippery Rock Creek, at Buhl's Mills and in other widely scattered points throughout the county produced no appreciable quantities of oil. While several of these wells successfully produced oil in small quantities, none of them was financially able to continue operations and all were soon abandoned. During that year, 1865, Julius Klinger and John Berg completed what was probably the first successful trans-Atlantic shipment of oil. These men purchased and shipped one thousand barrels of oil consigned to England.

"While repeated failure rewarded the efforts of Butler county citizens to harvest the bonanza they were sure existed in the county, developments in Venango County were proceeding apace. The strike at Pithole in 1865 produced nationwide excitement. Other important fields in the vicinity caused a stampede of speculators and operators to the Venango district. Tidioute, Petroleum Center, Pleasantville, Pithole, and other towns rose, some to become permanent, others to vanish, but all to take their place in the rapidly developing saga of petroleum.

## C. D. Angell Theory

"Among the many intelligent operators who flocked to the Venango county field was one C. D. Angell. His observation and theories were to play an important part in the ultimate

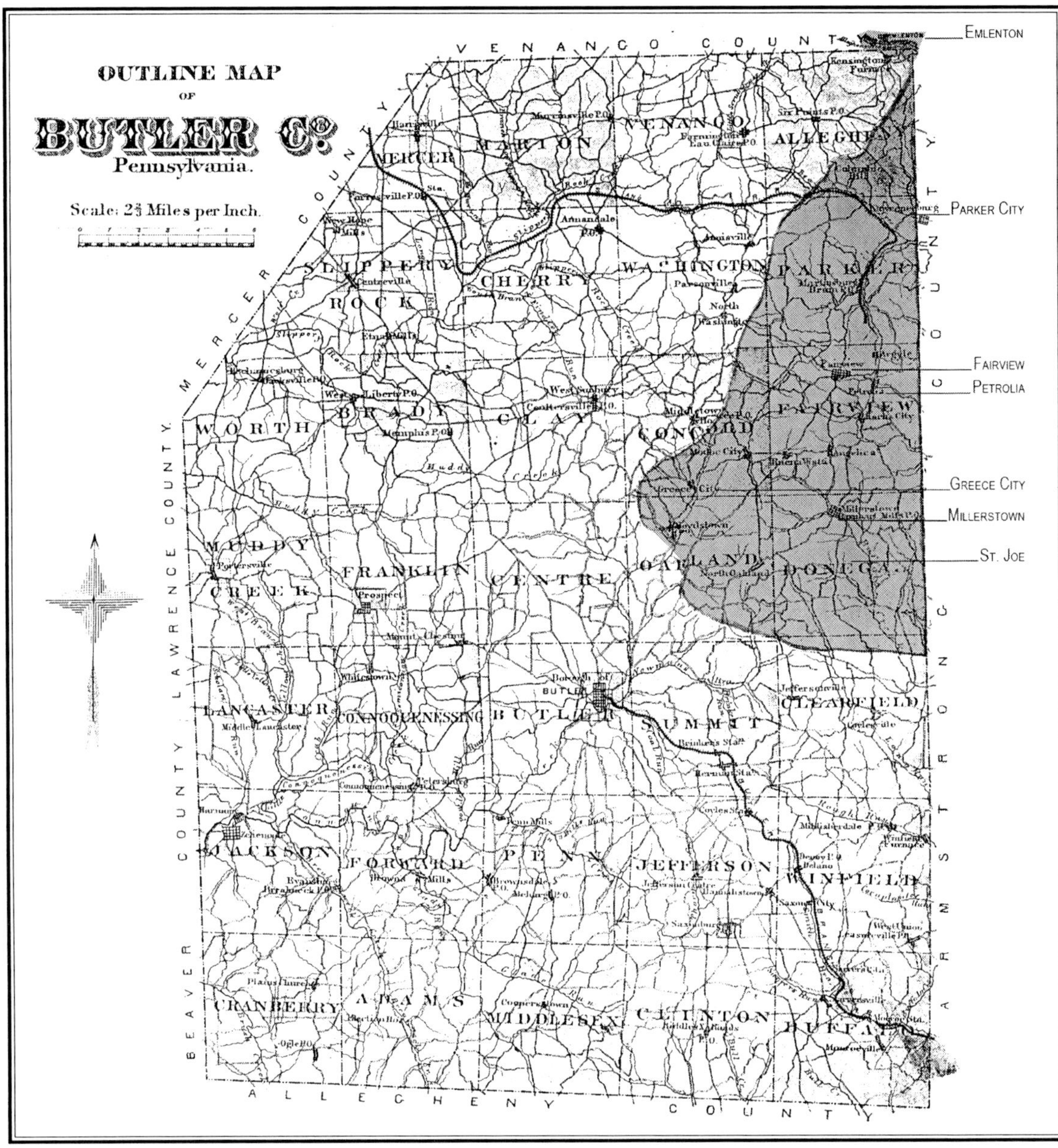


***Butler County, Pennsylvania. This 1874 map shows the general location in Butler County of the oil belt continued on through from Clarion County. Refer to page 77. This oil territory was expanded but never with the original oil boom excitement. Later producers moved south in Butler County. (Courtesy of the Butler Public Library)***

*Butler County oil territory, farm land now oil land, 1876*

development of Butler County as an oil producer. Mr. Angell, after considerable observation and surveying, advanced the theory that oil lay in continuous lines or belts. According to the Angell theories, two such belts existed, a northern belt extending through the Venango field, and southern line or belt which followed, he believed, a general direction of 16 degrees east of north.

"Extended south, the line passed through Scrubgrass and St. Petersburg and followed a course which eventually led to Butler County. While not universally accepted by drillers and explorers, the Angell theories were the best then obtainable. In accord with the Angell beliefs, the drillers gradually advanced their operations south through St. Petersburg and toward Parker City, just outside the limits of Butler County.

"While the Angell theories, first expounded in 1867, were being explored, Jacob Zeigler and other Butler county operators had again organized themselves into a company known as Jacobs Oil Company. The Jacobs company, organized in 1868, immediately took over some of the now-abandoned leases of the Butler County Oil Company.

"Jacobs Oil Company started a well, known as Jacob's Well near Martinsburg in the fall of 1868. In February of 1869 the well was completed with a small showing of oil. The superintendent, J. Q. A. Kennedy, after an examination of the well, decided that it was too deep and ordered the tools raised.

## First Paying Well in Butler County

"At the point designated by Mr. Kennedy the well inauspiciously produced three barrels per day. A 60-barrel production was reached after the well was shot. Jacobs' Well never exceeded a 60-barrel production and in the light of later developments was not a large producer. The importance of the well lies in the fact that it has generally been considered the first paying well within the borders of Butler County.

## Parker, Queen City of Oil

"The exploration and development of the Southern Angell belt reached Parker during the summer of 1869 and the spring of 1870. Parker quickly developed as the "Queen City" of oil. An oil exchange was immediately established. The expansion of the Parker oil field was phenomenal. In July 1869 it is reported that there were not more than twenty-five wells in the entire Parker district. The influx of oil men during the summer of 1869 brought on a drilling fever of remarkable proportions. By the middle of November the number of wells had increased from twenty-five to ten hundred and fifty eight.

"While the Parker field was by no means limited to Butler county, the lands lying between Bruin and Parker and within Butler county were considered a part of the Parker oil district. During the years 1869, 1870 and 1871 a general leasing of that part of Butler county took place and a considerable number of productive wells were drilled.

*Early photograph of Parker City – 1870-72*
*This photograph was before the iron bridge was built across the Allegheny River. The Wallace House, Spencer's Opera House and H. Fuller lumber yard and feed store are pictured. The large barge in the center is believed to be the famous Ben Hogan's "Floating Palace of Pleasure." The area above "Parker's Landing" was called the "Bluff." Two derricks are shown sticking up above the tree tops, one is the "Mt. Vernon," first well on the "Bluff," and the other is the "Eagle Well."*

"Jacobs' Well, lying near Martinsburg, strangely enough was not considered an extension of the Angell belt south and west of Parker, but rather was thought to mark the southern limit of the Angell line. Drilling, therefore, particularly during 1869 and 1870 progressed away from, rather than toward, the principal field later to be developed within the county. It was not until the latter part of 1871 that the more daring operators began cautiously to feel their way southward again along the Angell line.

## Scene Shifts

"In November of 1871 the A. L. Campbell farm near the present site of Petrolia became the front of operations. Several paying wells were drilled on that farm during the fall of 1871 and the spring of 1872. Around these wells the village of Argyle, later included within the boundaries of Petrolia borough, sprung up.

"Lying immediately south of the A. L. Campbell farm lay the Blaney and Jamison farms. To these two adjoining farms came George H. Dimmick. He, together with his business partners, George H. Nesbitt and William Larkin, leased both farms. They immediately located a well on the line adjoining the two farms and began drilling. The well, known as the "Fanny Jane," was drilled in during April of 1872. A flowing well producing 250 to 300 barrels a day resulted.

"The success of the "Fanny Jane" spread as if by magic throughout the oil country. Within four months of the Fanny Jane strike, the village of Petrolia had appeared from nowhere and was a flourishing oil field metropolis. Petrolia, built around the site of the "Fanny Jane," quickly swallowed up the village of Argyle and launched itself as one of the most open and lawless towns of the oil country. Within its boundaries the criminal element that infested every oil town flourished under the leadership of such notorious characters as Ben Hogan, whose temerity reached the ultimate height of running as a candidate for the office of burgess at the first election held in Petrolia borough. An oil exchange was quickly established which for a time controlled the price of crude oil for the entire industry.

"The "Fanny Jane" was quickly followed by other wells. The Lightfoot, the Maple Shade, the Walnut Shade, Argyle wells one, two, three and four and many other wells of considerable productivity came in rapid succession. By June of 1872 the field of operation had progressed beyond Petrolia to the McClymonds' farm upon which a major portion of Karns City is now erected. Cooper Brothers and 'Dunc' Karns had leases on or in the vicinity of the McClymonds farm and drilling was in full swing.

## Belt Abandoned

"While the more conservative drillers continued to pursue the general line established by Mr. Angell in 1867, the 'Fanny Jane' strike brought on a rash of wildcatting. Wells were drilled away from the now established Angell belt, which had by this time been determined to extend through Parker, Bruin, and Karns City in a general north 22 degrees east direction, rather than north 16 degrees east direction as originally fixed by Mr. Angell.

"Among the wildcat wells resulting from the "Fanny Jane" discovery was one drilled by David Morrison on the farm of his father-in-law, S. S. Jamison, in the southern part of Concord township about seven miles from Petrolia and well outside of the accepted oil belt. This well was drilled during August of 1872. To the utter amazement of experienced oil men, the Morrison well produced 700 barrels of oil its first day and averaged 250 barrels a day over the first six months of its existence. The surprising success of the Morrison well brought about an oil excitement similar to that which resulted at Petrolia from the 'Fanny Jane' strike. Greece City sprang up on the site of the Morrison well almost overnight.

## Greece City

"During its brief career of about 13 months, Greece City grew into a thriving community of banks and hotels, alive with the bustle of an oil town and prosperous from the wealth which poured from the land surrounding it.

"The success of the Morrison well did not long remain a mystery. Wells drilled in the vicinity soon established the existence of a new belt of oil, later known as the 'Cross Belt' or Greece City line. Unknown to anyone, Morrison had accidentally tapped the fabulous 'Fourth Sand' which was ultimately to exert the most depressing influence on oil men known in Butler County. The harvest of oil from the Fourth Sand, together with the productivity of the Third Sand in the main belt, eventually forced the price of oil to a low of 40 cents per barrel.

## Troutman

"Upon the heels of Greece City other oil towns sprang up. Modoc City, Troutman and Buena Vista quickly followed. Each new town marked a step in the development of the Cross Belt from Greece City toward Petrolia and Karns City. Large operators flocked to the district. The spring of 1873 saw the Old Troutman well come in at 800 to 1,000 barrels per day. Other fabulously productive wells were brought in with regularity. It was not until September of 1873, more than a year after the drilling of the Morrison Well at Greece City, that the existence of a different oil bearing strata from that which the Angell belt wells were drawing was definitely established. While several different explanations of the discovery of the new sand, known as the 'Fourth Sand,' as distinguished from the Angell or Main belt strata known as the 'Third Sand,' have been advanced, the most generally credited is the explanation give by A. L. Campbell, whose pioneer development at Argyle had been the forerunner of the Petrolia excitement.

## Old Hickory

"According to Mr. Campbell's account, Daly and Company, including one Charles C. Stewart, had begun the drilling of a well known as 'Old Hickory' on the Scott Heirs' farm west of Karns City and near the corner of the MacLear farm where Tack, Morehead and Company were in the process of drilling their 'MacLear Number One' well. Mr. Campbell was superintendent of drilling on the MacLear Well for Tack, Morehead and Company. The accurate record of the MacLear well which Mr. Campbell kept enabled Mr. Stewart of Daly and Company to keep a close check on the progress of the MacLear well.

"Daly and Company, after the usual drilling depth was reached, concluded that their 'Old Hickory' well was down and dry. Mr. Stewart, checking the progress of MacLear Number One, which was a producing well when completed, concluded that the two wells had been completed in different strata. The refusal of Daly and Company to drill further down induced Mr. Stewart to purchase 'Old Hickory' and drill further. He resumed drilling and finally finished the well in the Fourth Sand, sixty-nine feet under the Third Sand to which the well had originally been drilled.

"Simultaneously with the completion of 'Old Hickory,' Tack, Morehead and Company completed MacLear Number Two well in the Fourth Sand. All of these wells flowed when they reached the Fourth Sand. Realizing that the knowledge in their possession was worth a fortune, both Mr. Stewart and Tack, Morehead and Company paid each of their men a $100 bonus to keep silent. Despite this precaution word soon spread throughout the district of the discovery.

"Drilling had gone on apace at Petrolia during the development of the Cross Belt from Greece City through Modoc City, Troutman and Buena Vista. The Petrolia-Karns City wells, all drilled in the Angell or Main Belt, unlike the Cross Belt wells, were Third Sand wells.

"Richard Jennings, probably the largest individual producer in the district, 'Dunc' Karns, Cooper Brothers and other substantial operators of the Petrolia-Karns City district, upon hearing of the findings of Mr. Stewart, quickly pulled their Third Sand Wells and drilled them down to the Fourth Sand. This maneuver was frequently successful.

## Becomes Frenzy

"What had theretofore been an excitement now became a frenzy. The Fourth Sand seemed to be everywhere. For a time it seemed as if no one could miss. Wells that had been fair producers, or poor producers in the Third Sand, suddenly flowed when the Fourth Sand was tapped. It was quickly ascertained that the Main Belt and the Cross Belt intersected between Karns City and Petrolia. Producers in that district, drawing from both sands simultaneously, reaped a rich harvest in oil. The land lying between those towns became the richest field discovered up to that time in the county.

"'Dunc' Karns was reported to have an income of $5,000 per day. He built a fabulous mansion near Freeport, owned a pipe line and railroad. Richard Jennings of Queenstown brought in the largest single well in the vicinity early in 1884. This well, the Jennings No. 19 on the Daugherty farm produced 4,000 barrels of oil during its first 24 hours. Many legendary wells including the Rob Roy, the Eureka and others were brought in during this period of development.

"The eastern extension of the Cross Belt was opened in the spring of 1874. Just over the county line in Armstrong county, D. S. Criswell brought in the 'Boss' well, a 2,500 barrel producer. This well was located about two and one half miles east of Petrolia. The 'Lady Hunter,' a well whose production equaled that of the 'Boss' well and other fabulous wells were drilled to the Fourth Sand east of Petrolia during 1874. By the summer of 1874, production in the Petrolia and Cross Belt areas had reached its peak. The highest production from the field during a single day was 42,000 barrels. Average production in the district for the month of July of 1874 was 28,424 barrels per day.

"As a result of the fabulous Fourth Sand discovery, the price of crude oil, despite the strenuous efforts of operators who united to restrict production, dropped at one time to 40 cents a barrel. Over the years, however, the price of oil held up remarkably well. From 1870 to 1875, years that saw the Parker excitement, the Petrolia frenzy, the Fourth Sand development and the Millerstown boom, the average prices over a year's time ranged from a high of $4.34 in 1871 to a low of $1.17 in 1874. By 1875 the price was again on the way up and averaged $1.31.

## Chicora Field Opens

"While operations in the Petrolia, Karns City and Greece City districts were rapidly moving toward their peak in 1874, the Millerstown (Chicora) field came into being. This production area, actually an extension of the Main Belt field south from Petrolia and Karns City came into its own early in 1873 with the completion of Divener No. 1, a 1,400 barrel producer. The first well in the district known as the 'Schreve' well, a small producer of 85 barrels, was drilled in during April of 1873 and was followed quickly by the Divener well and other large producers.

"Millerstown quickly mushroomed from a sleepy hamlet to a bustling town of about 5,000 persons. The drilling of the Meade well in Donegal township in 1875, the Lambing wells and others tended to extend the Millerstown field. The booming oil town of St. Joe marked the southern extremity of the field. Plummer, Danville, and Greer were other towns that developed from the Millerstown excitement.

## The End Is Near

"By the beginning of the year 1876 the Butler County oil excitement was beginning to wane. The exploratory development period had, for the time being, run its course and the then developed territory within the County was in the process of complete drilling development. While the oil fever had, to a considerable extent, abated, production continued to rise in the Butler-Armstrong territory. It achieved its highest point during the year 1877 when the area as a whole produced 5,500,000 barrels of crude oil.

"The recent strike at Bradford caused an exodus of drillers from Butler County. The rush to Bradford and Tarpot, which drained off many large operators from the county, halted exploratory operations in the county only momentarily. A rash of drilling in the Six Points area in 1877 and 1878 produced new wells, but no particularly noticeable excitement. By 1880 the drilling fever in the northern townships had expended itself. The few remaining pioneers who

*Butler County Oil Territory*

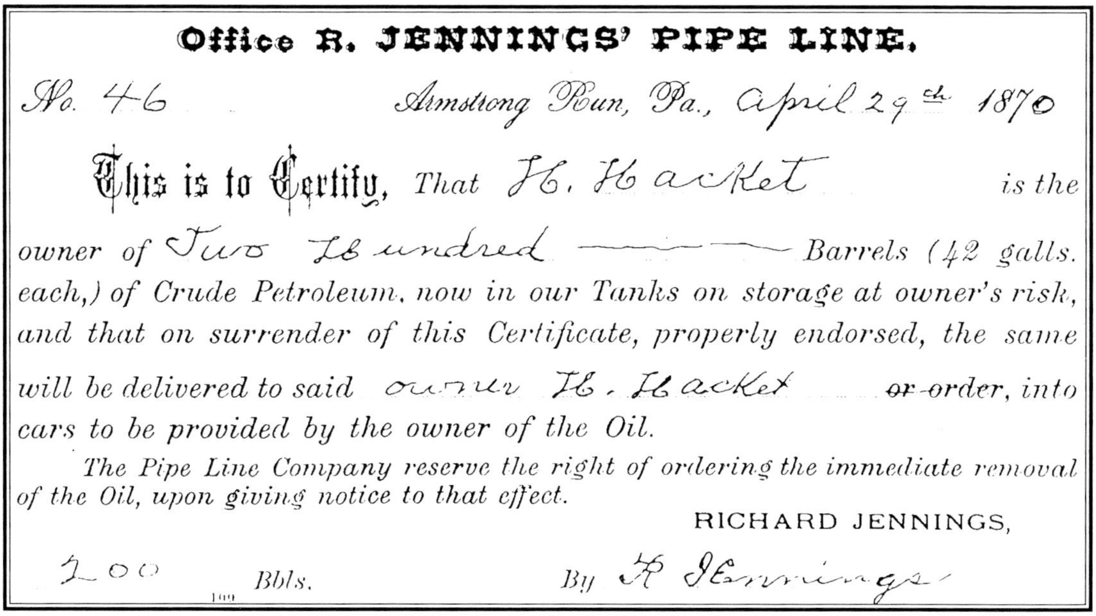

Office R. JENNINGS' PIPE LINE.

No. 46 Armstrong Run, Pa., April 29th 1870

This is to Certify, That H. Hacket is the owner of Two Hundred ——— Barrels (42 galls. each,) of Crude Petroleum, now in our Tanks on storage at owner's risk, and that on surrender of this Certificate, properly endorsed, the same will be delivered to said owner H. Hacket ~~or order~~, into cars to be provided by the owner of the Oil.

The Pipe Line Company reserve the right of ordering the immediate removal of the Oil, upon giving notice to that effect.

RICHARD JENNINGS,

200 Bbls. By R Jennings

***THE ARMSTRONG RUN OIL FIELD***
***The Armstrong Run Oil Field was near Armstrong Run about four miles north along the river from East Brady. On April 18, 1870, the Armstrong Run Oil Company began drilling on the Mildren farm. Among the dozen owners were Richard Jennings of Queenstown, Colonel W. D. Slack of the iron company, and Nick Keener, a butcher from East Brady. They completed a 1,262-feet-deep well in the fourth sandstone that developed a large flow of oil and gas. It caught on fire, burning for three days, and contractor Ed Swan, driller Bryan Swan, tool dresser John Walley, and a spectator, C. Clark, were severely burned. Initial production was estimated at 1,000 barrels a day. After it had produced for 45 days, it was still flowing 250 barrels daily. (Butler County Historical Society, Butler, Pennsylvania)***

had not gone to the Bradford field began to search for new production areas in the Butler County field."

The proceeding article was published in the Butler Eagle's sesquicentennial edition in July 1950, which marked the 150th anniversary of the founding of Butler County. The article was researched and written by William C. McElvain. (Butler Eagle, Butler, Pennsylvania, July 1950)

## Parker City
## An Oil Town of Rapid Phenomenal Growth

The discovery of oil was one of the most important events in the history of Armstrong county. The northern portion was the area most affected by the mid to late 1860s oil excitement. After the discovery of oil, land that was agriculturally worthless became the old farmer's "Golden Egg." About 1786, John Parker, a surveyor, left his home in Washington County, Pennsylvania, and ended up in the wilds of western Pennsylvania. Both John and his brother surveyed much of the land east and west of the Allegheny River, which included the northern portions of Armstrong and Butler Counties.

A small, fast flowing creek empties into the Allegheny River near Parker's Landing and became known as Bear Creek. Along the river, near Bear Creek, was located a large Indian village. A man by the name of Elliot was part of the surveying party and he shot and killed an old Indian

*Atlas of Armstrong County, Pennsylvania; Pomery, Whitman and Company, Philadelphia, 1876*

at the mouth of Bear Creek. The surveyors were greatly alarmed and fled. The Indians followed the men a very long distance with the intent of killing them, but finally gave up the pursuit and the Indians gave the white men no further problems. Sometime prior to 1797, a river traveler by the name of Baird made a deal with the Indians and purchased the "flat" where the Indian village was located. This site was then sold to John Parker and became known as Parker's Landing.

Parker's Landing, which later grew into the great oil town of Parker City, is situated on the western bank of the Allegheny River, 82 miles up river from Pittsburgh, Pennsylvania. It was named after the honorable John Parker who had been granted several hundred acres for his services in surveying and around 1797 built a house and settled here. John Parker was a true pioneer and was appointed one of the first associate judges of Butler County. He was considered a true and trustworthy friend who promoted the settlement of the surrounding territory. His son, Fullerton Parker, was a leading operator and general speculator during the "oil boom." William Parker, the father of

ALLEGHENY VALLEY R. R.

—AND—

CONNECTIONS.

*Three Through Passenger Trains Daily*

EACH WAY BETWEEN

PITTSBURGH & CORRY,

Passing through the important cities of

TITUSVILLE, CORRY, TIDIOUTE,

Oil City, Franklin, Foxburg,

Parker's Landing, Brady's Bend,

AND MAKING

**DIRECT CONNECTIONS**

TO AND FROM

Buffalo & Niagara Falls

—AND—

ALL IMPORTANT POINTS NORTH.

J. J. LAWRENCE,

*PITTSBURGH, PA.* *Gen'l Superintendent.*

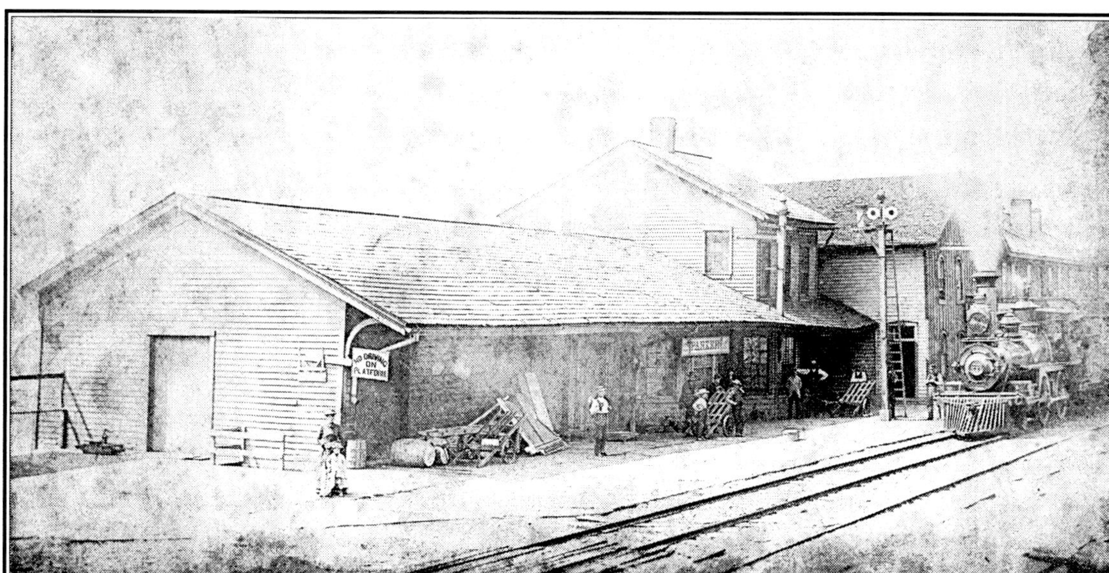

*The Allegheny Valley Railroad Station at Parker's Landing. All equipment, supplies, and people came by rail through this station. The sign at the left reads, "No Driving on Platform." At the right is the Phillips House, Wilkins and Thompson Proprietors. Old Steamer No. 37 sets in front of the hotel. The station was located opposite Parker's Landing along the Allegheny River. Photograph taken 1876. From this station the train could cross the river on the new iron bridge into Parker City and on into the Butler County oil region.*

PARKER CITY
HOVEY TWP.
Scale 25 Rods to an Inch
1ST WARD
2ND WARD
BUTLER CO
WASHINGTON
PARKER CITY TROTTING PARK
Union Pipe Tanks
Bear Creek
Geo Parker
Dr. Eggert
Schultz Brewery
Union Hotel
Grant Pipe Co.
U. P. Co.
Spring
Grist Mill
Abrams
R. Dawson
A. Duncan
J. Gruble
School
Parker, Cooper & Co.
Morgan Cooper & Co
T. Donnelly

Judge Parker, moved from Washington County, Pennsylvania, about 1798 and built a mill on Bear Creek. It was built of logs and contained the rudest machinery, but it still provided a valuable service to the early settlers. This mill was the first erected in that section of the county. Another valuable business was started about three-fourths mile up Bear Creek. The Bear Creek Furnace was built by Whiting and Stackpole around 1820 and produced about 75 tons of pig iron per week. The furnace was fired with charcoal and steam, but finally ceased operations in 1840 due to the scarcity of timber.

## The Village of Lawrenceburg

Lawrenceburg was laid out by Judge Parker about 1819. This village was brought into existence by the Bear Creek Furnace, and consisted mainly of crude dwellings occupied by employees of the company operating the furnace. The number of businesses increased as the village grew and eventually there were a few log homes, three stores and numerous taverns which brought in people from miles around. The village of Lawrenceburg was at the top of the river hill, whereas Parker's Landing was at the bottom of the hill along the river. There was a large amount of traffic and travel upon the Allegheny River by means of canoes and keel-boats, and all who had business to transact at Parker's Landing naturally came to Lawrenceburg. At that time there were few buildings and businesses along the river.

The closing of the Bear Creek Furnace in 1840 delivered the death blow to the village of Lawrenceburg. It suffered a steady decline until it became only a name with few businesses, few people, and a very bleak future—that is, until the commencement of the oil excitement and by 1870 thousands of people had located there, either as permanent or transient residents. The surrounding oil fields were also heavily populated.

Old John Marshall was born in 1802 and first came to the village of Lawrenceburg in 1825 and purchased 20 acres of land at one-dollar per acre. For years, John thought and complained that $20 was too much. That is until 1870 when his land was found to be valuable oil territory and he was offered $45,000, plus royalties. Old man Marshall politely thanked them for the offer, but refused and slowly walked away.

The name "Boom Town" is quite often associated with gold, silver, or oil. No one who has not witnessed the rapid up-building of towns in the oil region can form an adequate idea of the growth of Parker's Landing. The importance of the discovery of oil near Parker's Landing was not fully realized until the summer of 1869, and this date marks the beginning of Parker's Landing, the chief emporium of the lower oil region.

## Parker's Landing, Indian Village, Steamboat Stop

Parker's Landing in the early years was an unimportant station in the wilderness visited occasionally by pioneers in canoes and keel-boats. A store, owned by W. R. Robinson, a blacksmith shop run by Samuel Craig, a boarding house and lumber storage station were located there, but a village never took root. Thomas M. Parker ran a ferry which provided a valuable service by transporting supplies, teams, and families across the Allegheny River. In 1824 Judge Parker built a large building and used it as a warehouse. This old structure has passed unscathed through many fires and floods and is still standing as the oldest house in Parker's Landing. It has been converted into a hotel and is now called the Parker House. This hotel and bar is still in existence in 2007.

*Parker City street scene. The Fuller House is on the left and W. T. Graham, Attorney and Counsellor-at-Law, is on the right. 1870*

A few buildings were all that the "flat" was comprised of in 1869. In just a matter of a few years, this area became the heart of a busy city. Petroleum caused great changes to take place resulting in a bustling city, a railroad, daily newspapers, an iron bridge across the river, banks, hotels, and hundreds of flourishing businesses. Lawrenceburg was swallowed up by its former tiny neighbor and Parker City was the exciting end result.

## Parker City

The first oil discovered in Armstrong County was pumped from the Clarion well, No. 1, on the Robinson farm, one mile north of Parker City, on October 10, 1865. The Clarion No. 1 produced 18 barrels a day, but it was not until the latter half of 1869 that a genuine oil excitement revealed the importance of the region surrounding Parker's Landing and Lawrenceburg. The number of drilling wells and the leasing of oil lands by many outside parties including the Venango County oil men, was explosive. In July 1869 there were 25 oil wells producing 310 bar-

LAWRENCEBURG

**SAVINGS BANK**

**Second Ward, PARKER CITY, PA.**

DIRECTORS.

| | | |
|---|---|---|
| John Scott, | M. S. Adams, | Robert Balph, |
| W. F. Messimer, | Wm. Morgan, | A. B. Gibson. |

OFFICERS.

| | |
|---|---|
| John Scott, | President. |
| M. S. Adams, | Vice President. |
| T. M. Joy, | Cashier. |

STOCKHOLDERS.

| | | |
|---|---|---|
| John Scott, | H. E. Picket, | W. F. Messimer, |
| M. S. Adams, | Z. B. Shepard, | A. L. Campbell, |
| A. B. Gibson, | Amos Young, | Robert Storey, |
| T. M. Joy, | E. R. Marshall, | E. H. Randolph, |
| J. W. McFarland, | Robt. Balph, | A. M. Hoover, |
| Wm. Morgan, | A. B. Rice, | W. H. Bradley, |
| W. J. Parker, | Jno. B. Campbell. | |

Stockholders Individually Liable.

Transact a general Banking business. Buy and sell Gold, Silver and Coupons. Make collections and prompt remittances at lowest rates of exchange. Sell drafts on England, Ireland, Scotland and all parts of the European Continent. Also passage tickets to and from the Old Country.

*Interest at 6 Per Cent. Paid on Time Deposits.*

rels a day, but by November of 1869, over 1,056 wells were either completed or in the process of drilling. Other wells that were drilled near Lawrenceburg and located in Butler County proclaimed the value of the Butler County oil territory. With the boom town of Parker City being so close to Armstrong, Clarion and Butler Counties, it became quite evident that Parker City would become an important base of operations for producers and operators. Parker City attracted operators of every class and condition. Speculators, gamblers, and adventurers thronged its busy streets. It had according to newspaper reports the "vim and vigor" of Oil City, Rouseville, Petroleum Center, and Pithole done up in a single package.

As word of the oil boom and possible easy riches began to spread, people began to pour in. The lawlessness and loose morals of the followers of the oil excitement did not fail to contaminate the new town. Ben Hogan and his partner, French Kate, were two of the most infamous residents of the city. Rapidly built shanties were thrown up over night and business was being conducted even before the echoes of the carpenters' hammers had faded away. Buildings were going up like "mushrooms in the dark," saloons, stores, machine shops, hotels, and businesses of every description. Every available building lot between the Allegheny River and the Bluff was sold for saddle and harness shops, ready-made clothing stores, furniture, liquor, dry goods, grocery stores, lumber mills, tobacco shops, and oil well supplies. Any goods that could be bought in Buffalo, Pittsburgh, Cincinnati, or New York City could now be bought in Parker City.

In a short time activity in and around the city increased. The population grew and great prosperity ensued. Refined gentlemen replaced the riff-raff. Fortunes were made and lost all in a day, and handsome residences were erected by men who were just days ago common day laborers. With a new magic in the air, imposing business structures lined the streets of the lower flats. Many large business establishments catered to the wants of the mushrooming population and every other business was either a saloon or an eating house. "Oil on the brain" was a condition shared by all, and all conversations led to petroleum. At one time the Parker City Oil Exchange did the largest amount of trading in oil of anybody in the petroleum fields. The Oil Exchange possessed a $5,000 library with lavishly decorated clubrooms. The traffic of people moving about from the Flat to the Bluff was so great that an elevator was constructed. It carried the weary speculators to their homes on the beautiful hilltop for only the price of a gallon of oil. (Refer to Page 6)

Many of the more respected citizens saw that there existed a need for a city government. The rapid growth of population and the belief that the good times would never end, naturally led the citizens to believe that Parker City was destined to become a large and important place. A directory of the oil region in 1875 placed the population of Parker City at 4,000. At the height of the boom there were probably 15 to 20 thousand residents with a transient population of about 5,000. A petition for a city charter was presented to the legislature and by an act approved March 1, 1873, Parker's Landing became officially the city of Parker. From 1870 to 1875, Clarion, Armstrong, and Butler Counties completely overshadowed all other oil fields. Parker City, Karns City, Petrolia, and Millerstown were major population centers in our Lower Region.

## Banking

The rapid growth and the increase of business soon rendered the establishment of a bank. The Parker City Savings Bank was opened in 1869 and failed in July of 1882 with large losses to its depositors. The Exchange Bank was established in 1871 and ceased operations in 1880. In October 1882, Parker Fullerton and Company revised the Exchange Bank until 1901 when it too failed.

## Waterworks

In 1872, the Parker City Waterworks was built by Miller and Vesey. Water was taken from the Allegheny River and pumped to an iron tank. The water was raised 280 feet straight up to the tank's location on the Bluff. The waterworks was managed by one man who proudly reported that it cost the city only $13 for repairs in the first eight years of operation. For fire protection hose carts were pulled to the fire by a volunteer force and hooked to a fire hydrant. The pressure was sufficient to control an ordinary fire. The water was suitable only for fire protection and was far from being desirable as a source of drinking water. The Allegheny River was shallow and fouled by oil, brine water, and dead animal carcasses the greater part of the year.

## Newspapers

Every oil town had its newspapers and Parker City was no exception. A number of unsuccessful newspapers originated in the city during its prosperous oil period. A daily paper was established by Johns and Jackson and published a short time in 1871 and 1872. Wilson, Clark and Co. published the "Oilman's Journal" several years but it eventually failed. These papers and several others were started but never became financially successful. Established by G. A. Needle in September 1874, the "Parker City Daily" had a very prosperous run. The Daily had a large circulation and was one of the most reliable and influential journals in the Lower Oil Region. The "Parker City Daily" was published as a rival to the "Oil City Derrick," and was the same size as the Derrick. The Daily contained the Associated Press dispatches and much general information, in addition to its careful digest of news from every part of the oil region. It was published as a morning paper until 1879 when a terrible fire befell the city and the Daily ceased to exist. The Drake Well Museum and the Butler County Historical Society have a collection of the "Parker City Daily."

## Fire and Hose Company

The fire department of Parker City responded to a large fire bell which rang when someone reported a fire. The department consisted of two volunteer fire companies, one in each ward. The members of the fire company were a well-organized group and each man knew his job in a fire situation. Fire meetings, picnics, and parades were a regular part of each volunteer's responsibilities. The Columbia Engine and Hose Company watched over the First Ward and John Tonks led the department. In the Second Ward, the Citizens Hook and Ladder and Hose Company was led by S. M. Brewer, foreman. During parade days at other local oil towns, competition between fire companies was very fierce. Many newspaper articles of these events tell the stores of "who fought who, who beat who, and who went home with a bloody nose and a black eye."

*Krosskop's Pharmacy*

## Other Endeavors

Parker City also boasted the Parker City Gas Works, which was built in 1877 and manufactured natural gas through a process known as Smiths Patent. Natural gas was delivered through miles of iron pipe to eloquent houses and businesses in the First Ward and to the residents on the Bluff. Oilwell machine shops thrived and the first was opened in 1869. It was operated by Bradley and Duff. A large oil well machinery repair business was carried on and created a large employment of men. Lumber yards were always a busy enterprise in a growing oil boom town with George Brown Lumber Yard and Wilkins and Fullerton being just two of them.

## Wightman Glass Company

As the oil business became less and less reliable, a group of investors organized the Parker City Glass Works in 1880. Twenty-six glass blowers were employed along with sixty-five laborers. This business helped Parker City employ many of its citizens well into the next century and long after the oil boom went bust.

## The Social and Moral Health of Parker City

Parker City maintained a fine school system. Churches were well attended. Social organizations such as the Masonic Lodge, the Royal Templars of Temperance, and others gave its citizens a variety of clubs. The race track and driving park was known throughout the eastern

United States and general health was taken care of by many fine doctors. And last but so important was the Grand Army of the Republic, Calvin A. Craig Post No. 75. This Civil War post was chartered April 21, 1880. Sixteen veterans appear on the charter and the Post grew to over 90 members. Many veterans returned from the Civil War and found employment in the oil region. Bless their souls and may they rest in eternal peace.

---

J. WIGGINS & CO.'S

DIRECTORY

OF THE

Parker Oil District,

1873-74.

CONTAINING THE TOWNS OF

Parker City, Foxburg,
St. Petersburg, Emlenton,
East Brady.

*COMPILED BY J. WIGGINS & CO.*

1873.
J. F. DIFFENBACHER, PUBLISHER AND PROPRIETOR,
78 FOURTH AVENUE, PITTSBURGH, PA.

W. B. M'CORD, STEAM BOOK AND JOB PRINTER, WELLSVILLE, OHIO.
12

## PARKER CITY

Parker City is situated on the west bank of the Allegheny river, and on the main line of the Allegheny Valley Railroad, about eighty miles above Pittsburgh, and fifty miles below Oil City. It is the principal town in the Parker Oil District.

The first oil produced here was in the year 1869. The further development of this territory proved it to be the most prolific in the history of oil production.

The tide of emigration began to flow in this direction, and the towns of Parker, St. Petersburg, Foxburg and Emlenton sprung into existence.

Parker City, previous to the year 1869, was only known as Parker's Landing. The station on the Allegheny Valley Railroad was on the east side of the river. The building up of a town at this place was very rapid, and in the fall and winter of 1872 it boasted of a population of about 3,000.

During the winter of 1872, the town was visited with two fearful calamities in the shape of fires, the first of which happened in the month of December when a greater part of the city was destroyed. This was soon followed by a second fire, which consumed the entire business portion and all that part of the town built along the river. This conflagration, though fearful in its effects upon the town, did not consume the energies of its business representatives. The importance of the locality for a town was such as to demand its speedy rebuilding, and during the summer the entire burnt district was replaced with more substantial and elegant buildings than those that were destroyed.

The consolidation of this town with Lawrenceburg—the latter place now being the second ward—and the increase from immigration, gives the place, at the present time, a population of nearly 5,000, and is still rapidly increasing. A costly bridge has been built over the river from Parker to the station on the Allegheny Valley Railroad.

There is one daily newspaper published here, the *Oil Man's Journal*, which is managed by Wilson, Clark & Co., and is devoted to the interests of the oil trade.

There are a number of elegant hotels which are kept in first-class style, livery accommodations are excellent, and the different monetary, commercial and manufacturing enterprises are conducted here on a large scale, and are well known throughout the oil country.

## A

Aaron Hannah, of I. Aaron & Co., Washington, second ward.
Aaron I., of I. Aaron & Co., Washington, second ward.
AARON I. & CO., furniture, Washington, second ward.
Abrasan M., oil operator, Washington, second ward.
Adams Edward, grocer, Washington, res. Jackson, second ward.
Adams Homer, oil dealer, Wayne, second ward.
ADAMS HOUSE, M. S. Adams proprietor, Washington, second ward.
Adams W. C., oil operator, Wayne, second ward.
ADAMS M. S., proprietor, Adams House, Jackson, 2d ward.
Adams M. S., oil operator, Jackson, second ward.
Ainstead George S., (colored,) laborer, Central avenue.
Aldinger H. D., boots and shoes, Washington, second ward.
Aldrich ———, carpenter, Central avenue.
Allen R. B., agent Dickey Torpedoe's, River ave., near Fifth.
Allen R. B., president Grant Pipe Co., River avenue.
Allen Thomas, teamster, Jackson, second ward.

ALLEN W. E.

Dealer in groceries, provisions, tinware, &c., Washington.
Allen W. F., harnessmaker, First.
Anderson J. R., jewelery, Washington, second ward.
ANGE J. E., Riverside Restaurant, River avenue.
Armon Robert R., benzine dealer, office, River and Tom's Run.
Arters John M., of Lyle & Arters, boards River avenue.
Atchison H. D., of Boyd, Atchison & Co., boards River avenue.

AYER & CRAWFORD,

Milliners, dealers in all kinds of ladies' furnishing goods, etc., Washington, second ward.

AYER O. D.

Restaurant, also dealer in tobacco, cigars and pipes, Washington, second ward.

## B

Babst William, drug store, Washington, second ward.
Bachman S., of S. Bachman & Bro., Wayne.
BACHMAN S. & BRO., clothing, corner Washington and Wayne, second ward.
Bachman Louis, Wayne, near Washington, second ward.
Bachman M., of S. Bachman & Bro., cor. Washington & Wayne.
Bachman M., painter, Wayne, near Washington, second ward.
Bacon Mrs. H., widow, Wayne, near Washington, 2d ward.
Badger Charles P., oil producer, River and Bluff avenue.
BAILEY E. H., farmer, Washington.
Bailey H. E., engineer, Washington, second ward.
Baristow W. A., photographer, Second, near River avenue.
Balph Robert, Justice of the Peace, Washington, second ward.
Banser J. H., blacksmith, Jackson, near Washington, 2d ward.
Bardsley Joseph, clerk, River avenue.
Barnes W. H., book keeper, River avenue.
Barrett William, tailor, boards River avenue.
Bartholic F., carpenter, Washington, second ward.
Baughman J. N., miner, Washington, second ward.
Beam E. F., of J. Beam & Son, Bluff avenue.
Beam W. M., of J. Beam & Son, Bluff avenue.
Beam J., of J. Beam & Son, Bluff avenue.
BEAM J. & SON., tank guagers, River avenue.
Bendheim L., boots and shoes, River avenue, near Third.
Bennett James, hardware and house furnishing goods, River avenue.
Bennett Thomas, clerk, boards River House, second ward.
Bishop John R., clerk, boards River avenue.
Black J. W., teamster, Washington, second ward.
Bleymiller Charles, cigar store, Washington, second ward.
Blemiller Gust., laborer, Washington, second ward.
Bligh William D., oil operator, Bluff avenue.
Boulge George, cigar store, River avenue.
BOOK FRISBEE, billiard hall, River avenue.
Boan Mrs. S., widow, Jackson, second ward.
Borland A., grocer, River avenue.
Borland Alex., clerk, River avenue, near Fourth.
Boughton F. W., of Boughton & Chandler, office, River avenue.
Boughton & Chandler, agents for Geo. W. Tift & Sons, Buffalo, New York, boilers, River avenue.
Bowyer Lee, laborer, boards Union House.
BOYD, ATCHISON & CO., saddles, harness and trunks, River Avenue, between second and third.



Boyd T. H., of Boyd, Atchison & Co., boards River avenue.
Boyles H. F., telegraph operator, River avenue.
Bracken P., oil operator, Washington, near Jackson, second ward.
BRADLEY & DUFF, machinists and blacksmiths, between Ludlow and Wayne, second ward.
Bradley William, of Bradley & Duff, Ludlow, second ward.
Brady E. H., agent Phillips Bros., office, River avenue.
BRAMAN E., agent J. H. William's livery, near Syracuse House, River avenue.
Brown Miss A., millinery, Washington, second ward.
Brown Charles, hostler, boards River avenue.
Brown George, teamster, River avenue.
Brown H. D., clerk, boards Phillips House.
Brown James, Washington, second ward.
Brown John, barber, River avenue.
Brown & Martin, hardware, near R. R. depot, Pollock, Pa.
Brown R. L., of Brown & Martin, Kittanning, Pa.
Brown T. B., of Brace & Brown, boards Florence House.
Brown T. P., books and stationary, Washington, second ward.
Brown W., clerk, Washington, second ward.
Bruce & Brown, dry goods, etc., Washington, second ward.
Bruce J. N., dry goods, boots, shoes, etc., Washington, 2d ward.
Buck A. D., book keeper, River avenue.
Buckas James, butcher, Washington, second ward.
Burgham Joseph, of Burgham & Odell, River avenue.
Burgham & Odell, bakers, River avenue.
Burk Morris P., clerk, boards Syracuse House.
Burnett Lawrence, hostler, boards River avenue.
Bushnell Joseph, oil producer and dealer, office, River avenue.
Butler W., barber, River avenue.
Byer Joseph, shoemaker, Washington, second ward.

## C

Caldwell P., clerk, boards River House.
Campbell William, carpenter, boards north Central avenue.
Castle Harry P., agent Erie Iron Works, River avenue.
Catlin O. L., clerk, Jackson, second ward.
Catlin R. N., clerk, Jackson, second ward.
Carriger H. C., grocer, Washington, second ward.
Carson William, carpenter, River Avenue, near Fourth.
Chambers John, painter, Ludlow, second ward.
Chandler W. F, of Boughton & Chandler, boards Phillips House.
Chapin Harry, oil producer, Washington, second ward.
Charters Joseph G., barber, River aveuue, near Fourth.
Christy J. H., carpenter, Washington, second ward.
CHURCH CATHOLIC, Washington, second ward.
Church Methodist, Washington, second ward.
Church Presbyterian, Jackson, near Washington, second ward.
CAURCH U. P., Ludlow, second ward.

Claffey Ealiler, blacksmith, boards Wayne, second ward.
Clattner Peter, butcher, Washington, second ward.
Clawson Charles F., clerk boards Bluff avenue.
Cleveland Pipe Co., office, River avenue.
Cluff William, boarding house, Central avenue,
Clyde James K., blacksmith, Jackson, second ward.
Coast Fred., dressmaking, Washington, second ward.
Cochrane C. C., of Boyd Atchison & Co., boards River avenue.
Cohen D. A., clerk, boards River avenue.
COHEN L., merchant tailor, second ward.
Coleman J. E. clerk, boards Adams House, second ward.
Colter J. P., Colter & Martin, boards second ward.
COLTER & MARTIN, attorneys at law, office River avenue.
Conn R. D., residence Petrolia.
Connoley Wm., livery stable, Washington, second ward.
Cook John, gate keeper, Central avenue.
Cooper Geo., oil operator, Ludlow, second ward.
Corothers James, oil operator, boards Adams House.
Coss A., paper hanger, Central avenue near third.
Coulter J. E., manager, James Bennett, residence Emlenton.
Cowan John, harness maker, Washington, second ward.
Crawford J. J., grocer, Jackson, second ward.
Crane J. G., clerk, River avenue.
Crell M., livery, River avenue.
Cribb J. W., policeman, first ward.
Critchlow S. E., book-keeper, Jackson, second ward.
CROSBY G. S., attorney at law, Washington, second ward.
Cross M. A., of S. B. Cross & Bro., Washington, second ward.
Cross S. B., of S. B. Cross & Bro., Washington, second ward.
Cross S. B. & Bro., flour and feed, Washington, second ward.
Crow James, laborer, River avenue.
Cunningham J. B., well burner, Washington, second ward.
Cunningham Robert, flour and feed, Washington, second ward.
Cunningham W. A., butcher, Washington, second ward.

## D

Dally C., bricklayer, River avenue.
Dally W. R., grocer, River avenue.
DAVIS S., surgeon dentist, office and house Washington and River avenue.
Davitt A. B., cashier Exchange Bank, boards Mansion House.
Day E., laborer, boards Union House.
Dean J. W., tobacconist, east Washington, second ward.
Dehen A., mason, Jackson, second ward.
Dellger Henry, laborer, Washington, second ward.
Dessing Michael, saloon, boards River avenue.
Dewoody George C., clerk, boards River avenue.
Devaux Miss E., dressmaker, Jackson, second ward.
Devling G. P., proprietor Florence Hotel, corner Washington and Wayne, second ward.
DIEHL J. J., of Dolan & Diehl, River avenue.
Dieselman L. W., clothes cleaner, Central avenue.
Dingman D. J., grocer, Washington, second ward.
Dolan Edward, of Dolan & Diehl, River avenue.
Dougherty Robert, commissioner, Parker Hill.
Dougherty William, tinshop, Washington, second ward.
Dresser S. R., oil producer, Bluff avenue.
Duff John, of Bradley & Duff, Wayne, second ward.

## E

English Alexander, harnessmaker, Washington, second ward.
Erickson O. E., laborer, Jackson, second ward.
Erly Phebe, widow, near Central avenue.
Ervin S. S. & Bro., furniture, Washington near Jackson, second ward.
Ervin S. J., of S. S. Ervin & Bro., Washington and Jackson.
Eshenbeaugh Thomas, blacksmith, Jackson, second ward.
Eavans L. E., of Eavens & Foster, Petroleum Centre.
Exchange Bank, A. B. Davitt, cashier, River avenue.

## F

Ferguson Charles, clerk, River avenue.
Findlay Abel, teamster, Water second ward.
Fisher Bro's., oil producers and dealers, River avenue.
Fisher Henry, of Fisher Bros., residence Oil City.
Fisher Jno. J., of Fisher & Bros., residence Oil City.
Fisher Jno. M., tailor, boards Washington, second ward.
Fleming R., of Mitchell & Fleming, Jackson, second ward.
Flickinger J. H., lumber dealers, Washington, second ward.
Flung Culver, pumper, River avenue.
FLORENCE HOTEL, G. P. Derling, proprietor, Washington and Wayne, second ward.
Floyd W. P., oil producer, Jackson, second ward.
Foster C. B., books and stationery, Parkers Station, Pa.
Foster J. A., clerk, Washington, second ward.
Foster Mrs. J. A., millinery, Washington, second ward.
Foster W. W., of Evans & Foster, River avenue.
Foster Wm. B., cashier, Parkers Savings Bank, River avenue, near Fourth.
Fowkes Alfred, clerk, boards Wilson House.
Fowler Francis, bar-keeper, boards River avenue.
Fowler Mrs. S., widow, Jackson, second ward.
Fowler W., engineer, Jackson, second ward.

Frederick Jacob, boot and shoe maker, River avenue.
Frew William, of Lockhart & Frew, Pittsburgh, Pa.
Frick J. W., carpenter, Water, second ward.
Frisbee F. H., of Book & Frisbee, River avenue.
Frost A., of Truby & Frost, River avenue.
Fuge John, coal diger, Washington, second ward.
Fulkerson A. C., restaurant, Washington, second ward.
Fulkerson J. C., machinist, Wayne, second ward.
Fuller H., lumber yard, River avenue.
Fullerton H. R., of Matson & Fullerton, River avenue.

## G

Gales J. L., of Rice & Gales, Washington, second ward.
Gales W. H., clerk, Washington, second ward.
Gales Walter, attorney, Washington, second ward.
Galey John H. oil producer, River avenue.
Gaskey Joseph of D. Robinson & Co., River avenue.
Gaskilt Joseph, tobacconist, River avuenue.
Gates Goe. W., watch maker, boards River avenue.
Geasey Henry, oil operator, Central avenue.
Gens Charles, south River avenue.
Gernand Wm., teamster, Jackson, second ward.
Gibson Thomas, teamster, Jackson, second ward.
Gillihand Mrs. T., widow, Parker Hill.
Goldsborough J. R., oil producer, Bluff avenue.
Goodwin W., tobacconist, River avenue near Fourth.
Goss John Jacob, baker, Washington, second ward.
Grace Peter, oil dealer, River avenue, residence Bluff avenue.
Grant Pipe Company, office River avenue.
Gray Samuel, notary public, office River avenue.
Gray Warren G., Fourth near River avenue.
Green M. E., of Matzell & Green, River avenue.
Green O. J., buyer and seller of oil, residence Bluff.
Griff Alex. C., oil producer, River avenue.
Groves A. W., carpenter, River avenue near First.

## H

Hackett Thomas, of Stewart & Hackett, River avenue.
Halm Charles, laborer, second near River avenue.
Halm Rebecca, widow, second near River avenue.
Haines Jno. H., of Parker, Thompson & Co., residence River ave.
Hall H. M., grocery and confectionary, River avenue, 3d ward.
Hammond Thomas, carpenter, Washington, second avenue.
Hampton W., laborer, Third and Central avenue.
HARRIGAN WILLIAM, manager Pacific and Atlantic Telegraph Company, boards Syracuse House.
HARMAN J. H., drug store, River avenue.
Harrington A. T., oil producer, Jackson, second ward.
Harrington A. S., oil producer, Fourth, residence Parker Hill.
Harrington James, clerk, River avenue.
Harrison S. B., carpenter, Jackson, second ward.
Harrop Wm., oil producer, office River avenue.
Hart W. W., clerk, boards Phillips House.
Headly Mrs. C., widow, River avenue.
Healed J. T., carpenter, south River avenue.
Heasly Mrs. C., widow, Jackson, second ward.
Henderson A. B., book keeper, boards Phillips House.
Henlen John, tinner, east Washington, second ward.
Henry Thomas, fruit stand, River avenue, near Third.
Henry Wm. Charles, superintedent Union Pipe Co., Water.
Hepler Levi, livery, Fourth and Central avenue.
Hershey Wm., saloon, River avenue.
Hery G., clerk, boards, River avenue.
Higgins J. C., clerk, Washington, second ward.
Higgins M., laborer, Jackson, second ward.
Hilfinger C., agent Roberts's torpedo, office Fourth, near River avenue.
Hipkins John, porter, boards Phillips House.
Hockenberry G. Z., of G. & W. Hockenberry, east Washington.
Hockenberry W., of G. & W. Hockenberry, east Washington.
Hogan James, proprietor Hogan House, Washington, 2d ward.
Hoover A. M., physician, office Ludlow, second ward.
Howard George, porter, boards Syracuse House.
Howard H. P., driller, Parker Hill.
Howard John, tailor, Centre avenue.
Huff G. W., oil buyer, office River avenue.
Huff J. H., pumper, Jackson, second ward.
Hughes Rev. S. A., Presbyterian, Jackson, second ward.
Hugh Samuel, hostler, Jackson, second ward.
Hullings M., oil dealer, River avenue.
HULLINGS & LORD, livery stable.
Hullings W. J., of Hullings & Lord, River avenue, north Fourth.
Humes W. H., east Washington, second ward.
HUMPHREY JOHN S., agent Union Pipe Co., River avenue.
Hunter A., carpenter, boards Washington, second ward.
Hunter John K., of Hunter & Patterson, Washington, 2d ward.
Hunter & Patterson, planing mill, Washington, second ward.
Hunter W. J., telegraph operator, boards Syracuse House.
Hutchinson Batchelder, oil producer, River avenue.

## I

Imus W. A., clerk, River avenue.
Irwin S. H., tank gauger, second ward.
Jackson William T., laborer, south River avenue.
Jenkins T., grocery, River avenue.
Jennings W. F., harness maker, boards First.
Johnson John, painter, River avenue.
Johnson John, carpenter, River avenue.
Jones J. K., barkeeper, boards Mansion House.
Joslyn Hopkins, blacksmith. boards Syracuse House.
Joy T. M., Cashier Lawrenceburg Savings Bank, boards Adams House, second ward.
June S. B., proprietor Union House, Jackson, second ward.
June Silas, pumper, Jackson, second ward.

## K

Kannard R., horse shoer, Wayne, second ward.
Karns S. D., oil producer, office River avenue.
Karns W. E., superintendent Cleveland Pipe Co., office River avenue.
Kelburg John, tailor, River avenue.
Kelly G. S., grocery, Washington.
Kelly P., boiler maker, Washington, second ward.
Kelly J., grocery, Washington, second ward.
Kendrick E. M., telegraph operator, Fourth, near River avenue.
Kennedy Joseph, grocery, Washington, second ward.
Kern A. L., oil operator, Fourth, near River avenue.
Kerns J. L., clerk, boards River avenue.
Kesselman Wm., machinist and blacksmith, Washington.
King F., book keeper, River avenue.
Klugh Charles, stonemason, Wayne, second ward.
KLUGH GEO. A., books and stationery, Washington, 2d ward.
KLUGH GEO. A., proprietor Revere House, corner Washington and Wayne, second ward.
Krosskop W. B., druggist and chemist, River avenue.
Krumpe John, notions etc., Washington, second ward.
Kuhn Adam, of Kuhn & McLain, Parker Road, second ward.
KUHN & McLAIN, boots and shoes. Washington, 2d ward.
Kuhne Chas, clerk, boards River avenue.

## L

Langan Harris, teller Exchange Bank, River avenue.
Lawacheck J. H., drayman, Jackson, second ward.
Lawrence P. H. case cutter, Wayne, second ward.
LAWRENCEBURG SAVINGS BANK, Washington, 2d ward.
Laurie Charles, hostler, boards Union House.

Lehmal Fred., carpenter, River avenue.
Lehman N., boots and shoes, River avenue.
Lemon Robt., teamster, Washington, second ward.
Levens N. O., contractor, Bluff avenue.
LESLIE H. W., physician, 10 River avenue.
Levi David, clothing, 12 River avenue.
Lewis V. laborer, boards Union House.
Liblong Joseph, bartender, River avenue.
Linden James, laborer, near brick yard, second ward.
Loan Thomas, oil operator, Jackson, second ward.
Lockhart Charles, of Lockhart & Frew, Pittsburgh, Pa.
Lockhart & Frew, oil dealers, River avenue near fourth.
Lord George, of Huling & Lord, Bluff avenue.
Lyle & Arters, engines and boilers, River avenue, first ward.
Lyle H. R. of Lyle & Arters, River avenue.

## M

Mack John, laborer, Union House, Jackson, second ward.
Magee J. M., proprietor Syracuse House, River avenue.
Mahaffy R. F., carpenter, north Pike.
Manifold S. H., of Wilson & Manifold, Bluff avenue.
Mann R., photographer, Washington, second ward.
Manroe James, clerk, boards Parkers Station.
Mansfield Charles P., clerk, Parkers Station.
MANSION HOUSE, Dolan & Diehl, proprietors, River avenue.
Mapes ———, tinner, boards River avenue.
Marlin H. A., book keeper, Exchange Bank, boards Syracuse House.
Marlow Augusta, widow, Wayne, second ward.
Marsh Frank, paper hanger, Water, second ward.
Marsh B. G., laundry, Washington, second ward.
MARSHALL E. R., of Randolph & Marshall, Wayne, 2d ward.
Martin D. W., of Brown & Martin, res. Kittanning, Pa.
Martin Wallace, of Colter & Martin, boards second ward.
Mason Levi, agent Robert's torpedoes, Fourth, near River ave.
MATZELL & GREEN, New York Restaurant, River avenue.
Matzell A. D., of Matzell & Green, River avenue.
Maultcsch Henry, butcher, River avenue.
McBride Jefferson, Washington, second ward.
McCarty James laborer, Water, second ward.
McChesney Joseph, clerk, boards Cammie Shear House.
McConnell T., Jr., sup't. Grant Pipe Co., River avenue.
McCoy Michael, oil hand, Washington, second ward.
McCRACKEN J. H., books, stationery, etc., River avenue, second ward.
McCune John, oil operator, Jackson, second ward.
McFARLAND J. W., Mayor, Wayne, near Washington, second ward.
McGunn F., blacksmith, Wayne, second ward.
McGlaughlin Bros., books and stationery, River avenue.
McGlaughlin Mrs. J., millinery, north River avenue.
McGlaughlin William, of McGlaughlin Bros., River avenue.
McGlaughlin John, of McGlaughlin Bros., River avenue, near Bridge.
McIntire James, blacksmith, Gourley House.
McKee S. M., telegraph operator, boards Syracuse House.
McKellip James, teamster, Water, second ward.
McKellvey W. C., book keeper, Bluff avenue.
McKeloy H., oil dealer, office, River avenue.
McKeloy W. C., teller, Parker Savings Bank, River avenue.
McKinley David, carpenter, Jackson, second ward.
McKissic M. S., carpenter, Washington, second ward.
McKnight Mrs. M., Millinery, Washington, near Jackson second ward.
McLain John, of Kuhn & McLain, near Washington, 2d ward.
McLaughlin John, of McLanghlin & Roof, near D. Karn's.
McLaughlin M., laborer, Jackson, second ward.
McLaughlin & Roof, furniture, 11, River avenue.
McLaughlin William, boarding house, 11 and 12, River avenue.
McLean Alexander, oil operator, Wayne, second ward.
McManus Felix, proprietor McManus House, River avenue.
McManus House, River avenue.
McManus P., clerk, boards McManus House.
McNickel James, coal miner, Washington, second ward.
McNay W. J., oil operator, Jackson, second ward.
McNeal H., clerk, boards River avenue.
McNeely, John, saddle maker, boards First.
McWILLIAMS N. B., carriage and wagon maker, Wayne, second ward.
McWilliams R. C., of Morrow & McWilliams, Water, 2d ward.
Mehurin John, painter, River avenue, above the Bridge.
Mehurin John, painter, River avenue, above the Bridge.
Messimer W. F., hardware, Washington, second ward.
Metzgar Fred., butcher, Washington, second ward.
Middleton J. P., book keeper, boards McBride House.
Miller Joseph, watches, clocks and jewelry, River avenue.
Miller Peter, watches, clocks, etc., River avenue.
Minneger A. C., laborer, Ludlow, second ward.
Minneger, W. C., pumper, Ludlow, second ward.
Mitchell & Flemming, tank builders, Jackson, second ward.
Mitchell James R., of Mitchell & Flemming, Jackson, 2d ward.
Mobley W. C., oil producer, River avenue, res. Bluff avenue.
Moffitt Charles A., watchmaker, Washington, second ward.
Morrow George, of Morrow & McWilliams, Water, second ward.
Morrow & McWilliams, grocers, Washington, second ward.
Morrow G. W., laborer, Water, second ward.
Morrison G. H., restaurant, River avenue.
Morse C. H., clerk, boards River avenue.
Matson & Fullerton, grocers, River avenue, opposite Bridge.
Matson James, clerk, River avenue.
Matson John, of Matson & Fullerton, River avenue.
Mos Fred., shoemaker, north Central avenue.
Mossinger Henry, saddler, Ludlow, second ward.
Mours Charles, Machinist, near Central avenue.
Mullin Thomas, hostler, Jakson, second ward.
Murdoch F. H., of Wynne & Murdoch, boards Adams House.
Mussinger Harry, harness maker, Ludlow, second ward.
Myer Henry, boots and shoes, Central avenue and Third.
Myers Sylvester, bootmaker, River avenue.

## N

Neff O. H., saddles, harness, etc., Washington, second ward.
Neill J. C., laborer, north Central avenue.
Newell J. P., livery, Central avenue.
Newell W. B., clerk, boards Phillips House.
Neyhart A., oil dealer and producer, office River avenue.
Nickerson C. W., carpenter, Water, second ward.
Nober J. A., barber, Washington near Ludlow.

## O

O'CONNOR JOHN, merchant tailor, River avenue.
O'Conner Patrick, painter, Water, second ward.
Odell John, of Burgham & Odell, River avenue.
O'Hara James T., hostler, boards Parkers Hill.
OILMAN'S JOURNAL, Clark Wilson & Son, proprietors, Washington, near Jackson, second ward.
Orton Mrs. S., widow, second ward.
Overheim Daniel, Jr., blacksmith, Washington, second ward.
Owens John, of Owens, Pixley & Co., Washington, second ward.
Owens, Pixley & Co., clothing, Washington, second ward.

## P

Pacific & Atlantic Telegraph Co., Fourth near River avenue.
Painter F. W., engineer, Water, second ward.
Palmer B. M., of Flickiner & Palmer, Washington, second ward.
Parker Mrs. E. H., widow, Bluff avenue.
Parker Fullerton, oil producer, River, residence Bluff avenue.
Parker & Karns, City Rail Road Co.'s office River, avenue.
Parker James P., President Parkers Saving Bank, River avenue.
Parker Saving Bank, River avenue.
Parker, Thompson & Co., oil dealers, River avenue.
Parker William, of Parker, Thompson & Co., Oil City.
Parker W. J., oil producer, River avenue.
Patterson R. L., physician, River avenue.
Patterson L., of Hunter & Patterson, Water, second ward.
Peck J. N., driller, boards south River avenue.
Perkins John, tailor, River avenue.
Peters T. D., book-keeper, Jackson, second ward.
Petersen O., bootmaker, boards River avenue.
Phillips Bros., oil producers, River avenue near Third.
PHILLIPS HOUSE, Wilkins & Thompson, Parkers Station.
Phipps John, justice of peace, River avenue.
Paine Joseph, pumper, Ludlow, second ward.
Pitts S. S., teamster, Jackson, second ward.
Pixley———, of Owens, Pixley & Co., Washington, second ward.
Plato F., druggist, boards Cammie Shear House.
Platt William, hostler, north River avenue.
Pollock P, O., Parkers Station.
Pool Rosel, teamster, Central avenue.
Porterfield W. H., residence Oil City.

Post Office, River avenue.
Potts John K., cashier Cleveland Pipe Co., River avenue.
Powers R. B., druggist, Washington, second ward.
Pratt Angeline, widow, Wayne, second ward.
Prest George D., dry goods, River avenue.

## R

Randolph—, of Randolph & Marshall, Wayne near Washington.
Randolph R., livery, Wayne near Washington.
Rawie Louis, baker, boards River avenue.
Reamer J. W., livery, Bluff avenue.
Reamer J., livery, Central, residence Bluff avenue.
Reedy John, hostler, boards River avenue.
Reicnert John, shoemaker, River avenue.
REVERE HOUSE, George A. Klugh, proprietor, corner Washington and Wayne avenue.
Reynolds Patrick, tailor, Washington, second ward.
Ribely George, baker, boards River avenue.
Rice A. B., of Rice & Gales, Washington, second ward.

RICE & GALES,
Variety store, Washington, second ward.

Richards Mike, butcher, River avenue.
Richardson C. S., laborer, Central avenue.
Riddle Charles E., oil operator, Bluff avenue.
RIEDLINGER J., barber, River avenue.
RIEFFEL BROTHERS, Revere House Stables, corner Washington and Wayne, second ward.
Riffel F. O., of Riffel Brothers, Washington and Wayne, second ward.
Riggle John, pumper, boards Union House.
RIVERSIDE RESTAURANT, River avenue.
Roberts J. B., stage line, Water, second ward.
Robertson William, shoemaker, Washington, second ward.
Robinson D., of D. Robinson & Co., River avenue.
Robinson D. & Co., tobacco and cigars, River avenue and Third.
Rodgers Wayne, hostler, boards River avenue.
Roof Geo., of McLaughlin & Roof, near D. Karns
Rosen S., clerk, boards River avenue.
Row Simon, watchman, Parkers Hill.
ROWE J. E., Proprietor McBride House, corner Washington and Jackson, second ward.
Rugh Levi, carpenter, boards Washington, second ward.
Rumbaugh J. B., grocer, Jackson, second ward.
Russel A. D., book-keeper, Parkers Savings Bank, River ave.

## S

Sabin N., tank builder, Jackson, second ward.
Scandrett Thos. Jr., clerk, boards River avenue.
Schulze F., tailor, Washington, second ward.
Scott H., shoemaker, Washington, second ward.
Scott John, President Lawrenceburg Savings Bank, residence Butler, Penn'a.
Sendker L., boots and shoes, River avenue.

SENDKER W. H.,
Practical boot maker, four doors south of the postoffice, River avenue.

Shaffer R. H., clerk, River avenue.
Sheers James, feed store, north River avenue.
Shepard Edward, agent engines, Fourth near River avenue.
Shingle William, pumper, River avenue.
Shirley Jno. T., of Shirley & Speer, River avenue.
SHIRLEY & SPEER, hardware, oil well supplies, etc., River avenue.
Shone Michael, barber, Washington, near Jackson.
Sickfut Benj., shoemaker, Ludlow, second ward.
Siegel F., of Siegel & Bro., boards Florence Hotel, second ward.
Siegel J., of J. Siegel & Bro., boards Florence Hotel, second ward.
SIEGEL J. & BROTHER, dry goods and notions, corner Washington and Wayne, second ward.
Sigel E., of Sigel & Co., Washington, second ward.
Sigel H., of Sigel & Co., second ward.
Siegfried B. F., shoemaker, Washington, second ward.
Simon A., clothing, River avenue.
Simon George, laborer, boards River avenue.
Simpson Thomas J., book-keeper, Syracuse House.
Sloan L. C., laborer, River avenue.
Smith Arthur, teamster, boards River avenue.
Smith & Co., oil producers and dealers, River avenue.
Smith Frank K., blacksmith, boards River avenue.
Smith Henry, of Smith & Co., River avenue.
Smith Mrs. J. B, laundress, River avenue.
Smith J. B, advertising agent, River avenue.
Smith J. B., photographer, Washington, Second ward.
Smith James, blacksmith, River avenue.
Smith John, of Smith & Co., River avenue.
Smith M. L., carpenter, Jackson, second ward.
Smith Wellington, driller, Jackson, second ward.
SMULLIN JOHN, attorney at law, office River avenue, boards Syracuse House.
Snyder A., wagonmaker, River avenue, near Fourth.
Snider Charles, bakery, Washington, second ward.
Solley James, justice of the peace, Wayne, near Washington.
Speer Geo. H, of Shirley & Speer, River avenue.
Sproull Robert, laborer, Water, second ward.
Stanley K. C., cooper, Jackson, second ward.
Starks Willis, barber, South, River avenue.
STEELE J. A., livery exchange, River avenue, near the Bridge.
Steele M. M., flour and grain, Parker Station.
Steen H., driller, Jackson, second ward.
Steirly Chris., paper hanger, Ludlow, second ward.
Sternberg M., clerk, boards River avenue.
Steveson Thomas, butcher, North, River avenue.
Stevenson N. W., oil dealer, Fourth, near River avenue.
Stewart Archie, expressman, Central avenue.
Stewart & Hackett, oil dealers, office River avenue.
Stewart John A., junk dealer, Washington, second ward.
Stewart W. L., photographer, Jackson, second ward.
Stewart Wm., of Stewart & Hackett, River avenue.
Stiver Samuel, carpenter, River avenue.
Stone Howard, porter, boards Adams House, second ward.
Stroup Peter, laborer, Wayne, second ward.
Swisher O. P., clerk, Jackson, second ward.
Syracuse House, J. M. Magee, proprietor, River avenue.

## T

TALMEY J. P., Catholic Church, Washington, second ward.
Tamer, Lee, tank builder, Jackson, second ward.
Tate Robert, clerk, Washington, second ward.
TATE T. W., grocer, Washington and Jackson, second ward.
Teerkes John, cigar maker, River avenue.
Theuret James, hostler, boards Parkers Hill.
Thomas John, agent sewing machines, Washington, 2d ward.
Thompson J. A., of Wilkins & Thompson, Parkers Station.
Thompson William, of Parker, Thompson & Co., res. Oil City.
Titus William, teamster, Jackson, second ward.
Truby H. S., of Truby & Frost, River avenue.
Turk John, pumper, Ludlow, second ward.
Twichell John, wagon maker, River avenue.



## U

Underwood William, carpenter, Bennertown, second ward.
Union Express Co., office, Parkers Station, Pa.
Union House, Jackson, second ward.
Union Pipe Co., office, River avenue.

## W

Waddell Mrs. N., Washington, second ward.
Walker A. C., teamster, boards McBride House.
Walker John, pumper, River avenue.
Walter George, tailor, Ludlow, second ward.
Warren G. G., book keeper, Bluff avenue.
Washabaugh Perry, attorney, office, River avenue.
Wassan A. A., grocer, second ward.
Wassan & Rumbaugh, grocers, Jackson, second ward.
Weaver & Son, restaurant, River avenue.
Welker George, clerk, boards Bluff avenue.
Wert T. H., stage driver, Water, second ward.
Widgeon Mrs. Jane, widow, Washington, second ward.
Wiesengthal Phillip, shoemaker, Washington, second ward.
Wightman C. D., Wayne, near Washington.
Wikmen Edward, boots and shoes, Jackson, second ward.
WILBER J. H., livery, back of the Syracuse House, River ave.
Wilcox W. W., mason, Washington, second ward.
Wilkins R. R., of Wilkins & Thomson, Parkers Station.
WILKINS & THOMPSON, proprietors Phillips House, Parker Station.
Wilson A. C., grocer River avenue.
Wilson Clark, of Clark Wilson, & Son, Washington, second ward.
WILSON CLARK, & SON, proprietors Oilmen's Journal, office, Washington, second ward.
Wilson Clayton L., Clerk, boards River avenue.
Wilson H. C., of Clark & Pierson, Washington, second ward.
Wilson J., blacksmith, River avenue.
Wilson John O., of Wilson & Manifold, Bluff avenue.
Wilson & Manifold, dry goods, notions, general merchandise, etc., River avenue.
Winder A. C., barber, River avenue.
Winthington H. W., teamster, Central avenue.
Wood I. H., clerk, River avenue.
Wright H. S., grocer, Washington, second ward.
Wright T. W., printer, boards Washington, second ward.
Wyant A. H., oil dealer, Water, second ward.
Wynne W., B , of Wynne & Murdoch, boards Adams House, second ward.
WYNNE & MURDOCH, physicians, office, cor. Washington and Wayne, second ward.

## Y

Yeanders G. W., carpenter and builder, Central avenue.
Yandors George, wagon maker, Bluff avenue, second ward.

## PARKER CITY BUSINESS DIRECTORY

**Attorneys.**

Colter & Martin, River avenue.
Crosby G. S., Washington.
Smullin John, River avenue.
Washabaugh P., River avenue.

**Bakery.**

Burgham & O'Dell, River ave.
Snider Charles, Washington.

**Bankers.**

Foster W. B., Fourth.

**Banks.**

Exchange Bank, River avenue.
Parker Savings Bank, River.

**Billiard Hall.**

Book & Frisbee, River avenue.
Mansion House, River avenue.

**Blacksmiths.**

Bradley & Duff, Ludlow.
Overheim Daniel, Washington.

**Books & Stationary.**

Brown T. P., Washington.
Klugh Geo. A., Washington.
McCracken J. H., River ave.
McLaughlin Bros., River ave.

**Boot and Shoe Makers and Dealers.**

Aldinger H. D., Washington.
Bendheim L., River ave.
Bruce J. N., Washington.
Frederick Jacob, River avenue.
Kuhn & McLain, Washington.
Lehman N., River avenue.
Myer Henry, Central and Third.
Sendker L., River avenue.
Neff O. H., Washington.
Wikmen E., Jackson.

**Butchers.**

Metzgar F., Washington.
Stevenson J. B., River avenue.

**Carpenters & Builders.**

Flickinger & Palmer, Washington.
Yeanders G. W., Central ave.

**Cigars and Tobacco.**

Ayer O. D., Washington.
Bleymiller Chas., Washington.
Boulger George, River avenue.
Robinson D. & Co., River ave.
Teerkes John, River avenue.

**Clothing.**

Bachman L. & Bro., Washington.
Levi David, River avenue.
Siegel & Co., Washington.

**Drug Store.**

Babst Wm., Washington.
Harman J. H., River avenue.
Krosskop W. B., River avenue.
Powers R. B., Washington.

**Dry Goods.**

Bruce & Brown, Washington.
Bruce J. N., Washington.
Prest Geo. D., River avenue.
Siegel J. & Bro., Washington.
Wilson & Manifold, River ave.

L. COHEN,
**MERCHANT TAILOR,**
AND DEALER IN
READY-MADE CLOTHING,
GENTS' FURNISHING GOODS,
HATS AND CAPS, NOTIONS &C.,
Washington St., Second Ward,
PARKER CITY, PA.

**Engines, &c.,**

Boughton & Chandler, River.
Lyle & Arter, River avenue.

**Furniture.**

Aaron & Co., Washington.
McLaughlin & Karns, River.

**Flour and Feed.**

Cross S. B. & Bro., Washing'n.
Cunningham Rob., Washing'n.
Sheers James, River avenue.
Steele M. M., Parker's station.

**Hats, Caps &.,**

Prest Geo. A., River avenue.

**Hotels.**

Florence Hotel, Washington.
Hogan House, Washington.
Revere House, Washington.
Union House, Jackson.
Mansion House, River avenue.
McManus House, River avenue.
Phillips House, Parker Station.
McBride House, Washington.
Syracuse House, River avenue.

**Grocers, &c.**

Adams Ed., Washington.
Adams M. S., Washington.
Allen Wm. E., Washington.
Borland A., River avenue.
Carringer H. C., Washington.
Crawford J. J., Jackson.
Dally W. R., River avenue.
Dingman D. J., Washington.
Hall H. M., River avenue.
Jenkins C., River avenue.
Kelly G. S., Washington.
Kelly J. G., Washington.
Kennedy Joseph, Washington.
Matson & Fullerton, River ave.
Morrow & McWilliams, Washington.
Tate T. W., Washington.
Wasson & Rumbaugh, Jackson.
Wilson A. C., River avenue.
Wright H. S., Washington.

**Hardware.**

Bennet James, River avenue.
Brown & Martin, Pollock, Pa.
Dougherty W., Washington.
Messimer W. F., Washington.
Shirley & Speer, River avenue.

**Harness-makers.**

Boyd, Atchisen & Co., River ave.
Neff O. H., Washington.

**Jeweller.**

Miller Joseph, River avenue.
Miller Peter, River avenue.

**Justice of the Peace.**

Phipps John, River avenue.
Solley James, Washington.

**Laundry.**

Marsh B. G., Washington.
Smith Mrs. J. B., River ave.

**Livery Stables.**

Connoley Wm., Washington.
Crell M., River avenue.
Hepler Levi, Fourth avenue.
Hulings & Lord, River avenue.
Petrolia Exchange, River ave.
Steele J. A., River avenue.
Wilber J. H., River avenue.

**Lumber Dealers.**

Fuller H., River avenue.

**Machinists.**

Bradley & Duff, Ludlow.
Kesselman Wm., Washington.

**Millinery.**

Ayer & Crawford, Washington.
Brown Miss A., Washington.
McLaughlin Mrs. J., River ave.

*Joseph O. Bullman, Civil War Veteran*
*Old Joe was well known by everyone in the area. He was born June 24, 1835, and died June 10, 1938, lacking only fourteen days of being 103 years old. His life span was undoubtedly the greatest one within the history of the Parker area. He worked as a blacksmith during the oil excitement and continued his trade until he was 80 years old. At the time of his death, Joseph was the oldest Civil War veteran and was the last surviving member of the Calvin A. Craig G.A.R. Post No. 75, Parker City. (Courtesy of Dr. Clinton R. Coulter, Parker City, Pennsylvania)*

## The Fancy New Iron Bridge

*New Iron Bridge crossing the Allegheny River at Parker City*

Parker City is in Armstrong County and across the Allegheny River is Clarion County. The Allegheny Valley Railroad was on the Clarion County side and until 1872 there was no means of reaching the Parker side of the river except by ferry. James P. Parker was an experienced river man and could handle with ease the trip back and forth. In 1872, three enterprising oil business men decided the days of the old ferry were numbered. S. D. Karns, H. R. Fullerton, and Fullerton Parker in partnership had a fancy iron bridge erected across the Allegheny River at a cost of $80,000. This tall bridge brought revenue to the owners and made Parker's Landing more accessible to all types of trade.

## The Dreaded Parker City Fire Bells Ring The Cry Was, "Fire! Fire!"

There are two basic ways that a fire gets started, one would be by accident, and the other intentionally. Intentionally-set catastrophies in oil towns throughout the upper and lower oil belts were an occurrence much too common. Fires raged and destroyed whole blocks of structures. New buildings were erected and business was resumed almost before the ashes were cold.

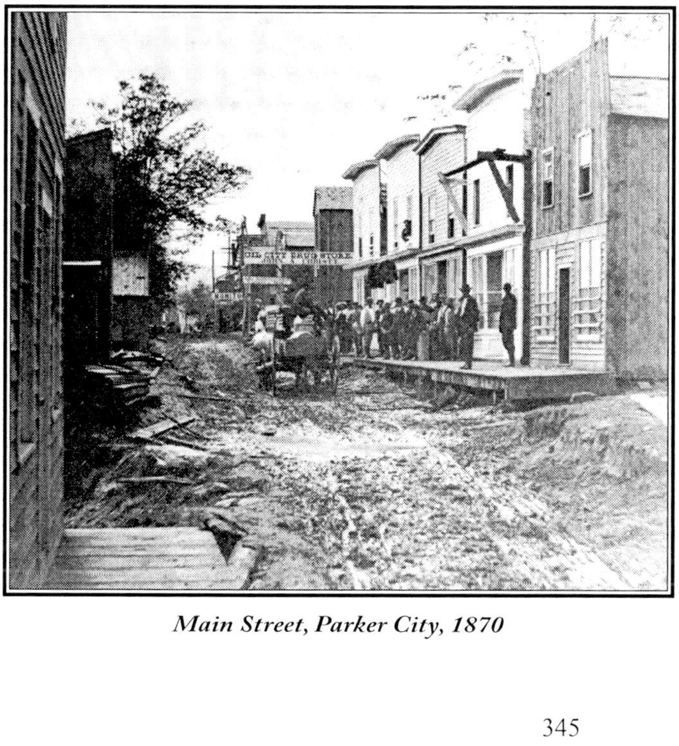

*Main Street, Parker City, 1870*

On November 4, 1870, there was a fire which destroyed the greater part of the buildings at the Landing. Again, in the same year, a terrible fire occurred on December 4, and the central part of Lawrenceburg was burned. On January 19, 1873, and again on the 4th of April, 1873, the business portion of the first ward was almost entirely destroyed.

Again, the excitement of day-to-day life was so great and business so good that fires counted for little. Carpenters replaced the wood buildings in a few days and everything went on as usual.

After the fire of April, 1873, the streets of the Landing were laid out on a new plan and gradually but surely more substantial and more costly buildings were set in place. River Avenue was built up on both sides and consisted of business houses for almost its entire length.

Parker City suffered financially as business slowed, wells became exhausted, and the population moved on by 1879. On October 30, 1879, River Avenue from Smullin's building to the site of the glassworks was entirely destroyed by fire and the financial depression was so great that little attempt was made to rebuild.

## It Was Announced April 23, 1974, Bear Creek Shut Down

The Great 1879 Bear Creek Pumping Station, one mile down the Allegheny River from Parker's Landing, for some time the nation's largest and a central point for a total of 21 lines connecting the eastern part of the United States, and west to the Mid-Continent Fields. At one time it had a storage capacity of 74 tanks, each 25,000 to 30,000 barrels. Original station was steam power with a battery of fifteen boilers.

The pumps used were Worthington and National Transit Company, the latter manufactured in Oil City, after shops were moved from Petrolia, Pennsylvania, April 5, 1880. Some of these huge pumps, known as Cline Triples, required seven railroad cars each to ship. Pump No. 4 at Bear Creek moved one barrel of oil every revolution. The station was electrified in the early spring of 1949.

*Bear Creek Pump Station*
*Len "Pithole" McCorry*
*Looking south down the Allegheny River. Built 1879-1880 by S. O. Company. It had a pumping capacity of 125,000 barrels daily, for some time the largest in the nation. Station abandoned May 1974. Photo May 5, 1949*

## Brown and McGee Lumber Yard

For some unknown reason, this yard was a popular place to be. There are about 50 men in this photograph. The sign over the office reads, "Brown and McGee Lumber Yard, Sash and Doors, Rig Timbers." One brave soul sits on top the derrick. Oil supplies and wood oil tanks are also built and sold here. This photograph shows pastureland with many cows grazing in the summer sun. A big black dog sits on the porch of the office. This photograph was taken in our lower fields but the location is unknown. (Butler County Historical Society, Butler, Pennsylvania)

*Oil Boom Lumber Yard*
*Brown and McGee, oil well supplies, tanks, doors, and rig timbers*

# Old Times in Oildom by Geo. W. Brown

*Being a Series of Chapters in which are Related the Writer's Many Personal Experiences During Fifty years of Life in the Oil Regions.*

George Brown, from Youngsville, Pennsylvania, wrote "Old Times in Oildom." The book was published by the Derrick Printing Company, Oil City, in 1909. The following information gives us, students of early oil history, insight into the times and personalities in the Western Pennsylvania oil region. One of George's favorite sayings was, "It is a long lane that has no turn." Many characters in this history hit rough turns in the lane and wrecked their horse and buggy, along with their business.

# The Lumber Business in Parker City From "Old Times in Oildom" by Geo. W. Brown

When this oil town was just getting a good start, your humble author was crossing the Allegheny River on the old chain ferry, owned by McLaughlin and Fullerton. And right here let me say that each of those men made a nice little fortune before the iron bridge was built spanning the river at that point, and connecting Clarion and Armstrong counties by this old ferry. Day and night, it was loaded with teams and passengers.

I leased a few square rods of swamp land of Old Fullerton Parker, as he was familiarly called, paying $600 a year rental. After renting the ground I was obliged to haul in about 100 loads of gravel before I could pile lumber on the soft land. Then an office appeared on the scene, and G. W. Brown ran the first lumber yard in the greasy city of Parker. For about two years my luck was the very best. I had for customers the best operators in the new field. They paid their bills at the end of each month. But, it's a long lane that has no turn. If any man did owe me during the first two years of my yard business in Parker's Landing, and happened to go into bankruptcy, he had just paid me off in full. This was rather a pleasant experience for me. But the trouble came in the turn of the lane. At the end of two years, when my individual profits had been about $15,000, oil took a downward plunge, and fell from about $3 a barrel to about 60 cents. Then it was that my heretofore good customers went into bankruptcy by the dozen, and I was kept busy for a few months going to Pittsburg to adjust claims with the register in bankruptcy. I became quite well acquainted with this genial gentleman. And this genial gentleman made more money out of this kind of business than I did. In the end I found that I had been throwing good money after poor money. I never received one dollar on my adjusted claims. I learned that when a Parker City oil operator went into bankruptcy, he went in to stay. What money he had in his pockets when the crash came, stayed there. I never heard of an assignee, register in bankruptcy, or any other officer of the law getting his hands on any of it. The creditor always paid his own railroad fare, hotel bills, and register's bills without aid from the debtor. The debtor generally started a little business of his own, as soon as he got his discharge from all his former obligations. This was my experience, at least, and I have yet to hear of a creditor who came out any better than I did. Old Times in Oildom were indeed slippery times. Of about 40 lumber yard men who ran lumber yards in these old times but two, to my knowledge, came out unscathed.

I'll give a couple of items now to prove the "slipperyness" of these times. One of my customers at that time, a carpenter, imbued with the spirit of the times, took the job of building an addition to the Phillips Hotel, owned by James E. Brown, the millionaire of Kittanning. The thrifty carpenter gave me the privilege of furnishing all the lumber for this addition, amounting to about $800. When the job was finished, said carpenter collected the money from James E. Brown for the whole job and forgot to pay G. W. Brown for the lumber.

Finally a draft came to the Youngsville Savings Bank with these words written on the back, "Give Brown my love, and tell him to draw again." I pocketed both the insult and loss. Many, many men like this were inhabitants of the oil regions and helped to make "old times in oildom" miserable.

My second item is one showing luck, and no luck. One Saturday evening I took the paltry sum of $2,000 insurance on my lumber yard, worth about $7,000. Sunday about half of the city burned, taking in my lumber yard. When the news was telegraphed to me Monday morning to my Youngsville home, I thought I had been lucky in getting the $2,000 insurance placed before the fire took place, but now comes the sequel. The company failed to come to time, and I got a tip from a reliable fire insurance agent that my company was a little "shaky" and all I ever received was $540.

## John Galey and The Robinsons

When the lumber business had become a little slow on Oil Creek, the author of this book transferred his rambling tread to the busy, busy precincts of Parker's Landing. There the oil business brought together a most motley crowd. No oil town produced a more mixed crowd. No oil town produced so many rich oil farmers as did Parker. The Parkers, Robinsons, Foxes, and scores upon scores of families were rich enough to live without oil, but when the oleaginous wealth was forced upon them, they very meekly accepted it. Among the richest, both after and before striking oil, were the Robinsons. There were three brothers, and each had a good, large farm, and every acre was good oil territory. The piling of riches on these good natured and contented men did not set them up above poor folks. They always dressed well and had a gentlemanly air about them not often found among ordinary farmers, and the striking of dozens of big wells on their farms, when oil was worth $3 a barrel, made not the least difference in their dress and actions. It was always a comfort for the author to visit with any of the brothers, before and after the finding of oil on their farms. Nearly everybody was their hobby, and one of these brothers had this for his hobby. Mr. Elisha Robinson paid his bill, but the odd cents on the bill Elisha would never pay. If the bill was $500.01, he would pay $500, and if the bill was $1.99, he would pay $1. Elisha always drew the line on cents. He would never pay only even dollars, but would never find any fault with any bill, either large or small. He was one of my best customers while I kept a lumber yard at Parker City. While operating his large oil farm his monthly lumber bills ran very high, and in my four years of business I never was obliged to present a bill for payment. On the first day of each month he would call for his bill and write his check for even dollars. If all my customers in Parker had been Robinsons, I would have been just about $10,000 ahead when I quit the lumber yard business. What a blessed world this would be if all the people were Robinsons.

## John H. Galey

John H. Galey was one of the many business men of Parker at the time I speak of. John's history from that time to the present is well worth a brief mention. He was an active boy, as the run of boys go, generally. He had his eyes open for some kind of an oil trade. For a while he did not have his mind made up as to what kind of a trade it would be. Finally, an operator put a well down on Stump Creek Island, a mile above Parker. The operator struck a very good paying well and offered the island to Galey for $10,000. John thought the matter over and made up his mind that $10,000 and more than that amount, could be pumped from that well, but to use his own words, "I had not the $10,000, but went to Pittsburg and borrowed it and paid for the property, and I have taken from that and another well which I put down on the same island, $125,000, and I have run the wells only 18 months. They are producing nearly as much as ever and they make a nice little property."

John went on making money hand over hand for a time, then he went far toward the setting sun and built a large, thriving town. One day, when he was away on business, the Indians came and wiped out his town. But John Galey was not the man to sit down and mourn over the loss of his wealth. The next time I heard of him he plunged into the wilderness at Haymaker, McKean County, Pa., and leased a large amount of land farther north than any oil company had thought of going. He took that well-known and moneyed firm, the McKinney Bros., in to the deal with him, and Galey was sailing over financial seas as gaily as ever before.

The next time I met John was several years afterwards at one of his boarding houses, or houses where he boarded, near Oakdale, Pa. This is the very strange story he told me on that occasion: "I came to McDonald a couple of years ago to lease gas territory for Guffey, Galey & Co. No oil had been found in this section at that time. But thinking that this was gas territory, I commenced at McDonald and made leases along this ridge for a distance of nearly three miles. Our strip is a little over a mile wide, on an average. Now every acre of it is the best oil territory in the McDonald field. I happened to keep right on the belt as near as if I had known just where the oil lay. A little side belt struck us occasionally but run out just as soon as it crossed us. We have the largest wells in this field, one of them producing 16,000 barrels a day when first struck (this is the old Matthews well). One flowed 30,000 barrels before we could control it. The oil rushed down the creek through Nobletown, but luckily did not take fire. It is the most strange thing to me imaginable to think how I followed this belt so far, and then stopped at the end of it when I was leasing gas territory with no thoughts of oil. I cannot but think it almost a miracle."

## Parker City

I take for my subject, in this eleventh chapter, Parker City. Who has not heard of Parker City? Certainly every oil man has become familiar with the name.

This little city is one of the "has beens." It never will be the great oil center that it once was, but it might have been one of the best of its size in Western Pennsylvania if not for the short sightedness of the original landholders. This remark applies particularly to The Flats or First Ward.

## Parker City's Big Mistake

When Mr. Fullerton Parker began to rent his land holdings along the river front, the oil business was so great that he could get nearly any price he put on it as rental. If one man did not give $10 a foot front as rent per year, another man would, and Mr. Parker did just what most any other man would in the same situation—put on a big price per year. He could get it, and it was worth a big price. His mistake was in not selling the lots and letting somebody beside himself pay part of the city taxes. This plan would have tied many business men to the young city, and they would have been residents today, instead of helping boom some other city. The selling of the lots would have been the best plan, as with the united efforts of the many owners, the city would have had a steady growth, thereby gradually increasing the value of city property.

This is no guess work. All travelers know that large towns are not found in bunches. As the traveler passes through the country at intervals of 40 or 50 miles, he sees large towns. The shadow of a large town keeps the little towns weak and spindling. A little town does not grow much with a large town just close by to take all the trade away from the small town. This is one of the uncontradicted facts. Kittanning is far enough down the river and Franklin is far enough up the river to give Parker City an open field. No shadows from any direction would dwarf the growth of "pretty little Parker City," with its magnificent view of the swift running old Allegheny.

Coal, oil, gas, timber and good soil are found all around the city. What more could be desired in the way of building up a large city? It is a fact, known to every Parkerite, that when rents were up to fever pitch and business booming, there was no grumbling, but when business began to adjust itself to the decreasing output of the oil wells of the vicinity, rents were not adjusted.

## An Expensive Swamp

The author speaks from personal experience. He paid $600 a year rent for several years on a few rods square of swamp land, as a spot to pile lumber on. The land was made usable by said author hauling many loads of gravel into the swamp. By the application of this gravel, the land was made firm enough to hold up lumber. This expense was borne by the renter. But, as I said before, Mr. Parker was not considered an extortioner at this time, as he could get the high rent from other parties, if your scribe had not frozen to the lot. But the trouble came when business fell off nine-tenths, and your humble servant plucked up courage enough to approach W. C. Mobley, the superintendent and son-in-law of Mr. Parker, and asked him, in view of the fact that the profits on sales of lumber would not pay the rent, to lower the rent a trifle. Mr. Mobley's answer was, "Not a cent." The result is easy to see. The lumber yard was obliged to close out business.

This was the case with many other branches of business. Instead of the motto being carried out, "Live and Let Live," thereby holding the population, the motto, "Die and Let Die," was carried out, and Parker City missed its great opportunity. I'll mention a few points that will not be new to the old residents of the city.

Before the bridge was built, John McLaughlin and 'Squire Fullerton bought a chain ferry, paying $8,000. It paid for itself in a few months and made a handsome fortune for the firm before the bridge took away their business. But with business foresight, McLaughlin and Fullerton took a good slice of the bridge stock, and again piled up money. 'Squire Fullerton is now dead, but his widow lives on the "Bluff," where she can overlook the place of her late husband's victories. John McLaughlin built the Globe Hotel and conducted it for several years, then sold it to his two sons, George and Will. Their father is connected with the natural gas business and lives at Murraysville, Pa.

## Mr. McKinney, Get Your Sleep

Who of the old operators but knew Fin Frisbee. Fin, together with Doc Book, built the Central Hotel. Oh, but that hotel did a great business. Just one little incident will prove this. I was a lodger one night, and by the noise below my slumbering place, I took it that the bar was doing quite a business. As I was used to noise, it did not deprive me of nature's sweet restorer—sleep. But Curt McKinney, of Titusville, did not fare so well. I stood in the office the next morning after the noise, when Curt (as he was called then, but now he is called Mr. McKinney), came down stairs and approached Fin, who stood in the office wearing one of his contented smiles, and addressed him in this manner, "Mr. Frisbee, if you allow so much noise about your bar every night, as you did last night, I will not stop over night with you again." Curt said this with his usual earnestness. Fin looked up very much unconcerned and replied, "I took in at the bar last night $500. I will not trade a noisy $500 at the bar for a quiet 70 cent lodging."

Poor Frisbee. After becoming proprietor of the great Kent House at Lakewood, he sold out and removed to Duluth, where he added very materially to his wealth, and bid a long farewell to his dollars and crossed over the river of death to try an unknown existence. If there are no hotels to be run in that other life, Fin will be unhappy.

## Elisha Robinson, 1905

I met Elisha Robinson on the street this morning. He is the same unassuming man of money that he was when his oil wells were forcing him to go to Pittsburg every few days to deposit his piles of cash. He is the same true Christian gentleman yet that he was when, 33 years ago, he would come as regular as a clock into my lumber office to pay his bills. If all my customers had been Elisha Robinsons, I would have $10,000 more money today than I have. And his brother, Sam, still clings to this terrestrial ball, which means that another good, honest, rich man still lives. Both brothers are tilling the soil the same as before that same soil poured forth rivers of oil. Elisha has his affections fixed this spring on a piece of hoarded land that he will clear up this coming summer and put in a state of cultivation.

## Fullerton Parker, 1905

Fullerton Parker, who was monarch of all he surveyed in this city in its palmy day, has, with many other pioneers, been gathered to his fathers, but his mansion on the "Bluff" still overlooks the city which bears his name. There are Parkers and Parkers here yet, but they are not of the old settlers. There is one in Oil City, William Parker, who is remembered here as the owner of the old "Rob Roy" well at Karns City, which produced nearly 150,000 barrels of oil and put nearly as many dollars into the pockets of its owner. There are very few "Rob Roys." None ever came and stopped with me. The "Rob Roy" spoken of above gave Oil City an ornament in the great brick mansion of Bill Parker. It is lucky for noted Oil City that such men as Mr. Parker gravitated in its direction, when they became too rich to stay anywhere else.

## Parker City, 1905

And now, let me close this article by saying that if Parker is not the Parker of old, it bears unmistakable signs of former prosperity in its five good churches, fine brick school building, water works, bank and many substantial buildings that were paid for when money was no object. For this and many other blessings the Parker of today has reason to be thankful.

## A Crowd of Dancers

I must mention one dance given in Parker City, soon after oil was struck. I was in the lumber business and occasionally visited the place before it was a city. At each visit I stopped at the same hotel. One evening I registered as usual and soon learned that a big ball was to come off that night. The proprietor of the hotel offered to find lodging for me outside the noise of his dance. I told him not to take the trouble, as the noise would not disturb my slumbers in the least.

This landlord had engaged two violinists from Brady's Bend. There was not a violinist at Parker City at that time. The musicians were promptly on hand, also a big crowd of dancers. About 11 p.m., one of the musicians received a telegram calling him home immediately on account of the death of a relative. He lost no time in catching the Pittsburg train that was just ready to leave the Parker depot. As the absent violinist was caller of the quadrilles, his partner was left in a bad shape. The band was also left in a bad fix, with two violins and but one player, and the player could not call one quadrille. I had played and called quadrilles for 25 years, but nobody in Parker City knew it.

The landlord told me that he would be obliged to pay a part of the money back to that large crowd if he stopped the program half finished. The milk of human kindness began to flow in my veins, and to save this clever landlord from making such a sacrifice, I told him that I could fill the place of the absent musician. To say that he was pleased would be putting it lightly. He smiled all over his face and I took up the absent man's work and saved a breakup of the ball. I lost my full night's sleep. I got a couple of hours sleep in the morning and that landlord charged me 50 cents for supper, 50 cents for lodging and 50 cents for breakfast, and I paid it without a word—just the same as he always had charged me when I had not saved $100 for him. However, this fiddler never stayed another night at that hotel after paying for the privilege of saving the collapse of the big ball.

## Something About Gas

In these chapters I have said but little about gas. In fact, I tell little in these chapters that would permit being called "Old Times in Gasdom," instead of "Old Times in Oildom." Just think a moment. Those of our readers who were on earth when the first big flow of gas was struck at Titusville on the Jonathan Watson farm. The first big flow of gas was not worth ten cents; not good for anything in fact but to scatter the nice flowing yellow oil to the four points of the compass. For years after that the gas from the many wells in the oil region was more of a nuisance than a benefit. It caused considerable expense. The owners of the wells were obliged to buy iron pipes to carry the gas to a safe distance from the well where it was burned to prevent the mischief it might do. And mischief it did do in hundreds of cases. It killed the lamented Henry R. Rouse and several others with him at the same time, besides disfiguring for life a score or more. Many lives have been lost and much property destroyed before this vapor was finally bridled by the ingenuity of man.

The operators in Butler and other sections soon found a safe way to destroy this terror. They laid a pipe, as described, touched a match to the gas, thereby destroying its power to kill. I stood on a high eminence in Butler County one night and counted 63 great gas torches high up in the air. It was a grand sight. But oh! the millions of money that was vanishing, all unawares to mankind. Even Cal Payne, who at that time lived in his big new house at the city of Butler, was doing his full share in destroying one of the best servants of mankind ever known. I guess he knows something about it now, as he sits on the throne and gives directions in regard to this vapor, as it lights millions of homes with a brighter light than oil, and softer light than electricity. And more than that—it cooks millions of meals, and good housewives have only to strike a match, and one match may even suffice for all winter.

My own little town of Youngsville would be in comparative darkness if not for this mischievous gas. Instead of a dim flickering street lamp, as in nights of old, we now have street lamps on nearly every corner and one bright light greets another all over the borough. The Forest Gas Company leads this once uncontrollable stuff, in iron pipes, from away over the Allegheny river, in the wilds of Forest County, to nearly every room in nearly every house in Youngsville. All stores and public places are a bright blaze of light.

Did you ever think of the triple benefit of this excellent illuminant? First, it saves you from straining your eyes while reading at night; second, it saves much hard work in cleaning lamps; and third, it saves a vast amount of wood chopping and whittling shavings every time a little fire is started. People within range of the gas are apt to forget to be thankful every day that gas was struck in their time and that it was not postponed until another generation. And I must not forget to say that another great benefit is derived from this source. The young timber, instead of being cut up for fire wood, is allowed to grow up into high priced lumber all over the gas producing region.

The great ruler of the universe, God, will provide for future generations. This is only one of the many benefits that will be vouchsafed to the millions of people who will come to fill our tracks after we have traveled that unreturnable journey. Great is gas, and it came from small beginnings.

## Iron Tanks

When oil was stored in large iron tanks to a greater extent than it is since the Standard Oil Company commenced the business of transporting it directly from the wells to the refineries, lightning played a conspicuous part in depleting the producer's bank account. Now and then a tank is struck by lightning, even yet, but a good share runs to the refineries or to the seacoast safely underground in iron pipes, free from danger from lightning.

I have seen a great many tanks burning after being struck by lightning and the most dangerous one of the lot that I have seen was one at Perryville, Clarion County, Pa., across the river from Parker City, about 28 years ago. Nearly half a dozen large tanks, of about 28,000 barrels capacity, stood on the left bank of the Allegheny River at Perryville. They stood on a side hill, about 40 rods from the railroad tracks and the river. One morning during a heavy thunderstorm, lightning struck one of the tanks and there was a wicked blaze immediately. It burned

*Large iron tanks at the tank yard near Perryville, Pennsylvania. Located across the Allegheny River from Parker City. Each tank held about 28,000 barrels of oil. These same tanks caused George Brown to nearly escape death after a lightning strike. (Photo courtesy of Sally Jordan Reed)*

all day and in the evening a carload of us traveled five miles in a chartered car on the Allegheny Valley railroad to see the tremendous big black blaze. At this time I was a reporter for the Erie Daily Dispatch, and I went with the crowd for the purpose of reporting this oil fire. A couple of hundred people, both men and women, had gathered about this great blaze and about 3 o'clock in the afternoon, the overflow that always comes when a full tank of oil burns about half way down, came, and rivers of burning oil started down the side hill. The volume before spreading was about four feet high.

For some unaccountable reason, I happened to be below, right in the way of this burning oil. All the others happened to be off at one side, where they easily got out of the range of the burning fluid. I was the only one who had a nip-and-tuck race with the flames. I ran slantingly across the hillside towards a piece of woods. I came to a rail fence which I climbed on double quick and dodged into the woods. As I went under the trees, the blaze from the burning oil struck the tops of the trees over my head with an ugly roar. As I ran, the heat struck my back with great force and I was quite strong in the belief that there would soon be one less reporter for the Erie Daily Dispatch. But as was my luck, when I struck the edge of the woods, I found a rise in the ground that turned the oil straight down the hillside, leaving a breathless correspondent sitting on a log, thanking God for a deliverance from sudden death.

*Geo. W. Brown*

This was a fire to be remembered, as it cleared a couple of acres of woodland between the oil tanks and the railroad. The burning oil poured down the hill, devouring green trees and everything it came to. It swept the Allegheny Valley railroad tracks, stopping trains for a day or two, burned a planing mill, a lumber yard, several dwelling houses, and a barn, then poured a great burning stream into the river—a stream which spread from shore to shore and floated toward Pittsburg. It was a strange sight indeed to see that broad expanse of fire towering high and moving down the old Allegheny on top of the water.

I began this chapter by speaking of gas. I am reminded by this Parker City fire of the Wilcox, Pa., burning well. Here was another gas freak. I and three others drove four miles from the Wilcox Hotel to see the famous burning well. We were amply rewarded for the trip. When we arrived, it was dark. Every seven minutes, without fail, the gas would throw the oil and water nearly twice as high as the derrick. Each time when the flow would come, a man with a long pole, having an oiled rag on the end of it, would reach out the full length of this pole and set fire to the oil and gas. The gas would throw an eight-inch stream far up into the air. The water would form itself into a barrel shape, and the gas and oil would go straight up this round tunnel, all ablaze, entirely encircled by the water. Then the water would spread and fall in beautiful spray, forming all colors of the rainbow. Such a sight taking place every seven minutes cannot be described by my weak pen. This free show has long since gone into "innocuous desuetude." It is doubtful if even a counterpart of this wonderful Wilcox well will ever be seen again. This is an age of wonders, and perhaps something will turn up in this picturesque line that will excel the wonders of the burning well at Wilcox. But, I say again, it is doubtful.

*Lawrenceburg Above Parker City*
*These mud streets were quite a challenge to the poor horses carrying loads up from the Allegheny River. The large iron oil tanks can be seen on the horizon belonging to the Bear Creek pump station and Standard Oil. (Courtesy Dr. Clinton Coulter, Parker City, Pennsylvania.*

## Terrible Fire At Karns City

The old times in the oil country were frequently enlivened by fire getting started in the tinder box houses forming all new oil towns. I was running a lumber yard at Karns City at the time that quite noted town went up in smoke. I owned an opera house, a building for a lumber office, and the Western Union telephone office, lumber yard and hotel, all located in the lower end of the town. Fire companies came from Millerstown, Petrolia, and Parker City just in time to stop the fire before it reached any of my property. In fact, nearly every building in town was burned except mine. I scored good luck for once.

The fire started in a little store. The proprietor was away from home. The story flew fast that the man, a Hebrew by the way, had been burned out three times and had been insured every time. When the man came home the next day, an angry crowd met him on the street and accused him of being the cause of their homes being in ashes. The man turned white with fear in a moment and protested his innocence. But his hearers were crazy mad and threatened him with lynching. A rope was procured and preparations made to hang him to the nearest tree. Just then the constable arrested him and started with his prisoner for the lock up. On the way to the lock up, an amateur cowboy sent a rope whirling through the air three times, but each time it failed to coil around the man's neck. I never saw a lynching bee, but I might have seen one if that rope had caught on around the man's neck, as lots of men were ready to grab the end of the rope and run for the nearest tree.

That night a trial was had before 'Squire Stewart and the store proprietor was sent to the Butler jail for safe keeping. He remained in jail a few days and then caused the arrest of his captors for false imprisonment. He proved that he was in Clarion all day on the day of the fire. He also proved that his wife's gold watch and chain and all of his best clothes were burned, and that one child, through fright, ran and concealed itself under a bed and was accidentally discovered just in time to save it from a horrible death. The would-be lynchers had to pay quite dearly for their cowboy play.

Although this fire swept the most of Karns City from the face of Butler county, no lives were lost. But after all this, when a hotel had been built near the depot of the Parker & Karns City railroad, a terrible loss of life took place when the hotel burned with seven perishing. The hotel people and two transient lodgers were burned. But this was not as bad as the hotel fire at Millerstown a short time before that when eight people were cremated in one hotel and many more injured. I was an eye witness to this holocaust.

If all the sudden and tragic deaths that have taken place in the oil region since the Drake well was drilled were mentioned, it would fill a large sized book. Of the hundreds and hundreds of lives lost by glycerine I will mention just one that took place over on Bear Creek when the stuff let go with a jar that broke plate glass windows in Parker City, two miles away over a hill. I accepted an invitation from a Mr. Stephenson to ride with him in his buggy to the place indicated by the sound. When we reached the spot, we found a great hole in the ground, about eight feet deep and eight feet in diameter. About one-half of a horse lay in the road unscathed, but as clean as it could be done with a knife. It was the forward part of the animal that lay in the road cut through the middle about half way between the fore legs and the hind legs. The hind quarters were nowhere to be seen. Only a little of the wagon could be found. Not a vestige of the well shooter could be discovered except a portion of the skin of his face. That was found hanging on a bush about 20 rods away. One wagon tire was driven through a hemlock tree a foot in diameter

I might describe many other scenes that I have observed in the torpedo line. I have stood in three different towns and have seen plate glass breaking and falling onto the sidewalks, all caused by glycerine magazines exploding—Butler, Willow Grove and Parker City. Death followed in the wake of all these explosions.

## Dunc Karns and Tom King

Noticing the announcement of the death of S. D. Karns brings very forcibly to my mind once again "Old Times in Oildom" in Parker City. I owned a lumber yard in that noted city at its inception. Dunc, as he was called by everybody, lived in his oil country house, next door to my lumber office. He was a lively oil operator and no mistake. All he had to do was to drill a hole in the ground and get a big gusher of five or ten hundred barrels a day. Oil was four dollars a barrel, and his income was simply immense. Everything he took hold of melted into great piles of money.

He bought, or leased, the McClimens farm, one mile and a half south of Petrolia, and aside from the gusher oil wells, he made a nice little pile in lots, on which to build Karns City. The writer of this bought ground for an opera house, a restaurant and hotel and a city office, where was located the lumber business and Western Union telegraph office, and land on which the lumber yard was situated. All these pieces of land were situated in the western part of the town, and when the town burned, a cross street was all that saved my property. My chief loss was simply some ornamental cut glass in the telegraphic part of the building, worth but a couple of hundred dollars. This would have been saved if the wild, noisy crowd of men had obeyed my instructions not to carry it out, but to let it take its chances, as there was a cross street between my building and the fire, and as the fire companies from Parker City, Petrolia and Millerstown each had a stream of water on the fire, I felt in but little danger. But they paid not the least attention to me. They were crazy, and they rushed past me as I stood in the front door, and a half dozen grabbed the frame work and carried the whole business out and into the middle of the street, where they dumped it into the mud, and the fire companies soon made mince meat of it.

After that fire, Karns City was partly rebuilt, but it never fully recovered from the damage. But it is today far from being a toadstool town. It is a pretty little farming village, reinforced by many old time wells, and it will be a success as long as good producing farms abound in that region.

But back to Dunc Karns. He was "it" in everything pertaining to the business of Parker City and vicinity. Any kind of business that did not have Karns attached to one or the other end of the names of it was considered "small potatoes." It was Parker and Karns City Railroad, Karns Bank, Karns Bridge, Karns City, Karns Pipe Line, in fact he was the main spring. Dunc was not one of those business men who went around with a troubled look on his face. No, indeed! He scattered sunshine every day. He mixed pleasure with business and had a good time generally. He was a young man, full of life and energy when his great piles of money fell upon him. The weight of his money did not crush him. He never put on money airs and made himself disagreeable to the average man. He would play a game of billiards with a respectable oil worker with as much enjoyment as he would have had playing with the owner of a thousand-barrel well. I recollect one little mark of pride on his part, however. He bought a billiard cue, finished with silver trimmings. The billiard man kept Mr. Karns's cue in a separate receptacle from the other sticks. Although this looked somewhat "uppish," Dunc bought it for what it was, a finely wrought plaything.

## The Mansion

The first sign of opulence on Dunc's part was the building of a fine brick mansion a few miles below Freeport, Pennsylvania. He graded a lawn all around it, large enough for a common sized farm, with a nice setting of evergreens all over it. This place was on the opposite side of the Allegheny river from his boyhood home, Karns Eddy. But the old saying is, "It's a long lane that has no turning." The lane was turned with Dunc, but turned the wrong way. He was not the first to dip in a little too deep, nor will he be the last. The oil country is paved with men who have made and lost money. The writer speaks from experience, having earned a small fortune in two years and lost it in one. You see by this that you can go down hill twice as fast as you can go up.

At a certain time when the oil business was at its height, there were 30 lumber yards in the lower oil regions. These lumber yards were all doing a good business when, lo and behold! oil dropped from $3 a barrel to 50 cents a barrel. Then the lumber men got it "in the neck." Oil operators by the score, who had been paying their lumber bills every 30 days, went into bankruptcy. One little bit of an example is this: One operator, who had been one of my best customers failed for $1,500,000. Then where was my $900 lumber bill? This kind of thing was not uncommon. Out of this number, 30 yards, only two came off unscathed, 28 having found the oil country a slippery place to do business in. Money slipped into the dealer's pocket easily, but it slipped out again much more easily and quickly than it came. You may ask 28 lumber yard owners if they do not agree with me. One of the surviving two is now dead, leaving but one witness in favor of the beauties of the lumber business in the lower oil regions.

## From a Business Point of View

The Parker City of today in a business point of view, is not a shadow of its former self. I feel inclined to give one incident of many that could be given to show the strenuous way the Allegheny Valley Railroad had of doing business.

The railroad being on the opposite side of the river from Parker City, within only a wire cable to guide a ferry boat across the stream, made it anything but easy to do business in the new and hustling town. Everything was "hurly-burly." The short side track at the Phillips House could not hold half the cars sent to the new oil town, and the cars would be switched off at Foxburg, three miles above, or at Bear Creek, one mile below, and they would lie there until a small opening on the switch at the Phillips House could make room for one or two cars. Then the cars destined for Parker City would be switched in. One day the freight train men put two car loads of lumber for my yard, lumber needed at once for oil well purposes, on the Bear Creek end of the side track. No room was made for cars for nearly a week, and no team could get near the cars to unload the lumber.

As "gondolas" were in great demand about that time, the railroad could not well spare its cars a week at a time lying idle on a side track. Consequently, Tom King, assistant superintendent of the road, came up from Pittsburg and ordered his track workers to pitch my lumber down the steep bank onto the gravelly beach. There it was, partly in the water and partly out of the water, before I was even aware of the King's decree. Well, my customers were in a great hurry for the lumber and I put my own team and a couple of other teams at work hauling that lumber nearly a half mile up that river beach over gravel and rocks and up a steep bank. A team could haul about one quarter of a load at a trip, and it was fully $100 damage to me considering the breakage and extra teaming, lost time and everything.

I had not time to sue the railroad company. Tom King had not time to lose with a law suit, and the most important of all the reasons for not suing the railroad was that I had a free pass over the road from year to year, and if I had resented Tom's mean action, it would have, in all probability, had a bad effect on the pass. And as the pass covered the Allegheny Valley road and its branches, and I was over the road very often, I figured that it would not take a great while to "deadhead" $100 worth of transportation in my kind of business. So, Tom King was not put in the Sheriff's hands to answer for his unheard way of doing unheard things. I hear that Tom King, with a big "K," is still railroading in the western country somewhere. If he takes such liberties with some of those cowboys as he did with me, he would have to excuse himself at the point of one of their playthings, a revolver. But what is the use of being named King if you can't be a King.

It was uphill and downhill business those days to run a lumber yard in the new Eldorado even when the hard-worked yard master could get a car on that little short side track at the Phillips House where wagons could reach it. It was no picnic to get the lumber to the yard on the other side of the river. When the teamster was lucky enough to get his team in through the crowd of teams, it required two men to load the wagon, one on the railroad car and another on the wagon. Then the brakes had to be put on good and hard to get down the steep river bank and onto Fullerton & McLaughlin's chain ferry boat. Then upon reaching the Parker side of the river, a steep bank had to be ascended before reaching the yard. It cost something to get lumber from that little crowded side track to the yard, the wages of two men and team and the tremendous ferry toll of Fullerton & McLaughlin on small loads.

And now I will give the other side for there were two sides to the business. I bought a raft of 100,000 feet of boards from the Weston mills, three miles above Olean, N. Y., on Monday morning, paying $10 per thousand feet for it in the water. Saturday evening it was all gone at $19 per thousand feet. That was without cost to me. The oil operators would drive their teams into the water beside the raft and load their wagons from the raft. I cleared $900 on that raft in one week without touching a board. Another $900 easy transaction took place a short time after the last mentioned. A bridge was built across the river and the contractor gave me the contract of furnishing the square hard maple timber to put under the bottom of the stone piers. I gave the contract to a Springcreek mill owner at his own price and cleared about $900 without touching a piece of timber.

*Office of Mutual Pipe Line Company - Foxburg, Pennsylvania, June 1873*
*Many important oilmen, dressed in their "Sunday Duds." The Mutual had over one hundred miles of iron pipe lines from the tank to the rail. Standing on the porch are from left to right: W. H. H. Fithian, Joseph Seep, R. M. Kerr, M. L. Lockwood; leaning against the porch post F. H. Langworthy; in front of window Mr. Pullman. In between the center posts are Edward Ogelby, M. Jackson and Alfred Smiley. Continuing after the third post are C. J. Eastwick, Mr. Armor, P. Whitmar, and Josh Norris.*

This looks like making money easily. But the old saying is, "Come easy, go easy." As stated heretofore in these chapters, there is no trouble about the reader seeing where the "go easy" came in. All the old operators will remember the method, if there was any method in it, of doing business. It was up and down and up and down again and again and all branches of business were in a state of chaos until the Standard Oil Company, through marvelous management, gradually brought things in the oil country into understandable shape.

Author's note: This last sentence is very powerful. Often John Rockefeller and the Standard received much criticism, but George saw the Standard as the oil region's Savior. Thanks, George.

## The Lower District Pipe Lines

The following is a very complete estimate of the Pipe Lines of the Lower Oil Field, comprising nine different lines, all of them gathering in the oil to places of shipment on the Allegheny Valley Railroad. The extent and ramification of these lines is surprising, and testifies to the importance of the transactions in that direction:

The Emlenton Pipe Line is fifty miles in length, carrying the oil from Upper Turkey Run.

The Antwerp Pipe Line runs from Upper Turkey Run, and points adjacent, and is about fifty miles in length.

The Mutual Pipe Company's lines consist of nine different branches, running from the following points: Upper and Lower Turkey Run, Clarion District, St. Petersburg, Antwerp City, and points as far as Beaver Creek, and is one hundred miles in length.

The Grant Pipe Line runs from Grant farm on the west side of the Allegheny, coming out at Parker's Landing, and is 20 miles in extent.

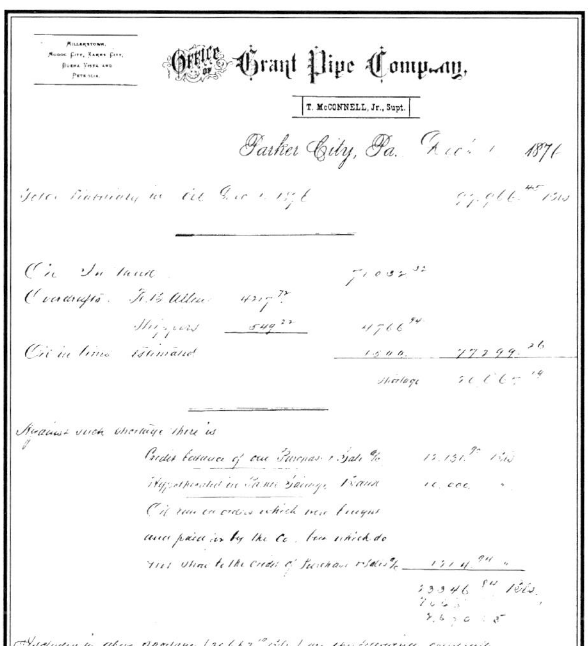

Office of Grant Pipe Company.

T. McCONNELL, Jr., Supt.

Parker City, Pa. 1871

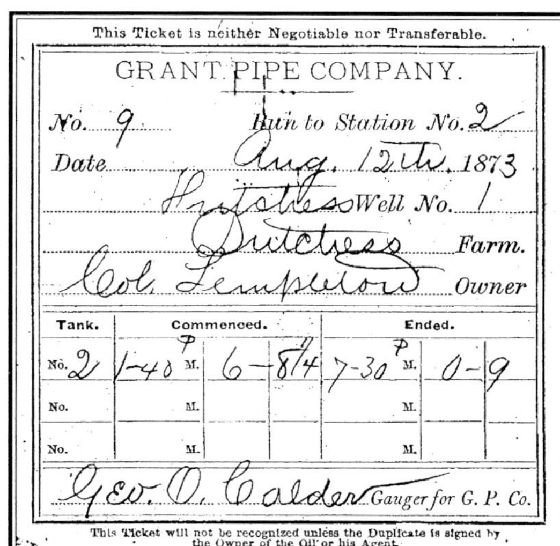

This Ticket is neither Negotiable nor Transferable.

GRANT PIPE COMPANY.

No. 9 Run to Station No. 2

Date Aug. 12th. 1873

Well No. 1

Dutchess Farm.

Col. Templeton Owner

| Tank. | Commenced. | | | | Ended. | | | |
|---|---|---|---|---|---|---|---|---|
| No. 2 | 1-40 | M. | 6 | 8/4 | 7-30 | M. | 0 | 9 |
| No. | | M. | | | | M. | | |
| No. | | M. | | | | M. | | |

Geo. O. Calder Gauger for G. P. Co.

This Ticket will not be recognized unless the Duplicate is signed by the Owner of the Oil or his Agent.

***Grant Pipe Company Flow Slip. Col. R. B. Allen of Parker was President of the Grant Pipe Line Company. (Col. Allen's biography is on page 535 of History of Petroleum, J. T. Henry, 1873)***

PARKER OIL EXCHANGE IN 1874.

TOP ROW— J. D. Emery. Warren Gray. —— Harris. E. Seldon. C. Seldon. Col. Sellers. Nelson Cochran. Unknown. Milo Marsden. W. A. Pullman. L. W. Waters. Lemuel Young. Chas. Archbold. Unknown. Unknown. Harry Parker. Hugh McKelvy. James Green. James McCutcheon. J. M'Donald. Dr. Thorn. Unknow Unknown

MIDDLE ROW— O. J. Greer. Fullerton Parker. Full. Parker, Jr. James Goldsborough. W. C. Henry. Thos. McLaughlin. Col. Brady. Sam. Morrow. Joseph Seep. Charles Hatch. John Barton. R. Moorhead. H. W. Batchelor. —— Gephardt. Shep. Morehead.

LOWER ROW— Capt. J. T. Chalfant. Thos. McConnell. Weston Howland. James Lowe. Chas. Riddell. Richard Conn. Rem Offley. Ken. Kerr. Harry Marlin. H. Beers. Jas. Garrett. Chas. W. Ball. Walter Fleming. Chas. J. Fra

The Iron Pipe Line runs from Bear Creek district, Sheakley, Argyle, Petrolia, Karns City, Millerstown, and Modoc City, and is in all, 125 miles in length.

The Cleveland Pipe Company (Vandergrift and Foreman) runs from Karns City and Greece City, and is 40 miles in length.

The Fairview Pipe Line, owned by Vandergrift and Foreman of Oil City, runs from Sheakley, Petrolia, Greece City, Millerstown and Modoc. Total length, 125 miles.

Relief Pipe Line runs from Story farm and Armstrong Run, and is some 12 miles in extent.

The Butler Pipe Line runs from Greece City, Modoc, Millerstown, and intervening points, to the Butler Branch Railway, and is sixty miles in length.

(History of Petroleum, J. T. Henry, 1873)

## Parker City, 1880
## Another Oil Town and Its Sad Decline From Prosperity

It does not require a long time for an oil town to fall from the height of prosperity to a position of comparative unimportance. The years 1878-79 were witness to a decline in the overall oil business. The reversal in fortune was a result of the partial exhaustion of the wells and the low price of oil. Farms, commercial and residential property depreciated in value to an alarming extent. In Parker City during the period of greatest prosperity, many costly and elegant residences were erected on beautiful sites near the crest of the Bluff. Their value is now reckoned in hundreds of dollars rather than thousands. Mansions that were built in 1870 for $15,000 sold in 1880 for $400.

Parker City, at the date of its incorporation in 1873 gave every evidence that she would one day stand in the class of the average metropolis of from 10,000 to 25,000 inhabitants. Even though Parker City holds the title of the "Smallest City in the United States," hope is now passed and she has only her charter as evidence of her once mighty population.

So, in the brief space of ten years, Parker City had seen the heights and depths of existence and had grown from a simple river landing place to a bustling city, and descended again to a minor village.

## Martinsburg

Now called Bruin, Martinsburg is situated twenty miles northeast of Butler, Pennsylvania, about four miles from Parker City. In 1820 Benjamin Fletcher built a log gristmill along the south branch of Bear Creek which began to attract settlers. In 1837 a village was laid out under the direction of John Martin and was named Martinsburg. Growth was slow with the building

*Martinsburg (Butler County Historical Society)*

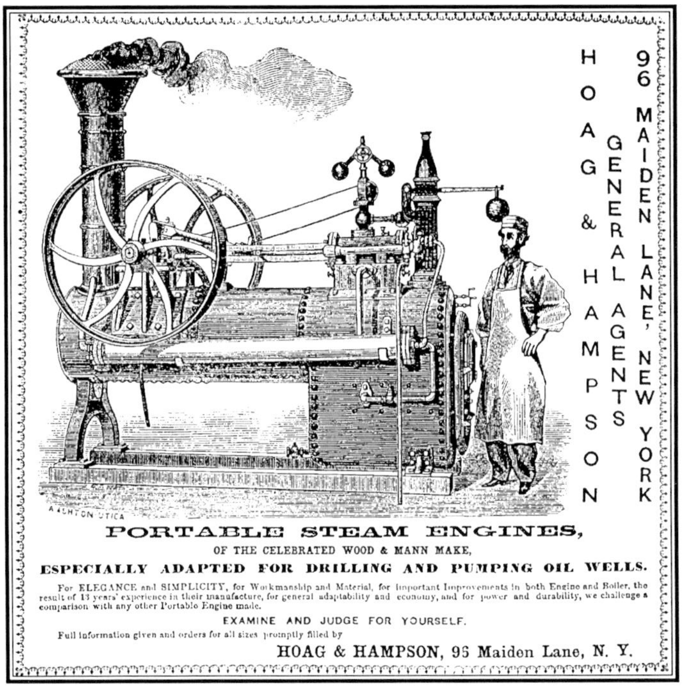


of a few houses, stores, taverns and places to stay the night. The post office was established in 1851 and at that time renamed Bruin, some say after all the bruins along Bear Creek.

The first well was kicked down by Jake Rieder near the Dudley Furnace. Due to the primitive methods, it took two years to complete. A second well was drilled by John Heiner just north of the Heiner trestle but both were mostly unsuccessful.

The village's growth was slow and stable until 1869 when the Jacobs Oil Company drilled the first successful oil well in the county near Bear Creek. In 1872, wild catters arrived and a tidal wave of oil operators, drillers, pumpers and supporting businesses arrived. Martinsburg was now at the front of the Butler County oil boom and the population rapidly increased from a couple dozen people to over 600. Hotels, machine shops, stores, taverns and other oil support enterprises arrived in town. The Parker-Karns City Railroad eventually played a major role in the transportation of people and freight in and out of the oil fields.

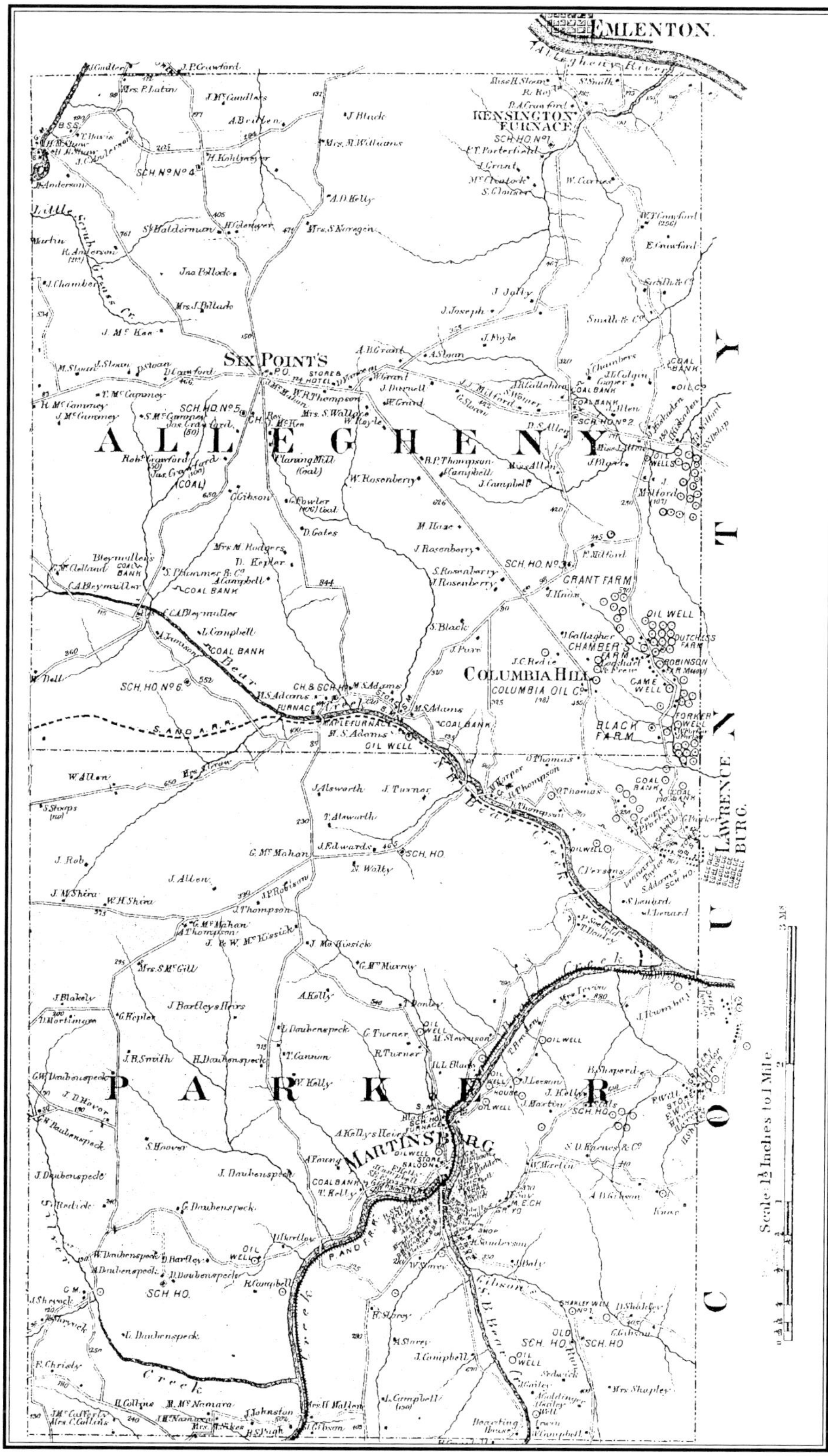


*Atlas of Butler County, Pennsylvania, 1874, Allegheny and Parker Townships, Six Points, Emlenton, Venango County, Lawrenceburg, and Martinsburg.*

Martinsburg never grew much in importance due to the fact that the oil operators were pushing the belt south. In the fall of 1877, production decreased and a smallpox epidemic in which five people died caused the population to move on.

## Argyle City

Argyle City was located in Butler County and just a short distance to the east of the great oil town of Petrolia. As with many oil towns, there is no visible evidence that Argyle ever existed. Today, the vicinity where the town was located is known as Hornertown. A. L. Campbell named Argyle after an old family home in Scotland.

Argyle City was founded and grew out of the wildcat operations of the Robert Campbell Oil Company. The Campbell Company was formed to test the sixteen degree oil sand belt theory of C. D. Angell and was located four miles south of Martinsburg on the A. L. Campbell farm. The company was put together by John Lamburg and drilled its first well in the spring of 1871. This first well came in at eighty barrels a day on November 19, 1871. The discovery caused much growth and excitement in the oil towns of Fairview and Martinsburg. The well that drew the greatest attention to the area was the "Ivanhoe" which came in May 1872. The "Ivanhoe" was drilled to the third sand and produced 228 barrels per day for over one month. This activity brought in great numbers of people and interest in oil operations grew.

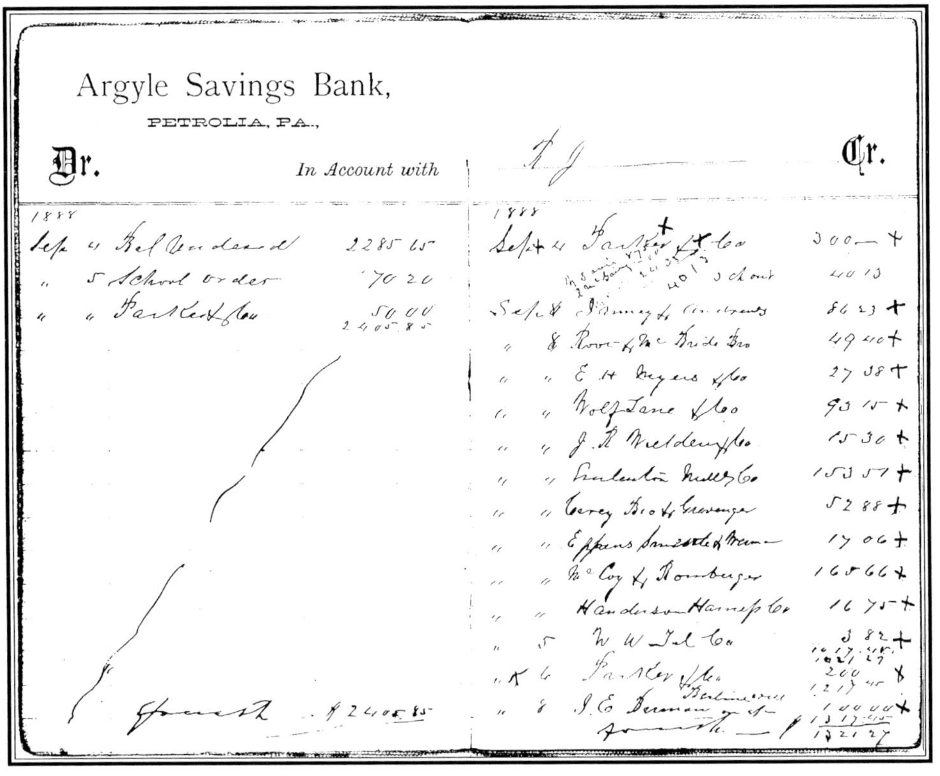
Argyle Savings Bank,
PETROLIA, PA.,
Dr. In Account with Cr.

*A page from a Argyle Savings Bank book. The Argyle Bank was originally located in Argyle City but moved to Petrolia as it grew to importance. This bank handled millions of dollars in its eighteen years, eight months of doing business. (Butler County Historical Society, Butler, Pennsylvania; Lenn G. McCorry, Butler County Oil History)*

*Argyle City, 1872, located near Petrolia, Pennsylvania. This oil town was laid out on the A. L. Campbell farm by its owner and named after their family home in Scotland. This photograph was done by Frank Robbins, Oil City, Pennsylvania. (Drake Well Museum, Titusville, Pennsylvania)*

A. L. Campbell leased lots in Argyle City for seventy-five cents to $2 per foot frontage and sold land on his farm for $800 to $1,000 per acre. With the sale of land and increased oil discoveries, the population of Argyle City increased to 600 residents. In the book, "The History of Petroleum," Argyle was described as a suburb of Petrolia. It took its rise from a well put down in the fall of 1871. At its best, Argyle had 600 inhabitants, and sported two hotels, two hardware stores and one or two machine shops. Argyle was described as but a "wreck of its former self, and is a sad comment on the changeable character of our new petroleum towns."

One of Argyle City's more lasting institutions was the Argyle Savings Bank. The need for a banking house was due to the sales of oil territory, increased population, and increased oil discoveries. The Argyle Savings Bank was established on August 22, 1872, and the bank's money turnover was one of the largest in Butler County. In its 18 years, 8 months, it handled millions of dollars and gave its patrons the finest banking practices in the lower oil region. The bank completed an office building at Petrolia and was moved in February, 1873, with H. L. Taylor, president, and E. A. Taylor, cashier. The bank ended its existence when the firm transferred its interests to a new company, under whose management the bank suspended on April 22, 1891.

Argyle City existed as a community until January 19, 1873, when the wells began to play out. The oil field was soon abandoned and the population moved to Petrolia, Karns City, or Butler. Beyond a few board shanties in hopeless decay, nothing now remains of Argyle City.

(History of Butler County, Pennsylvania, 1883, Waterman, Watkins and Company, Chicago, Illinois)

## Petrolia

The first wells in the Petrolia district were put down by Messrs. Dimick, Nesbitt and Co., in November, 1871, on the Sheakley farms and attracted considerable attention from operators. Early in April, 1872, Dimick, Nesbitt and Co. finished the "Fanny Jane" well which was the first big strike around Petrolia and caused the Petrolia district boom. The rush was on. A newspaper reporter, who rode down from Parker's Landing to Petrolia in the summer of 1872, passed en route some twenty loads of houses and household effects, including one two-story house forty feet square, which was undergoing removal for the fifth time.

## Petroleum Built Petrolia

In November, 1871, F. M. Campbell built the first house in Argyle—excepting farm houses. A few months later, Argyle presented the appearance of a very lively village with stores, machine shops, etc. But the crossroads farther to the south seemed a more desirable site for the future "city," and about this point buildings began to spring up as if by magic. The first house in Petrolia was erected by W. E. Clark in February, 1872. John Painter, of Brady's Bend, established the first grocery and saloon on the corner of Main and Jamison streets. This was quickly followed by a store, 12 x 18 feet, built by R. W. Cram, and used as a newsroom and office of the Penny Post. The oil excitement was at a high pitch, and in a few weeks Petrolia had grown into a flourishing town with stores, machine shops, hotels, saloons and shanties, all hastily contrived. George H. Graham made the first survey of lots and, in an incredibly short time, the Graham farm (then owned by the Reno Real Estate Company), and the farms of John B. Jamison, James Campbell and James Blaney had been partially absorbed in the fast-growing town.

## Burgess George H. Dimick

The morals of the new settlement were characteristic of oildom. Fully 3,000 people had their homes here, and among the population were many roughs, and men and women of shameless character. Drunkenness, prostitution and other forms of vice became so prevalent that the leading citizens deemed it advisable to secure the corporation of the town, in order that these evils might be checked. During the summer of 1872, measures were taken to form a borough government, and in December of the same year the organization was effected, and George H. Dimick elected Burgess. The borough was fully incorporated according to law, February 1, 1873. Mr. Dimick proved a very efficient officer and, aided by a strong police force, soon was able to maintain comparatively good order and establish justice.

Among the first business places opened were Robertson's machine shop, built by W. E. Clark; McBride's drug store, the Central Hotel, by Peter Christie & Co. A man named Ben Hogan erected a large building which he styled the Opera House. Under his management the house gained very unenviable notoriety, but after it passed into different hands it came to be well patronized, and many men of national reputation came hither in the role of lecturers.

Courtney built one of the first machine shops. In 1874, Ireland and Mr. Caughtry bought it. In 1878, the building was destroyed by fire, but was rebuilt in the fall of the same year. Mr. L. N. Ireland, in partnership with James E. Hughes, was connected with the management of the shop since 1874. In 1875 and 1876, from 18 to 28 men were employed in these works.

Petrolia soon came to be regarded as the very center of the lower oil region, and some of the largest operators accordingly made it their headquarters.

R. W. Cram was the first Postmaster. The office was established in 1873. Mr. Cram was succeeded in 1882 by William Gibson. At one time, the office was a large distributing office. In 1877, it was estimated that at least 5,000 adults received their mail through Petrolia Post Office. It became a third-class office in 1877, and paid a larger salary than any other office in the county.

The population of towns in the oil region is always an unstable quantity. Petrolia in its most prosperous days was probably the home of at least 6,000 people. The years 1875 through 1877 were by far the best in the history of the town. The rapid rise in the price of oil in 1876 stimulated the industrial growth of the place to the utmost. In 1879, business activity declined, and there was a consequent exodus of population. The census of 1880 showed over 1,100 inhabitants in the borough. Petrolia was very important oil town in the lower oil region.

Petrolia had a very efficient fire department, consisting of two hose companies with a membership of about fifty each. The United Hose Company No. 1 was organized by A. C. Beeson, in 1873, and consisted solely of employees of the United Pipe Line Company. This

*Main office of "The United Pipe Lines" at Petrolia, Pennsylvania, 1873.*

was followed by a citizens' company called the Petrolia Hose, which, while it continued, was a large and well-managed company. The Protective Hose Company was organized in 1874 by citizens of the borough, and was a most efficient organization.

For the protection of the borough against losses from fire, an ordinance was enacted April 7, 1874, granting to A. C. Beeson and Murat Compton the full and exclusive right to construct a water works. The right was transferred to E. O'Donnell and Frank L. Masson, who at once proceeded to construct works as a private enterprise. In June, 1880, the property was divided into a capital stock of $1,200, and mostly sold to citizens of the borough. The company organized with the following board of managers: Guerdon Williams, Chairman; D. C. Backus; T. T. Mapes; D. A. Cypher; F. L. Masson, Superintendent; S. W. Harley, Secretary; and E. A. Taylor, Treasurer. The works consisted of one water pump of twenty-horse power to supply the town with water; a stationary pump of forty-horse power for use in case of fire, and about five miles of hose.

The machine shop of Ireland & Hughes; the boiler works of J. C. Lyons, and of Frank Quinn & Co.; the cup and valve manufactory of A. W. Root; and the machine shop of A. C. Price were large employer's. The United Pipe Line Company formerly had large machine shops and an extensive foundry in Petrolia. These works were removed in 1880.

When the "Fanny Jane" was struck in April, 1872, the Jamison farmhouse was the only dwelling within the present limits of Petrolia Borough, but the new and rich strike was a sufficient incentive to set a number of people to building, and with the mushroom-like growth, only known in oil and mining regions, a village was formed in the little valley of Bear Creek. It seemed to the astonished farmers, who had lived for long years on the hills of Fairview Township, as if the bustling little village grew up in a single night. The little cluster of houses, shops and stores certainly did grow very rapidly and spread out over the lands west of the creek, sold by George H. Graham to C. D. Angell, and the Jamison farm and property of Adams & Scott east of the creek. It was incorporated as a borough in February, 1873, in response to a petition circulated in the fall of 1872 and signed by 103 citizens. Argyle, the older but lesser town, was swallowed up. Like all oil towns springing quickly into existence through the pressure of a suddenly developed need, Petrolia consisted entirely of light and flimsily constructed wooden buildings. They were put up hastily to meet the demands of the strange heterogeneous population which poured into the county. Hotel followed hotel, and all were crowded to their utmost capacity as soon as completed. The town quickly leaped to a population of three thousand and ultimately to six thousand. The lucky strikes in the 22-degree belt, and the rapid development of the territory, brought in all classes of people. The heavy capitalist, the experienced operator, the shrewd speculator, the penniless adventurer, the "man who had seen better days," the green novice, the curious tourist, the honest citizen, the common laborer, the tramp, beggar, gambler, thief, the courtesan, all were there, and jostled each other on the narrow sidewalks.

The sodden, aimless, broken-down wretches who form the human *flotsam* and *jetsam* of the ocean of life, depraved characters of every type and every degree of degradation, came upon the heels of the pushing men of business as a horde of camp followers straggling on after an army. Petrolia afforded a marked illustration of condensed and intense life. Six thousand people—a constantly changing population, made up of all grades and classes, good and bad, lived in a town which at a casual glance appeared scarcely large enough to hold as many hundred, and the majority of them crowded ten years of action into one of actual time. Business and pleasure and dissipation were carried on during the height of the great oil excitement with a rush, which is never equaled outside of a great center of oil production and oil speculation.

The better elements of society, however, were always dominant in Petrolia, and it never had as bad a reputation as some of the older oil towns in the upper region.

The town soon became the head center for some of the oldest and heaviest operators, and gigantic business interests had their inception there and were successfully conducted. In October, 1875, after the great cross belt development, the oil exchange was organized (the first in Butler County), with S. H. Smith as President. Speculation ran high, and at one time Petrolia made the market price for oil for the world. The borough passed through the usual ups and downs of oil town history, had its great fires, its record of quickly made fortunes and heavy failures of individuals, and finally, its own prosperity began to wane, as the oil production which had made it fell off.

*View of Petrolia City*

## A Destructive Flood

On Saturday, July 26, 1879, Petrolia was visited by one of the heaviest storms ever known in this region. About 11 a.m., so much rain had fallen that Bear Creek was converted into a rushing river and soon after, in the vicinity of Karns City, fences, small buildings and all other objects in the way were swept away and carried down stream. The railroad bridge above Petrolia stayed the onward rush of the debris for a time, but suddenly giving way without warning, Bear Creek and its heavy burden of floating timber was precipitated upon the town.

Jamison Creek had previously broken through buildings on the east side of the railroad track, and when the dam at the railroad bridge broke, it became evident that the business portion of the town was doomed to destruction; the coliseum, a large building 60 x 170 feet erected by the pedestrian association at a cost of $1,400, was washed from its foundations and floated upon the water like a toy. This huge bulk was driven against other buildings, and they in turn gave way, until along the entire course of the stream the town was bereft of buildings. Some fifty buildings were destroyed, and about the same number inundated. The wildest excitement prevailed, but fortunately there were no serious accidents.

Some very amusing incidents occurred, among others, the following: During the morning, a walking match was in progress at the Coliseum, and the champions were still contending when the flood struck the town. One athlete, in his walking suit, and with spectacles upon his nose, was seen on top of the Coliseum as it moved down the stream. When the building came in contact with the hose house, he leaped from the roof to the top of the latter building, and rode onward until his novel conveyance stopped, and he was enabled to land in safety.

The buildings destroyed were generally of little value, and probably would not average over $200 each. But the value of goods and furniture lost was very great. The Record, in its next issue, estimated the entire loss in Karns City and Petrolia at $75,000.

*View of Petrolia City, 1873*

Karns City was but slightly damaged. Among the largest losses in Petrolia, according to the Record's estimates, were the following: C. Barthold, clothing store, $8,000; R. J. Bottner, jewelry store, $5,500; Foote & Weeks, drug store, $6,000; M. N. Miles, law office, $1,100; M. C. Benedict, law office, $1,384; damaged United Pipe Line, buildings, etc. $5,000; Central Drug Store, $1,000. An entire solid block of buildings, standing over the creek on Railroad Street, was completely destroyed.

## Petrolia's Newspapers

The Petrolia Record was established in 1877 by its present editor and proprietor, Charles E. Herr. The first issue appeared October 27. The Record was then a folio of twenty columns. In April, 1878, it was enlarged to its present size, a twenty-eight-column folio, with pages 24 x 36 inches. The Record is all printed at home and typographically, it is one of the neatest local papers published. It is well conducted and prosperous.

Lerch & Mapes started the Advertiser a short time before the Record appeared. Only a few numbers were ever issued. In 1878, the same firm began the publication of the Producer's Free Press, a journal devoted to the oil interests. The Free Press was edited by P. C. Boyle, and had an existence of about one year.

## Fires and Instant Death

The first fire, other than that at the Jamison well, occurred in November, 1872. While it destroyed five or six houses and entailed heavy losses on the people of the town, it did not for a moment retard their chase after oil and the almighty dollar. The fire of January, 1873, originated at the Wilson well above the town, whence it spread to five wells in the vicinity, destroying the rigs and 5,000 barrels of oil. The oil, which flowed into the creek, was accidentally fired by one of the Dean brothers. The two brothers were severely burned for their carelessness, while the dwellings of Prof. Loomis and Parsons were destroyed and one man killed by the falling of a burning derrick. In April, 1873, the Evans & Jennings well on the Dougherty farm, near Petrolia, took fire, and 900 barrels of oil were destroyed.

On Sunday evening, December 10, 1876, occurred a terrible explosion of nitroglycerine on the outskirts of the town, which broke the windows of many buildings and blew to atoms David Smith and a teamster who had just brought the explosive from Titusville. A second nitroglycerine catastrophe took place near Petrolia October 28, 1878, resulting in the deaths of S. O. Gotham, H. R. French and John W. Fowler.

The fire of August 1, 1879, came to destroy what the flood left undisturbed. Lutzinger's barber shop, the Grand Central Hotel, Newman's candy shop, Harvey's store, Comerford's saloon, and the post office on the upper side, Dr. Mahneke's office and Spence's stable on the lower side, and the old Shirley store buildings across Main Street, were burned. The drug store of A. A. West was torn down by the mob under the pretence of preventing the spread of the fire. A notice on an outbuilding of Burgess Stone, with the motto, "Bread or Blood," distinguishable among the hieroglyphics, led to the supposition that incendiaries were at work.

(History of Butler County, Pennsylvania, 1883; Waterman, Watkins and Co., Chicago, Illinois.)

## Saved By Grace

During the terrible flood of July 26, 1879, along Bear Creek in Butler County, Major Wilson, an old oilman, was trapped in his workshop. The foundation lifted and the structure began to float and move away. Major Wilson was an old gentleman who liked to participate in oil country amusements, amusements his preacher would call "drinking freely from the flowing bowl."

Major Wilson fell to his knees and in a loud voice, as the small shop was hurled down the rushing stream, asked for divine help to save him from a watery grave. At this very moment old Major Wilson pledged himself to spend the rest of his life in the Lord's work. His prayers were soon answered and what was left of Major Wilson's ark was soon carried to shore.

How long he kept his promise to the Lord is questionable and later that day poor old Major Wilson was seen at Hogan's Seminary taking a "long cool drink from a tall dark bottle." (Oil and Gas Man's Magazine, Butler, Pennsylvania, 1907-1918; Drake Well Museum, Titusville, Pennsylvania)

# Dilks, Tank Car Loading Terminal

Dilks was the name given to a railroad siding located then on the A. Frederick farm and was one of the main oil pipeline and rail terminals in the Butler oil region. There was more freight business transacted here than in the entire town of Butler, Pennsylvania. Freight was hauled in and out of the oil fields and trains with fifty tank cars each loaded here.

The Butler Pipe Company was started in 1873 with the office in the First National Bank building in Butler, and was a branch of the Union Pipe Company. The rail oil tank cars and all connections were known as the Empire Transportation Company. This operation was under the direction of the Pennsylvania Railroad Company. The name Dilks came from General Superintendent J. H. Dilks, builder of the oil terminal and the Butler Pipe Company. The Humes Pump Station pumped the crude oil to the Dilks terminal through a three-inch pipe line, constructed in 1873, two years after the railway was opened for business.

Dilks was part of the transaction when Standard Oil Company bought the Union Pipe line. Standard Oil Company bought the Union pipe line and the Dilks terminal on December 1, 1877.

One of the most disastrous Butler County oil terminal fires occurred at Dilks on July 20, 1876. Lightning struck a group of three 25,000 barrel oil tanks and 64,973 barrels of petroleum went up in smoke. The oil market that day quoted the price of $2.28 per barrel of crude. This would have been quite a loss of money for the producers in 1876. (Butler County Historical Society, Butler, Pennsylvania; Lenn "Pithole" McCorry, Butler County Oil Historian)

***Early Refinery at Petrolia City***
***Barrels of refined crude oil setting on the loading docks of the W. H. Daugherty and Son Refining Company, Petrolia, Pennsylvania. Railway ran through the refinery. (William Kraustok, Butler County Historical Society)***

*Bird's Eye View of Petrolia City 1875*
*Butler County Historical Society, Butler, Pennsylvania*

## Nervy If Nothing More

"I understand sir, that you have recently discovered oil and thereby you now possess a swollen fortune," said the gay young youth.

*John D. McKee, oil tank gauger, going to work. A gauger kept a record of the amount of oil a producer ran into the pipeline system. The producer was given a flow slip that recorded the amount of petroleum the producer would be paid for. (1877, Butler County Historical Society, Butler, Pennsylvania)*

"Well," gruffly answered the beautiful girl's father. "What is that to you?"

"Oh, nothing," said the gay young youth, "I thought I would give you fair notice of my intention to help you take the swelling out of it. Myrtle and I are going to be married!"

(Oil and Gas Man's Journal, Vol. 6, No. 2, Butler, Pennsylvania, 1911)

# PETROLIA CITY.

This thriving and enterprising town has sprung into existance since the spring of 1872. The Gibson well No. 1, and the Blaney well, were the first successful wells put down here, and when struck flowed at the rate of about one hundred barrels per day. There are many pumping weils now in operation in and around Petrolia. The site of the village comprises about fifty acres of ground, and was occupied by Mr. George H. Graham, for farming purposes. This point being the nearest to East Brady and Parkers, was soon the scene of an active and busy village. There are some very prominent business houses here. Also large manufacturing establishments for the manufacture of oil well supplies.

The Karns City Rail Road will pass through the centre of the town. One banking house represents the monetary interests, and by reference to the advertisements in the accompanying Directory of the place, it will be seen that the commercial enterprises are largely represented. The "Central" is the principal hotel, but there are others which offer good accommodations.

## A

Aken G., proprietor Forest House, Fairview.
ARGYLE SAVINGS BANK, H. S. Taylor president, Main.

## B

Backus D. C., of Backus Morrison & Co., office Main.
BACKUS MORRISON & CO., hardware and oil well supplies, Main.
Bahney Jos., of Jos., Bahney & Co., Main.
Bahney Jos., & Co., clothing, furnishing goods &c., Main.
Baker ———, superintendent oil wells, boards Jefferson.
Baker Lewis, bartender, Fairview.





Barker B. H., engineer, Main.
Barns John, driller, boards Jameson.
Barns Nellie, restaurant, Fairview.
Barthold C., clothing furnishing goods, Main and Argyle.
Beam I. & Son, tank gaugers, Main.
Beeson A. C., superintendent Fairview Pipe Line, boards Main.
Berwald Levi, boots and shoes, Main.
Bills Samuel, laborer, boards Jameson.
Black W. C., of Dean & Black, Argyle.
Blystone George, contractor, Jameson.
Blystone Henry, laborer, Jameson Farm.
Boyles G. W., bookkeeper, Blaney Farm.
Boyle Wm. J., laborer, Main.
Bullock C., insurance agent, Argyle.
Burmore George, clerk, boards Main.
Buzzard James, grocer, Main.

## C

Cameron John, engineer, boards Jameson.
Campbell George, harness maker, boards Main.
Carlin R. L., of R. L. Carlin & Co., livery, Fairview.
CARLIN R. L. & CO., livery stable, Fairview.
Carlin Thomas, of R. L. Carlin & Co., Fairview.
Clark R. P., teamster, Main.
Clark William E., merchant, Jameson.
Cleary James, saloon, Fairview.
Cohen H., clerk, boards Main.
Collins Sue, laundress, Blaney Farm.
Conn R. D., oil dealer, Main.
CONWAY A., boot & shoe maker, Jameson.
Cook Charles, clerk, boards Crittenden House.
Corson Smith, pumper, Fairview.
Courneen James, clerk, Fairview.
Crittenden House, Fairview.
CRAM R. W., books and stationery, and post master, Main.
Crittenden C. A., proprietor Crittenden House, Fairview.
Currier Peter, blacksmith, Main.

## D

Dale S., machinist, boards Jameson.
Dale Samuel, blacksmith, boards Central House.
Dauberspeck John, teamster, Fairview.



Davis A. J., restaurant, Fairview.
DEAN & BLACK, grocers, Argyle.
Dean John C., stage driver, Fairview.
Dean William, of Dean & Black, Argyle.
Delap S. N., oil operator, Fairview.
Deutsch Ignatius, tailor, Fairview.
Deutsch Jos., tailor, Fairview.
Devinney L., of Shepley & Devinney, Jameson.
Dillen Wm., butcher shop, Main.
Dimich G. H., oil producer, Fairview.
Divlier Benjamin, laborer, Jameson.
Dougherty B., grocer, Fairview.
Dougherty J. B., of Moorehead & Dougherty, Main.
Dunbar Ambrose, barber, Main.

## E

Ebbert D., of R. P. Levalle & Co., Petrolia restaurant.
Elliott James, baker, boards Main.
Elsworth C. S., clerk, Fairview.

## F

Fairview Pipe Line, office Main.
Fisher Brothers, oil dealers, Main.
Furman B. J., of Furman & Wisener, Fairview.

## G

Gardner Samuel, carpenter, Main.
Gardner & Tyler, Misses, dressmaker, Main.
Gibboney G. W., boots and shoes, Argyle.
Glethill Andrew, tailor, Fairview.
Goodman Ellis, of Joseph Bahney & Co., Main.
Gratton Mike, wagon maker, boards Petrolia House.
Grazier W. P., salesman, boards Main.
Griffith Charles, barber, Main.

## H

Hamilton A., peddler, Jameson.
Harlan G. W., book keeper, Main.
Hatch C. R., agent, Fairview.
Hazelwood Oil Co., Fairview.
Herr F. F., clerk, boards Main.
Hicksey Jas., machinist, Argyle.
Hileman Henry, engineer, Jameson.
HILL J. B., grocer and dry goods, &c., Jameson, res. East Brady, Pa.
Hill W. B., baker, Fairview.



Holderman Jacob, laborer, Main.
Holderman R. H., of Levine & Co., Main.
Holman H. K., livery, Fairview.

## I

Island Roley, Oil Dealers, Argyle.
Ireland L. N., blacksmith, Main.

## J

Jones J. Glancy, restaurant, Argyle.
Jones J. W., Argyle.
Jacobs Isaac, peddler, Main.
Jacobs L., cigars and tobacco, Main.
Jarvis Robert C., carpenter, Jameson.
Jenkins David, tool dresser, Jameson.
Jersey William, blacksmith, Main.

### K

Kennedy Charles, restaurant, Fairview.
Kilkenstein George, tailor, boards Fairview.
King Charles, sup't. oil wells, boards Jameson.

### L

Lanahan J. C., proprietor Valley House, Main.
Lardin Wm. M., of Nesbitt & Lardin, Blaney Farm.
Lattin C. B., pumper, Fairview.
Lattin J. M., pumper, Fairview.
Lawrence S. E., cigars, &c., Fairview.
Lauders Michael, butcher, Main.
Leckey Robert, book keeper, Nesbitt & Lardine's.
LeRoy E., watches, clocks, &c., Main.
LEVALLE R. P., of R. P. Levalle & Co., boards Petrolia Restaurant.
LEVALLE R. P. & CO., proprietors Petrolia Restaurant, Argyle.
Levin M., of Levin & Co., Main.
Levin & Co., confectioner, Main.
Lewis James H., attorney, Jameson.
Lewis William, driller, boards Jameson.
Lucas William, laborer, near Blaney Farm.

### M

Mackey Edward, musician, Jameson.
Maloney Timothy, stage driver, Fairview.
Maloney Tim., tailor, Fairview.
Manning Patrick, laborer, Blaney Farm.



Mames Henry, driller, off Jameson.
Martin Thomas, driller, boards Jameson.
Martin William, painter, boards Jameson.
McBride Edward, Main.
McBRIDE SAMUEL, drug store, Main.
McBride Thomas S., rig builder, Jameson.
McBride Thomas, wagon maker, Main.
McCall E. M., pumper, Blaney Farm.
McCantlish M. J., blacksmith, Main.
McCarthy O., saloon, Main.
McCartney Alex., lumber dealer, Fairview.
McCartney Wm., teamster, Fairview.
McClure D. W., photographer, Jameson, residence Fairview, Pa.
McCray Samuel, teamster, boards Jameson.
McCreary Wm., manager Union Pipe Line, Main.
McCrum Samuel, book-keeper, boards Main.
McCue John, druggist, boards Jameson.
McCullough T. J., of Sheffield & McCullough office, Main.
McCurdy John, carpenter, Fairview.
McDonald ——, machinist boards Central House.
McDonald T., book-keeper, boards Forest House.
McFee Hugh, shoemaker, Fairview.
McGuire Edward, stage driver, Fairview.
McVey John, billiard room, Main.
Meyers Chas., of Rothstein, Bros & Co., Main.
Meyer Isidor, clerk, boards Argyle.
Monroe A. M., baker, Fairview.
Moore D. L., book-keeper, Main.
Moore Frank, driller, boards Jameson.
Moorhead A. J., of Moorehad & Dougherty, Jameson.
Moorhead & Dougherty, restaurant, Main.
Moorhead Milton, oil operator, boards Crittenden House.
Moorhead S. W., oil operator, boards Crittenden House.
Morad D. C., musician, Jameson.
Morgan ——, blacksmith, Fairview.
Morrison David, of Backus, Morrison & Co., office Main.
Morrison George, salesman, Jameson.
MORRISON JAMES, Wines, Liquors, Cigars, etc. Main.
Murray Patrick, boarding house, Argyle.

### N

Neff E. N., telegraph operator, boards Crittenden House.
Nelson Sauford, laborer, Fairview.
Nesbitt G. H., of Nesbitt & Larden Blaney Farm.
Nesbitt & Lardin, oil producers, Blaney and Jameson Farm.

### O

O'Dell Misses, dressmakers, Jameson.
Olliver David, painter, Fairview.

### P

Pacific & Atlantic Telegraph Co., Main.
Parker G. W. D., barber, Main.
Peaks S., telegraph operator, boards Main.
PETROLIA RESTAURANT, R. P. Levalle & Co., proprietors, Main and Argyle.
Postoffice, R. W. Cram P. M. Main.
Primer J. J., barber, Fairvew.

### R

Reed D. V., helper, Main.
Reedy Frederick, butcher, Main.
Reiber G. W., livery stable, Main.
Reiber George W., restaurant, Main.
Richards E. C., saddles and harness, Main.
Richardson T., manufacturer of drilling tools, Main.
Riddley W. B. clerk, boards Crittenden House.
Ringleader John, baker, Main.
Robertson D. W., blacksmith, Main.
Robertson George D., machinist, Main.
Robertson J. N., manager Pipe Line Shops, Main.
Robertson James N., machinist, Main.
ROBERTSON Wm., machinist, Main.
Rogers & White, tank builders, Fairview.
Ross J. O., fruits and vegetables, Main.
Rothstein Bros. & Co., dry goods &c., Main.
Rothstein K., of Rothstein Bros. & Co., Main.
Rothstein M., of Rothstein Bros. & Co., Main.
Rowley George H., blacksmith, Jameson.

### S

Saalfield Henry, cigars and tobacco, Main.
SACAEN ALBERT, boots, shoes, clothing, &c., Main.
Satterfield John, oil dealer, Main.
Scott Thomas B., carpenter Main.
Seaton J. H., shoemaker, boards Wilson House.
Shae Patrick, helper, boards Main.
Shearman E. I., book-keeper, boards Main.
Sheffield Frank, of Sheffield & McCullough, Main.
Sheffield George, of Sheffield & McCullough, boards Main.
Sheffield & McCullough, hardware and oil well supplies, Main.
Shepley C. H., of Shepley & Devinney, Jameson.
Shepley & Devinney, druggists, Jameson.
Simpson Alex., clothing and gents' furnishing goods, Main.
Smith A. H., restaurant, Fairview.
Smith Robert, carpenter, Jameson.
Spencer John, clerk, boards Thompson House.
Stahl George W., agent Fairview Pipe Line, Main.
Steinberger F., clerk, boards Main.
Steinberger W., tailor, Main.
Sweeny August, clerk, Fairview.

### T

Tack Augustus, oil operator, boards Crittenden House.
Tack F., oil operator, boards Crittenden House.
Tarr Minnie, Jameson.
Taylor E. G., book-keeper, Argyle Savings Bank, Main.
Taylor H. L., president Argyle Savings Bank, office, Main.
Tew James, driller, boards Jameson.

Thompson House, James Thompson proprietor, Main.
Thompson James, proprietor Thompson House, Main.
Thompson C. M. C., clerk, boards Thompson House.
Thompson S. D., engineer, Main.
Thumm George W., cashier Argyle Savings Bank, Main.
Thumm S., bookkeeper Fairview Pipe Line.
Tubbs David W., grocer, Jameson.
Tupper Thomas, assistant sup't. Fairview Pipe Line, Main,
Tyler Harlow, contractor, Fairview.

**V**

Valley House, J. C. Lanahan proprietor, Main.
Vashon George, Waiter, boards Petrolia restaurant.

**W**

Wanamaker G. B., clerk, boards Main.
Warden Deloss, blacksmith, Argyle.
Warwick William, blacksmith, Brady.
Weeks James K., blacksmith, Gibson Farm.
Wesson B. S., engineer, Fairview.
Westerman Fredrick, pumper, Fairview.
Western Union Telegraph Co., Main.
White James, billiards, Main and Argyle.
Wilcox Laugdon, tinner, boards Main.
Williard Dr. L. M., physician, office Main.

**WHITE S. W., DEALER IN ALL KINDS OF GROCERIES, JAMESON.**

Wilson Joseph, boarding house, Main.
Wisener J. J., of Furman & Wisener, blacksmith, Fairview.
Woodberry H. C., clerk, boards Main.
Woodberry Henry, Fairview.

## PETROLIA BUSINESS DIRECTORY.

**Banks.**

Argyle Savings Bank, Main.

**Billiard Halls.**

McVey John, Main.
White James, Main and Argyle.

**Books & Stationary.**

Cram R. W., Main.

**Boot and Shoe Makers and Dealers,**

Berwald Levi, Main.
Conway A., Jameson.
Gibboney G. W., Argyle.
Bahney Joseph, & Co., Main.
Sachen Albert, Main.

**Butchers.**

Dillen Wm., Main.
Lauders Michael, Main.

**Cigars and Tobacco.**

Cram R. W., Main.
Jacobs L., Main.
Lawrence Mrs. S. E., Fairview.
Morrison James, Main.
Saalfield Henry, Main.

**Clothing.**

Bahney Joseph, & Co., Main.
Barthold C., Main and Argyle.
Sachen Albert, Main.
Simpson Alexander, Main.
Slimberger W., Main.

**Dressmakers.**

Gardner & Tyler, Misses, Main.
O'Dell Misses, Jameson.

**Drug Stores.**

Hill J. B., Jameson.
McBride Samuel, Main.
Shepley & Devinney, Jameson.

**Dry Goods.**

Rothstein Bros., & Co., Main.

**Groceries.**

Buzzard James, Main.
Dean & Black, Argyle.
Dougherty B., Fairview.
Hill J. B., Jameson.
Tubbs David W., Jameson.
White S. W., Jameson.

**Hardware.**

Backus, Morrison & Co., Main.
Sheffield & McCullough, Main.

**Hotels.**

Central House, Main.
Crittenden House, Fairview.
Thompson House, Main.
Valley House, Main.

**Livery Stables.**

Reiber G. W., Main.

**Machinist.**

Robertson Wm., Main.

**Notions, &c.**

Rothstein Bros., & Co., Main.

**Oil Well Supplies.**

Richardson T., Main.
Backus, Morrison & Co., Main.
Robertson Wm., Main.
Sheffield & McCullough, Main.

**Saloons and Restaurants.**

Barnes Miss Nellie, Fairview.
Kennedy Charles, Argyle.
Jones J. Glancy, Argyle.
Levalle R. P., & Co., Argyle.
Moorhead & Dougherty, Main.
Reiber George W., Main.

## Petrolia February 1, 1873
## Central Hotel

Christie and Orr, Proprietors – first class new livery attached. The leading house and the best accommodations in the city. Sample room and good stabling. Charges moderate. Stages leave the hotel for Parker City, Brady, Greece City, Butler and other points in the oil region.

*Top, Richard Jennings; Left, George Nesbit; Right, Samuel D. Karns; and bottom, George Dimick, victor in the mayor's battle between Dimick and Ben Hogan in Petrolia City. (Sketches in Crude Oil, John J. McLaurin)*

## November 22, 1873
## The Jennings No. 19

On November 22, 1873, on the Daugherty farm between Petrolia and Fairview, a well owned by Richard Jennings absolutely reached the fourth sand at an estimated depth of 1,600 feet. This well, the "Jennings Number 19," in its first twenty-four hours produced 2,500 barrels of oil, and in its first month, produced 30,000 barrels. The Jennings well was one of the largest in Butler County and the first of hundreds that would eventually be drilled in the "cross belt" of the "lower region." (Oil City Derrick, Oil City, Pennsylvania)

## PETROLIA'S FRENCH MARY

It must be noted that Petrolia already had a Madam who had established a business just before Ben Hogan hit town. It was operated by Mary Marshall who became known as "French Mary." Her girls walked the streets of Petrolia from sun up to sun down soliciting customers. After Hogan arrived in Petrolia, Mary joined the female seminary crowd and soon became another of Ben's "lady friends," and Ben gave her his pet name, "French Mary."

The following interview was recorded between Aryl Ann Bashline and Lenn McCorry in 1977 by Vanity Press, Butler, Pennsylvania. Lenn G. McCorry was a well-known western Pennsylvania oil history expert. Lenn was nicknamed "Pithole" by his friends and was born February 22, 1889, in the St. Joe oil field near Butler, Pennsylvania. Mr. McCorry received many citations from the Pennsylvania Historical Commission for his oil field research and authored a book still sold by the Butler County Historical Society titled "Beautiful, Bountiful Butler County." This book is a must read for all Western Pennsylvania oil students. Lenn died in 1988 at the age of 99. "Peace to his ashes." The following is an interview about what Lenn remembered about "French Mary."

***Could this be Petrolia City's infamous "French Mary?" Even today, 2007, the hill out of Petrolia is called "French Mary Hill."***

Bashline: Tell me about Ben and "French Mary."

McCorry: Yes, Ben had a lady friend there in Petrolia. Now, I remember her and also as a young man saw her. She died, I don't know how many years ago. But, when I first went up in the Petrolia area, I can remember her. She lived in her house which is standing today right as you go into Petrolia on the left going in, a little home, there, handy the railroad depot. She was Ben Hogan's lady friend. And when he would put on a show of some kind there at his building, they called it "Hogan's Theater." Why, she would ride a horse through town. And she was well dressed, good looking lady and wore leather boots, spats, ya know. And she would ride around and mention about the show and other types of personal entertainment in the local oil fields. She done the advertising (laughs), and I remember she died and the County buried her in Butler. I don't know if she had any people left or anything here. But she was up in the eighties when she died. But that was back in the early twenties when she died. But I remember of seeing her; never got talking with her or anything. But this Hogan, he was supposed to be the toughest guy in the world. He was a powerful man. He was in every place where there was any money like that.

Bashline: Yeah.

McCorry: Like in the boat there, at Parker and those places. Yes, he was into everything. Yeah, this "French Kate," the one he knew up in Pithole, and then "French Mary" here at Petrolia (laughs). – End of interview

The hill going out of Petrolia, Pennsylvania, towards Karns City, is still called "French Mary Hill."

It has been reported that late on warm, moon-lit, summer nights, the lonely spirit of the beautiful "French Mary," although long gone, still rides side-saddle on her horse through the streets of Petrolia evoking her long lost loves. (The author)

## HOGAN'S SEMINARY

Benedict Hogan was born in Wurttemburg, Germany, in 1841 and with his parents came to America when he was four years old. Ben led a rough but exciting life and his fame among oil men was second only to that of Colonel Edwin Drake.

In the upper oil field Hogan got his start in the saloon and prostitution business at the famous Pithole City in Venango County, Pennsylvania, and in the lower fields at Parker City, Armstrong County. Ben had no formal education, had numerous scrapes with the law, spent time in jail, was a champion bare knuckle fighter and a spy, a gambler and proprietor of many well-known brothels, a vaudeville entertainer, hustler, con artist and finally in later years an evangelist. Now we know why it was reported that Hogan led a fast and exciting existence and to top things off, add a dash of French Kate, Kitty Bowers, Lizzie Toppling and French Mary.

Ben Hogan knew the ins and outs of making a fast dollar and concentrated his financial endeavors in providing therapy for the lonely, hard-working men who after the Civil War had been attracted to the western Pennsylvania oil fields. Hogan knew that where the oil activity was feverish, there were men with money and no problems spending it.

It was the "Fannie Jane" well that came in at 200 barrels on April 17, 1872, and caused Petrolia to spring up overnight. Within a year Petrolia's population had reached 2,000 and the boom town had developed the most notorious reputation of any place in Butler County. Each day brought on more gambling, drunkenness, shootings, gun fights, stabbings and brawls. When Hogan arrived in Petrolia he purchased a small shack-like house from a Mr. Lauterbrick for the total sum of three hundred dollars, added a bar, stocked in the liquor, and put his girls to work. As business picked up, Hogan spent $11,000 and built a larger building with fifteen bedrooms, a large dance floor, and a large well-decorated bar. This brothel was built at the head of Main Street and became known by several different names, including Hogan's Ranch, Hogan's Opera House, and Hogan's Female Seminary. It is written that on a trip to Petrolia, an old farmer's wife noticed the presence of several young ladies in the large new building and asked Hogan what sort of place it was. One of Hogan's girls overheard the question and before Hogan could speak up, replied, "This is a female college, a seminary." Thus Hogan's house of ill repute was thereafter known as "The Female Seminary."

Another time Hogan's Seminary of ill fame came under fire from local law enforcement and Hogan was hauled off to county court. In his own defense, Hogan explained the situation to the judge that he was only running "a gymnasium where different members of both sexes may enjoy wholesome exercise, using the different parts of the body in such a way as to bring all the muscles into play." He informed His Honor solemnly that if people would spend a few hours each week in his gym, it would do them "more good than all the pills in the world." The good judge must have been a fan of wholesome exercise, seemed satisfied with Hogan's defense, and fined him only a small fee.

# BEN HOGAN—THE LEGEND IN BUTLER COUNTY

The oil boom history of Butler County would not be complete without the mention of good old Ben Hogan. After his appearance in Parker City in 1870, Hogan self proclaimed himself "the man from Hell." In Parker City, Armstrong County, Ben's escapades were well documented in all the oil region newspapers with the grand "Floating Palace of Pleasure" leading the way in wickedness. Ben Hogan was ditched by his old partner in crime, "French Kate," and now joined in the sale of "wine, women and song" by the lovely, but street-wise, Kitty Bowers. Kitty was Ben's right-hand girl in his Butler County business days. Ben Hogan often remarked that in Butler County the "oil was more plentiful, the whiskey was stronger, and his girls were the prettiest."

As the oil belt was expanded south through Butler County, more oil towns sprang up. Martinsburg, Petrolia, Karns City, Modoc City, Troutman, Buena Vista, Greece City, Argyle, Millerstown and St. Joe. In 1872, Hogan moved his operations to Petrolia which was at that time the most lawless patch of oil soaked dirt in the new field. Petrolia in 1872 was the roughest oil town and Hogan's place was the roughest, vilest, toughest dive in town. The St. Petersburg Progress dated November 22, 1872, reported that "acts of violence are becoming all too frequent in Hogan's bawdy house, and he should be imprisoned or maybe even lynched as punishment for his acts."

## Mayor Ben Hogan

Instead of lynching Ben, the rough elements, the hoodlums and saloonkeepers ran Hogan as a candidate for Burgess when Petrolia officially became a borough in February, 1873. Ben lost the election and charged that the vote count had been rigged. Still onward Hogan and Bowers pushed their houses of ill repute. In 1873 they opened a wine room in Greece City and two variety theaters, one in Petrolia and one in Millerstown. About this time Hogan became disgusted as new laws were passed against his kind of monkey business and police were becoming more common. Ben and Kitty formed a variety troupe and toured the middle west, but they missed the oil fields and in 1876 returned only to lose his money when Kitty abandoned him for a saloonkeeper in Petrolia.

Ben Hogan ran against George H. Dimick, a well-known oil operator who was secretary to the famous oil man Henry R. Rouse. Rouseville, above Oil City in Venango County, was named after him and Henry was burned to death in an oil well fire near Rouseville in 1861.

***NEW SHOES***
*A dining-room girl at Petrolia in 1873 stood on top of a derrick to win a new pair of shoes being offered by a jovial oilman to the first fair maiden entitled to the prize. Note the maiden nearing the top of the derrick and can you imagine what the young girl's mother must have exclaimed upon hearing the good news!*

*Bird's Eye View of Petrolia City*

Today, in 2007, the Henry Rouse estate still provides income to run the Rouse Estate and Personal Care Center near Youngsville, Pennsylvania, and provides personal care to Warren County's less fortunate.

## Hogan's Lost Race

The well-known oil region newspaper man, John J. McLaurin, however, declares that Hogan really won the contest but was counted out by the election board. A letter from Hogan appeared in the Oil City Derrick a few days later:

"It was my desire, if elected, to reform and become a good citizen and law-abiding man, and to bring myself before the public to show the people at large that a pugilist's mind does not run entirely to the art of self defense. I have studied human nature for the last twenty years, having had a great amount of experience in my travels in Europe and all over America, with both men and women. I am defeated for the present, which has been many a political man's case. It is evident there was foul play, for in counting the votes at 4:30 in the afternoon I was forty votes ahead. I return thanks to my friends and fellow citizens for their liberal support. The number of votes I got goes to show that there are men of good judgment in Petrolia. It is acknowledged that this election caused more excitement than any ever held in Butler County."

For more information, read "The Golden Flood" by Herbert Asbury, 1942.

## Row at the Butler County Castle
## A Revolver Drawn by a Prostitute
## Ben Hogan Shot, His Injury Not Dangerous
## Oil City Derrick, May 7, 1874

"A row occurred at the Butler County Castle in Petrolia at one o'clock this morning which resulted in the shooting of Ben Hogan, the famous pugilist. The trouble originated in consequence of a dispute between a prostitute and a male lodger, whose loud words soon brought another of its inmates, known as 'Vic' Woodhull, to the scene of the disturbance, in which she took sides against the degraded masculine specimen. The rumpus soon grew so boisterous that Ben's innocent slumbers were disturbed and he appeared upon the scene just in time to prevent the murder of the young man. Vic held in her hand a seven shooter, already cocked and aimed at the object of her fury. Ben rushed in and placed himself between the infuriated parties and received the shot which would perhaps have sent into eternity the pimp for whom it was intended.

"The ball entered the left side of the abdomen, about two inches above the pansits ligament, the physicians in attendance say a down and inward course. The wound caused but little pain and no fears are entertained by the medical attendance as to his recovery.

"The girl who committed the crime is a resident of Kittanning and has only recently left her home. She appeared before the Burgess this afternoon and was fined ten dollars for disorderly conduct."

## Petrolia
## Oil City Derrick, May 8, 1874

"Ben Hogan's new opera house is rapidly approaching completion, and will be formally opened some time this month should the gay and frisky proprietor not in the meantime be dispatched to the Kingdom Come by some row like that narrated in another column. The building is much the largest in the settlement and will be finished throughout in good style. The ground floor contains a big sample room and several dozen sleeping apartments, while the upper story is entirely devoted to the purposes of an opera house. A commodious stage occupies one end, with a gallery at the other extremity and sitting room between for over a thousand persons."

## Ben Survives Shooting
## Petrolia
## Oil City Derrick, May 22, 1874

"Ben Hogan has recovered from his little shooting."

## Petrolia's 1,100 Seat Opera House Opens
## Oil City Derrick, May 22, 1874

"The opening of the Opera House on Monday evening was one of the events of the season. The services of a talented company had been secured for the occasion and by the appointed hour over a thousand persons comprising the best citizens and many of the leaders of Petrolia and vicinity, had convened in the spacious auditorium.

"To those who had not previously visited the hall, the spectacle was a gratifying surprise. Eight hundred chairs occupied the body of the grand apartment, besides which the gallery at the east end contained seats for three hundred people. From the walls and ceilings numerous chandeliers and bracket lamps were suspended, affording excellent light in every quarter of the room. The large stage was amply supplied with handsome scenery which attracted universal admiration. The entertainment was highly enjoyable and successfully giving full satisfaction to the finest and most respectable audience ever assembled in this section of the country.

"The building is two stories in height, thirty five feet wide by one hundred forty feet in length and situated near the center of town. The lower floor is divided into a large sample room, twenty sleeping apartments, several parlors and two dressing rooms connecting with the stage, the whole furnished in elegant style. Its erection is a credit to the progressive borough to whose varied attractions it must prove a valuable addition. It is designed to provide standard entertainments for the benefit of the community and to this end fresh entertainment will be presented.

"A party will be given at the close of the performance tonight, under the auspices of leading citizens who have very properly determined to reward in a fitting manner the common table enterprise that has bestowed upon their town the most inviting and extensive place of amusement in the lower oil districts." (John J. McLaurin, Correspondent)

## PETROLIA
## Oil City Derrick, May 25, 1874

"The new opera house was filled last night with delegations from all the neighboring oil towns, Karns City especially being represented in force. In point of number enthusiasm and genuine appreciation the proprietors have been rewarded with the success they so much deserved. Great and beneficial changes were made in the program which may be deemed the sole cause of success. 'Sketches in India' the opening comedy, was well rendered and elicited much applause. It would be unjust to praise any particular actor of this play as they all maintained their parts as well as the nature of the piece would warrant. The 'Egyptian Box Mystery' took many a fellow down who supposed the matter could be easily solved. The Henri family were most conspicuous by their excellent dancing and we are free to admit that Miss Flora Henri can well make the famous Jennie Benson look to her laurels on a 'heel and toe jig.' Frank Giblers, although a fine gymnast, particularly on the trapeze, did not have an opportunity to show many of his feats, the ceiling not being sufficiently high, but what he did do was enthusiastically received. The concluding piece, 'Servants by Legacy,' was extremely comic, and produced more laughter than any of the programme.

"From the great numbers of ladies that witnessed the performance and their appreciation of all that was said or done, we can safely predict a financial success to those undertaking this hazardous scheme."

## The Opera House – Petrolia
## Oil City Derrick, June 2, 1874

*The famous Ben Hogan*

"Hogan still continues a success with his splendid combination troupe, although the houses are not so well filled, in our opinion, as they would be with other than such strict temperance dramas as 'Ten Nights in a Bar Room' on the bills. Oildom at best is not a rigid temperance community and the morals of such places are seldom if ever appreciated as they should be.

"Last night Ben took a benefit and received a splendid testimonial. An extra train was run from Parker, which brought out many ladies and gentlemen, who were highly pleased with the performance in every particular. 'Twenty Years of a Drunkard's Life' came first on the program and the elegant manner in which it was rendered received great applause. Every member of the company did well, and were highly appreciated by the audience. Most of the company will be engaged for the coming week, when many new attractions will be presented."

## Petrolia, Pennsylvania
## The Derrick - June 2, 1874

"In Petrolia, gin mills are getting as thick around the Parker and Karns City railroad depot and vicinity as bed bugs in our boarding houses. It was reported that at one saloon the owner had eight assistants and was still having trouble keeping his patrons happy and their glasses full. Each day forty lager beer kegs are shipped through the express office of the Parker and Karns City Railroad at Parker City.

"To ensure that if the kegs went dry, S. W. Seldner's wholesale liquor house advertised that the following stock had just arrived. Twenty-five barrels of Seldner's celebrated Derrick whisky; 15 barrels of S. W. S. whiskey, 5 barrels of sweet Catawba wine, and last but not least, 5 barrels of tasty port wine.

"Mr. Ludwig Mayer, a wholesale beer distributor, advertised that, 'My customers will please take notice that I am now able to fill their orders in large or small quantities of beer, in quart and pint cases, one eighth, one fourth, and one-half barrels. Parties wishing beer for the coming fourth of July will please send their orders three or four days in advance. Also, return cases as soon as empty in order to fill orders promptly for the coming Fourth of July. Sole wholesale agent of Philadelphia's Best and Milwaukee Beer.'"

# Derrick Specials
# Nitroglycerine Horror, Another Torpedo Victim
# July 7, 1874

"Another nitroglycerine explosion occurred today, adding still one more to the long list of victims by this explosive. The man killed was named Alonzo Taylor, a moonlight torpedo agent from St. Petersburg. He was unloading an old shell this afternoon at Troutman, when it exploded, breaking one of his legs, cutting his head, and otherwise injuring him. He lingered three hours, suffering intense pain, when death relieved him." (John J. McLaurin)

Alonzo Taylor was laid to rest and "sleeps the sleep that has no awakening." Alonzo is buried in the Union cemetery at St. Petersburg, Pennsylvania. "Peace to his ashes." (John J. McLaurin)

# Young Man Shot – Fire at Petrolia
# July 14, 1874

"Fire was discovered issuing from Carlin Brothers stable on Jamison street at one a.m. this morning. The alarm was promptly sounded by the United Fire Company, which was not long in making its appearance. The water was supplied from the creek, by a pump owned by the United pipe line. The horses and the harnesses were all saved. A small blacksmith shop adjoining the stable was partially saved.

"Loss on both buildings will not exceed five hundred dollars. The efficiency of the fire company may be estimated from the fact of the fire extending no further than the building wherein it started. This was the work of an incendiary.

"A young man named Ira Peterson was accidentally shot today, the wound being in the right side. Its extent is not exactly known, but Drs. Willard and Oldfield seem to think it will prove fatal." (Oil City Derrick)

# LOADED UP

Frederick E. Hinman of Petrolia, Pennsylvania, manufactured millions of pounds of nitroglycerine and shot thousands of wells. He manufactured in his 86 years no less than 2,400,000 pounds of the mighty explosive and he lowered it into wells scattered over the oil regions with no fewer than 5,500 shots.

A good many men were blown up shooting wells and one reason was that shooting wells at night is always more dangerous than shooting in the day. If you were not a Roberts Torpedo Company employee, it was illegal to shoot wells on your own, thus many wells were shot at night in order to not be caught. Another problem that added to the danger of the job was some shooters drank too much and became careless.

Mr. Hinman was asked one day if torpedo men would go to shoot while under the influence. To this Frederick replied, "I have known many to go out pretty dam well loaded up." Some shooters went home at the end of the day in one piece, others did not. (Oil and Gas Man's Journal, Butler, Pa., 1910)

## Picnics and Prize Fights

Ben enjoyed visiting the Brady's Bend picnic grove, located up the Dewey farm road on the hill overlooking the river. There, a large wooden platform was used for public picnics and dancing. The grove drew large crowds and a fourth of July celebration often lasted several days.

Ben was a champion bare fist boxer and according to the late Butler historian, Lenn "Pithole" McCorry, a ring was roped off and a prize fight was held in May of 1874 in Shakey Grove, Petrolia. The notorious Ben Hogan, a national champion, was the referee and "Big Red Mullen," a driller, fought "Stop Cock" Jim Cummings, a pipe gauger who was also known as "Lightning Jim." There were over 1,500 people in attendance and more than $50,000 had been wagered on the fight. After four hard-fought rounds, "Lightning Jim" Cummings was declared the winner by referee Ben Hogan.

The fight started over little Johnny Moore, a crippled newspaper boy known as "Rusty." Rusty sold newspapers at the Central Hotel in Millerstown, and was struck and knocked down by Red Mullen in the presence of "Lightning Jim." The fight wagers brought "Lightning Jim" enough money to buy part interest in a new refinery at Titusville and send poor little crippled Rusty to a doctor. Later "Lightning Jim" married Rusty's sister, "Sis" Moore, and moved to Titusville to manage the refinery. (Oil and Gasman's Magazine, May 1911, Butler, Pennsylvania)

## One New Year's Night, A Story of the Early Days of Petrolia, Pennsylvania The Oil and Gas Man's Magazine, Butler, Pennsylvania 1909

Ben Hogan's Female Seminary building in Petrolia was built for the sole purpose of conducting immoral purposes. With the striking of a gusher field at Bullion, Ben had flitted there and Summit City became Hogan's place of business. Here, Ben allured the young from the straight and narrow.

*The Female Seminary*
*Petrolia, Pennsylvania*

*"Take that!"*

*Fairview Hotel, 1877*
*Mains Street Fairview, Pennsylvania*

***Sitting Bull and Buffalo Bill, 1885.***
***Buffalo Bill is holding a 1873 Winchester rifle.***
***Courtesy of the Buffalo Bill Historical Society,***
***Cody, Wyoming, Vincent Mercaldo Collection***
***"Show your cards Bill"***

## Buffalo Bill in Petrolia City

His Petrolia building was remodeled into an opera house and Ben played all the leading acts there. One night Buffalo Bill and his show played at Hogan's Opera House to an overflowing crowd. The place was packed with oil men, their wives, and excited wide-eyed children.

Buffalo Bill was no stranger to the "wild and wooly" side of life. After the show, Ben Hogan and his gambling friends invited Buffalo Bill to a friendly game of poker and it was said that he, Buffalo Bill, dropped the entire receipts of the evening in the poker match with the Petrolia gamblers that very same night.

## Petrolia, June 4, 1875
## The Derrick

"Everyone loves the fourth of July with the crack, crack, crack of the fireworks. This special reporter reminds all the citizens of Petrolia, Karns

City and Millerstown that one month from today it will be lawful to shoot fire crackers, get drunk, and pay your fine."

*A Butler County oilman being lured into the arms of "Crazy Nell."*

## All Your Hardware and Grocery Needs Filled At Tinker and Duncans The Peoples' Cash Grocery and Hardware Corner of Main and Jamison Streets, Petrolia, Pa.

"We carry one of the oil region's largest stocks of goods and grocery needs. Owing to extensive trade our goods are always fresh and new.

"We also carry at a bargain second hand spring wagons, stoves and tinware, iron and nails, table cutlery, oil stoves, and ice chests.

"Stop at Tinker and Duncans, Petrolia, Pa."

## "Crazy Nell"

Nell Robinson, in the spring of 1869, hired Ben Hogan as a bouncer and bartender at her resort at Petroleum Center. Nell later resurfaced in 1872 in Antwerp City, Clarion County. After Antwerp burned in 1873, Hogan rented his resort at Petrolia City to Nell Robinson of Petroleum Center in the summer of 1873 and opened his wine room at Greece City, a boom town some eight miles south of Petrolia, which was founded, grew to fifteen hundred population, and vanished all within thirteen months. "The place was purely a wine room," said Hogan's biographer. Ben soon tired of the wine room and sporting women and returned to the prize ring. (See page 48)

## Ben's Conversion

*Could this actually be Mary Stockman, Ben's beautiful wife? It would have taken a very intelligent and beautiful woman to tame Ben.*

During the summer of 1878, Hogan was visiting New York City where he entered a crowded theatre thinking it to be a place of amusement. To his surprise, it was a religious service, a revival meeting, conducted by Charlie Sawyer. Never had Hogan been a church-goer. He thought long and hard and finally the Lord entered his heart. The "Wickedest Man in the World," Ben Hogan renounced his evil ways, took the pledge against strong drink, and became a born-again Christian.

Ben became a traveling evangelist and toured the nation telling his story of conversion from sin to the path of glory and salvation. Sometime between 1878 and 1881, at the age of 40, Hogan married Mary Stockman, aged 29, who helped Ben in his new way of life. By 1886 records show he and Mary were running Ben Hogan's Mission in Chicago, a far cry from Ben Hogan's Female Seminary. The Hogans continued to preach, taught Bible School, counseled alcoholics, and fed the hungry. Ben died in Chicago on November 1, 1916, age 75. "Peace to his ashes."

## Fairview, Butler County

Fairview is situated one mile south of Petrolia on a commanding position, affording a delightful prospect of the country for miles around, from which circumstances the place derived its name. Fairview is of old date, a settlement having established itself some seventy years ago, when the country around was nothing but a howling wilderness. Most of the buildings were located on one street, and generally presented the appearance of considerable age.

Fairview had one good hotel, a tavern or two, one bank, a church and schoolhouse, and numbered in population about 800.

This pretty country village, sober, quiet and unpretentious, dates its foundation as far back as the year 1826, when James McElwee started a small grocery at the crossroads. He sold whiskey; the whiskey produced fits, as it invariably does, and made business for the Squire. In 1830, Thomas McCleary, Esq., and William Hawk caused town lots to be laid off on portions of their farms, and soon a small but prosperous village sprang into existence. He kept the first tavern. When he engaged in this business, he had not much to start with, not even a glass for his customers to drink out of. When a man called for a drink of whisky, a flask was handed to him, and he gauged the drink according to his own liking.

A number of destructive fires visited the town, destroying two hotels and other buildings.

Fairview grew but slowly until 1872, when the oil developments began to lend impetus to its prosperity. Previous to the date mentioned, probably the population of the village was at no time over 200; in 1876, it was estimated that there were 1,000 people living in the town. During the oil period, there was great business activity and commendable progress in all direc-

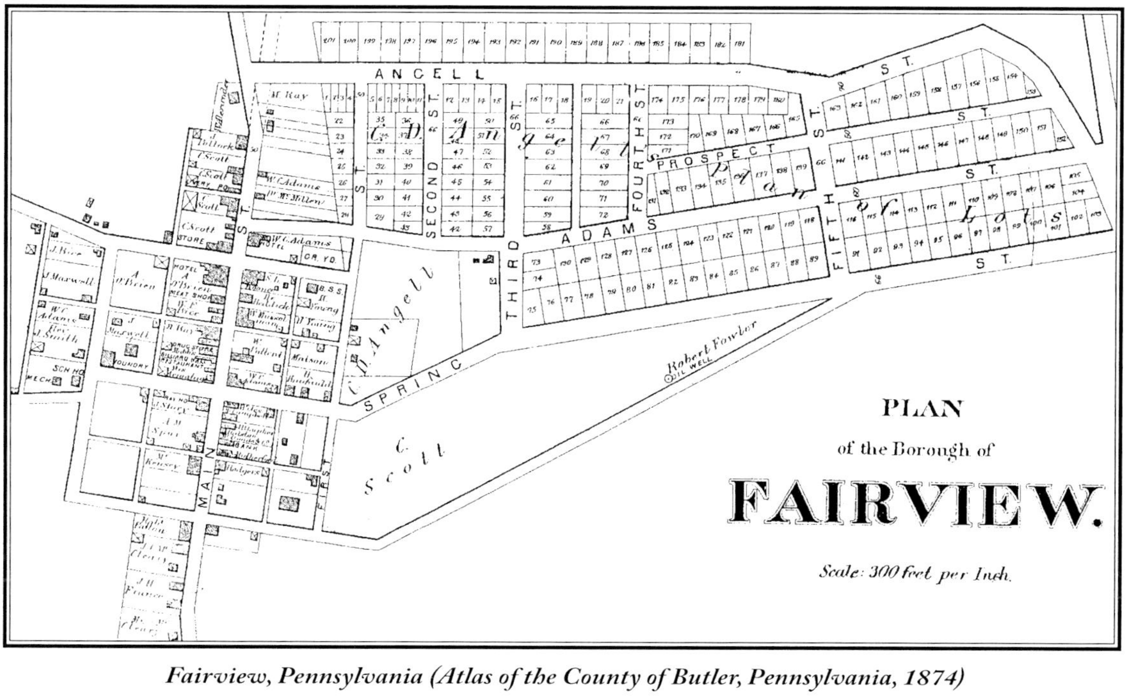


*Fairview, Pennsylvania (Atlas of the County of Butler, Pennsylvania, 1874)*

tions. Many business houses appeared, and a large number of tasty and elegant residences were erected. Nearly all of the old buildings of the town disappeared and were replaced by new ones of a better kind.

In 1872, the borough of Fairview was organized, and John Pollock elected the first Burgess.

Natural gas was used here for light and domestic purposes in 1872. It was among the first in the nation. Gas was from the Indian Spring well, one mile west of town on the Campbell farm.

The only attempt to start a newspaper in this place was made in 1872, when the Fairview *Reporter* made its appearance. Col. Samuel Young was proprietor, and R. W. Criswell, now engaged with the Cincinnati Enquirer Company, editor. The *Reporter* survived about one year.

In June, 1872, the "Fairview Pipe Line," a two-inch, from Argyle pump station to East Brady, was five miles long and there were eight storage tanks. A line to West Monterey, four miles, four storage tanks on the east side of river. On September 2, 1872, it was extended to Greece City, seven and one-half miles.

The Fairview Deposit and Savings Bank was established August 15, 1872. It was a private bank, managed by Ralston, McQuaide & Co.; D. A. Ralston, President, and R. W. McKee, Cashier. This bank closed March 15, 1882; liabilities about $225,000, and assets very small. This was the most damaging blow ever received by Fairview.

The situation of Fairview, away from railroads, and the springing up of towns near by, and the decline of the oil business, contributed to diminish the prosperity of the town. (History of Butler County, Pennsylvania, 1883; Waterman, Watkins and Co., Chicago, Illinois)

# FAIRVIEW.

This town is one of the oldest in Butler county, and possesses residential advantages superior to any of the towns in the oil region. Being situated on an eminence overlooking the valleys in which the towns of Petrolia and Karns City are built, commands a view of the surrounding country, which presents a scene picturesque and beautiful; and being located contiguous to the heart of the oil producing region, the advantages for business purposes are good.

The commercial pursuits are represented by several flourishing stores, while the manufacturing departments are doing a thriving business.

The hotel accommodations here are not to be overlooked. The Ray House and McFann are well kept hotels, and well known in this county. The Adams House is a new building, and will be opened for the reception of guests this fall. It will be the most commodious hotel in this part of the country.

## A

ADAMS HOUSE, W. C., Adams proprietor, Main.
Adams J. M., stock dealer, Main.
Adams John, teamster, Harrisville Road.
ADAMS W. C., proprietor Adams House, Main.
Alexander Alex. wagon maker, Main.
Alexander C. C., books and stationery, Main.
Alexander Wm., wagon maker, Main.
Angell C. D., oil operator, office Main.
Angell E. A., office Main.
Arthurs Richard J., livery stable, near Main.
Avery S. S., attorney, office Main.



## B

Badger J. L., butcher, Harrisville Road.
Badger Sylvester, laborer, Harrisville Road.
Baker Albert, copper and tinsmith, boards Main.
Baker Jennie, dressmaker, Main.
Baird Byron, teamster, Harrisville Road.
Bemus George H., attorney, office Main.
Bemus W. M., student, Main.
Beatty Dr. J. F., physician, Main.
Bice Abram, carpenter, boards Harrisville Road.
Bice John, carpenter, Harrisville Road.
Bice W. F., saddles and harness, Main.
Bice W. N., clerk, Main
Birchard H. C., physician, Main.
Black Henry, teamster, Main.
Bloomingdale & Marks, clothing and furnishing goods; Main.
Bloomingdale Sol., of Bloomingdale & Marks, Main.
Brayton M. D., clothing and etc., Main.
Brown J. F., cigars and tobacco, Main.
Bruce W. J., painter, Main.
Bullock W., oil operator, boards Adams House.
Byers J. M., wagonmaker, Petrolia.

## C

Campbell J. G., dress maker, Main.
Campbell J. G., of Wiles & Campbell, Bradys' Bend.
Carmichael J., clerk, boards Ray House.
Carter Thomas, painter, boards Adams House.

Chew Thomas, blacksmith, Harrisville Road.
Christy John A., drug store, Main.
Clark C. B., pumper, Angell avenue.
Clarke John, machinist, Sunbury.
CLARKE & MAXWELL, foundry and machine shop, near Main.
Clarke Peter, of Clarke & Maxwell, near Main.
Coates J. C., watchmaker, Main.
Collins Eugene, bootmaker, Main.
Conway G. P., clerk, boards Main.
Cowen Alexander, carpenter, boards Adams House.
Crocker T. B., stage driver, Sunberry.
Cross B. R., photographer, near Main.



## D

Dean Isaac, contractor, Angell avenue.
Dufford J. H., bootmaker, Main.

## E

Eden John, carpenter, boards Adams House.
Ervin John, furniture, Main.
Ervin William, laborer, Harrisville Road.
FAIRVIEW DEPOSIT BANK, of Ralston, McQuaide & Co., Main.
Fairview Hotel, A. O'Brien, proprietor, Main.
Farrell Boston, mechanic, Main.
Fleming Wm., oil operator, Main.
Folsom M. J., dressmaker, Main.
France O. P., carpenter, rear U. P. Church.

## G

Genther Charles F., New York Tonsorial Rooms, Main.
George J. B., clerk, boards Main.
Gibson Harrison, moulder, Main.
Graham John R., carpenter, Main.
Griswold Burt, driller, Angell avenue.
Griswold Grant, oil producer, Angell farm.
Griswold Harry, oil producer, Angell farm.
Griswold Orr, oil producer, Angell farm.
Grow A., of G. N. Grow & Bro., Main.
Grow G. N., of G. N. Grow & Bro., Main.
Grow G. N. & Bro., bakers, Main.
Guthill Louis, barber, Main.

## H

Haskell A. M., superintendent Fairview Coal Mine, boards Adams House.
Hart Charles H., operator, east of Main.
Hites J. L., carpenter, rear U. P. Church.
Hopkins Dr. T. W., office Main.
Humphries D., grocer, Main.

## J

Jacobs Lemned, driller, Harrisville Road.

## K

Kaufbold Henry, carpenter, east of Main.
Kelly Edward of W. Kelly & Bro., Main.
Kelly W., of W. Kelly & Bro., Main.
Kelly W. & Bro., boots and shoes, Main.
Keys George, carpenter, Harrisville Road.
King Frank, blacksmith, boards Main.
Knox James M., pumper, Angell avenue.
Kuntz Elijah, carpenter, Harrisville Road.

## L

Larsen J. M., office Main.
Lauders George, butcher, Main.
Lecky Wm. M., oil operator, office Main.
Lockridge D. M., book-keeper, Main.
Loudon David, B., carpenter, east of Main.
Lovelace C. R., of Main & Lovelace, Main.
Luce Willis, restaurant, Main.
Lupher J. D., hardware and oil well supplies, Main.
Lupher Lee, teamster, Main.
Lupher & Pitcher, boilers, engines, &c., Main.

## M

Mabbs H. W., billiard saloon, Main.
Madison Thomas, east of Main.
MAHAFFY A. G., lumber dealer and justice of peace, Adams, residence Third.
Mahaffy J. C., teamster, Adams.
Main S. B., of Main & Lovelace, Main.
Main & Lovelace, livery stable, Main.
Marks H., of Bloomingdale & Marks, Main.
Maxwell G. C., of Clarke & Maxwell, Sunbury.
McCLEARY J. J., dry goods, groceries, etc., Main.
McClure S. M., dress making, Main.
McCoy David, coal digger, Angell avenue.
McCrackin Joseph, teller, Fairview Deposit Bank, Main.
McFANN HOUSE, Main.
McKee R. W. cashier, Fairview Deposit Bank, Main.
McKinsey George, teamster, Main.
McMILLAN T. M., physician and sergeon, Main.
Miles T. D., carpenter, rear U. P. Church.
Miller A. S., laborer, near Main.
Miller George C., stage driver, Sunbury.
Mills Joseph, blacksmith, Harrisville Road.
Moore Harper, manager coal banks, Main.
Montgomery Alexander, livery stable, Main.
Morrison John, barkeeper, boards McFann House.
Musselman W., restaurant, Main.

## N

Neill David, wagon maker, boards, Main.
Nelson P. K., proprietor McFann House Main.
Nichelson Andrew, carpenter, Angell avenue.
Nutting Alonzo, stage driver, Sunbury.

## O

O'Brian A., proprietor Fairview Hotel, Main.
O'Brian George, clerk, boards Main.
O'Brian James, clerk, boards Fairview House.

## P

Palmer James M., harness maker, Main.
Patton ——, carpenter, Main.
Patton D. L., clerk, boards Ray House.
Patton L. D., livery stable, Main.
Patton Robert, Sr., Main.
Patton Robert, Jr., farmer, Main.
Perkins W. W., oil operator, office Main.
Pettit S. L., laborer, Main.
Pitcher E. .L, of Lupher & Pitcher, res. Parker City.
Pollock John, oil operator, Main.
Pollock William J., carpenter, Angell Farm.
Post Office, Main.

## R

RALSTON McQUAIDE & CO., bankers, Main.
RAY HOUSE, Thomas H. Ray proprietor, Main.
RAY THOS. H., proprietor Ray House, Main.
Ray Matthew, Justice of the Peace, Main.
Reineman Isaac, oil operator, office Main.
Riddle J. R., photograph gallery, near Main.
Rimbey Louis, clerk, boards Main.
ROBERTS A. W., of Roberts Bros., Main.
ROBERTS BROS., cigars and tobacco, Main.
ROBERTS G. H., of Roberts Bros., Main.
ROBERTS' HALL, Main.
Roberts John, teamster, boards Sunbury.
Robinson William, carpenter, Harrisville Road.
Rodgers A., blacksmith, Main.
RUTHERFORD JAS., hardware and house furnishing goods, Main.

## S

Saeger Ferry, carpenter, near Main.
Scott C., dry goods and groceries, Main.
Scott Wm. H., clerk, boards Main.
Selven A. C., surveyor and conveyancer, officer Main.
Selven C. C., oil operator, boards Fairview House.
Shull W. H., cigars and tobacco, Main.
Skillman G. F., butcher shop, Main.
Smith J. C., lumber dealer, near Main.
Smith John Rev., Sunbury Road.
Smith & Van Worner, lumber, near Main.
Sperr Captain, oil operator, Main.
Staley George, blacksmith, boards Sunbury.
Stewart D. J., grocery, Main.
Stevick Frank, coal miner, east of Main.
Stillwell E. L., restaurant, Main.

## T

Thorn John N., laborer, Angell Farm.
Thorn William, wagon maker, Angell Farm.
Tiffany O. E., telegraph operator, Main.

## V

Vandergrift W. K., oil operator, office Main.
Varner John, moulder, boards Sunbury.

## W

Weaver Noah, driller and contractor, Angell avenue.
Weaver Thomas, Angell avenue.
Western Union Telegraph Company, Main.
Wheeler George B., book-keeper, boards Ray House.
Wheelock, N. S., teamster, Harrisville Road.
Wick J. H., teamster, Harrisville Road.
Wiles & Campbell, dry goods and groceries, Main.
Wiles R. J., of Wiles & Campbell, Main.
Waynard J. V., manager, Sunbury.
Wood J. A., clerk, Main.

## Y

Young Hugh, blacksmith, rear U. P. Church.

# FAIRVIEW BUSINESS DIRECTORY.

**Attorneys.**

Avery S. S., Main.
Bemus George H., office Main.

**Bakery.**

Grow G. N., & Bro., Main.

**Banks.**

Fairview Deposit Bank, Main.

**Barber Shop.**

Genthner Charles F., Main.

**Bath Room.**

Genthner Charles F., Main.

**Billiard Hall.**

Mabbs H. W., Main.

**Books and Stationery.**

Alexander C. C., Main.

**Boot and Shoe Makers and Dealers.**

Bloomingdale & Marks, Main.
Kelly W., & Bro., Main.

**Butchers.**

Lauders George, Main.
Skillman G. F., Main.

**Cigars and Tobacco.**

Brown J. F., Main.
Roberts Bros., Main.
Shull W. H., Main.

**Clothing.**

Bloomingdale & Marks, Main.
Brayton M. D., Main.

**Drug Stores.**

Christy John A., Main.

**Dry Goods.**

McCleary J. J., Main.
Scott C., Main.
Wiles & Campbell, Main.

**Engines, &c.,**

Lupher & Pitcher, Main.

**Furniture.**

Ervin John, Main.

**Grocers.**

Humphries D., Main.
McCleary J. J., Main.
Scott C., Main.
Stewart D. J., Main.
Wiles & Campbell, Main.

**Hardware.**

Rutherford James, Main.

**Hotels.**

Adams House, Main.
Fairview Hotel, Main.
McFann House, Main.
Ray House, Main.

**House Furnishing Goods.**

Rutherford James, Main.

**Justice of the Peace.**

Mahaffy A. G., Adams.

**Livery Stables.**

Arthurs Richard, Jr., near Main.
Main & Lovelace, Main.

**Lumber Dealers.**

Mahaffy A. G., Adams.
Smith J. C., near Main.

**Machinists.**

Clarke & Maxwell, near Main.

**Oil Well Supplies.**

Lupher J. D., Main.
Rutherford James, Main.

**Photographers.**

Cross B. R., near Main.
Riddle J. R., near Main.

**Physicians.**

Beatty Dr. J. F., Main.
Birchard Dr. H. C., Main.
Hopkins Dr. T. W., office Main.
McMillan Dr. T. M., Main.

**Saloons and Restaurants.**

Luce Willis, Main.
Musselman W., Main.
Roberts Bros., Main.
Stillwell E. L., Main.

**Wagon Makers.**

Byers J. M., Petrolia near Main.

**Watchmakers.**

Coates J. C., Main.

# Haysville

The development of the oil resources of this township began in 1871, and in the five or six succeeding years great social and financial changes were wrought. During the excitement, many of the old residents disposed of their farms and removed to other parts of the country. Others remained and reaped great financial gains, thus laying the foundation of future prosperity, while not a few embarked upon the treacherous sea of speculation and lost the earnings of long years of constant toil.

This little village, on the farm of Thomas Hays, was built by the oil developments in its vicinity. At one time the place consisted of fifty or more houses, two hotels, stores, shops, machine shops, etc.

The first store in the place was conducted on a small scale by John McCorkle. In 1879, W. G. Hays was the only merchant in the place. He began business in 1875.

N. W. Krouse's machine shop and store for oil well supplies is the most important business establishment in the place. Mr. Krouse came to Haysville in 1876. The village was then at its best.

The farm home of Senator Thomas Hays stands to the left. In this area and on the Hays land 125 oil wells were drilled. Senator Hays's town mansion was in Fairview.

On December 15, 1873, Johnson and Shaw hit the fourth sand and a well named the "Kitty Ann" was the largest in the field here.

After 1879, Haysville declined considerably, though there was still considerable business transacted there.

(History of Butler County, Pennsylvania, 1883; Waterman, Watkins and Co., Chicago, Illinois)

*Haysville, Butler County Oil Town*
*Named after Senator Hays, his farm house is located at the left of this photograph, 1875.*
*(Butler County Historical Society)*

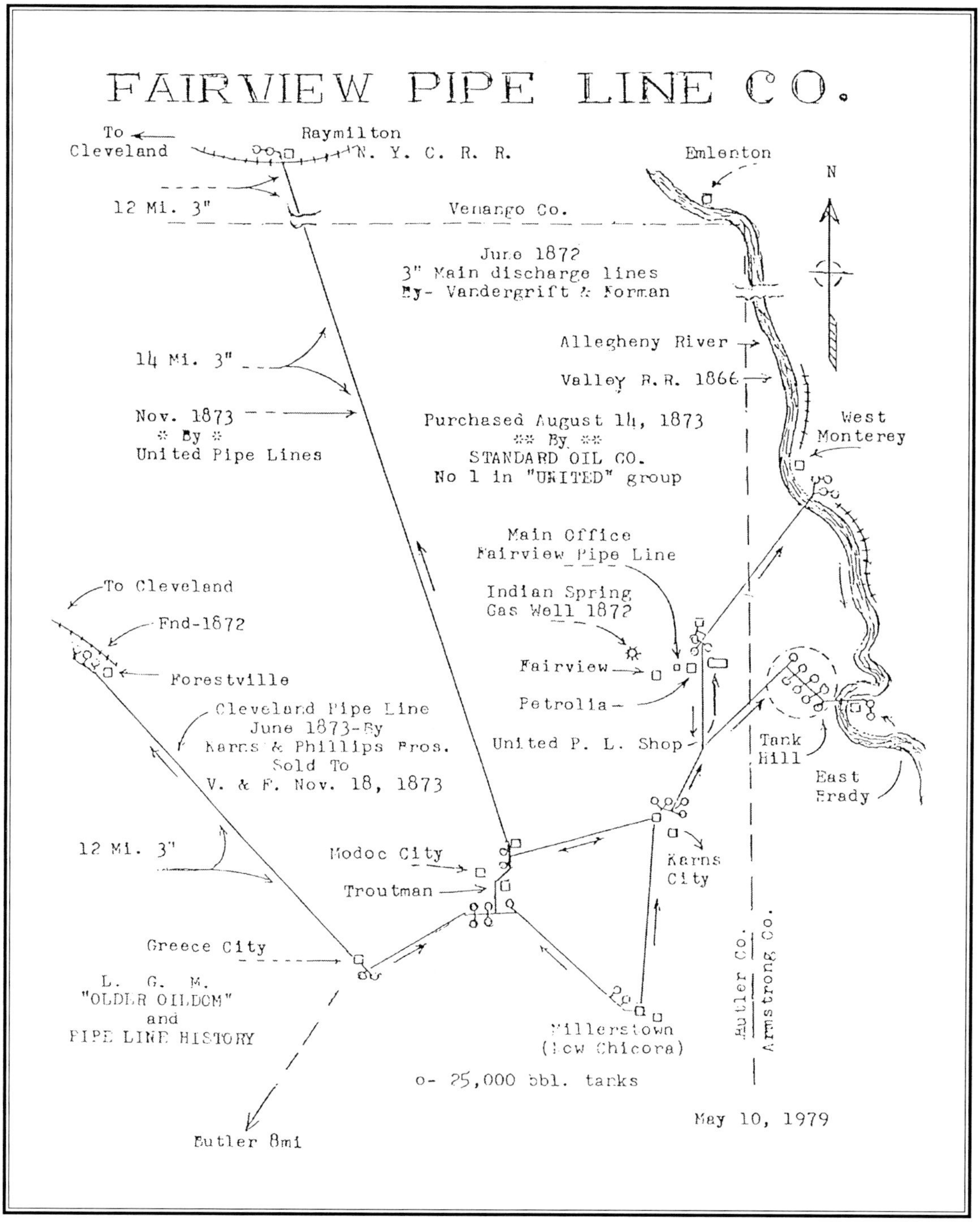


*Courtesy Lenn G. McCorry, Pipe Line History*
*Butler County Historical Society*

*Fairview Oil Line Pump Station, 1874*
*The large iron tank to the right has United Pipe Line lettered on it. This station pumped east to "Tank Hill" at East Brady and west to Hilliards. The man standing in the doorway is James O'Hara, local oilman from Petroleum Center. As wells became more numerous in less accessible areas of the townships, it became necessary to find a cheap, efficient way to get petroleum to a transportation system, thus pipe lines.*
*Cal Stevens, Karns City, Penna.*

## Fairview Pipe Line Pump Station

Fairview Pipe Line pump station was located at Karns City. It was begun in June 1872, purchased by Standard Oil Company August 14, 1873. The large steam engines ran pumps that forced the petroleum through pipe lines to the tank yards. The Fairview Pipe Line Company was Butler County's first oil pump station. Also, this was Standard Oil Company's first venture in the transportation of oil, much to John D. Rockefeller's delight. Vandergrift and Foreman were the builders. (Lenn "Pithole" McCorry, Butler County Historical Society, Butler, Pennsylvania)

Between the two stacks at the top is "The Colonel Woodward" Well No. 2. It was "Fourth Sand" and came in at six hundred barrels a day on January 10, 1874.

Note the little house to the left of the stacks. This was the Connelly home and this is an often told story of the "oil patch." Mr. Connelly was a machinist and left his wife and five children for employment with Farrar and Trefts, oil field steam engine manufacturers, Buffalo, New York. He told the Mrs. he would send for them but he was never heard from again. Poor Mrs. Connelly took on washing oilmen's clothes for the rest of her life.

*Fairview Pipe Line office. Mr. Vandergrift and Mr. Foreman were the builders of the line and are standing in the doorway. 1872*

The narrow gauge railroad tracks of the Parker and Karns City Railroad ran its first train to Karns City on April 4, 1874. The railway station was located where the Penn-Drake Refinery tower now stands. The tracks were eventually pushed on to Millerstown and then on to Butler, Pennsylvania. (Courtesy of J. P. O'Hara Sr. family, Karns City Oil Pioneer, Cal Stevens, Karns City, Butler County Historical Society)

## Karns City Borough

When, in December, 1871, the Cooper Brothers leased fifteen acres of land from Hugh P. McClymonds and fifteen acres from Samuel L. Riddle, these farms were destined to be the site of a remarkable oil town. The first well was located in the valley on the McClymonds farm, near the west line, and was named the "Shasta." In June 1872, it was producing 120 barrels a day. On May 29 of that year, S. D. Karns, having great faith in this venture of the Cooper Brothers, leased the entire 214 acres of the McClymonds farm at a bonus of $200 per acre and one-eighth oil royalty, the owner reserving the Cooper lease and fourteen acres surrounding his farm buildings. On June 1, he leased, on the same terms, 204 acres from Samuel L. Riddle, who reserved the lease given to Cooper Brothers and ten acres about his farm buildings. Oil was then selling at $4 per barrel, and the new strike was attracting the alert producer. A fierce dispute arose among several operators as to the possession of the McClymonds farm, but a compromise was effected, and June 18, 1872, he sold his farm, reserving the farm buildings and surface of fourteen acres, for $60,000, the purchasers being O. G. Emery, S. D. Karns, William Thompson, William Parker and John H. Haines. Excitement was now intense. Derricks were looming up on every side and buildings were rapidly erected.

The borough was incorporated January 4, 1875, in answer to a petition of September 12, 1874, with L. D. Akin, burgess; A. J. Elliott, J. C. Shay, J. E. Criswell, H. Rodgers, John S. Kane and Dr. S. H. Pettigrew councilmen, and T. P. Brown, clerk.

Previous to the organization of a borough government, the usual lawlessness of new oil towns prevailed. At one time there were 29 places where liquor was sold, and drunkenness and fighting were things of every-day occurrence. Sundays, especially, the roughs that infested the town carried their revels to a shameless extent. But all this soon passed away, and in its place law and order were established.

*Old No. 16 on the Pittsburgh and Western Railroad, 1881, formerly the Parker and Karns City Railroad.*

In honor of S. D. Karns the new town was named "Karns City." Lots on the McClymonds farm sold at $100 to $550, while many were leased for an annual rental of $25 to $150. Dwellings, stores of all kinds, hotels, restaurants, livery stables and machine shops lined the streets. Business was booming, and during this year, 1872, Ralston, McQuaide & Company established a banking office and transacted a large business until they suspended in 1880. With the advent of the "fourth sand gushers," the excitement was intense, and the rapidly accumulating fortunes of many oil producers were amazing.

Believing that a permanent town would be built, the owners of the McClymonds farm, in February, 1874, laid out lots on their land, and on January 4, 1875, the borough was incorporated, the boundaries embracing all of the McClymonds farm and about ten acres of the Riddle farm. Karns City being the terminus of the Parker and Karns City railroad, it became the depot of supplies for points at the "front." On the arrival of trains passengers were greeted with boisterous invitations to be transported by all kinds of vehicles to Barnhart's Mills, Greece City, Angelica, Troutman and Modoc. On June 13, 1874, the McClymonds farm was sold to James H. Abrams for $30,000, and since that time has been owned by various persons, but on June 18, 1879, it was purchased by E. E. Abrams.

Of the five remaining oil wells on this farm, one is the famous "Rob Roy," which has produced more than 200,000 barrels of oil, and is yet pumping about four barrels per day. The total value of oil from this farm is estimated to be over $1,500,000, much of the oil having been sold at four dollars per barrel. The Riddle farm was not sold, but the leases and producing wells formed large transactions. It was on this farm that in 1878 E. O. Emerson drilled an exhausted third sand well in search of the fourth sand, and struck the "Teaser," which flowed 500 barrels per day and surprised every person. Mr. Emerson sold this well and others on the Riddle farm to H. L. Taylor & Company for $90,000.

In 1876 Karns City had a population of over 2,000, with an excellent fire department, good schools, churches, crowded stores, glittering saloons, fountains of oil and a general whirl of excitement, but the exodus to the "upper country" rapidly decreased the population. In 1877 an opera house was built, which was afterwards torn down and moved elsewhere. In 1880 less than half the population of 1876 remained, and in 1890 the total was 427, which is about the present population.

The National Transit pump station forced oil through the various pipe lines, there being five large lines near it. P. D. Sherwin's Enterprise Coal Works shipped several cars of coal daily over the Pittsburgh & Western Railroad.

## Fire and Flood

The fire of May 26, 1873, which destroyed much of the Karns City property and threatened the new oil town, was followed in June of that year by a second fire, when a 1,200-barrel tank, belonging to S. D. Karns, in the Stone House neighborhood, burst and the oil caught fire, destroyed the Union Line pump house, two rigs owned by a Butler syndicate and a large quantity of oil. Refer to page 404.

The fire of December 2, 1874, burned 64 buildings in the business center; and in September, 1876, another fire destroyed 16 buildings between the present post office and the machine shop, including P. R. Burke's two buildings.

The burning of the Bateman House, March 5, 1877, was the most terrible in its results. Mrs. Bateman and her four daughters, Minnie, Lou, Belle and Winnie, lost their lives and N. Brown, a New York salesman, one of the guests, perished in the flames, while F. E. Bateman, his son and another guest died of burns within a few hours. Notwithstanding this series of destructive events, the citizens renewed their exertions so that, by the close of the fiscal year in June, 1878, a sum of over $2,913 was received for taxes, together with various sums for local improvements.

The flood of July 26, 1879, which wrought such havoc at Petrolia, was proportionately unmindful of the interests of Karns City. At 11 o'clock in the forenoon the waters of Bear Creek began to claim the valley for their bed, and as the rain increased in volume they attacked the small buildings and then Luther's stable. The old Conduit Pipe Company's office, occupied by the Toy family, Mrs. Emerick's house, the kitchen of the Primrose Hotel and A. D. Hall's farm house were moved from their foundations, and some of them carried down stream to mingle with the wreckage.

The burning of the tank belonging to the National Transit Company, June 1, 1884, was one of the awe-inspiring scenes of this field. The tank contained 26,000 barrels of oil, and though much of it was drawn off, what remained supplied fuel to keep the flames alive for 24 hours.

The first paper published here was the "Karns City Item," which had an existence of about one year. At first, it was printed at Millerstown, but afterward the office was established here. The Item was a small paper and The Rev. A. S. Thorn was the editor.

In 1878, the Karns City Telephone was started by J. Borland. The Telephone had a fairly successful career of about four years.

(History of Butler County, Pennsylvania, 1883; Waterman, Watkins and Co., Chicago, Illinois)

## An Act of Daring and Bravery

Just one more tale about the great flood of July 26, 1879, on Bear Creek. It was reported that an act of bravery was performed by one of the Lerdecker brothers.

An old couple, Mr. and Mrs. Daniel Manning, were in danger of being drowned and their home was in the low lands below Karns City. Any moment the old couple's homestead could have been swept away. To the rescue came young Lerdecker who stripped void of all his clothes, dove in naked, and rescued, one at a time, the helpless couple.

This was all done in his "Birthday Suit" amid great cheering from the large crowd of spectators on shore. (Oil and Gas Man's Magazine, Butler, Pennsylvania, 1907-1918; Drake Well Museum, Titusville, Pennsylvania)

## The Old Oil Driller Said

Never let a cold run. Take it in time. All that is necessary is to procure a bottle of "Sellers Cough Syrup" and be cured.

*Karns City, 1874*

## Oil Deals on Cash Basis

Ever wonder where the saying, "Cash on the barrelhead" came from? Much cash was exchanged in the oil fields and with no desk to work on, the good old wooden oil barrel made a good desktop. Thus the term, "I want cash on the barrelhead."

## Oil City Derrick
## May 7, 1874
## Karns City and Chicken Thieves

"A certain few among the population of this city have been on watch for a considerable period of time over the fact that some of our citizens are stealing chickens. The late night raids on hen roosts have been reported and their owners were on the trail of the offenders. One of our more esteemed townsmen concealed himself one evening under a bed in the house of a suspected chicken thief. From his hiding place that night he saw the occupant of the house return with four well-developed chickens. Yesterday morning the offender was arrested and sentenced by Squire Ellis to pay a fine and undergo two days imprisonment.

*An old Dutch farmers personal "Oil Patch"*

## Karns City, Butler County, Pennsylvania

"Karns City is about fifteen months old, pleasantly situated on a large plateau of ground about a mile long, and a quarter of a mile broad. Prior to the oil excitement, the site of Karns City presented nothing to distinguish it from the average of farming lands. The McClymonds' homestead and an old sawmill were its chief erections. The first well here was struck in June, 1872, and the second in September, and from the later period the place continued to improve rapidly. At the present time it has a population of about 1,000 inhabitants. A number of important business houses have established branches at this point. There are three large hotels (the Empire House, Apollo House, and Exchange Hotel), one bank, and the customary number of business houses of all varieties. The town is named in honor of S. D. Karns, Esq., a fortunate and extensive operator." (The History of Petroleum, Vol. I and II, J. T. Henry, 1873)

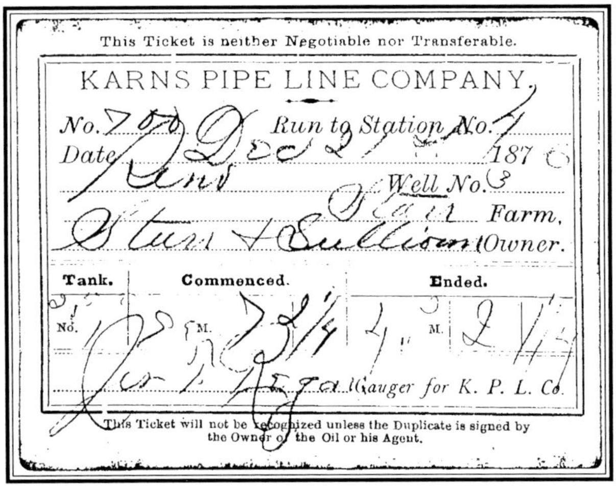
This Ticket is neither Negotiable nor Transferable.

KARNS PIPE LINE COMPANY.

No. Run to Station No.

Date 187

Well No.

Farm.

Owner.

| Tank. | Commenced. | Ended. |
|---|---|---|

Gauger for K. P. L. Co.

This Ticket will not be recognized unless the Duplicate is signed by the Owner of the Oil or his Agent.

*Flow Slip, Karns Pipe Line December 21, 1876. These slips were as good as "Green Backs." (Butler County Historical Society, Lenn "Pithole" McCorry, Butler County Oil Historian)*

## Karns City, May 8, 1874 Empire House

"A. J. Salisburg, proprietor, Headquarters of Oilmen, the largest hotel and the best accommodations in the lower oil district. The restaurant is attached to the house. Charges moderate." (Oil City Derrick)

## Karns City, May 22, 1874

"New buildings are going up in all directions until Karns City begins to look quite metropolitan in its appearance and dimensions." (Oil City Derrick)

*This large petroleum tank burst at Karns City, spilling oil and creating a mess. This iron tank was owned by S. D. Karns and gave way when rivets popped and 1,200 barrels of petroleum caught fire and ran down the creek. Karns City 1873 (Courtesy of Venango County Historical Society, Personal Notebook of John J. McLaurin, Franklin, Pennsylvania)*
*Refer to page 401.*

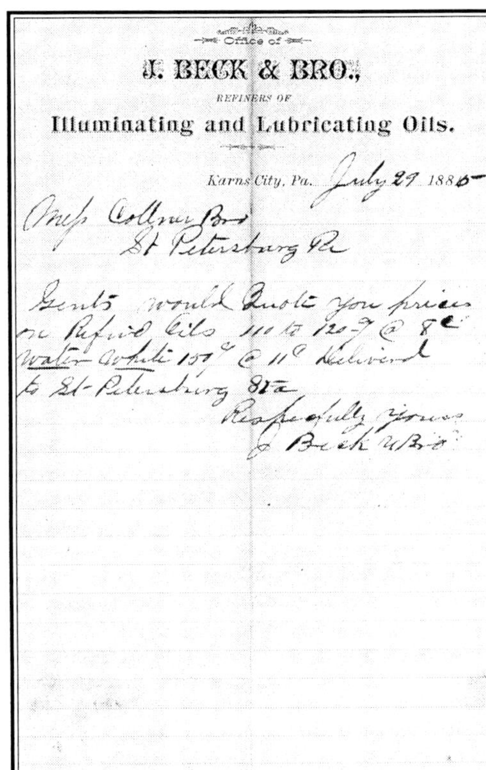
Office of

J. BECK & BRO.,

REFINERS OF

Illuminating and Lubricating Oils.

Karns City, Pa. July 29 188

Messrs Collner Bros
St Petersburg Pa

Gents would quote you prices on Refined Oils 110 to 120 @ 8c Water White 150 @ 11c Delivered to St Petersburg Sta

Respectfully yours
J Beck & Bro

*J. Beck and Brothers, Oil Refiners Karns City, Pennsylvania*

*Karns City, Pennsylvania, 1876*

## The Parker and Karns City Railroad

The importance of having access to further railroad service became apparent to the citizens, oil men and business men of Parker City. A railroad was needed to move machinery and goods into and out of the new Butler oil region. Many investors came together to form the railway but when the severe financial panic of 1873 swept over the country, many of the stockholders were forced to forfeit their stocks. The Parker and Karns City Railroad Company would not have raised the capital needed and succumbed to the pressure of the panic. Thankfully, four Parker City citizens came to the rescue, throwing their private means and their energies into the project, Mr. Fullerton Parker, Mr. Samuel D. Karns, Mr. W. C. Mobley and Mr. H. R. Fullerton. These men carried the Parker and Karns City Railroad to a successful completion.

In August 1873, the work of building the Parker and Karns City narrow gauge railroad was undertaken. The road was completed as far as Petrolia early in 1874 and by April 4, 1874, the train began running to Karns City.

The railroad was pushed on into Butler, Pennsylvania, in 1876 and largely increased the business of the road. The new iron bridge at Parker City was used by the Parker and Karns City Railroad to connect with the Allegheny Valley railroad. This road from Butler to the Allegheny Valley Railroad gave direct access to the length of the Butler County oil towns and fields.

In 1879 the Parker and Karns City Railroad became financially embarrassed. In 1881, the road became part of the Pittsburgh and Western and was made standard gauge in 1887.

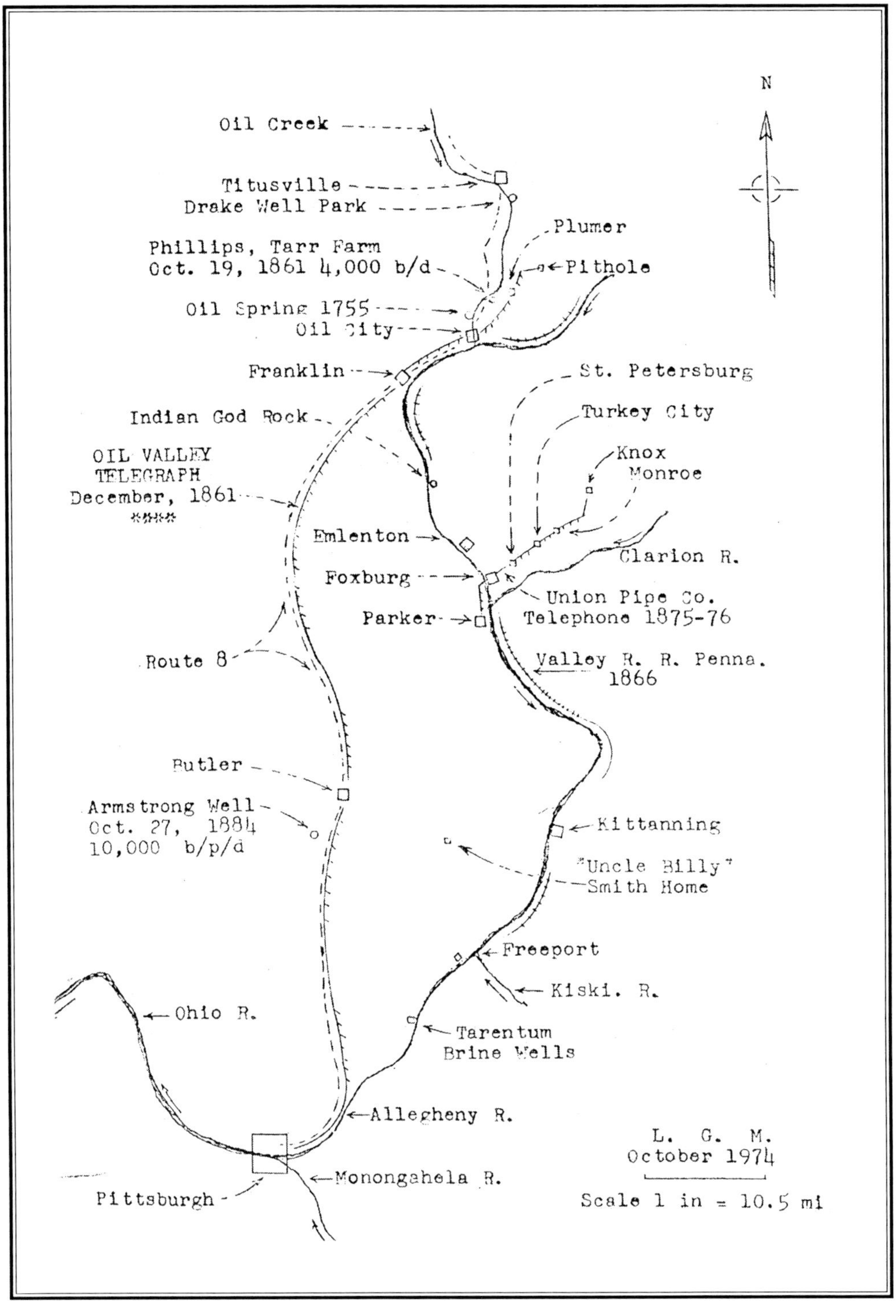


*Map drawn by Lenn G. McCorry showing historical sites from Titusville, Pennsylvania, to Pittsburgh. Courtesy Butler County Historical Society.*

# Parker & Karns City Railroad and Karns City & Butler Railroad.

**No. 3. TIME TABLE. No. 3**

(FOR THE GOVERNMENT AND INFORMATION OF EMPLOYES ONLY)

*TO TAKE EFFECT MONDAY DECEMBER 4TH, 1876, SUPERSEDING TIME TABLE NO. 2.*

**Eastward Bound Trains.**

| Third Class, Freight 17 | 15 | 13 | Second Class Mixed. 11 | 9 | First Class Passenger. 7 | 5 | 3 | 1 | NAMES OF STATIONS. |
|---|---|---|---|---|---|---|---|---|---|
| A. M. | P. M. | A. M. | P. M. | P. M. | P. M. | P. M. | P. M. | P. M. | |
| | | | | 4.35 | | 5.40 | | 10.35 | AR *Parker J* DE |
| 7.30 | 9.40 | 11 00 | 8.50 | 4.20 | 10.20 | 5.20 | 1.10 | 10.25 | *Parker* |
| 7.20 | 9.27 | 10.52 | *8.40 | *4.10 | *10.12 | *5.10 | *1.02 | *10.17 | *Donnelly* |
| 7.10 | 9.15 | 10.42 | *8.30 | * 4.00 | *10.07 | *5.00 | *12.52 | .10.07 | *Stone House* |
| 7.00 | 9.05 | 10.35 | ‡8.20 | ‡3.50 | ‡10.00 | ‡4.52 | 12.45 | ‡9.57 | *Dudly Siding* |
| 6.50 | 8.55 | 10.30 | *8.15 | 3 45 | 9.58 | 4 47 | 12.40 | 9.53 | *Martinsburg* |
| 6.43 | 8.45 | 10 25 | 8.10 | 3.37 | 9.53 | 4.40 | 12.35 | 9.48 | *Campbell* |
| 6.30 | 8.30 | 10.17 | *8.05 | *3.30 | •9.48 | *4.35 | *12.30 | *9.43 | *Argyle* |
| [illegible] | 8.25 | [illegible] | 7.58 | [illegible] | 9.45 | 4.30 | 12.25 | 9.38 | *Petrolia* |
| 6.08 | 8.05 | 10.00 | *7.55 | *3.10 | *9.42 | •4.27 | *12.22 | 9.35 | *Hazlewood* |
| [illegible] | [illegible] | 9.47 | 7.45 / 7.45 | [illegible] | 9 35 | 4.20 | 12 12 | 9.27 | *Karns City* |
| 5.35 | 7.17 | 9.37 | *7.37 | *2.37 | *9.27 | *4.12 | *12.05 | *9.20 | *Angelica* |
| 5.27 | 7.10 | 9.30 | *7.30 | *2.30 | *9.22 | *4.05 | *12.00 | *9.10 | *Moore Farm* |
| [illegible] | [illegible] | 9 20 | 7.20 | 2.20 | 9.15 | 3.52 | 11.50 | 9 00 | *Millerstown* |
| 4.47 | 6.27 | A. M | P. M. | P. M. | *9.05 | *3.42 | 11.37 | *8 45 | *Summit* |
| [illegible] | [illegible] | | | | 8 55 | 3 32 | 11.25 | 8.35 | *St. Joe* |
| 4 20 | 5 45 | | | | *8.48 | *3.27 | *11 22 | •8.30 | *Durkins* |
| 4.10 | 5.35 | | | | 8.37 | 3 17 | 11.15 | 8.25 | *Carbon Cent'r* |
| 4.00 | 5.15 | | | | *8.30 | *3.10 | *11.10 | *8.20 | *Jeffersonville* |
| 3.50 | 5.05 | | | | *8.20 | *3.09 | *11.00 | *8.13 | *Bonnie Brook* |
| 3.42 | 4 55 | | | | *8.12 | •2.55 | *10.52 | •8.05 | *McClure.* |
| 3.30 | 4 45 | | | | 8.00 | 2.40 | 10.40 | 7.55 | DE *Butler.* AR |
| A. M. | P. M. | | | | P. M. | P. M. | A. M. | A. M. | |

**Westward Bound Trains.**

| NAMES OF STATIONS. | First Class, Passenger. 2 | 4 | 6 | 8 | Second Class Mixed. 10 | 12 | Third Class, Freight 14 | 16 | 18 |
|---|---|---|---|---|---|---|---|---|---|
| | A. M. | A. M. | P. M. | P. M. | P. M. | P. M. | A. M. | A. M. | P. M. |
| AR *Parker J* DE | | 10.55 | | 6.00 | 12.10 | 5.05 | | | |
| *Parker* | 7.50 | [illegible] | 1.50 | Sup. [illegible] | 12.20 | 5 20 | 6.00 | 9.00 | 10.35 |
| *Donnelly* | *7.58 | *11.12 | *2.00 | *6.50 | *12.27 | *5.27 | 6.10 | 9.07 | 10.45 |
| *Stone House* | *8 07 | *11.22 | *2.10 | *7.00 | *12.37 | *5.37 | 6.20 | 9.17 | 10.55 |
| *Dudly Siding* | ‡8.15 | ‡11.32 | ‡2.17 | ‡7.07 | 12.45 | ‡5.45 | 6.27 | 9.25 | 11.10 |
| *Martinsburg* | 8.18 | 11.37 | 2.22 | 7.12 | 12.50 | 5.52 | 6.35 | 9.30 | ‡11.15 |
| *Campbell* | *8.23 | *11.45 | 2.27 | 7.20 | 12.55 | 6.00 | 6.43 | [illegible] | 11.30 |
| *Argyle* | *8.30 | *11.52 | *2.32 | *7.25 | •1.00 | *6.05 | 6.50 | 10.00 | 11.35 |
| *Petrolia* | 8.32 | 12.00 | 2.38 | 7.30 | 1.05 | 6.10 | 7.00 | 10.03 / 10.30 | 11.40 / 12.00 |
| *Hazlewood* | *8.35 | *12.03 | *2.40 | *7.32 | *1.08 | *6.13 | 7.05 | 10.35 | 12.05 |
| *Karns City* | 8.40 | 12.12 | 2.48 | 7.43 / 7.43 | 1.15 | 6.20 | 7.15 | 10.45 / 11.00 | 12.13 / 12.30 |
| *Angelica* | •8.47 | *12.20 | *2.58 | *7.55 | *1 22 | *6.27 | 7.25 | 11.15 | 12.40 |
| *Moore Farm* | *8.53 | *12.27 | *3.05 | *8 05 | *1.27 | *6.37 | 7.32 | 11.30 | 12.50 |
| *Millerstown* | 9.00 | 12.35 | 3.12 | 8.15 | 1.35 | 6.47 | 7.47 | [illegible] | [illegible] |
| *Summit* | *9.15 | *12.45 | *3.23 | *8.32 | P. M. | P. M. | A. M. | 12.22 | 1 32 |
| *St. Joe* | 9.27 | 12 55 | 3.32 | [illegible] | | | | [illegible] | [illegible] |
| *Durkins* | *9.32 | *1.00 | *3.37 | *9.00 | | | | 1.10 | 2 05 |
| *Carbon Cent'r* | 9.40 | 1.10 | 3.45 | 9.10 | | | | 1.30 | 2.20 |
| *Jeffersonville* | *9.45 | *1.17 | *3.50 | *9 15 | | | | 1.37 | 2.28 |
| *Bonnie Brook* | *9.55 | *1.27 | *4.00 | *9.25 | | | | 1.47 | 2.35 |
| *McClure.* | *10.02 | *1.35 | *4.08 | *9.35 | | | | 2.00 | 2.45 |
| DE *Butler.* AR | 10.15 | 1.50 | 4.20 | 9.45 | | | | 2.20 | 3.00 |
| | A. M. | P. M. | P. M. | P. M. | | | | P. M. | A. M. |

Trains Nos. 3, 6, 9 and 10 run daily. All other train daily except Sunday. Train No. 10 will have the right to road over Train No. 3. Sstop for supper. ‡ Trains not stop. *Trains stop only on signal.

W. C. MOBLEY, Sup't.

## Oil City Derrick June 17, 1874

"Business on the Parker and Karns City Railroad passenger account is very strong. Much machinery is being shipped on the Allegheny Valley Railroad from Titusville and Oil City. This road is splendidly managed and is being improved daily. Road gangs are constantly improving the tracks and replacing hemlock with oak ties. The conductors and employees are all courteous and study the wants of the public, giving universal satisfaction."

## Stephen Duncan Karns

Not so fortunate was perhaps the most fabulous of all the producers who found liquid gold in the hills of Butler County.

Stephen "Dunc" Karns was one of the leading oil producers in Western Pennsylvania because of his interests in the Karns City field and the cross belt region. It is estimated that his income exceeded $5,000 per day in 1872.

Samuel Karns divided a field into building lots. He charged a minimum of $100 and as much as $550 for his lots. The lots sold well, and soon the area had developed into a town having all types of commercial establishments. This new oil town was named Karns City and on January 4, 1875, the town was incorporated as a Butler County borough. Being the last major community south of Parker's Landing, Karns City became the terminal and supply depot of the narrow gauge Parker-Karns City Railroad, serving the surrounding communities and oil fields.

The railroad itself was organized by S. D. Karns on August 20, 1872, with an issue of $22,750 worth of stock. In May, 1873, he issued additional stock worth $100,000. In August, 1873, with the available capital, construction began; however, in December Karns had to issue $250,000 additional stock in order to complete the road. Officially the opening was on April 4, 1874. The offices for the Parker-Karns City narrow gauge railroad and the Karns Oil Pipe Line Co. was a red brick building located next to the Parker House in Parker City.

Along the Allegheny River at Freeport, Pennsylvania, Karns was the owner of one of the finest mansions in Western Pennsylvania, had a string of expensive race horses, and made many trips to Europe where he entertained royalty on a lavish scale. Refer to page 409.

In addition to his vast oil interests, Karns controlled at least one bank and had built the later famous Parker bridge.

But all that he had acquired vanished in 1874 when a sudden spiral forced crude oil to 45 cents a barrel. "Dunc" Karns made a terrible mistake by dealing in oil speculation. He purchased oil at a higher price and when the bottom dropped out of the market, he lost money and was wrecked financially.

Practically penniless, Karns went west and while managing a cattle ranch in Colorado, hit a small gusher. Returning east he bought and published a newspaper in Pittsburgh for a few

*Parker and Karns City Railroad trestle over Bear Creek near Parker City. Frank Robbins, Photographer, Oil City, Pennsylvania, 1874. Note the two brave men. This railway was torn up in the 1980s and no longer exists.*

*Railbed of the Parker and Karns City railroad near the Bear Creek trestle. Frank Robbins, photographer, Oil City, Pennsylvania, 1876. Pushing the oil belt south. Note the solid stone along the railway.*

years. But "Dunc," as he was known to the oil pioneers of Butler County, never regained the fortune that once had been his. (Freeport Public Library, Freeport, Pennsylvania)

Oil City Derrick
As of June 23, 1874

The locations and tankage for the lower belt are as follows:

General Summation

| Location | No. Tanks | Ag. Capacity |
|---|---|---|
| Raymilton | 16 | 286,000 bbls. |
| Emlenton | 2 | 22,000 |
| Edenburg | 1 | 17,000 |
| Fullerton | 7 | 94,000 |
| Foxburg | 18 | 165,850 |
| Tarrentown | 10 | 147,150 |
| Parker | 26 | 376,400 |
| Monterey | 4 | 49,300 |
| Argyle | 6 | 58,000 |
| Petrolia | 3 | 43,100 |
| Karns City | 10 | 161,800 |
| Modoc | 3 | 54,700 |
| Troutman | 3 | 161,200 |
| Greece City | 3 | 44,800 |
| Butler | 3 | 44,800 |
| Dilks | 6 | 113,900 |
| Millerstown | 8 | 168,300 |
| Sarah Furnace | 7 | 119,300 |
| Brady | 10 | 118,140 |
| North Washington | 2 | 12,000 |
| Clintonville | 1 | 6,000 |
| | 154 | 2,260,040 bbls. |

*Samuel Duncan Karns*
*Oil operator and leading capitalist, Samuel Duncan Karns was described in these terms: He was a man of light build, nervous temperament, prepossessing appearance. He was cordial, frank, and manly. As a businessman he was without blemish and his word was his bond. He was generous and conscious of his own power and rapid in execution in whatever he may undertake. He was a man of a will of his own.*

*Mansion lost to financial ruin*
*This is a photograph of the famous Samuel Duncan Karns Mansion along the Allegheny River near Freeport, Pennsylvania. The mansion was built of red brick and a real showplace, a most beautiful home. This is the only known image to exist. The mansion mysteriously burned down in the 1960s. This home was located on the present-day River Forest golf course. (Courtesy of Butler County Historical Society, Butler, Pennsylvania, Lenn "Pithole" McCorry)*

JOHN RALSTON. J. G. M'QUAIDE. D. ALEX. RALSTON.

**RALSTON, McQUAIDE & CO.,**

**Karns City Banking Office,**

**KARNS CITY,**

Butler County, Pa.

*JAS. R. JOHNSTON, Cashier.*

Money Received on Deposit; Government Bonds Bought and Sold. Collections made on all accessible points in the United States.

---

S. H. PETTIGREW, M. D.,

DEALER IN

**DRUGS AND MEDICINES**

*Paints, Oils, Fine Toilet SOAP, Perfumery and Brushes; Wines and Liquors for Medical Use; Glass, Putty, &c., &c.*

***Main street, KARNS CITY, PA.***

*Physicians' prescriptions carefully compounded.*

---

**COX & KANE,**

DEALERS IN

**Drugs, Medicines, Chemicals,**

PERFUMERIES, FANCY ARTICLES, WINES, LIQUORS, ETC.,

PAINTS, OILS, &c.

**KARNS CITY, PA.**

*Physicians' prescriptions carefully compounded at all hours.*

## KARNS CITY.

Karns City, Butler county, Pa., is located in the Modoc Oil District, about three miles west from Petrolia, one-and-a-half miles south of Fairview, and five miles north from Millerstown.

The business carried on here does not differ in character from that conducted in the other oil towns. Hardware, dry goods, groceries, &c., are well represented; while the machinist, tank builder, &c., are doing a thriving business.

The Karns City Railroad has its starting point here. This road, when completed to important points in the oil region, will give a fresh impetus to the growth and business interests of the city.

There are several good hotels here: The Empire House, the Exchange Hotel, the Apollo House and others,—all of which are well kept. The city contains upwards of 1000 inhabitants.

### A

Adams J. S., of Lard & Adams, Main.
Adriance Wm. H., billiard hall, Main.
Akin L. D., proprietor Exchange Hotel, Main.
ALLEN J. K., grocer, boards Empire House.
Andrews Phillip, agent Empire House.
APOLLO HOUSE, A. J. Elliott, proprietor, Main.
Arteman D. K., clerk, boards Main.

### B

Bacon O. J., butcher, Main.
Ball ———, driller, boards Apollo House.
BARBARY MICHAEL, of Barbary & Prowse, Main.
BARBARY & PROWSE, proprietors Devonshire house, Main.
Bell George W., carpenter and builder, Main.

**A. W. SPEASE,**

WHOLESALE & RETAIL DEALER IN FINE

**Groceries & Provisions,**

**KARNS CITY, PA.**

*Goods Delivered.*

---

SCOTT & TRAVERS,

DEALERS IN

*Clothing, Fine Furnishing Goods,*

**Tobacco, Cigars, Notions, &c.**

*KARNS CITY, PA.*

---

D. F. KELLERMAN,

DEALER IN

**Dry Goods and Groceries.**

ALSO DEALER IN

BOOTS AND SHOES, CROCKERY, QUEENSWARE,

**And a large stock of Notions generally.**

"QUICK SALES AND SMALL PROFITS" is my motto.

**Main Street, - - - - - - KARNS CITY, PA.**

---

J. K. ALLEN,

DEALER IN

**FAMILY GROCERIES,**

*Flour, Provisions, Queensware & Crockery,*

Foreign and Domestic Fruits, Butter, Cheese, Canned Fruits, Pickles,

**PRESERVES, MAPLE SUGAR, etc.**

*Next door to Empire House, KARNS CITY, PA.*

J. WIGGINS' BUTLER OIL REGION DIRECTORY.

**D. T. PAPE,**

**DRY GOODS HEADQUARTERS, BUTLER, PENN**

Bell John, oil operator, boards Empire House.
Bennett J. N., contractor, boards Exchange Hotel.
Berline John, teamster, boards Atlantic House.
Bice E. H., clerk, boards Apollo House.
Binaerd A. T., carpenter, Main.
Bishop Robert A., oil producer, Fairview Road.
Bishop Robert L., oil operator, boards Empire House.
Bradley J. R., barber, Main.
Bramle S., driller, boards Apollo House.
Brek Eliza, millinery, Main.
Brown & Martin, hardware, oil well supplies, Main.
Brownell J. D., tank builder, boards Exchange Hotel.
Bryan Hugh S., laborer, Fairview Road.
Bachtell James, driller, hill west of town.
Burk P. R., clerk, boards Exchange Hotel.
Bussell Gillmor S., oil producer, Fairview Road.

## C

Calvert Robert, carpenter, boards Exchange Hotel.
Carmoody C. S., rod ironer, boards, Main.
Case John B., saloon, boards Apollo House.
Casper Phillip, teamster, boards McKee House.
Chapman W. H., boiler maker, boards, Apollo House.
Clark Isaac, saloon, boards Apollo House.
Conard S. D., pumper, Main.
Conner James, carpenter, boards Apollo House.
Core John, billiard room, Main.
Corts Joseph J., engineer, Fairview Road.
Cox J. H., of Cox & Kane, Main.
COX & KANE, drug store Main.
Criswell J. E., of J. E. Creswell & Co., Main.
Criswell J. E. & Co., dry goods, etc., Main.

## D

Davis, ——— tank builder, Main.
Daughbach Edward, driller, boards Apollo House.
Deutsch Ed., bootmaker, boards Main.
Dravo A. E., laborer, Main.
Dufford George, carpenter, Main.

## E

ELLIOTT A. J., proprietor, Apollo House, Main.
EMPIRE HOUSE, A. J. Salisbury, proprietor, Main.
Erns Morris, carpenter, Main.
Evans James, helper, boards Main.
Exchange Hotel, L. D. Aken, proprietor, Main.

## F

Ferguson S. W., oil operator, boards Empire House.
Flick Jos. of Jos. Flick & Son, Main.
Flick Jos. & Son, livery stable, Main.
Flick Robt., of Jos. Flick & Son, Main.
Fohover Frank, oil operator, boards, Empire House.
Foster J. A., teamster, Main.
Fox Henry, dry goods, etc., Main.
Frothingham Thos., oil operator, boards Empire House.
Frazer Washington, carpenter, Main.

## G

Geider M. C., clerk, boards Empire House.
Gibbs Samuel, boots and shoes, Main.
Goldrick Owen S., manager pipe line, on hill, west of city.
Gowdy R. T., oil operator, Main.
Grant Ira, carpenter, Main.
Gravistin John, bootmaker, boards Main.
Green Benjamen F., driller, Main.
Green David, driller, boards Apollo House.
Green R. M., barber, Main.
Green William, tool dresser, Main.

## H

Hall George, teamster, Main.
Harper D., physician, Main.
Harpst Andy, restaurant, Main.
HARVEY E. S., books and stationery, Main.
Harons Thomas, pumper, boards Exchange Hotel.
Hollobaugh Mary, millinery goods, Main.
Hoskins Charles, salesman, boards Exchange Hotel.
Houghton P. S., machinist, boards Main.
Hudson O., carpenter, boards Apollo House.
Hudson P. E., barkeeper Apollo House.
Hutchison Foster, bookkeeper, Main.

## I

Inebaugh Joseph, driller, boards Apollo House.
Ireland Isaac, laborer, Main.

## J

Jones C. L., boards Exchange Hotel.
JOHNSTON JAS. R., cashier Karns City Banking Office, boards Empire House.
Johnston John, wagon maker, Main.
Jones David, teamster, Main.

## K

Kane J. S., of Cox & Kane, Main.
Kane J. S., restaurant, Main.
KARNS CITY BANKING OFFICE, of Ralston McQuaide & Co., Main.
KELLERMAN D. F., dry goods and groceries, Main.
Kennedy Daniel, tool dresser, Main.
Keough David, blacksmith, boards Main.
Kern William, operator, boards Empire House.
Kilroy James, of Kilroy & Wersh, Main.
Kilroy & Wersh, bakers, Main.
Kiser Wm., teamster, boards Main.
Knoebler John, restaurant, Main.
Knittle Geo., teamster, Main.

## L

Lawson Thomas M., blacksmith, boards Main.
Lard & Adams, grocers, Main.
Lard W. H., of Lard & Adams, Main.
Lee Thomas, shoemaker, Main.
Lenox M. W., telegraph operator, boards Exchange Hotel.
LENTZ GEORGE W., clerk, Main.
Leslie Charles, oil operator, boards Empire House.
Leslie James, contractor, boards Empire House.
Lewis John, oil operator, boards Empire House.
Lize A. A., clerk, boards Main.
Logue Patrick, tank dealer, Main.

## M

Manning Daniel, laborer, Fairview Road.
Marshall S. S., carpenter, boards Apollo House.
McAdoo James, hostler, Main.
McCabe John, butcher, Main.
McCann James, blacksmith, Main.
McClymond H. P., oil producer, Fairview Road.
McCorkel Robert, pumper, Main.
McCane John, boarding house, Main.
McGuire John, driller, boards Exchange Hotel.
McKee David, proprietor McKee House, Main.
McKenzie Wm., machinist, Main.
McLain Daniel, machinist, boards Exchange Hotel.



McLean William, teamster, Main.
Miller John, operator, boards Empire House.
Montgomery William, carpenter, Main.
Moody G. A., salesman, boards Main.
Morrison Rezelle, carpenter, Main.
Murdock Jerry, driller, hill west of town.
Murta Francis, livery, Main.
Myers Joseph, carpenter, Main.

## P

Parker N. B., contractor, Fairview Road.
Parker Norman, oil operator, boards Empire House.
Parks George, pumper, boards Exchange Hotel.

Parks George, pumper, boards Exchange Hotel.
Paterson Henry, pumper, Main.
PETTIGREW, S. H., M. D., druggist, Main.
Pickard William, stage driver, Main.
Pollock J. T., boards Exchange Hotel.
Pollock Herman, blacksmith, Main.
PROWSE JAMES, of Barbary & Prowse, Main.
POST OFFICE, E. S. Hardy, P. M., Main.

## R

Rathburn Adam, driller, boards Exchange Hotel.
Ragan John, machinist, boards Main.
Redford H. H., book-keeper, boards Empire House.
Regan Michael, laberer, hill west of city.
Relief Pipe Line, office Main.
Rhinehart Abraham, tank builder, west side of town.
Rice William, teamster, Main.
Rigney E. V., butcher, Main.
ROCKENSTEIN JOSEPH, saddler, Main.
Rogan Captain, helper, boards Main.
Rogers Henderson, lumber dealer, boards, Exchange Hotel.
Rose Charles L., oil producer, Fairview Road.
Russell———, blacksmith, Main.

## S

SALISBURY A. J., proprietor Empire House, Main.
Salisbury Jonathan, oil producer, Fairview Road.
Scott Alexander of Scott & Travers, boards Empire House.
SCOTT & TRAVERS, clothing, gents furnishing goods, &c., Main.
Segerdahl D., blacksmith, Main.
Segerdahl F. A., blacksmith, Main.
Shaffer Nicholas, boot and shoemaker, Main.
Sheffield & McCullough, hardware and oil well supplies, Main.
Shay J. C., machinist and blacksmith, Main.
Shumway William, helper, boards Main.
Sickles Charles, oil operator, boards Empire House.
Sisson Rollin, clerk, boards Exchange Hotel.
Sloan J., teamster, Main.
Smiles Alfred W., telegraph operator, boards Empire House.
Smith J. S., operator, boards Empire House.
SNYDER H. N., attorney and notary public, Main.
SPEASE A. W., grocer, Main.
Spence W., livery stable, Main, near Exchange Hotel.
Sproul William, harness maker, boards Apollo House.
Staffoger Jacob, bootmaker, boards Main.
Stanly William, teamster, boards Exchange Hotel.
Stanly P., tank builder, boards Exchange Hotel.
Stetson Wm. H., carpenter, hill west of town.
Sweeney James, teamster, hill west of town.
Sypher David, gauger, hill west of town.

## T

Thompson T. P., oil operator, office Main.
Titus Meyers, machinist, boards Main.
Titus William, book-keeper, boards Empire House.
Tomblinson———, gauger, boards Exchange Hotel.
Travers F., of Scott & Travers, boards Empire House.
Traviston John, bootmaker, boards Main.

## V

Vanliew T. G., restaurant, Main.

## W

Wathey Henry, oil operator, McClymonds farm.
Weed William, book-keeper, boards Empire House.
Welsh A. M., carpenter, Apollo House.
Welsh Robert M., carpenter, Apollo House.
Wersh John, of Kilroy & Wersh, Main.
West A. A., druggist, Main.
Western Union Telegraph Office, Main.
Wheaton C. Warren, carpenter, Fairview Road.
Wheeler George, photographer, boards Apollo House,
Whitney———, boards Empire House.
Wick John R., clerk, boards Exchange Hotel.
Woods Mrs. R. G., millinery goods, Main.
Worthington John, book-keeper, Main.
Wright Lewis, carpenter, boards Apollo House.
Wynings A., carpenter, boards Exchange Hotel.

## Y

York Thomas, engineer, Apollo House.
Young John, painter, Apollo House.



## KARNS CITY BUSINESS DIRECTORY.

**Attorneys.**

Snyder H. N., Main.

**Banks.**

Karns City Banking Office, Main.

**Bakery.**

Kilroy & Wersh, Main.

**Billiard Hall.**

Adriance Wm. H., Main.
Core John, Main.

**Blacksmiths.**

Shay J. C., Main.

**Books and Stationery.**

Harvey E. S., Main.

**Boot and Shoe Makers and Dealers.**

Gibbs Samuel, Main.

**Butchers.**

Bacon O. J., Main.

**Carpenters & Builders.**

Bell George W., Main.
Killerman D. F., Main.
Lard & Adams, Main.
Spease A. W., Main.

**Gents' Furnishing Goods.**

Scott & Travers, Main.

**Harness-makers.**

Rockenstein Joseph, Main.

**Hardware.**

Brown & Martin, Main.
Sheffield & McCullough, Main.

**Hotels.**

Apollo House, Main.
Devonshire House, Main.
Empire House, Main.
Exhange Hotel, Main.

**Livery Stables.**

Flick Joseph, & Son, Main.
Spence William, Main.
Murta Francis, Main.

**Machinists.**

Shay J. C., Main.

**Notary Public.**

Snyder H. N., Main.

**Cigars and Tobacco.**

Scott & Travers, Main.

**Drug Stores.**

Cox & Kane, Main.
West A. A., Main.

**Dry Goods.**

Criswell J. E., & Co., Main.
Fox Henry, Main.
Kellerman D. F., Main.

**Grocers.**

Allen J. K., Main.

**Oil Well Supplies.**

Brown & Martin, Main.
Sheffield & MCullough, Main.

**Photographers.**

Wheeler George, Main.

**Physicians.**

Harper D., Main.

**Saloons and Restaurants.**

Harpst Andy, Main.
Kane J. S., Main.
Knoebler John, Main.
Vanliew T. G., Main.

**Furniture.**

Fleegers S., Main.
Smith G. C., Main.

**Grocers.**

Aldinger H. C., Slipperyrock.
Bernard James J., Slipperyrock.
Boyle P. M., & Sons, Slipperyrock.
Clemenger & Cleland, Slipperyrock.
Fredericks B., & Co., Main.
Kormer John, Mill.
Rumbaugh D., Slipperyrock.
Ryan P., Slipperyrock.
Westerman H. L., Main.

**Millinery.**

Gibson Miss M., Kittanning.

**Newspapers.**

Daily Sand Pump, Slipperyrock.

**Oil Well Supplies.**

Hersh & O'Brien, Main.
Norton & Murphy, Slipperyrock.
Reid & Durant, Main.
Lupher J. D., Slipperyrock.

**Photographers.**

Freedley & Conner, Main.

**Hardware.**

Fredericks B., & Co., Main.
Norton & Murphy, Slipperyrock.
Reid & Durant, Main.
Westerman H. L., Main.

**Hotels.**

Central Hotel, Main.
Eureka Hotel, Main.
Hanlen House, Main.
Reiber's House, Main.
Sink's Hotel, Main.

**Justice of the Peace.**

Fleegers S., Main.

**Livery Stables.**

Barton Alexander, Main.
Flick Abraham, Hanlen House.
Robinson C. C., Main.

**Lumber Dealers.**

Clark & Meeker, Slipperyrock.

**Machinists.**

Hirsch & Forst, Slipperyrock.

**Merchant Tailors.**

Hersh & O'Brien, Main.

**Physicians.**

Bell & Leffingwell, Main and Green.
McMichael Josiah, Main.
Tyler ——— Main.

**Queensware.**

Westerman H. L., Main.

**Real Estate.**

Newell Z., Main.

**Saloons and Restaurants.**

Heaton Peter, Slipperyrock.
McFarland John A., Main.
Mulligan Patrick, Main.

**Stoves.**

Simpson A. H., Slipperyrock.
Temple & Ingersoll, Main.

**Tank Builders.**

Booth A. P., Slipperyrock.
Marshall & Welch, Slipperyrock.

**Wheel Factory.**

McKee W. J., Water.

*Millerstown, Main Street shown, the front of the Central Hotel, 1874. (Courtesy of Butler County Historical Society and Lenn C. McCorry, Butler County Oil Historian)*

## Millerstown

Millerstown was incorporated as a borough in 1855, and until the discovery of oil in 1873 was a quiet, rural town, each day and year succeeding one another with hardly a ripple of excitement. At that stage of agricultural development, it had nearly reached the maximum of business the country would support, and but few changes or improvements took place. It then numbered about 32 voters, and contained four mercantile houses conducted by H. L. Westermann, Barnhart Frederick, Andrew Barnhart and Michael Dieter; two hotels, the Hoch and Schreiber Houses; two churches; one furniture store; two blacksmith shops and a few minor establishments.

In 1873, E. Shreve put down a 250-barrel oil well on the farm of Adam Stewart, and immediately the scene changes from one of comparative apathy to one of the most astonishing activity. People became almost wild with excitement, and everything immediately advanced to most extravagant prices, while thousands visited the new oil field with the expectation of becoming wealthy. Land that had been worth from $30 to $40 per acre was leased at the extravagant figures of from $100 to $200 per acre, and the land-holders retained a one-eighth royalty of the

*Atlas of the County of Butler, Pennsylvania, G. M. Hopkins and Company, Philadelphia, 1874)*

*Millertown 1874. Note Central Hotel at far left. Courtesy of Butler County Historical Society and the Drake Well Museum.*

production. At this time the railroad was not built to this place, and all kinds of vehicles were used to convey anxious speculators to the new field of operations. Buildings went up as if by magic, stores and business houses sprang up almost in a night, and before one year had elapsed a population of over 3,000 people claimed this place as their temporary home. The streets were filled with teams laden with oil well apparatus, merchandise, etc., so as to fairly blockade the way, while the sidewalks swarmed with an excited populace. The land where the Central House stood was leased for a term of ten years by Martin Hoch to Dr. W. P. Book at an annual rental of $1,200, and he built a three-story hotel on the ground, and still could not begin to accommodate the people who lived in any place where they could find covering for their heads. Hundreds of rough board shanties were built and occupied by those accustomed to aristocratic homes.

A twelve-foot front where Barnhart Frederick's store stood was rented for $60 per month. The rental for land usually ran from $12 to $28 per foot according to location, the latter price generally obtaining. The post office, from a small country office, soon became a prize worth contending for, the Postmaster's salary being $2,800. Farmers on whose land oil was found ceased to cultivate it, and merchants were pushed to their utmost capacity to furnish supplies for oil producers, and so neglected the farmer's trade that the attachment to plows and other farm implements frequently could not be obtained in the borough. Everybody was impressed with the importance of oil in all its bearings and ramifications.

## Fires

Everything was at the flood tide of prosperity when a terrible calamity befell the place, which practically swept the business portion out of existence. On the night of April 1, 1874, a fire broke out in the Book House, which originated from natural gas used for heating and lighting purposes that was not properly protected. The flames spread with fearful rapidity, and the

*Millerstown, Main Street in center of photograph with the Central Hotel at the top. 1874. Courtesy of Butler County Historical Society, Butler, Pennsylvania*

borough being destitute of fire protection, the people could only stand and see their property devoured by the fierce destroyer, utterly impotent to stay its ravages. The people became panic-stricken, especially those whose tenure of citizenship was dependent upon the result of speculation, and who had hazarded their all upon the single cast of the die. The mere mention of fire sent them rushing through the streets, and as they saw their property disappear through the lurid flames and murky smoke, their very impotence to stay the flames destroyed their self-possession so needful at such times, and as a result eight persons perished in the flames.

The property destroyed extended to the Frederick blacksmith shop and Reiber Hotel on the north; to Buffalo Creek on the south. It also destroyed nearly all the buildings on the west side of the street. The destruction of property was enormous, and the loss placed at $229,000 in the local papers, with only $64,000 insurance.

Hardly had the ashes cooled, however, before new buildings began to be erected, and in a few months, the evidences of the fire had largely disappeared, and in a majority of instances the new buildings were superior to those destroyed. Such was the recuperative powers of the place, under the large production and high price of oil, that very largely those who had lost their all in the flames immediately rebuilt and commenced life again with redoubled energy, and the population continued to increase.

On the night of April 11, 1875, another fire destroyed $100,000 worth of property. The fire started between two stores, located about where the Herald office stood, and swept down the west side of Main Street, to the German National Bank, where its progress was stayed; but it swept westward destroying about forty houses. A local paper says: "The most important losses were those sustained by McKinney Bros., and Galey's building, oil producers, the German National Bank, S. McBride's store, A. Talmo's paint store, and the Hogan Opera House," the other structures "being principally saloons and small tenement buildings." The lots laid bare were soon covered with buildings, and business was again resumed as actively as ever before. But fate seemed against the business men, for Millerstown was again visited by the dread destroyer on the night of December 6, 1877, which laid waste over two blocks of the business portion of the borough, entailing a loss of about $125,000. As the insurance covered only about one-fifth this amount, the business interest received a severe but not death-like shock.

The origin of the fire is to this day clouded in mystery, and the only solution to the enigma is the theory advanced that it resulted from the explosion of a lamp left burning in the store of C. F. Aldinger, situated on the east side of Main Street. The fire was first discovered by G. B. McCalmont, whose office was located in the second story of the Aldinger building.

*Union Oil Pump Station, Millerstown, Pennsylvania, 1874*
*This pump station was located in Millerstown, Pennsylvania, about one-half mile from the Central Hotel, Main Street. The Union Pump Station was owned by the Union Pipe Company and is the white building with smoke, steam and a high smoke stack near the center of the photograph. (Courtesy of Lenn "Pithole" McCorry, Butler County Historical Society, Butler, Pennsylvania)*

The alarm was promptly given, but the flames had made too much progress to be easily stayed, and they swept southward from the store of C. F. Aldinger to the drug store of C. D. Aldinger, adjoining, and soon it was enveloped in flames and totally destroyed. The Book House, a large three-storied building, was also fated and the place where it once stood was marked by a pile of ashes. By the utmost persistency, the fire was stayed from crossing the street and laying waste the property on the south of the Book House and the west side of Main Street. It however extended northward and in addition to other buildings destroyed F. Schweiger's shoe store. Among the heavy losers could be mentioned Barney Forst, who lost some $10,000, and B. Frederick, whose hardware and dry goods store fell a prey to the flames, his loss being some $14,000, only a part of which was covered by insurance. The flames were not stayed until the Schreiber House and the dwelling house of A. Henshaw, and blacksmith shop of H. Myers beyond had been destroyed.

The energy and pluck that marked the business men of this place was again brought into requisition, and soon the burned and blackened traces of the fire was hid from view by buildings erected to replace those so recently burned.

In 1874, after the destructive fire, the absolute necessity of providing some adequate means of fire protection presented itself to all, and at an informal meeting held by the business men, they decided to raise money by subscription for the purpose of building water works.

## Newspapers

The first paper in Millerstown was established by O. P. Jackson in 1874. It was a diminutive sheet, about eight by ten inches, rejoicing in the name of "The Sand Pump." It was as short-lived as it was small-sized. The "Millerstown Review," a six-column paper, established by Rev. A. S. Thorn, made its appearance in 1875. In 1879, the proprietor discontinued publication and removed to Atwood, Kansas, where he established the Atwood Pioneer.

The "Weekly Herald" was started in September, 1876, by S. J. Small, who carried it on until May 19, 1877, when it was purchased by the present proprietor, P. A. Rattigan.

## Oil Exchange

The Millerstown Oil Exchange was organized in October, 1882, with Thomas Dorsey, president, and Dr. S. D. Bell, vice-president. The Oil Exchange room was in the Central Hotel and this exchange influenced occasionally the oil market of the world. Many oil men speculated and what made fortunes for the members at one time wrecked many of them financially in the end.

## Oil Territory Declines

Millerstown began to decline as the oil supply became exhausted, and from a population variously estimated from 5,000 to 6,000, it had at the last census but 1,250, exclusive of Iron City, a suburban place, which consisted almost exclusively of board shanties built during the excitement. Deserted buildings are seen on every hand, and a fine residence costing about $4,000 was sold for $450, which gives an idea of the depreciation of property. A large number of buildings have been torn down and taken to other places.

(History of Butler County, Pennsylvania, 1883; Waterman, Watkins and Co., Chicago, Illinois)

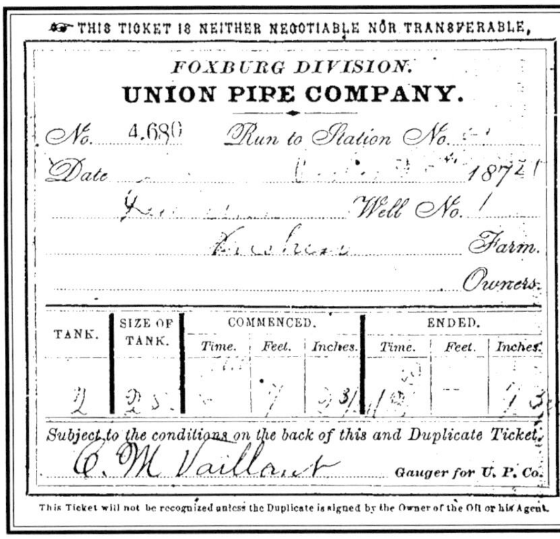

THIS TICKET IS NEITHER NEGOTIABLE NOR TRANSFERABLE,

FOXBURG DIVISION.

UNION PIPE COMPANY.

No. 4,680 Run to Station No.

Date 187

Well No. 1

Farm.

Owners.

| TANK. | SIZE OF TANK. | COMMENCED. | | | ENDED. | | |
|---|---|---|---|---|---|---|---|
| | | Time. | Feet. | Inches. | Time. | Feet. | Inches. |
| 2 | [illegible] | | 7 | 3 | | | |

Subject to the conditions on the back of this and Duplicate Ticket.

C. M. Vaillant Gauger for U. P. Co.

This Ticket will not be recognized unless the Duplicate is signed by the Owner of the Oil or his Agent.

*Union Pipe Company oil flow certificate, 1874. Courtesy Lenn "Pithole" McCorry, Butler County Historical Society*

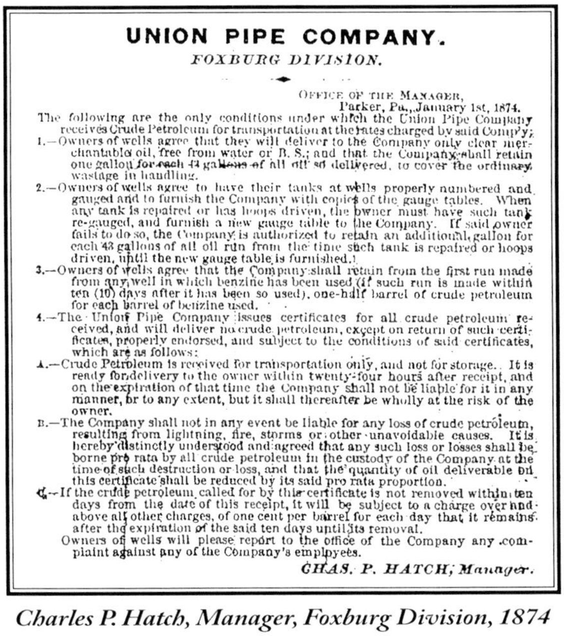

UNION PIPE COMPANY.

FOXBURG DIVISION.

OFFICE OF THE MANAGER,
Parker, Pa., January 1st, 1874.

The following are the only conditions under which the Union Pipe Company receives Crude Petroleum for transportation at the rates charged by said Comp'y;

1.—Owners of wells agree that they will deliver to the Company only clear merchantable oil, free from water or B. S.; and that the Company shall retain one gallon for each 43 gallons of all oil so delivered, to cover the ordinary wastage in handling.

2.—Owners of wells agree to have their tanks at wells properly numbered and gauged and to furnish the Company with copies of the gauge tables. When any tank is repaired or has hoops driven, the owner must have such tank re-gauged, and furnish a new gauge table to the Company. If said owner fails to do so, the Company is authorized to retain an additional gallon for each 43 gallons of all oil run from the time such tank is repaired or hoops driven, until the new gauge table is furnished.

3.—Owners of wells agree that the Company shall retain from the first run made from any well in which benzine has been used (if such run is made within ten (10) days after it has been so used), one-half barrel of crude petroleum for each barrel of benzine used.

4.—The Union Pipe Company issues certificates for all crude petroleum received, and will deliver no crude petroleum, except on return of such certificates, properly endorsed, and subject to the conditions of said certificates, which are as follows:

A.—Crude Petroleum is received for transportation only, and not for storage. It is ready for delivery to the owner within twenty-four hours after receipt, and on the expiration of that time the Company shall not be liable for it in any manner, or to any extent, but it shall thereafter be wholly at the risk of the owner.

B.—The Company shall not in any event be liable for any loss of crude petroleum, resulting from lightning, fire, storms or other unavoidable causes. It is hereby distinctly understood and agreed that any such loss or losses shall be borne pro rata by all crude petroleum in the custody of the Company at the time of such destruction or loss, and that the quantity of oil deliverable on this certificate shall be reduced by its said pro rata proportion.

C.—If the crude petroleum called for by this certificate is not removed within ten days from the date of this receipt, it will be subject to a charge over and above all other charges, of one cent per barrel for each day that it remains after the expiration of the said ten days until its removal.

Owners of wells will please report to the office of the Company any complaint against any of the Company's employees.

CHAS. P. HATCH, Manager.

*Charles P. Hatch, Manager, Foxburg Division, 1874*

# Millerstown, Butler County, Pa.

A village has existed here for upwards of half a century—and previous to the advent of the derrick and drill, presented that dead-and-alive appearance common to towns situated at considerable distances from railway connections, manufacturing or mining industries. Its silent streets wore a Sunday aspect. It lived in quiet peace, and could boast of a couple of taverns, a number of stores, tradesmen's shops, a church, school-house, and less than 200 men, women and children. Many of the dwellings gave evidence that the owners were well-to-do, as they presented a comfortable appearance.

In March of 1873, the first well, "The Shreve," was struck, which gave ample proof of the value of the Millerstown territory, and so attracted many of our operators. Millerstown soon became crowded with people from all parts of the region. New wells were struck, and with each strike the town swelled in population and importance. At the present writing it has two banks,

*Men of vision to whom the world owes much. Sixty-five of the world's most important pioneer oilmen at Millerstown, Pennsylvania, 1874.*
*This old photograph needs no introduction to oil men. It has been reproduced on more than one occasion for their benefit, but it probably will carry fresh interest for the layman. The men who posed for this picture were pioneers in the true sense. They were blazing the trail which lighted the world. They had faith in the permanence of petroleum as a servant of man. Among them were John D. Archbold, Daniel O'Day, Capt. J. J. Vandegrift, Michael Murphy, J. M. Guffey, Charles Pratt, Platt Crosby, Jonathan Watson, M. L. Lockwood, Curt McKinney, Richard Jennings, John H. Galey, Joseph Sibley, Major John Steel, Peter M. Grace, John Preston, Washington Emery, Thos. W. Phillips, G. T. Braden and N. V. V. Franchot—all well-known in the oil world. The chief value of the picture, possibly, is its reminder that were it not for men of vision we would make no real progress. We are too prone to judge the future by the present, and to scout the suggestion that tomorrow may be brighter than today. Civilization is advanced by the men who can see through the clouds of pessimism and who have the courage to hope for better things. The petroleum industry is essentially progressive. As well try to stop a flood with a pile of sand as to attempt to restrain its development. Nature has given petroleum to the human race for use, and it is human nature to extract from it the maximum service possible. But this will be done, not by those who see the horizon of human experience and say this is the boundary of our estate, but by men of broader faith who push forward in the van of progress, marking the way which their critics must follow.*

two pipe line offices, four hotels, four livery stables, a number of hardware and drug stores, and some six or seven grocery and dry goods houses, and has a population of 2,000 to 2,500. The town, as it now stands, is well built, and has the look of permanence. It is situated on a side-elevation overlooking the valley, and is surrounded by a beautiful and fertile farming country—its countless hills and dales forming a picturesque landscape of rare beauty. "The Sand Pump," a daily newspaper, is published here.

(History of Petroleum, Vol. I and II, J. T. Henry, 1873)

## Men of Vision, 1874

One of the largest gatherings of "Oil Dignitaries" in the world at Millerstown, Pennsylvania, in 1874. On the left is the Butler County Bank, established in 1873 by H. L. Taylor and Company, John Satterfield, President, George G. Stiles, Cashier, and H. J. Hoyt, Teller.

Other officers are Joseph Bushnell, dealer in petroleum for Standard Oil, The Union Pipe Line, United Pipe Line, Relief Pipe Line, Columbia Conduit Pipe Line, Karns Pipe Line, Grant Pipe Line and the Keystone Pipe Line. Prices for the world's crude oil market were decided and posted here.

The second gentleman from the left is Joseph Overy, a "Good Joe" and a big oil producer. He also was the builder of the Keystone Pipe Line, which ran from St. Joe to Freeport, Pennsylvania. (Photograph courtesy of Butler County Historical Society, Lenn "Pithole" McCorry, Butler County Oil Historian) Refer to page 419.

## The Great Divener Well
## Millerstown, Pennsylvania, 1874

One morning at the Central Hotel on Main Street, Millerstown, Fred Plumer overheard two drillers conversing about a big well coming in west of town and some talk about the Divener farm, which was one and one-half miles to the southwest. Plumer immediately ordered his horse saddled and galloped to the Divener farm home. He made the elderly couple an offer of $25,000 and the customary one-eighth royalty for the farm, which after some discussion was accepted.

John Lee and Frederick Plumer were in business together. They began the oil business at Franklin in 1860 and followed it through south of Millerstown to the great Divener. John Lee and Plumer started immediately to drill their test well and reached the pay sand on February 28, 1874. It proved to be a prolific producer, which averaged more than 1,000 barrels daily for some time. Just eight days later the well paid for itself. On March 7, 1874, it was sold to Vandergrift and Forman with a 107-acre lease for $92,500.

A quote from an Oil City paper on July 15, 1875: "The Hope Oil Company well at Troutman and the Divener have produced more oil than any other wells in the Butler County field." The Divener had produced an aggregate amount of 150,000 barrels of oil by November 23, 1875.

Lenn C. "Pithole" McCorry, Butler County oil historian, reported that up to 1962 the great Divener well had seven different owners in the past 94 years. Up to 1962 it had produced 451,370 barrels of oil which would represent the value of $1,354,110 and is producing at a profitable rate today, 1962. The great Divener well was drilled by Lee and Plumer to the depth of 1,700 feet. (Refer to pages 240-241.)

When the Butler and Karns City Narrow Gauge Railroad was built in November of 1876, a depot was established at the north end of the wood trestle near the well and the depot was named "Divener." It served for many years, as long as passenger trains were in service. In 2006 the Divener wood trestle was mysteriously burned down during a strike at a local refinery. (Lenn C. "Pithole" McCorry)

---

JNO. J. SATTERFIELD, President. GEO. G. STILES, Cashier.

**BUTLER COUNTY BANK,**

**Of Millerstown, Butler County, Pa.**

STOCKHOLDERS INDIVIDUALLY LIABLE.

Collections made on Fairview, Petrolia, Karns & Greece Cities at lowest rates.

Interest allowed on Time Deposits, 4, 5 and 6 per cent.

*Buys and sells U. S. Bonds, Gold, Coupons, Collects Interest, Sells Exchange, and does a General Banking Business.*

DIRECTORS:

JNO. SATTERFIELD, . . . . . . . . Millerstown, Pa.
JNO. PITCAIRN, JR., . . . . . . . . Oil City, Pa.
H. L. TAYLOR, . . . . . . . . . Petrolia, Pa.
J. J. VANDERGRIFT, GEO. V. FORMAN, } Of Vandergrift & Forman, . Oil City, Pa.

CHAS. DUFF, Pres't. W. G. STOUGHTON, V. Prest. JAS. C. SCOTT, Cash'r.

**Millerstown Savings Bank,**

**Millerstown, Butler County, Pa.**

**Paid in Capital Stock, $50,000.**

*Stockholders Individually Liable.*

DIRECTORS:

Chas. Duffy, W. G. Stoughton, John M. Thompson.
Chas. McCardless, Andrew Barnhart, Henry Westerman,
Martin Hoch.

***A General Banking Business Transacted.***

## MILLERSTOWN.

Previous to the oil excitement in this section, the greater portion of the ground upon which this town is built, was used for agricultural purposes, and the place was known as "Barnhardt's Mills," which was and still continues to be the Post Office address.

The character of the improvements made here have more of a tendency toward permanency, than is noticeable in many of the towns throughout the oil country, and with a little enterprise in the way of a Railroad, either to Butler, Pa., or to connect with the Allegheny Valley Railroad, would have the result of making this point the metropolis of the Lower Oil District, as the advantages offered here as a disbursments centre is favorable.

The town is principally built on two business streets, Main Street running north and south, and Slipperyrock, from east to west, on which are many flourishing stores, which are stocked with goods of every description, for the wholesale and retail trade.

The Hotel accommodations here are not to be excelled by any town in this part of the country.

### A

Adget ——— carpenter, Main.
Aldinger C. A., clerk, Slipperyrock.
Aldinger C. D., boots and shoes, Main.
Aldinger C. F., tobacco and cigars, Main.
Aldinger David, carpenter, Slipperyrock.
Aldinger H. C., dry goods and groceries, Slipperyrock.
Allegheny House, H. S. King, proprieter, Slipperyrock.
Anthony Frank, livery, Reiber House.
Anthony William H., hostler, boards Reiber House.
Barber Harris, driller, Slipperyrock.
Barnhard Andrew, Main.
Barnhart David F., wagonmaker, Mill,
Barnhart David, Sr., laborer, Mill.
Barnhart E. W., shoemaker, Mill.
Barnhart Simeon J., shoemaker, Mill.
Barnhart W. H., shoemaker, Mill.
Barry John, operator, Water.
Barton Alexander, livery, Main.
Becker John, teamster, Mill.
Bell Andrew, hostler, boards Reiber House.
Bell & Leffingwell, physicians, Main and Green.
Bell S. D., physician, office Kittanning.
Bernard Jas. J., grocer, Slipperyrock.
Bixby Laura, dressmaker, Slipperyrock.
Black Adam, teamster, boards Parker House.
Blystone J. W., grocer, Slipperyrock.
Book Charles L., barkeeper, boards Central Hotel.
BOOK W. P., proprietor Central Hotel, Main.
Booth A. P., tank builder, Slipperyrock.
BOYER EDWARD, proprietor, Sinks Hotel and Restaurant, Main.
Boyle P. M., of P. M. Boyle & Sons, Slipperyrock.
Boyle P. M. & Sons, groceries and dry goods, Slipperyrock.
Boyle J. W., of P. M. Boyle & Son, Slipperyrock.
Boyle H. F., of P. M. Boyle & Sons, Slipperyrock.
BRADFORD & HARDENBURG, livery, Main.
BRADFORD W. H., of Bradford & Hardenburg, Main.
Braman Phillip, constable, Water.
Brown Wm. S., miner, Mill.
Brownell Samuel, clerk, boards Main.
Brownell J. H., clerk, Slipperyrock.
Buckley C. W., miner, Kittanning.
Burford E., clerk, boards Central Hotel.
BUTLER COUNTY BANK, Main.
Byers S. B., carpenter, Main.
Byers Susan, widow, Mill.

### C

Canning Wm., shoemaker, boards, Allegheny House.
Cannon Amos, rig builder, Green.
Capsy J. H., carpenter, Kittanning.
CENTRAL HOTEL, W. P. Book, proprietor, Main.
Chestnut & Healy, boots and shoes, Main.
Chestnut Samuel, of Chestnut & Healy, Main.
Cleland W. J., of Clemenger & Cleland, Slipperyrock.
Clemenger F. J., of Clemenger & Cleland, Slipperyrock.
Clemenger & Cleland, grocers, Slipperyrock.
Clark Daniel, of Clark & Meeker, Slipperyrock.
Clark & Meeker, lumber yard, Slipperyrock.
Cohen D., Boots and shoes, Main.
Colby House, O. D. Colby, proprietor, Slipperyrock.
Colby O. D., proprietor Colby House, Slipperyrock.
Coners Martin, carpenter, Green.
Coners Thomas, carpenter, Green.
Conlan James, hostler, Mill.
Connor M. W., painter, boards Parker House.
Cooper Thomas, blacksmith, Kittanning.
CONNER W. F., of Freedley & Conner, Main.

Cramer Phillip, policeman, Water.
Cress John, blacksmith, Main.
Cumberland John, blacksmith, boards Parker House.

## D

DAILY SAND PUMP, O. H. Jackson, prop., Slipperyrock.
Dandyke John, contracter, Mill.
Dawson J. G., agent, Slipperyrock.
Dettie Louis, of Fairbairn, Dunlap & Co., Main.
Derling W. A., carpenter, boards Parker House.
Dick George, laborer, Green.
Dixon A. M., barber, Green.
Dorsey J. P., cook Sink's Hotel.
Drummond David, restaurant, Main.
Dudley C., barber, Slipperyrock.
DUNLAP SAMUEL R., of Fairbairn, Dunlap & Co., Main.
Dunlap William, carpenter, Frederick Farm.
Durant H. J., of Reid & Durant, boards Central Hotel.
Dye Orlando, carpenter and builder, boards Reiber's Hotel.

## E

Eckel Michael, tank builder, Slipperyrock.
Elsworth V., hostler, boards Reiber's Hotel.

## F

Fairbairn, Dunlap & Co., druggists, Main.
Fairfax Harriett, laundress, Green.
FETZER G. F., of Fetzer & Myers, Slipperyrock.
FETZER & MYERS, Millerstown flouring mills, Slipperyrock.
Filer Thomas, carpenter, Slipperyrock.
Fleegers S., furniture, Main.
Flick Abraham, livery, Slipperyrock.
FORST B., of Hirsch & Forst, boards Main.
Forquer Frank, driller, Green.
France E., tank builder, boards Slipperyrock.
Fredericks A., shoemaker, Green.
Fredericks Edward, blacksmith, Main.
Fredericks B., of B. Fredericks & Co., Main.
Fredericks B., & Co., dry goods, groceries, &c., Main.
Fredericks H., blacksmith, Slipperyrock.
Fredericks John, shoemaker, Green.
Fredericks J., blacksmith, Main.
Fredericks P. F., boots and shoes, Slipperyrock.
Fredericks P. G., of Fredericks & Co., Main.
Fredericks Wm., blacksmith, Main.
FREEDLEY J. J., of Freedley & Connor, Main.
FREEDLEY & CONNOR, photographers, Main.
Frisbee J. C., barkeeper, Central Hotel.
Fry S. S., hostler, boards Reiber's Hotel.

## G

Grisford J. C., of Grisford & Winger, Main.
Grisford & Winger, billiard room, Main.
Gallagher Michael, cabinet maker, Slipperyrock.
Gibson Minnie, millinery and dressmaking, Kittanning.
Gilespie John, driller, Main.
Goldstein Henry, cigars, tobacco and furnishing goods, Main.
Green Frank, clerk, boards Slipperyrock.
Greenwood P. W., contractor, Mill.
Griffin David, barkeeper Reiber's Hotel.
Grove Simon, teamster, Green.
Grove Willard, teamster, Green.

## H

Haislet John, driller, boards Parker House.
Haight Henry, carpenter, boards Parker House.
Hall Wm. F., baker, Slipperyrock.
HANLEN HOUSE, Jos. Hanlen proprietor, Main.
HANLEN JOS., proprietor Hanlen House, Main.
HANLEN W. A., barkeeper Hanlen House.
Hardenburg H. A., of Bradford & Hardenburg, Main.
Hang Edward, engineer, Green.
Haynes D. B., clerk, Main.
Healy B., of Chestnut & Healy, Main.
Heaton Peter, restaurant, Slipperyrock.

Hemfry ——— saloon, Main.
Herrmann John, mason, Green.
Hersh J. D., of Hersh & O'Brien, Main.
Hersh & O'Brien, machinist and oil well tools, Main.
Hills W. S., grocery, Slipperyrock.
HIRSCH & FORST, merchant tailors, Slipperyrock.
HIRSCH S., of Hirsch & Forst, boards Central Hotel.
Hoch Adam, brewer, Main.
Hoch Adolf, of Hoch Bros., Slipperyrock.
Hoch Agustus, of Hoch Bros., Slipperyrock.
HOCH BROS., livery, Slipperyrock.
Hoch Gotlieb, brewer, Main.
Hoch Martin, livery, Slipperyrock.
Hoffman William H., oil operator, Mill.
Hogan Martin, driller, boards Parker House.
Holcomb D. P., painter, Main.
Holland Austin, clerk, boards Main.
Hollan Josiah, teamster, Green.
Holmes David, bootmaker, Main.
Hubard R. O., hostler, Central House.
Huccard Phillip, carpenter, Green.

## I

Ingersoll H. D., of Templer & Ingersoll, Main.
Irons B. O., oil operator, Main.
Irons J. F., oil producer, Main.
Irons J. B., livery, Green.

## J

Jackson O. H., proprietor daily sand pump, Slipperyrock.
Jacobs Stephen, carpenter, Main.
Johnson Alexander, miner, Kittanning.
Johnston C. H., blacksmith, Kittanning.
Johnson Chasles H., of Johnson & White, baker, Mill.
Johnson Samuel, miner, Kittanning.
Johnson & White, bakery, Mill.

## K

Keil W. P., stewart Reiber's Hotel.
Kelly D. J., carpenter, boards Parker House.
Kelly Washington, butcher, Slipperyrock.
KENNEDY Wm. T., manager Reid & Durant, Main.
King H. S., proprietor Allegheny House, Slipperyrock.
Kormer John, grocer, Mill.
Kostinger Cristopher, hostler, Reiber's Hotel.
Kunz George, porter Sink's Hotel.

## L

Lahay Michell, contractor, boards Parker House.
Langworthy S. A., telegraph operator, boards Reiber's Hotel.
Larway S., pumper, Main.
Lavelle Charles, painter, Green.
Lawrence Henry, carpenter, Green.
Ledyard H. C., dentist, Main.
Leffingwell J., of Bell & Leffingwell, Green.
Leicht Christian, wheelwright, Kittanning.
Lemme A., tank builder, boards Slipperyrock.
Little Frank, laborer, Mill.
Lockhart Henry, of Russell & Lockhart, Main.
Long Joseph, miner, Kittanning.
Loudon Amos, helper, Main.
LOW W. A., clerk, Hanlen House.
Lupher J. D., oil well supplies, &c., Slipperyrock.

## M

Manks Joseph, waiter Sink's Hotel.
Manning William, miller, Mill.
Marks A., of A. Marks & Co., Main.
Marks A., & Co., clothing, dry goods, &c., Main.
MARSHALL J. O., attorney at law, Main.
MARSHALL KENNEDY, attorney, Main.
Marshall S. S., of Marshall & Welch, boards Main.
MARSHALL S., law student, Main.
Mashall & Welch, tank builders, &c., Slipperyrock.
McBRIDES DRUG STORE, Main.
McBRIDE P., drggist, Main, boards Central Hotel.
McCabe James, boarding house, Slipperyrock.
McCann James, blacksmith, boards Parker House.
McDowell A. D., rig contractor, Mill.



McELROY BROS., butchers, Slipperyrock.
McElroy Henry, of McElroy Bros., Slipperyrock.
McElroy James, of McElroy Bros., butcher, Slipperyrock.
McFarland John A., Eureka Hotel, Main.
McGuire Dennis, pumper, boards Parker House.
McKEE W. J., carpenter and builder, planing mill and wheel factory, Water.
McKee D. F., manager J. D. Lupher's, Slipperyrock.
McLaughlin Wm., books and stationary, Main.
McMichael Josiah, physician, Main.
McMillan William, grocer, Water.
Meeker Louis, of Clark & Meeker, Slipperyrock.
Micheal J., physician, Green.
Miller J. W., bookkeeper, boards Main.
Miller R. P., clerk, boards Central Hotel.
Millerstown Flouring Mills, Fetzer & Myers, Mill.
Millerstown Pipe Line, Main.
MILLERSTOWN SAVINGS BANK, Main.
Mordan E. D., barber, Main.
Morgan G. S., agent Millerstown Pipe Line, boards Central Hotel.
Moore James, policeman, Kittanning.
MUDGE ALFRED G., attorney at law, Main.
Mullin James, B., druggist, boards Central Hotel.
Mulligan Patric,k saloon, Main.
Murphy S., of Norton & Murphey, Slipperyrock.
Myers John G., of Fetzer & Myers, Green.

## N

Newel James, butcher, Slipperyrock.
Newell Z., real estate agent, Main.
Norton D. E., of Norton & Murphy, Slipperyrock.
Norton H. D., blacksmith, boards Reiber's Hotel.
Norton & Murphy, hardware and oil well supplies, Slipperyrock.

## O

O'Brien M., of Hersh & O'Brien, Main.
Olin George, tank builder, boards Slipperyrock.

O'Neal John, hostler, boards Slipperyrock.
Otto ———, tank builder, Slipperyrock.

## P

Pacific & Atlantic Telegraph Co., office Main.
Parker William H., teamster, Water and Main.
Paul Eliza, widow, Slipperyrock.
Poff Valaentine, butcher, Mill.
Pohl Rev. A. F. W., German Lutheran, Main.
Pope Henry, driller, boards Parker House.
Post Office, Main street, W. McLaughlin, post master.
Power Charles M., clerk, Main.

## R

Rawson William, boards Parker House.
Rawson Morris, Main.
Ransew William, miner, Main.
Reard John, carpenter, Green.
Reiber George, clerk Reiber House.
Reiber Henry, butcher, Slipperyrock.
REIBER WILLIAM, proprietor and manager Reiber House.
REID & DURANT, hardware oil well supplies, Main.
Riger John, of John Riger & Co., Green.
Riger John & Co., wagonmakers, near Main.
Ripley H. A., manager billiard room, Central Hotel.
Roach David, teamster, Water.
Robinson C. C., Livery and Exchange Stable, Main.
Robinson John, butcher, Slipperyrock.
Rumbaugh D., groceries, &c., Slipperyrock.
Russell A., livery, Main.
Russell R. H., clerk, boards Main.
Russell & Lockhart, drug store, Main,
Russell T. H., of Russell & Lockhart, Main.
Ryan John, clerk, boards Slipperyrock.
Ryan P., grocer, Slipperyrock.

## S

Salik Adam, hostler, boards Slipperyrock.
Salvage E. J., boots and shoes, Slipperyrock.
SATTERFIELD JOHN, Pres't Butler County Bank, boards Central Hotel.
Seibert Reuben, wagonmaker, Slipperyrock.
Short John, baker, boards Slipperyrock.
Shuster Andy, blacksmith, Slipperyrock.
Sierrel John, laborer, Mill.
SIMPSON A. H., tinware and stoves, Slipperyrock.
SINKS' HOTEL AND RESTAURANT, Main.



SINKS J. H., Agent Sinks' Hotel, Main.
Slesman J. M., carpenter, Green.
Slesman W. F., carpenter, Green.
Sloan S. S., carpenter, boards Parker House.
SMALLEY J. D., flour, feed and groceries, Slipperyrock.
Smith G. C., furniture dealer, Main.
Smith Henry, driller, boards Parker House.
Smith Maxwell, driller, boards Parker House.
Smith N. A., clerk, boards Slipperyrock.
Snively Charles, driller, boards Slipperyrock.
Snyder Andrew, brewer, Main.
Snyder T. K., shoemaker, Mill.
Spatrz Morris, Centre Avenue, house Main.
Splain Wm., telegraph operator, Main.
Stahl H., of Stahl & Scharbach, Main.
Stahl & Scharbach, clothing and furnishing goods, Main.
Steele J. A., livery, Mill.
Steirly Charles, machinist, boards Sinks Hotel.
Stern M., of A. Marks & Co., Main.
Stewart W. A., laborer, Slipperyrock.
STILES GEO. G., Cashier Butler County Bank, boards Central Hotel.
Summers Joseph, confectionary, Slipperyrock.
Sutherland James, hostler, boards Reiber House.
Swank D., tankbuilder, Slipperyrock.
Scharbach C., of Stahl & Scharbach, Main.
Schnitzius John, boarding house, Mill.
Scott D. H., bookkeper Millerstown Savings Bank, boards Central Hotel.
SCOTT JAMES C., Cashier Millerstown Savings Bank, boards Central Hotel.

## T

Taylor H. L & Co., oil producers and dealers, Main.
Taylor Lot, teamster, Water.
Temple F., of Temple & Ingersoll, Main.
Temple & Ingersoll, saloon, Main.
Threde Albert, oil operator, Mill.
Troxall Jacob, pumper, Green.
Turner John, rig builder, Main.
Tyler ———, physician, Main.

V

Varry John, operator, Water.
Voss Edward, tinner, Slipperyrock.

W

Wagner ——, widow, Green.
Wares F. P., dressmaker, Slipperyrock.
Waugh John H., Teller Butler County Bank, boards Central Hotel.
Welch A. M., of Marshall & Welch, boards Main.
Weld Desilver, dry goods and notions, Main.
Wells J. C., of I. C. Wells & Bro., Main.
Wells I. C., of I. C. Wells & Bro., Main.
Wells I. C. & Bro., books and stationary, Main.
Westerman C. J., Teller Millerstown Savings Bank, Main.
Westerman H. L., dry goods and groceries, Main.
Western Union Telegraph Company, office, Main.
White N. S., of Johnson & White, Mill.
WHITESTONE D., clothing and gents' furnishing goods, Main.
Wilcox Harrison, hostler, boards Reiber House.
Williams Thomas, contractor, Main.
Winger J. J., of Gaisford & Winger, Main.
Wise Luke, helper, Main.
Wolford Eli, blacksmith, Slipperyrock.
Wright J. H., tool dresser, Green.
Wright W. A., carpenter, Green.

Y

Young James B., hostler, Slipperyrock.
Young James, livery, Mill.

## MILLERTOWN BUSINESS DIRECTORY.

**Attorneys.**

Marshall J. O., Main.
Mudge Alfred G., Main.

**Bakery.**

Johnson & White, Mill.

**Banks.**

Butler County Bank, Main.
Millerstown Savings Bank, Main.

**Barber Shop.**

Dudley C., Slipperyrock.

**Billiard Hall.**

Central Hotel, Main.
Grisford & Winger, Main.

**Books and Stationery.**

McLaughlin Wm., Main.
Wells J. C., & Bro., Main.

**Boot and Shoe Makers and Dealers.**

Aldinger C. D., Main.
Chestnut & Healey, Main.
Cohen D., Main.
Fredericks P. F., Mill.
Salvage E. J., Slipperyrock.

**Butchers.**

McElroy Bros., Slipperyrock.
Poff Valentine, Mill.

**Cabinet Makers.**

Gallagher Michael, Slipperyrock.

**Carpenters & Builders.**

Dye Orlando, Reiber's Hotel.
McKee W. J., Water.

**Cigars and Tobacco**

Aldinger C. F., Main.
Goldstein Henry, Main.
Whitestone D., Main.

**Clothing.**

Hirsch & Forst, Slipperyrock.
Marks A., & Co., Main.
Stahl & Scharbach, Main.
Whitestone D., Main.

**Dentist.**

Ledyard H. C., Main.

**Drug Stores.**

Fairbairn, Dunlap & Co., Main.
McBride's Drug Store, Main.
Russell & Lockhart, Main.

**Dry Goods.**

Aldinger H. C., Slipperyrock.
Boyle P. M., & Sons, Slipperyrock.
Fredericks B., & Co., Main.
Marks A., & Co., Main.
Weld Desilver, Main.
Westerman H. L., Main.

**Engines, &c.,**

Lupher J. D., Slipperyrock.

**Flour and Feed.**

Smalley J. D., Slipperyrock.

*Central Hotel, Millerstown, Pennsylvania*
*This brick building replaced the original wood structure. The Central no longer exists.*

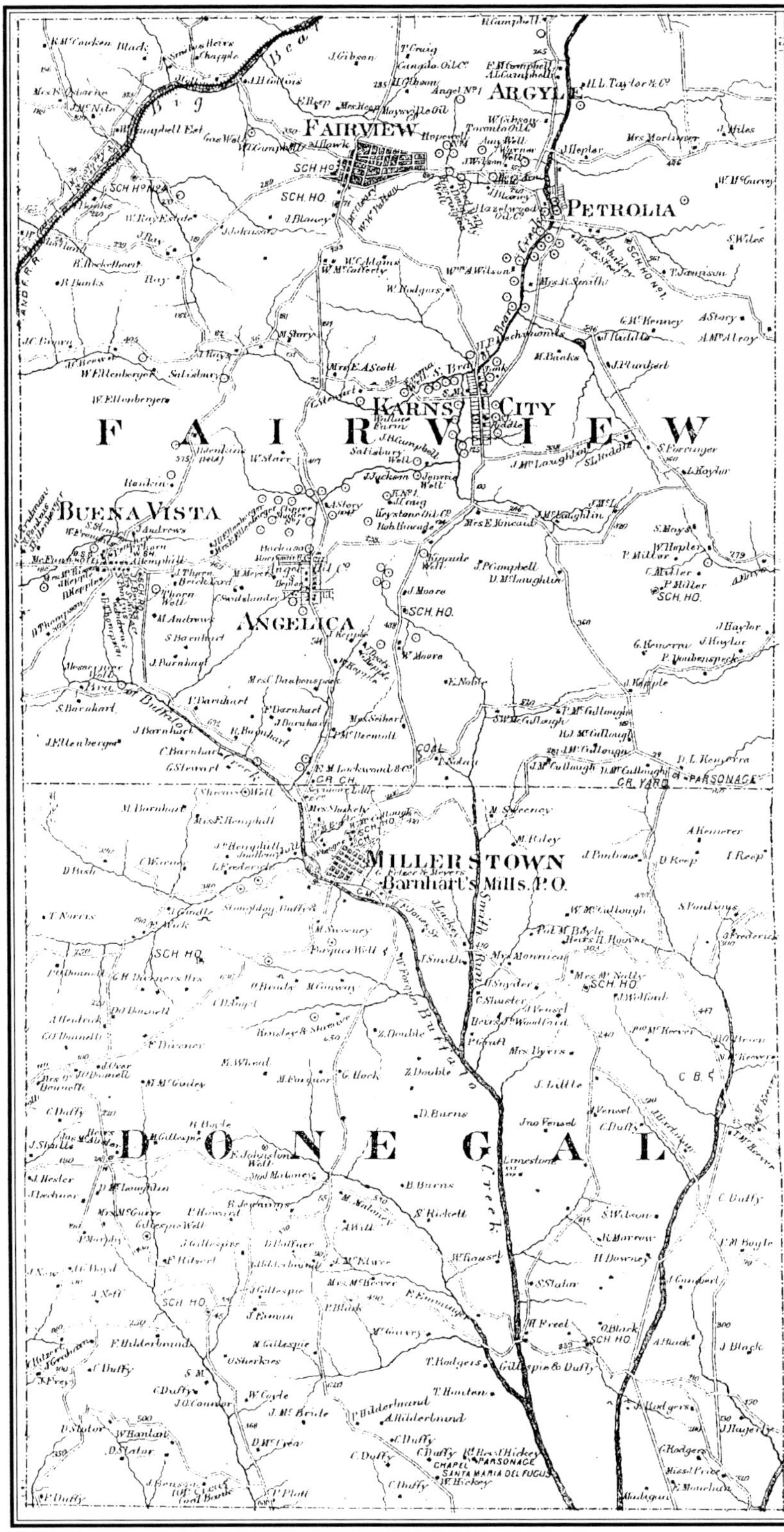


*Atlas of the County of Butler, Pennsylvania, 1874, Fairview and Donegal Townships, Argyle, Fairview, Petrolia, Karns City, Buena Vista, Angelica and Millerstown.*

## Millerstown, June 13, 1880
## Runaway and Accident
## Special to Derrick

"John Campbell left here for his home in Butler last night, and his horses got frightened at the top of the Butler hill. They turned and came dashing down the hill, throwing Campbell out. He was not injured. One horse struck an engine in front of Dale and Larkin's Shop and was instantly killed.

## Pipe Line and Railroad War
## 1874

Samuel Van Syckel has been given credit for building the first oil pipe line in 1865. This first pipe line was four miles in length and ran from Pithole to the railroad at Miller's farm. Van Syckel hired armed guards to prevent teamsters, who hauled oil in barrels by horse and wagon, from tearing up the pipeline. The teamsters saw pipe lines as the enemy which threatened their livelyhood.

## Columbia Conduit Pipe Line

On October 10, 1873, the Oil City Derrick reported that a new pipeline from Butler County oil fields to Pittsburgh is projected. Work is expected to begin soon with over one hundred fifty new wells being drilled in the Millerstown district. On April 1, 1874, a survey was undertaken to determine the right-of-way and on June 19, 1874, they started to lay a three-inch pipe from Millerstown to Fairview near Pittsburgh, a distance of 35 miles. On November 28, 1874, the Columbia conduit was completed and began to pump 3,500 barrels a day. This was the first oil pipeline to reach the city of Pittsburgh.

*Transferring oil across the railroad tracks, 1874. A sign on each side of the tracks reads, "Look and Listen!" (Frank Robbins Photograph, Oil Field Photographer, Oil City, Pennsylvania)*

Everything was running smoothly until the railroad company realized that just like the Pithole teamsters, the railroad would no longer haul petroleum and would lose revenue for the Pennsylvania Railway.

The Columbia Conduit Pipeline Company had no right-of-way across the tracks and the railroad had no intention of giving them one. The oil pipeline was laid along the bed of a stream but where it crossed the tracks it was torn up by the railway employees.

## The Solution To A Problem

The petroleum pipeline was relaid by a large force of men, and again destroyed by the railroad company. Finally, J. G. Benton of Titusville, Pennsylvania, came up with the startling solution. To overcome the difficulty, the oil was transferred on the public highway crossing the tracks in tank wagons of 25-barrel capacity, which transferred 8,000 barrels per day. It took only two minutes to fill the horse-drawn tank wagon from a six-inch fill pipe leading from a storage tank. To empty the oil back into the pipeline, the tank was drained into a trough leading to another large iron tank from which it was pumped into the line. Everyone was happy but the railroad. The Oil City Derrick announced that on July 19, 1875, the Columbia Conduit pipeline established their new office at St. Joe.

## Greece City

The drilling crept south through Parker, Bruin, Karns City and Chicora. This narrow strip running north and south was the third sand and was thought to be the only strata containing oil. New wells were being put down everywhere and among them was one put down by David Morrison on the farm of his father-in-law, S. S. Jamison, in the southern part of Concord Township, about seven miles from Petrolia, in a southeasterly direction. This well, on August 22, 1872, struck the great fourth sand deposit, which was literally to "pour forth rivers of oil," to bring into existence several new towns, among them Greece City, and ultimately to have the most powerful depressing effect upon the petroleum market ever known in the history of the trade.

*Hauling oil across the Pennsylvania Railroad right-of-way near Pittsburgh. This photograph was taken from the top of the receiving oil tank. (Views of the Pennsylvania Oil Region, Frank Robbins, Photographer, Oil City, Pennsylvania, 1874)*

The "wild cat" known as the Morrison well surprised even the most sanguine, and after making a spurt of 700 barrels the first day, flowed at the rate of 300 for some time. Her average for the first six months was 250 barrels per day. Territory in the vicinity, and especially between the new well and Petrolia, was eagerly leased, and a number of wells were soon located. Theodore Huselton, who owned the farm south of the Karns branch of the

Connoquenessing and adjoining the Jamison farm, on which the lucky strike had resulted, immediately laid out a portion of his land in village lots, and the owners of property on the north side followed suit.

A village sprung up as if by magic and grew with astounding rapidity, being incorporated as a borough a few months after its origin. Greece City, as it was called, has now almost entirely disappeared, but for a brief period during the years 1872, 1873 and 1874, it was one of the prominent towns of the oil region and exhibited great bustle and business stir. It had several good hotels, three banks and a large number of shops, stores and dwellings. That portion of the town north of the creek was twice burned down and rebuilt. In this respect, it shared the usual fate of the other lower region oil towns—Petrolia, Karns City, Modoc, Buena Vista, Martinsburg and Millerstown all having disastrous fires, and some of them being several times visited. Greece City was probably, during the heyday of the oil excitement, the hardest town in the lower country. Saloons, gambling places and bawdy houses abounded; the town was thronged with brazen and depraved characters, and drunkenness, profligacy and crime held high carnival, much to the disgust of the many steady-going, substantial class of citizens who were compelled by business interests to make their home there. (History of Butler County, Pennsylvania, 1883; Waterman, Watkins and Co., Chicago, Illinois)

## Greece City, Butler Co.

On August 22, 1872, the first well was struck on the Jamison farm and early in September the first building made its appearance. Greece City rose rapidly and fell with the same speed. It had, at the height of its glory, about 1,200 inhabitants, and had four or five hotels, three banks, and the full variety of stores.

*Greece City, Butler County, Pennsylvania, 1873*
*Photograph courtesy of Leroy Kuhn, Don Litzenberg, and the Butler County Historical Society.*

The "Greece City Review," a daily newspaper, was edited by Rev. W. A. Thorne, and could be purchased in the hotel lobbies. On December 3, 1873, fire destroyed a large portion of the town at a loss of $90,000.

Twenty buildings were destroyed in the disastrous fire of December. Greece City was comprised of a general store, a hardware, a billiard hall, a restaurant, four hotels, a clothing store, two drug stores, two business buildings, two banks, a boarding house, a blacksmith shop, a livery stable, a well and tank shop and a barber shop. As production declined the newspaper wrote, "The city is fast disappearing from the face of the earth."

*Oil Exchange Hotel, Greece City, 1873*
*Drake Well Museum, Titusville, Pennsylvania*

For a time Greece City was considered the center of the oil excitement in Butler County. The town had every possibility of becoming an important city. At one time, the post office had a greater cancellation than the Butler office. Maps of 1872 show the proposed railroad from Parker to Butler was routed through Modoc and Greece City. From there it would have followed the gentle slope of the creek to Butler. Had the railroad become a reality and the oil boom lasted, things would have been very different. Today Greece City is only a few houses scattered along Route 38. (History of Petroleum, Vol. I and II, J. T. Henry, 1873)

## Greece City
## Parker City Daily News, May 15, 1874

It was reported by the local correspondent for the Oil City Derrick that after a long night of drinking freely from the ever "flowing bowl" at the Oil Exchange Hotel on Main Street, Greece City, young Will Warden developed a situation. His situation was explained in a rather unique way: Young Will "became a rampageous lot of whisky with a man outside of it," and was lodged in the cooler Monday.

OIL EXCHANGE HOTEL.

JOHN F. COLLINS,

PROPRIETOR.

MAIN STREET,

GREECE CITY, PA.

# GREECE CITY.

Greece City is located about eight miles from Butler Borough, in Butler County, has received its present growth since September 1872, and owes its existence and importance to the development of oil territory in that vicinity.

About the first of October, 1872, the Well known as the "Old Morrison Well" was struck, which flowed about 500 barrels daily.

Other wells were put down and a good yielding territory developed. The town at present contains about 1200 inhabitants, and a goodly number of stores, commission houses, &c. There are two Banks, the Greece City Bank and the Producers' Bank of Butler county. There are four stores for the sale of Hardware, Oil well supplies, &c. The other branches of commerce, such as Drugs, Dry Goods, Groceries, Cigars and Tobacco, and Gentlemen and Ladies' Furnishing Goods, are all represented here and in a flourishing condition.

There are five Hotels in the place, which offer to the traveling public as good accommodation as any town in the oil region.



## A

Abrahamson S., candy stand, &c., Main.
Addle James M., attorney at law, Main.
Aiken & Russell, grocers, Main.
Aiken W. A., of Aiken & Russell, Main.
Andrews J. A., hostler, Main.
Andrews John, of Andrews & Nelson, Main.
Andrews & Nelson, hardware and oil well supplies, Main.
Andrews Samuel, butcher shop, Hutchinson.
Andrews William, clerk, Main.
Armstrong S. P., oil broker, Main.

## B

BACKUS, MORRISON & CO., Hardware and oil well supplies, Main.
Bagley Alex., helper, Main.
PRODUCERS' BANK OF BUTLER COUNTY, Main.
Barnes J. R., cigars, tobacco &c., Main.
Barr & Craig, drug store, Main.
Barr M. A., of Barr & Craig, Main.
Bartold C., of Bartold & Haas, Main.
Bartold & Haas, clothing &c., Main.
Belles & Krumpe, blacksmiths, Thompson.
Belles Sipe, of Belles & Krumpe, Modoc Road.
Birchard W. B., boarding-house Thompson.
Bladen Aaron, cook, Thompson.
Boardley E. H., barber, Thompson.
Boardley Wm., barber, Main.
BOOK W. P., of W. P. Book & Co., Main.
BOOK W. P., & Co., druggists, Main.
Boundon W., saloon, Main.
Boughton C. V., of Boughton & Chandler, and Notary Public, Main.
Boughton & Chandler, general agents, Geo. W. Tift, Sons & Co., engines, Main.
BRAWLEY ED., proprietor Brawley House, Main.
BRAWLEY HOUSE, Ed. Brawley, proprietor, Main.
Bright Christopher, teamster, Thompson.
BRISCOE A. E., of W. P. Book & Co., Main.
Bristoll Thomas, hostler, Main.
Buckles ———, boarding-house, Main.
BUTLER HOUSE, Main.

## C

Chaffee P. P., clerk, Main.
Clark A. L., tinsmith, Main.
Cleveland Charles, blacksmith, Main.
Cleveland E. C., butcher, Main.
Cluxton Dr. C. F., physician, Main.
Cochran C. G., superintendent oil wells, Main.
Cochran Charles, justice of the peace, Main.
Cochran L. M., clerk, Main.
Cochran Sulton, pumper, Main.
COLLINS JOHN F., proprietor Oil Exchange Hotel, Main.
Conn James L., deputy sherriff, Main.
Coobau Frank, barber, Main.
Coobau Frank, painter, Thompson.
Cooper J. M., agent United Pipe Line, office head of Main.
Cornell I. C., salesman, Main.
Cottrall J. W., boarding-house Main.
Coxton Robert, supt. Collins wells, boards Hillside Hotel.
Craig J. S., of Barr & Craig, Main.

## D

Devinger John, porter, Main.
Dorsey Patrick, oil operator, boards Hillside Hotel.
Downey James, driller, Main.
Durham W. M., telegraph operator, Main.

## E

Eldridge Charles, hostler, Main.
EUROPEAN HOTEL, D. STEWART PROP. MAIN.

## F

Fleming E. G., clerk, Main.
Frazier Simon, Main.
Frost Amon, of Frost & Truby, Main.

## G

Gardner T. C., clerk, Main.
Glenn J. S., wagon maker, Main, res. Second.
Gordon J., of Thompson & Gordon, High.
Grant Wm., porter Brawley House.
GREECE CITY BANK, Main.
Grossman John A., of Sullivan & Grossman, Main.

## H

Haas H., of Bartold & Haas, Main.
Hair & Waggor, cigars, tobacco, &c., Main.
Hale Harry, laborer, Thompson.
Hall B., of B. Hall & Co, Main.
Hall B. & Co., saloon, Main.
Hall George F., of Hibbard & Hall, Main.
Hambly William, of McFall & Hambly, Main.
Hanna John, baker, Main.
Hare & Co., druggists, Main.



Hare F. N., of Hare & Co., Main.
Hart Samuel, salesman, Main.
Hibbard A., of Hibbard & Hall, Main.
Hibbard & Hall, dining hall, Main.
Hillside Hotel, MissesM. & M. Thompson, Hutchinson.
Hillyard —, insurance agent, boards at Hillside Hotel.
HINDMAN G. S., furniture, Main.
Hines C., harness maker, Main.
Hoadley J. N., candy store, Main.
Hogan David, contractor, Thompson.
Holmes E., hostler, Main.
Hopkins Samuel, clerk and teamster, High.
Hosford C. F., clerk Oil Exchange Hotel.
Howe H., assistant cashier, Producers' Bank of Butler, Co., Main.
Hunt Hiram, blacksmith, Main.
HUSELTON G. W., cashier Greece City Bank, Main.
Hutchinson Forgus, carpenter, Hutchinson.
Hutchinson George, butcher, Hutchinson.
Hutchinson Miss M. J., dressmaker, Hutchinson.
Hutchinson Wm., butcher, Hutchinson.

## J

Jones Charles, bootmaker, Main.

## K

King J. L., physician, Main.
King Thomas, of Weaver & King, Main.
Krumpe W. A., of Belles & Krumpe, boards Main.
Krug Charles, butcher, Main.
Krug Louis, butcher, Main.

## L

Lacey J., of Hare & Co., Main.
Larimore John, bootmaker, Main.
Lesseig David, tool dresser, Thompson.
Levy J., dry goods, notions &c., Main.
Lindsay Jacob S., hostler, Main
Lonsberry J. H., stencil cutter, boards Hillside Hotel.
Luce W. G., gents' furnishing goods, &c., Main.
Lyons Frederick, oil operator, Main.

## M

Marks & Levy, dry goods, &c., Main.
McAllen John, carpenter, Main.
McCandless R. J., boarding house Main.
McCrum Samuel, bookkeeper, Main.
McDermott Charles, machinist, Main.
McDONALD, T., livery stable, Main.
McFall & Hambly, druggists, Main.
McFall James, of McFall & Hambly, Main.
McGilvey A., boarding house Main.
McPherson Wm., agent Farrar & Treet, Main.
MELLINGER J. W., clerk Butler House, Main.
MELLINGER J. A., proprietor Butler House, Main.
Melville James, barkeeper, Main.
Miller Christopher, butcher, Main.
MITCHELL H. J., teller Greece City Bank, Main.
Moore Nelson, foreman United Pipe Line, office head of Main,
Morrow J. H., of Smith & Morrow, Main.
Mount Archy, of Mount & Kugler, Main.
Mount & Kugler, oil well supplies, Main.
Munson Hiram, driller, Main.
Murray A., engineer, Main.
Myler James, of Waugh & Myler, Main.

## N

Nelson J. C., of Andrews & Nelson, Main.
Norris H. W., blacksmith, Main.

## O

O'Donell J. J., cigars, tobacco, &c., Main.
OIL EXCHANGE HOTEL, Main.
O'Rourke J. T., contractor, Thompson.
Oughton James, grocery, Main.

## P

Pacific & Atlantic Telegraph Co., office head of Main.
Paxton John, hostler, Main.
Peck C. N., telegraph operator, office head of Main.
Peters Addison, blacksmith, Thompson.
Phillips H., of Seidenberg & Co., Main.
Pollock M. A., bakery, Main.
Post Office, Miss Kate Hair, P. M., Main.
Price Anna, laundress, Main.
Purdon A., watchmaker, Main.

## R

Reisdorph F. C., saloon, Main.
Retcher W. C., barber shop, Main.
Richey J. A., groceries, &c., Main.
Roach Michael, helper, Main.
Robinson Calvin, of Robinson & Russell, Main.
Robinson L. C., clerk, Main.
Robinson & Russell, livery stable, Main.
Rusch B., engineer, Main.
Rush J., baker, Main.
RUSSELL A. E., of Tadder & Russell, Main.
Russell Elmer, of Robinson & Russell, Main.
Russell John W., clerk, Main.
Russell W. G., of Aiken & Russell, Main.

S

Samuels Charles, dry goods, notions, &c., Main.
Seidenberg A., clerk, Main.
Seidenberg & Co., clothing, gents' furnishing goods, &c., Main.
Seidenberg M., of Seidenberg & Co., Main.
Sexton Charles; barkeeper, Main.
Sherry Charles, barber, Thompson.
Shutt A. A., manager Steele's stables.
Slaven Michael, tool dresser, Thompson.
Smith G. W., of Smith & Morrow, Main.
Smith & Morrow, wagon makers, Main.
Smith S. M., blacksmith, Main.
Steele W. J., teamster, Main.
STEELE J. A., livery stable, Main.
Stevens James, barkeeper, Main.
STEWART A. P., groceries and provsions, Main.
STEWART D., proprietor European Hotel, Main.
Sullivan C. C., of Sullivan & Grossman, Main.
Sullivan & Grossman, boots and shoes, Main.
Sutherland Edward, manager B. Hall Co., Main.

T

TADDER & RUSSELL, billiard hall, Main.
TADDER Wm., W., of Tadder & Russell, Main.
Thompson E. N., of Thompson & Gordon, High.
Thompson & Gordon, flour, grain and feed, High.
Thompson Mattie, Hutchinson.
Thompson Mary, Hutchinson.
Thompson J. M., oil operator, boards Hillside Hotel.
Truby Henry, of Truby & Frost, Main.
Truby & Frost, livery stable, Main.

U

United Pipe Lines, head of Main.

W

Wagger Mrs. M. L., assistant P. M., Main.
Wallace J. K., attorney, Main.
WAUGH & MYLER, hardware and oil well supplies, Main.
Waugh R. M., of Waugh & Myler, Main.
Weaver James, of Weaver & King, Main.
Weaver & King, boots and shoes, Main.
Western Union Telegraph Co., office Main.
Wilson Charles, supt. McDonald's stables, Main.

## GREECE CITY BUSINESS DIRECTORY.

**Attorneys.**

Addle James M., Main.
Wallace J. K., Main.
Seidenberg & Co., Main.

**Drug Stores.**

Barr & Craig, Main.

**Bakery.**

Pollock M. A., Main.
Rusch J., Main.

**Banks.**

Greece City Bank, Main.
Producers' Bank, of Butler, Co. Main.

**Barber Shop.**

Retcher W. C., Main.

**Billiard Hall.**

Tadder & Russell, Main.

**Blacksmiths.**

Belles & Krumpe, Thompson.

**Boot and Shoe Makers and Dealers.**

Durham Wm. M., Main.
Sullivan & Grossman, Main.
Weaver & King, Main.

**Butchers.**

Adrews Samuel, Hutchinson.

**Cigars and Tobacco.**

Barnes J. R., Main.
Durham Wm., Main.
O'Donell J. J., Main.

**Clothing.**

Bartold & Haas, Main.
Book W. P., & Co., Main.
Hare & Co., Main.
McFall & Hambly, Main.

**Dry Goods.**

Levy J., Main.
Samuels Charles, Main.

**Engines, &c.,**

Boughton & Chandler, Main.
McPherson Wm., Main.

**Flour and Feed.**

Thompson & Gordon, High.

**Gents' Furnishing Goods.**

Bartold & Haas, Main.
Luce W. G., Main.

**Furniture.**

Hindman G. S., Main.

**Grocers.**

Aiken & Russell, Main.
Richey J. A., Main.
Stewart A. P., Main.

**Hardware.**

Andrews & Nelson, Main.
Backus, Morrison & Co., Main.
Waugh & Myler, Main.

**Hotels.**

Brawley House, Main.
Butler House, Main.
European Hotel, Main.
Hillside Hotel, Hutchinson.
Oil Exchange Hotel, Main.

**Livery Stables.**

McDonald T., Main.
Robinson & Russell, Main.
European Hotel, Main.
Steele J. A., Main.
Truby & Frost, Main.

**Notions &c.**

Levy J., Main.
Samuels Charles, Main.
Sullivan & Grossman, Main.

**Oil Well Supplies.**

Andrews & Nelson, Main.
Backus, Morrison & Co., Main.
Mount & Kugler, Main.
Waugh & Myler, Main.

**Physicians.**

King J. I., Main.

**Pipe Line.**

United Pipe Line, office head of Main.

**Saloons and Restaurants.**

Frazier Simon, Main.
Hall B., & Co., Main.
Hibbard & Hall, Main.
O'Donell J. J., Main.
Reisdorph F. C., Main.

**Wagon Makers.**

Glenn J. S., Main
Smith & Morrow, Main.

# Modoc, Butler County

Modoc City, a short distance north of Greece City, also in Concord Township, attracted the attention of enterprising operators to an uninviting tract of land about midway between the many roads leading to Fairview and Buena Vista respectively. The property was secured, and the spot chosen on which to put down a well. The place was low and forbidding in its appearance, surrounded with a dense growth of underbrush, having a small run and extended ravine contiguous, and in no respect possessing any natural features likely to please an aesthetic taste. Neither did it seem destined to reward the adventurers with abundance of the greasy fluid of which they were in search, as the "belt" was generally believed to run in a direction that would

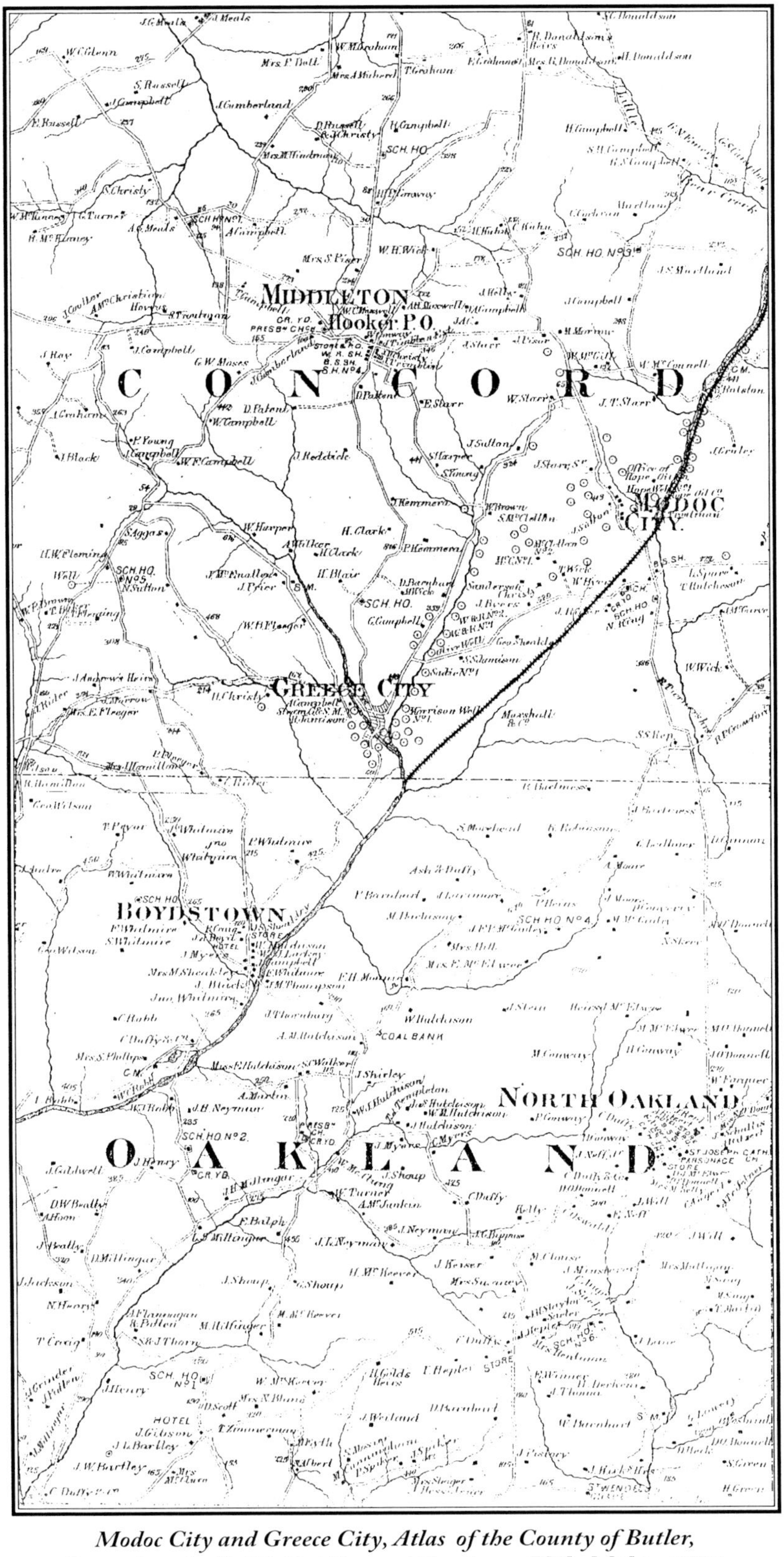


*Modoc City and Greece City, Atlas of the County of Butler, Pennsylvania, G. M. Hopkins and Company, Philadelphia, 1874.*

leave the entire locality completely "out in the cold." A rig was put up during the fall and winter, drilling quietly went on, and on March 22, 1873, the biggest strike since "sixty-five," was announced to the outside world. Such was the origin of the renowned Troutman Well, the most prolific yet discovered south of Oil Creek.

At the outset, the new "spouter" flowed for a period of seven months before the drilling tools were removed from the hole and the well averaged 1,000 barrels daily. It required two pipes to convey the product from the casing to the tanks, of which a number were provided in the quickest manner known to the business.

The City of Modoc contained a population of about 1,000. The place grew very fast, and was an important point in the oil region. The town had a number of commodious store rooms, shops, etc, and several good hotels. (1879)

## Hope Oil Company

In 1873 a small office building stood along the road which would later be the main street of Modoc City. Modoc City was created by the Hope Oil Company. This company was formed by William Vandergrift, Joseph Bushnell, William G. Warden and J. A. Bostwick, who had likewise secured a large tract of land in the immediate vicinity. The famous "Troutman Well" hit and for weeks the geyser continued to flow without serious diminution, frequently giving the pipe line lively exercise to convey the oil to Greece City, exactly three miles distant. Up to this date, it has yielded nearly 90,000 barrels of petroleum, thus ranking it with the most productive wells ever struck in any portion of Oildom.

Operators and speculators flocked to the latest Oil Dorado, intent on acquiring some interest in the neighborhood. Rigs multiplied rapidly, and the nucleus of the town appeared as if by magic. The Starr and the Sutton farms, below the Troutman property, were generally considered with special favor, and upon these Modoc City was at once located. Building lots were staked off that found ready purchasers or lessees at high figures, balloon houses were quickly erected along both sides of the principal street, even the fearful condition of the roads did not deter crowds from seeking this second Pithole, and within a month a town bristling with business of every description occupied the vacant fields of four weeks before. The embryo city received an Indian epithet, and henceforth Modoc became a familiar name in petroleum circles. Many oleaginous sages wisely shook their heads at the spectacle of activity everywhere presented.

Modoc contained one leading hotel, two or three hardware stores, a dozen groceries, a livery stable, two banks, and a score of unlicensed local-option dispensaries. The place is built without special regard to architectural arrangement, its appearance and surroundings are decidedly picturesque, and the newest tyro would not be long in discovering he had "fetched up" in a genuine oil city. With the exception of the three old homesteads of the Troutman, Starr and Sutton farms, everything about the town is of the modern date of the present season. The muddy streets are crowded with travelers from all quarters, and no one who has not visited Modoc can be said to have "done" the specimen oil town of the period.

On March 17, 1874, a terrible fire destroyed a large section of Modoc City's business district. The fire destroyed three hotels and other property with a loss of $100,000. Only one hotel remained, the American House.

# Weakened Market

A number of wells in this field produced 500 to 1,000 barrels each day. Big strikes were a daily occurrence and on July 31, 1873, Purse and Beck hit the "Bottom Dollar." This well came in at 450 barrels a day. These strikes weakened the market. On July 25, 1873, oil was $1.85 a barrel, August 1, 1873 a barrel was $1.55, and by November 12, 1873, it lowered to 47 1/2 cents a barrel. (History of Butler County, Pennsylvania, 1883; Waterman, Watkins and Co., Chicago, Illinois)

## MODOC CITY.

The Modoc district, in which Modoc City is located, is one of the most prolific in the oil Region, and is now yielding more oil than has been produced in any territory since the discovery of oil by artisian wells.

The first well put down in this district was the enterprise of Mr. William K. Vandergrif, early in the spring of 1872. This well was put down under many discouragements, owing to the difficulties met with in getting machinery, tools, &c., on the ground.

However, the third sand was struck, on the 23d of March 1872, and it proved to be the first flowing well in the district, averaging 600 barrels per day when it was struck, and is familiary known as the Hope or Trautman well.

Other wells were struck in this vicinity which yielded as follows:

| Name of Well. | When struck. | No. Bbls. |
|---|---|---|
| Captain Jack | July 29th. | 800 |
| Modoc | July 9th. | 600 |
| Morris & Dean | August 11th. | 1,500 |
| Walt & Thompson | August 12th. | 1,000 |
| Spotted Tail | August 20th. | 500 |
| L. H. Smith | August 20th. | 600 |
| Percy | August 1st.. | 400 |
| Brawley | August 5th. | 700 |
| Captain Grace No. 1 | August 1st. | 900 |
| Captain Grace No. 2 | August 15th. | 400 |
| S. P. Boyer | August 1st. | 500 |
| Miller | September. | 700 |
| Phillips | September 2d. | 500 |

This exhibit does not represent the productions of these wells, at this date. The total number of barrels produced by the above wells is about 40,000 daily.

### A

Akins C. F., contractor, boards at Live Oak Restaurant.
AMERICAN HOTEL, W. A. McDonald, proprietor, Main.
Anderson J. B., operator, boards American House.
Anderson Jospeh, tool-dresser, boards Beatty House.
Andre Samuel, butcher, Main.
Andre William J., butcher, Main,
Argrave H., miller, boards Meadville House.
Arrowsmith A., tank builder, Main.

### B

Barr T. C., miner, boards Beatty House.
Baker ——, of Ramsey, Baker & Co., McClellan Road.
Bateman F. E., proprietor Philips House, Main.
Baughman S., Sup't oil wells, boards Live Oak restaurant.
Beeck Jacob, shoemaker, boards Meadville House.
Bell W. T., operator, boards American House.
Betzhoover S., U. S. Marshall, boards American House.
Bliss A. O. druggist, Main.
Bliss D. L., druggist, boards, Live Oak Restaurant.
Bly E. P., tobacconist, Main.
Bole Henry, druggist, boards American Hotel.
Booz William H., clerk, boards Meadville House.
BRACKEN JOHN N., of Parkison & Bracken, Main.
Breedin ——, carpenter, boards Phillips House.
Brooks ——, operator, boards Phillips House.
Brothers Edwards, restaurant, Main.
Brown James, saloon, boards Meadville House.
Burlin B., of Burlin & Co., Pleasantville, Pa.
BYERS J. E., of Roessing & Byers, Main.

### C

Campan James T., pumper, boards Oil Exchange.
Campbell George, butcher, Main.
Cardenbergh O. F., saloon, Main.
Case R. W., physician, boards Live Oak Restaurant.
Casey J. W., restaurant, Main.
Cavebry Richard, driller, boards, Beatty House.
Central House, Valentine Shell, proprietor, Main.
Chambers Thomas J., carpenter, Main.
Chaner Lee, driller, boards Meadville House.
Chappell John, teamster, boards Ameican Hotel.
Charters John Z., barber, Main.
Chase A. G., physician, office Main.
Clifford Henry, barber, boards American House.
CONRATH ISRAEL, of Conrath & Woodward boards,, Phillips House.
CONRATH & WOODWARD, oil operators and rig builders. Parker City, Pa., branch at Modoc.
Corbett Thomas, tool-dresser, boards Beatty House.
Coulter Alexander, blacksmith, McClellen Road.
Coulter D. D., driller, boards Meadville House.
Crager ——, laborer, boards Phillips House.
Crittenden A. B., hardware, boards American Hotel.
Curtis Thomas, tool-dresser, boards Meadville House.

### D

Davis ——, carpenter, Main.
DAVIS W. H., proprietor Fullerton Oyster Bay, Main.
Davis William S., grocer, Main.
Drake Frederick, engineer, boards Meadville House.
Driscoll Frank, blacksmith, boards American Hotel.

### E

EICHER DAVID, of Longabaugh & Eicher, Main.
Elsasser Max, tobacco and clothing, Main.
Evans E. J., operator, Main.

### F

Feeney Lawrance, baker, McClellen Road.
Fields Joseph, driller, bords Meadville House.
Flood James, engineer, boards Beatty House.
Follett P. M., blacksmith, boards, American Hotel.
Fowler A., of Fowler & Skinner, Main.
Fowler & Skinner, Billiard Hall, Main.
Frost D., contractor, boards Meadville House.
Furgeson ——, expressman, boards Phillips House.

### G

Galvin M. T., operator, boards Phillips House.
Garrett C. S., lumber dealer, boards American Hotel.
Gibny James, operator, boards Phillips House.
Gilson G. B., lumber, boards American Hotel.
Glass J., carpenter, boards Meadville House.
Gordan ——, operator, boards Meadville House.

Grace Captain, operator, boards Phillips House.
Grayson John, carpenter, Main.
GRAYSON IRA W., proprietor Spring House Hotel, Main.
Griffin W. M., Billiard Hall, Main.
Grossman John, driller, boards Meadville House.

## H

Hamsher Wesley, manager Columbia Oil Company, boards American House.
Hearon Lee, driller, boards Meadville House.
Heath Alexander, contractor, boards Phillips House.
HECKETT J. F., proprietor Modoc Oil Exchange, Main.
Hilberd William, driller, boards Meadville House.
Hindman J. L., Main.
Holt Thomas, contractor, boards Meadville House.
Hooks Blare, driller, boards American Hotel.
Hoy Samuel, barber, boards American Hotel, Main.
Hultem Peter, tank builder, McClellen Road.

## I

Irven Oscar, tank builder, boards Skeets House.

## J

JACK E. H., of Waugh, Jack & Co., hardware, Main.
JACK M. J., of Waugh, Jack & Co., hardware, Main.
Jack Modoc, saloon, corner Main and Cherry alley.
Jamison E. C., painter, boards American Hotel.
JESSOP W. S., of Wiles & Jessop, Main.
JAMES ROBERT C., machinist, boards Hillside House.

## K

Karns S. D., banker, Main.
Keefe Michael, guager, boards Beatty House.
Kennedy Charles, sample rooms, Main.
Kennedy N., pipe layer, boards Meadville House.
Kennedy S. A., tank builder, boards Meadville House.



## L

Leigh A., livery sale and feed stables, Main.
LIVE OAK RESTAURANT, Longabaugh & Eicher, proprietors, Main.
LONGABAUGH S. S., of Longabaugh & Eicher, Main.
Loree H. B., butcher, Main.
Lynch James A., contractor, boards Meadville House.

## M

Markhan John, operator, boards American Hotel.
Marr W., saloon, Main.
Matteson T. B., of Burlin & Co., boards American Hotel.
McCabe John, tank gauger, boards Phillips House.
McCalla Geo. W., tank gauger, boards American Hotel.
McCool John, tool dresser, boards Beatty House.
McCormick Thos., carpenter, Main.
McDonald Daniel, ice dealer, boards American Hotel.
McDonald James L., carpenter, Main.
McDONALD W. A., proprietor American Hotel, Main.
McGill James P., confectionary, Main.
Meadville House, Charles Means, proprietor, Main.
Means Charles, proprietor Meadville House, Main.
McQuade John J., clerk, boards American Hotel.

Mercer ——, operator, boards Phillips House.
Millroy ——, superintendent oil wells, boards American Hotel.
Munch Valentine, driller, boards Meadville House.
MODOC OIL EXCHANGE, J. H. Heckett proprietor, Main.
Monniea Charles, grocer, Main.
MORARITY TERRY, manager Modoc Oil Exchange, Main.
Morrison R., operator, boards Phillips House.
Morrison S., driller, boards Meadville House.
Mulnix J. W., proprietor Beatty House, Main.

## N

Nawborg Albert, tank builder, boards McClellen Road.
Nichols J. D., laborer, boards Oak Shade Saloon.
Nichols Winfield, proprietor Oak Shade Dining Saloon, Main.
Norton Thomas, tank builder, boards Beatty House.
Noxon George, contractor, boards Phillips House.

## O

O'Brian John, laborer, boards Beatty House.
O'Donnell Lemuel, driller, boards Meadville House.
O'Hare John, driller, boards American Hotel.
Olsson C. A., shoemaker, boards Meadville House.

## P

PARKISON & BRACKEN, grocers, Main.
PARKISON JOHN, of Parkison & Bracken, grocer, Main.
Patterson J. M., driller, boards Meadville House.
Percy & Beck, oil operators office, Main.
Phillips House, F. E. Bateman proprietor, Main.
Pierce Zak., carpenter, boards Beatty House.
Pitts James, lunch room, Main.
Plante Henry, boots and shoes, boards Meadville House.
Purdum R. W., telegraph operator, boards Live Oak Restaurant.

## R

Ramsay A. P., of Baker & Co., Tidioute.
Ramsay, Baker & Co., hardware, McClellen Road.
Reed Charles, miner, Main.
Reed Hiram, carpenter, Main.
Reilly Daniel, tank builder, boards Beatty House.
Richey J. N., driller, boards Meadville House.
Roberts ——, operator, boards Phillips House.
Robinson Bros., tank gaugers, Main.
ROESSING A. C., of Roessing & Byers, Main.
ROESSING & BYERS, grocers, Main.
Ruff Martin, barber, boards Beatty House.
Ryan A. J., gents' furnishing goods, &c., Main.

## S

Say H. W., coal dealer, Main.
Say J. D., teamster, boards Oak Shade Saloon.
Saunders S., tank builder, boards Meadville House.
Shealds R. P., engineer, boards American Hotel.
Shell Frank, clerk, boards Central House.
Shell Valentine, proprietor Central House, Main.
Sinclair James, Sup't pipe line, boards American Hotel.
Shepard J., livery, Main.
Skeet James, boarding house, Main.
Skinner James, of Fowler & Skinner, Main.
Skeet Boarding House, James Skeet, proprietor, Main.
Slattery Mat., driller, boards, Beatty House.
Sloughter John, driller, boards Meadville House.
Smith A. H., restaurant, Main.
Smith Nelson, teamster, Main.
Smith W., driller, Main.
SPRING HOUSE HOTEL, Ira H. Granger, proprietor, Main.
Starr J. H., clerk, boards Live Oak Restaurant.
Starr William, oil well supplies, Main.
Statt E. K., of E. K. Statt & Bros., Main.
Statt E. K. & Bros., tank builders, Main.
Stewart W. P., operator, boards American House.
Stevens H. K., tank builder, McClellen Road.
Steveson R. C., driller, boards American Hotel.
Suplee W. J., of Ramsey, Baker & Co., McClellen Road.
Sutton Jacob, real estate agent, Main.
Sutton James, real estate agent, Main.
Sweeny T. J., deputy sheriff, boards Oil Exchange Hotel.



## T

Tawson Thomas, engineer, boards American Hotel.
Taylor ——, tobacconist, boards Meadville House.
Taylor Z., cigars and tobacco, boards Meadville House.
Taylor Bros., cigars and tobacco, Main.
Thompson Charles, bartender, Main.

## W

Waugaman B. C., tank builder, Main.
Warmingham John, jeweler, Main.
Warningham John, boards Live Oak Restaurant,
Waugh, Jack & Co., hardware, tubing and casing, Main.
WAUGH R. M., of Waugh, Jack & Co., Main.
Wright C. B., news room, Main.
WILES H. J., of Wiles & Jessop, Main.
WILES & JESSOP, grocers, Main.
Wilkins A., driller, boards Meadville House.
Wilson George, contractor, boards Meadville House.
Wineman James, blacksmith, boards Beatty House.
WOODWARD JOHN, of Conrath & Woodward, Parker City.

## Z

Zuver Andrew, laborer, Main.
Zuver William, carpenter, Main.

## MODOC CITY BUSINESS DIRECTORY.

**Barber Shop.**

Charters John Z., Main.

**Billiard Hall.**

Fowler & Skinner, Main.
Griffin W. M., Main.

**Blacksmiths.**

Coulter Alex., McClellan Road.

**Boarding Houses.**

Hindman J. L., Main.
Skeet James, Main.
Stevens H. K., McClellen Road.

**Boot and Shoe Makers and Dealers.**

Plante Henry, Main.

**Butchers.**

Campbell George, Main.

**Carpenters & Builders.**

Conrath & Woodward, Parkers and Modoc.

**Cigars and Tobacco.**

Bly E. P., Main.
Elsasser Max, Main.
Taylor Bros., Main.

**Drug Stores.**

Bliss A. O., Main.

**Gents' Furnishing Goods.**

Elsasser Max, Main.
Ryan A. J., Main.

**Grocers.**

Davis William S., Main.
Monniea Charles, Main.
Wiles & Jessop, Main.

**Hardware.**

Starr William, Main.

**Hotels.**

American Hotel, Main.
Beatty House, Main.
Central House, Main.
Meadville House, Main.
Phillips House, Main.
Spring House Hotel, Main.

**Jeweller.**

Warmingham John, Main.

**Livery Stables.**

Leigh A., Main.
Shepard J., Main.

**Lumber Dealers.**

Conrath & Woodward, Starr Farm.

**Physicians.**

Case R. W., Live Oak Restaurant.
Chase A. G., Main.

**Saloons and Restaurants.**

Casey J. W., Main.
Live Oak Restaurant, Main.
Oak Shade Dining Saloon, Main.
Modoc Oil Exchange, Main.
Ritts James, Main.
Smith A. H., Main.
Cardenbergh O. F., Main.

**Tank Builders.**

Arrowsmith A., Main.
Statt E., Main.

**Tank Gaugers.**

Robinson Bros., Main.

# Troutman

The major part of the town was laid out on the Troutman farm. Bear Creek runs through the town and a small bridge crosses the creek on main street. The Hope Well Company office, which opened this oil field, was located near the bridge. There were 12 large oil tanks located in the Modoc and Troutman area. Several 25,000-barrel tanks were located very near the town of Troutman and owned by the Fairview Pipe Line Company.

Joseph Seep owned an oil farm near the town and he had his office in Oil City. His firm handled much of all the oil produced and posted the daily crude oil market prices for New York, Pennsylvania, Ohio, West Virginia and Kentucky. Troutman was the third oil boom town in the township but on April 19, 1877, three large iron tanks holding 46,000 barrels of oil caught fire by lightning and burned. The burning oil ran down Bear Creek making it a blazing river and destroying 29 buildings, many wells, derricks and smaller oil tanks. The loss to Troutman and surrounding areas was over $160,000. Today two or three houses and barns are all that is left and the little bridge still carries traffic on Main Street. Refer to pages 306 and 307.

*Troutman oilfield General Store built 1876. Courtesy of Lenn "Pithole" McCorry and the Butler County Historical Society. Photograph taken May, 1956. The last oil era commercial building left in Troutman. This building no longer exists.*

## Petrolia
## Special to the Derrick – June 27, 1880
## Boiler Explosion, Buried Alive, A Boy Hurled Many Paces

"A terrific boiler explosion occurred at a well in Troutman this evening, which will probably cause the death of one of the victims. While a man named Harvey Rathbun was putting on the belt and Mr. Rogers was assisting him, and his son Willie was in the engine house, the boiler suddenly exploded, tearing everything to pieces. Mr. Rogers was buried under the falling debris, but managed to crawl out after the lapse of some minutes. Willie Rogers was picked up twenty paces away so seriously injured that his life is despaired of. Mr. Rathbun escaped with comparatively little harm."

## Buena Vista
## Oil City Derrick, June 20, 1874

"Buena Vista is pleasantly located on a sloping hillside about midway between Modoc and Millerstown, whence a magnificent view is afforded of the surrounding country. Though the village had existed half a century, not until last year, 1873, did the humble settlement attain any considerable degree of importance. The oil developments in the immediate neighborhood gave tasteful promise of a profitable return during the early summer, attracting scores of adventurous operators and business men to the quiet hamlet. Houses multiplied very rapidly, in a few weeks so completely metamorphosing the place as to render it almost a new creation. But the glowing expectations of the enterprising immigrants were not destined to be realized, the presence of enormous quantities of salt water quickly dissipating the roseate visions of prospective riches to the four winds. The superior inducements of Millerstown combined with the remarkable developments in the fourth sand district around Petrolia, proved fatal to the continued growth of the ephemeral town. Many of the best buildings were never occupied, people no longer dreamed it essential to their eternal salvation to locate in the pleasant village, and the Buena Vista of today presents a strong contrast to the stirring locality of eight brief months ago. But with oil at lively prices, the place will certainly revive; it is situated in a central position, has a good agriculture and coal region surrounding it, and some day must undoubtedly acquire a fair measure of prominence." (John J. McLaurin, Reporter)

## Buena Vista

Buena Vista was laid out about 1847 by John McKisson and is located four miles from Petrolia on the road to Greece City. It is pleasantly situated at the intersection of roads leading in all directions. It was formerly quite an important oil town, and contained six stores, two drug stores, machine shops, hotels, etc., all of which have disappeared. A large hotel, erected by Alexander Storey soon after the oil excitement commenced, was destroyed by fire in August 1875. At the same time 37 buildings were destroyed, including four stores and the heart of the business portion of the place. Numerous buildings have since been torn down and removed, so that now but little remains to tell what the village once was.

The most prominent gentlemen engaged in developing this territory are S. D. Karns, C. D. Angell, Parker, Thompson & Co., Lambing Brothers, Campbell Bros., Fisher Brothers, Tack Brothers, Moorehad and Ripley, Robert Leckey, H. W. Scott, F. F. A. Wilson, Marcus Brownson, Dimick, Nesbitt & Co., Jno. Preston, Jno. L. and J. C. McKinney, and Jno. H. Galey, Vandergrift & Foreman, Phillips Bros., H. L. Taylor, Jno. Satterfield, Tarbell & Hess, Orange Noble, William D. Robinson, A. N. Perrin and Col. R. B. Allen. (History of Butler County, Pennsylvania, 1883; Waterman, Watkins and Co., Chicago, Illinois)

---

## BUENA VISTA.



### A

Ashin John, driller, boards Continental Hotel.

### B

Burns John, rig builder, boards Butler.
Byers John L., shoemaker, Brady.
Black James, driller, Brady.
Blight Heman, tool dresser, Brady.
Boyd C. M., hotel, Butler.
Boyd J. H., hotel, Butler,

### C

Centre House, McFann L. R., proprietor, Butler.
Cloak Nicholas, teamster, Butler.
Crawford Samuel, carpenter, Butler.

### D

Davis Alexander, carpenter, Butler.
Dickey J. A., agent Weed Sewing Machine, Brady.
Dillon ——, butcher, boards Centre House.
Dirling William, rig builder, Butler.
Dorman John, clerk, boards Centre House.

### F

Forcht William, driller, Butler.

### H

Hall William N., carpenter, Brady.
Heller Henry, miner, Butler.
Hemphill Adam, miner, Brady.
Henderson L. D., billiards, boards International Hotel.

### I

International Hotel, Robert S. Millinger, proprietor, Butler.



### K

Kepple Isaac, farmer, Brady.
Kepple Wm., farmer, Brady.

### L

Losh O. P., carpenter, Butler.

### M

Malvin James, driller, boards Continental Hotel.
McCann J. C., saloon, Butler.
McCoy Robert, grocer, Brady.
McDonald David, lumber yard, Butler.
McFann L. R., livery, Brady.
McFann L. R., proprietor Centre House, Butler.
McGarvy William, driller, Brady.
McKisson Ann, widow, corner Brady and Butler.
McNees Walter, pumper, boards International Hotel.
Miles G. W., painter, boards International Hotel.
Miller Jacob, hostler, boards Centre House, Butler.
Miller Martin, carpenter, Butler.
Millichamp S., of Millichamp Bros, tank builders.
Millichamp Bros., tank builders, Brady.
Millingar Abraham, engineer, Brady.
Millingar David, carpenter, boards International Hotel.
Millingar Robert S., proprietor International Hotel, Butler.
Mooney Robert E., carpenter, Butler.
Morris W. H. H., contractor, Brady.
Murtland J. H., sewing machine agent, Brady.

### N

Nicholas Carroll, grocer, Brady.

### O

Osborne Leslie, driller, Butler.

### P

Phillips John, of W. Phillips & Co., Brady.
Phillips William, of W. Philips & Co., Brady.
Phillips Wm. & Co., druggists, Brady.
Pontius Nicholas, dry goods and groceries, Brady.
Porter Geo., driller, Brady.
Pulling George, carpenter, boards International Hotel.

### R

Retta & Whitsay, machinists, Butler.
Retta George, of Retta & Whitsay, Butler.
Robinson Cal., livery, Butler.
Rumberger C. C., physician, Brady.

### S

Smith John T., hotel, Brady.
Steel William, driller, Butler.
Stewart Myrl, teamster, Butler.
Shook John, blacksmith, Butler.
Shyock L. D., pumper, boards International Hotel.
Sutton J. J., post master, Butler.

T

Taylor Hiram, teamster, Brady.
Thompson R. J., of Thompson & Turner, Butler.
Thompson & Turner, tobacconists, Butler.
Thompson William, contractor, Butler.
Thorn Peter, wagonmaker, Brady.
Turk Samuel, carpenter, Butler.
Turner R. J., of Thompson & Turner, tobacconist, Butler.

W

Wallace Sarah, widow, Brady.
Whitsay John, of Retta & Whitsay, boards International Hotel.
Whitzell William, newsroom, Butler.
Wick Miller, carpenter, Butler.
Wimer Christopher, carpenter, Butler.
Wolford J. D., clerk, Brady.

Y

Young M. C., carpenter, Butler.

# UNCLE BILLY SMITH OIL'S FORGOTTEN MAN

"Here lies the blacksmith of Tarentum, "Uncle Billy" Smith, who made at his own anvil the tools he used to drill the world's first oil well at Titusville, Pennsylvania, completed August 27, 1859 giving the world petroleum in abundance for light, heat, power, and lubrication. The petroleum industry's millions of workers salute the memory of the original oil well driller, William A. Smith, in tribute to his loyalty, courage, and craftsmanship." These words are carved on "Uncle Billy's" stone.

William Andrew Smith was born February 5, 1812, at Hubb Run in Butler County, a tributary to Little Buffalo Creek near the head-waters in Winfield Township, on the farm of his grandfather, Jeremiah Smith, who was a pioneer and a native of County Down in Ireland. He had come to America in 1792.

William Smith was a middle-aged blacksmith when the events first began moving which would ensure his place in history. Billy Smith had drilled wells for Samuel Kier from Pittsburgh, Pennsylvania—not oil wells, but salt wells. Kier had a salt business at Tarentum, Pennsylvania, and his salt wells produced a little oil as well as salt water. Sam Kier sold the oil and went on boiling down the water for salt.

Billy Smith preferred farm life and had no plans of being a driller all his life. He mostly worked at his forge, making tools for others to work and drill with and was considering giving up blacksmithing and devoting his energy to farming.

## Colonel Edwin L. Drake

In the fall of 1858, Drake and his associates entered Smith's life. Drake knew little about drilling and his only uniform had been that of a railroad conductor. But when a group of promoters talked him out of his last $200 for stock, then sent him to Pennsylvania to dig for oil, his associates wanted to put up a big front. They addressed his mail to "Col. Edwin L. Drake," and "Col. Drake" he remained to his dying day.

"Colonel Drake" persuaded Smith and his son, Samuel B. Smith, to drill for $2.50 a day and to tackle the job of sinking the first oil well. Smith agreed to try it. They discussed the tools needed for the drilling and decided that Drake would furnish the materials and Smith would make the tools in his own blacksmith shop before setting out for Titusville.

An entry made by Uncle Billy in a memorandum book on May 14, 1859, shows that the full drilling outfit cost Drake $76.50. Total weight of the tools was a little more than 100 pounds.

In an 1880 interview, Uncle Billy said to the Titusville Herald, "I never saw such a time as there was at that Drake well. Hundreds of people every day, and the fool questions they would ask were awful. I thought it funny at first, but soon got sick of answering questions."

The well was completed on August 27, 1859, and as told by "Uncle Billy," he worked there and pumped the well up to April 20, 1862, when he returned to Butler County and his farm on the south bank of Rough Run.

## "Struck Oil," Was the Cry Heard 'Round the World

It was a Butler Countian's cry that brought world fame to Titusville, Pithole, Oil City, Parker's Landing, Petrolia and hundreds of other hamlets, many of which have now gone the way of ghost towns. It was Uncle Billy's cry that in 1871 caused men to draw from the strata of Butler, Clarion and Armstrong Counties one and a quarter million barrels, or nearly 30 per cent of the entire U. S. production, of oil for that year.

"Uncle Billy" Smith's cry in 1859 not only echoed through the streets of Titusville and Pennsylvania, but all over America and around the world. It brought thousands of fortune seekers, even some who had rushed to California 10 years earlier, to the Oil Creek Valley and eventually to the rich oil fields of Butler and surrounding counties.

*Drake Well, 1866*

*William Smith, "Uncle Billy" sits on the wheelbarrow with the rebuilt Drake Well in the background. The first derrick and engine house burned down October 7, 1859, after "Uncle Billy" gazed down the well head with an open flame. Also in the photograph from left to right are "Uncle Billy's" brothers, Eldridge, Lock, Smith's son William, and his son James. Seated near the engine house are Ellen and Adeline, his young daughters. William Smith passed on July 27, 1890, and is buried in the Pape Cemetery near Hannahstown, Butler County. "Peace to his ashes." (John J. McLaurin, Courtesy of Drake Well Museum, Titusville, Pennsylvania.)*

Both Drake and "Uncle Billy" had financial problems and in 1880, "Uncle Billy" faced the loss of his farm in a sheriff's sale. However, the Oil City Exchange raised money to pay off his debts and when he visited the Titusville area that year, he was interviewed by a number of local papers and the only known formal portrait was taken of him in 1880.

He spent the last days of his life at his farm, deaf, ill much of the time, and almost penniless. In 1886, while suffering from a general debility, he made his own medicine by burning wheat grains like coffee beans which he said cured him. "Uncle Billy" died July 27, 1890 and his widow buried him between his two former wives in the Pape Cemetery at Hannahstown. His grave remained unmarked and he was almost forgotten by history.

## Monument Erected In 1952

However, in 1952, as a part of Pennsylvania Week ceremonies, the Commonwealth's petroleum industry erected a tall granite monument over his grave. Many dignitaries from all parts of the nation attended this dedication. Some of the main speakers were the Honorable Oscar L. Chapman, United States Secretary of the Interior; Frank M. Porter, president of the American Petroleum Institute; also, Dr. Paul D. Foote, scientist and vice-president of Gulf Oil Corporation; W. S. Zehrung, president of Pennzoil Company of Oil City; R. D. Whitely, representative of Socony Vacuum Oil Company, and many other prominent citizens and relatives of the Smith family. (Courtesy of Butler Eagle, Butler, Pennsylvania)

## "Uncle Billy" Smith's Anvil

What ever happened to the anvil used by "Uncle Billy" to hammer out the tools used to drill the Drake Well? The answer to this question lies in not one dented, rusted anvil but two, and both are worn by many years of long, hard usage. After much research, it seems that two anvils claim the title of the historic Drake Well anvil.

It is written in Oil Region Reminiscences, a souvenir book published in 1907 by the Oil Men's Association of Butler County, that Charles H. Oliver, of Butler, owns the true anvil. The authenticity is reported to be without a doubt. Loading his tools on a wagon, "Uncle Billy" started on his journey for Watson Flats. Eight miles out of Butler City, he stopped at his brother's farm, W. L. Smith. "Uncle Billy's" anvil had the horn broken off so he exchanged anvils with his brother. This borrowed anvil was returned to W. L. Smith about six months later and after the Drake Well had been completed.

Charles H. Oliver, a gauger for the National Transit Company, purchased the anvil and its history from the Smith family.

***"Uncle Billy" Smith Dressed In His Finest** The basic technology used by Smith became the mainstay of the oil industry. William Smith was born, lived, and passed on in Butler County, Pennsylvania. In the year of our Lord 1859, "Uncle Billy" took on the job of drilling an oil well specifically for petroleum. Photograph taken in 1880. (Drake Well Museum, Titusville, Pennsylvania)*

## The Second Drake Well Anvil

Mrs. Mae Sheehan lived in Titusville at 702 West Elm Street, and died October 16, 1953. Mrs. Sheehan's father was Samuel B. Smith, who came to Titusville in May, 1859, as a boy of fifteen to work with his blacksmith father, "Uncle Billy." Stored in Mae Sheehan's tool house was the original anvil which had been stored for decades. As reported by the Butler Eagle Newspaper, Butler, Pennsylvania, dated Tuesday, February 2, 1954, "The anvil, termed a large one by veteran oil well drillers, is somewhat rusted but appears as it must have when Smith and his sons, Samuel, James and William, were working on it." No identifying marks link the anvil with its distinguished history.

Howard M. Levy, of Titusville, purchased all Mrs. Sheehan's personal property and turned the anvil over to the Drake Well Museum. The original drilling tools made to drill the Drake Well were donated in 1949 on the 90th anniversary by Mrs. F. Wayne Fesenmyer, of Bradford, Pennsylvania, a descendant of Senator Lewis Emery.

Today, both anvils are at the Drake Well Museum and salute the memory of the original oil well driller.

## Saxonburg and John A. Roebling

In the early Pennsylvania canal system, John Roebling had been working with heavy manila rope lines which were hard to handle when wet. This no doubt started him to experiment with steel wire which was woven or twisted into rope form. In 1842, he produced a wire cable scientifically constructed, which was made in Saxonburg. With Julius Riedel and John A. Knoch as his assistants, he secured a patent on his invention this same year. From this point on he began the work that made for him a lasting world-wide fame and fortune, besides entirely revolutionizing the art of bridge building throughout the civilized world.

*World's first oil well drilled with Roebling wire cable in Butler County near St. Joe. John Roebling built the Brooklyn Bridge and his son, Washington Roebling, was general manager of the Mercer Automobile Company in the early 1900s, Mercer County, New Jersey. (Butler County Historical Society, Butler, Pennsylvania)*

Roebling Brothers, formerly of Saxonburg, Pennsylvania, furnished a wire cable as an experimental project, which was used on the first well in the nation to be drilled with anything other than a manila line or cable. The well came in on November 3, 1875, at 160 barrels daily and was located on the Peter Duffy farm, Donegal Township, Butler County. The well was drilled by Stouchton and Weiser and called the No. 1 Peter Duffy.

As previously mentioned back at St. Joe, where the Roebling wire cable was first introduced in October of 1875 for oil well drilling, they are used universally in all parts of the world today for cable, tool, and rotary type drilling to swing the heavy drill pipe to depths of more than 15,000 feet. (Lenn "Pithole" McCorry)

*The sign on the tree reads, "St. Joe 1 Mile"*

## St. Joe

The existence of St. Joe is due wholly to the discovery of oil in the surrounding country in 1874. Joseph Overy was owner of the first well called the "Meade," drilled in on June 30, 1874, at 400 barrels. St. Joe was named after oilman Joseph Overy, with Main Street running north and south and the east-west street known as Butler Street.

The first well completed on the farm of Joseph Graham produced 200 barrels of oil and brought a large number of oil speculators and producers to this section, and many wells were

*The Joseph Graham Farm*
*Mr. Graham has followed farming principally, though he kept hotel for two years in the village of St. Joe. This oil town, once quite a stirring place, but now nearly extinct, was built upon Mr. Graham's farm. Since 1875, Mr. Graham has been engaged in oil production, in addition to the management of his farm. He is a progressive farmer, and is making improvements. St. Joe was known for some time as the southern front of the oil industry.*

*Oil men, dressed in their finest at St. Joe, Butler County, Pennsylvania, November, 1874. Just three days later the town was burned to the ground. This photograph was taken on the steps in front of the drug store in the oil town of St. Joe. Top row from left to right: John J. McLaurin, newspaperman; Tom King; Pratt S. Crosby, oil broker at Parker and Oil City; Fred Wheeler; Capt. John K. Barbour, Head of Standard Oil Company's right-of-way department; James Orr, broker in the Producers Exchange at Bradford; Samuel R. Reed from the Clarion field. Bottom row from left to right: James Amm, who did well in the Bradford field; Joseph Seep, the "King Bee" of good fellows, who purchased most of the oil in Pennsylvania for Standard Oil; L. H. Smith, who got his start at Pithole, did good at Petrolia, and was a big man at 300 pounds; and Thomas McLaughlin, oil buyer. St. Joe was host to a countless number of leading oil men during its short existence. (Courtesy of Butler County Historical Society, Lenn McCorry)*

drilled. Such a large influx of population necessitated the erection of stores and hotels to accommodate them. From a quiet rural community, far removed from the busy scenes of commercial life, it was transformed within the space of a few months, almost as if by magic, into a busy little mart of 1,000 inhabitants with nearly all the necessary adjuncts of a place of this size, including three large hotels, grocery and dry good stores.

All was life, activity and bustle in St. Joe and people were full of bright anticipations, for oil, upon which this wonderful prosperity was based, was found in large quantities and was bringing a remunerative price. Suddenly the scene changes, for nearly the whole village was blotted out of existence by fire one night in November of 1874. It can be said that the entire property was practically uninsured, and the loss a total one. This, however, did not discourage

***Wells in the St. Joe Oil Field***
*On July 8, 1874, a 200-barrel-per-day well was brought in. By September, St. Joe had some 250 buildings, houses, and stores. The pumper in the photograph is Alphonse Hinterlang and his son, Martin. 1877 (Courtesy of Martin Hinterlang, Lenn McCorry, Butler County Historical Society)*

***May 1877***
*The original St. Joe post office was moved to the John Ritzert farm and used for a harness repair shop. The mail slot can still be seen in the door. May 1956. (Courtesy of Lenn G. McCorry, Butler County Historical Society, Butler, Pennsylvania)*

some of the owners from rebuilding, and in an almost incredible short space of time it was rebuilt. A telegraph office was opened in November, 1874, but not until May, 1877, was a post office established with William Durham as Postmaster, which office he still retains, also the telegraph office, he being the first operator.

For about five years, St. Joe was at the floodtide of prosperity, and many fortunes were here made and lost. About this time, many "dry holes" were struck, and the product of those producing having materially decreased, and in some instances failed entirely, the people began leaving for the new fields of operation. From this time on, the village rapidly melted away until today it contains but two stores, one kept by William Durham and the other by the Reisner Bros., and one hotel. The land once covered with stores and houses is now used for agricultural purposes. Still, considerable oil is produced in this locality. (History of Butler County, Pennsylvania, 1883; Waterman, Watkins and Co., Chicago, Illinois)

*St. Joe Station oil field. Located one mile south of North Oakland in Oakland Township, Butler County, and up the hollow from the St. Joe railroad station, 1879. (Courtesy of Lenn G. McCorry, Butler County Historical Society)*

## Carbon Center, 1877

This village and oil pump station is located in the northwest corner of Clearfield Township. As the oil fields moved in a southern direction in 1874, Robert Thompson laid out Carbon Center in lots and streets and eventually a post office was established here.

The Carbon Center pump station was built in 1874 by United Pipe Lines and was bought by the Standard Oil Company. Oil was pumped to Freeport and down the river to Pittsburgh through a four-inch discharge line which was the first in the nation. Stations such as this were the heart of the oil pipeline system. The larger building at the right was the boiler house where

*Carbon Center oil line pump station, 1877*

***Boiler Room, Carbon Center pump station, 1879 These men fired the boilers and kept the petroleum flowing to the refineries. Note early 1890s electric lights. (Lenn McCorry, Butler County Historical Society)***

giant boilers produced steam which drove the pumps. The smaller building at the left was the pump house. John Page, who was the first engineer, and his wife are standing between the two buildings.

Surrounding the pump station were oil tank farms which were often struck by lightning causing ruptured tanks and terrible fires. In the background is the oil town of Carbon Center. Some of the buildings are the Blaney Hotel, the Sherman Hotel, Bob Moffert grocery store, Myers and Shultz meat market, the railroad station, and the Finnesy Brothers blacksmith shop. Lenn McCorry, Butler County oil historian, wrote on the back of this photograph, "We were 20 years locating this picture." This station was wired for electric lights in the early 1890s. (Lenn "Pithole" McCorry, Butler County Historical Society)

## Criswell City

Refer to page 452.

***Office of the McKinney Well***

***James Curtis McKinney, John Luke McKinney, unknown. Great wealth came from small beginnings. (John Mather, Drake Well Museum, Titusville, Pennsylvania)***

The famous "Lady Hunter" well is in the upper center of this Robbins photograph. This view of Criswell City in June, 1875, shows a telegraph office, stores, and boarding houses. The big strike that opened up this territory occurred on July 18, 1874, on the James Parker farm, southeast of Petrolia City. The Boss well was drilled to a depth of 1,600 feet and first produced 3,000 barrels per day. The "Lady Hunter," drilled on October 12, 1874, was 100 yards from the "Boss Well," and produced 8,000 barrels per day and averaged more than 150 barrels per day for the next six months. By March 1, 1875, there were over 600 producing wells yielding 24,000 barrels of oil per day on the "Cross Belt." Criswell City was named after David S. Criswell, one of the investors of the famous "Boss Well," also Satterfield and Taylor were owners. The "Cross Belt" and a small part of it known as the "Sucker Rod Belt," due to the narrowness of the belt stretching from Greece City to Criswell City. The "Cross Belt" was estimated to be one quarter mile in width, while the "Sucker Rod Belt" was no more than 500 feet at its widest. The wells on the "Cross Belt" added greatly to the total production of the 1870's Pennsylvania oil field production.

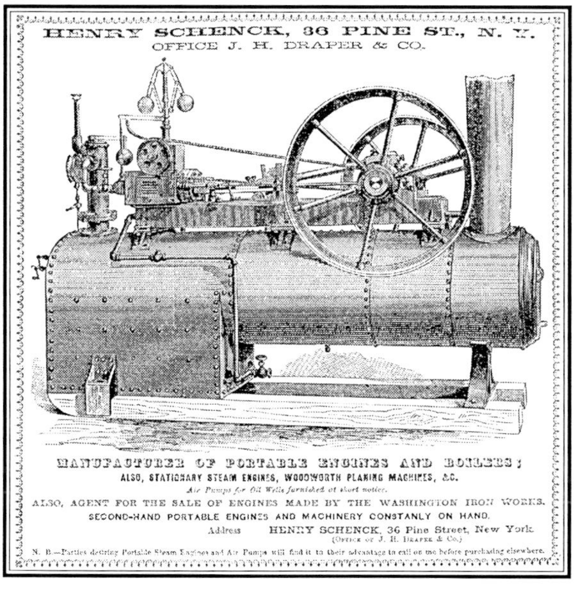


The "Cross Belt" was given its name because of the overlapping of the third and fourth sands at right angles between Petrolia and Karns City. The fourth sand crossed seventy-five feet below the third sand.

The production from the fourth sand was so great that it was necessary to establish an Oil Exchange in Petrolia. The exchange was part of the Chicago Board of Trade, and for a period of time set the price of crude oil until 1880 when the building was destroyed by fire. The abundant production from the "Cross Belt" caused the price of oil to drop from one dollar and seventeen cents to forty cents per barrel. The price remained forty cents for a period of time and caused the local operators

*CRISWELL CITY*

*Gas Well, Butler County, Pennsylvania*
*Some wells produced enough gas to illuminate a city of 10,000 inhabitants. (Frank Robbins, Photographer, Oil City, Pennsylvania)*

a great profit loss. To prevent further decline in price, the local operators formed a cooperative. A vote of the cooperative was taken and production was stopped for ninety days in hope of increasing oil prices. The stoppage was successful, and the price of oil increased to one dollar and thirty-five cents. The price was also affected by decreasing production from the fourth sand, and by February, 1874, oil was selling for one dollar and sixty-seven cents per barrel. Never in the history of oil in Pennsylvania had so many wells been so successful in such a short period of time and in such a small area as in the "Cross Belt" region. (History of Butler County, Pennsylvania, 1883; History of Petroleum, Vol. I and II, J. T. Henry, 1873; Oil City Derrick's Handbook of Petroleum, 1898; East Brady Historical Society's "If These Hills Could Talk")

## Gas Well Hit on October 5, 1875

There are, and have been, many gas wells in the county, but none of them could compare with the Delemater and Duffy well on the Lowry farm. It is in close proximity to St. Joe, which was struck in 1875, and the noise of the escaping gas could be heard for 10 miles. It is estimated that the volume of gas per hour was sufficient to supply the city of Philadelphia two days and nights with all the gas needed. By accident or design, it was set on fire, and the flame, which from the force of the gas was fifty feet from the ground, extended upward some 300 feet. The heat was so intense that the grass for several acres surrounding it was kept green and grew through an unusually severe winter and afforded pasturage for several calves. Visiting parties came here in sleighs and picked dandelions and violets during the winter months, while the grasshoppers and other insects found a paradise here and could be found in very considerable numbers.

After a time, the gas was utilized in drilling and operating other wells for domestic lighting purposes in St. Joe, and for the pump station at Carbon Center, etc. It is estimated that from 300 to 500 miles of piping, including main and lateral lines, extended from this truly wonderful well. Owing to the failure of gas, it was practically abandoned in 1881.

Up to the present time when this field is plowed its burned red clay can still be seen. The waste of natural gas throughout the oil region was appalling. (History of Petroleumm Vol. I and II, J. T. Henry, 1873)

*The Oil Man's Friend*
*This old outhouse, with its half moon cut in the door, was a part of every oil town. Hundreds, thousands of these old "privies" existed but like the old oil man, most if not all are gone. A common name was the "two holer" but in Turkey City, I actually saw a "five holer." God Bless America! (The Author)*

## Great Oil Belt City

As the search for oil continued in Jefferson Township, Butler County, David Kirk and Richard Dilworth, who were producers from Pithole in Venango County, purchased the Gottlieb Wolf farm and laid out the place in town lots. This village was started in 1870 during the completion of the Western Pennsylvania Railroad. It was at this time that it received the name Great Belt because supposedly it was situated on or near the Great Butler County Oil Belt. The great oil boom never materialized.

## Butler, Butler County, Pennsylvania

The first settlement of whites in what is now Butler County, was made in 1796, when the four families of Harbison, Holt, Fulton, and Kennedy, located in Middlesex Township. The town is pleasantly situated, and the surrounding country is rich in soil and mineral products. Butler is laid out with tolerable regularity, the streets being intersected at right-angles, and possessing ample width. A large proportion of the buildings are of brick, and the sidewalks paved with the same material. Iron furnaces once existed here, to which is due much of the growth of the town. It was made a borough in 1817. The first Court House was built in 1807, the present building in 1853. The Presbyterian Church was built in 1822.

**Changes in the Price of Oil**

The following table, furnished by W. C. Coles & Co., 61 Broadway, New York, N. Y., gives the highest and lowest price paid for oil, each year, since 1860:

| Year | Highest | Lowest | Year | Highest | Lowest |
|---|---|---|---|---|---|
| 1860 | $20 00 | $ 2 00 | 1888 | 1 00 | 71⅝ |
| 1861 | 1 75 | 10 | 1889 | 1 12½ | 79½ |
| 1862 | 2 00 | 10 | 1890 | 1 07⅞ | 60¾ |
| 1863 | 4 00 | 2 00 | 1891 | 81⅝ | 50 |
| 1864 | 14 00 | 3 75 | 1892 | 64⅛ | 50 |
| 1865 | 10 00 | 4 00 | 1893 | 80 | 62⅞ |
| 1866 | 5 00 | 1 65 | 1894 | 96¾ | 78½ |
| 1867 | 4 00 | 1 60 | 1895 | 2 69 | 94½ |
| 1868 | 5 50 | 1 80 | 1896 | 1 50 | 90 |
| 1869 | 7 00 | 4 25 | 1897 | 96 | 65 |
| 1870 | 4 90 | 2 75 | 1898 | 1 19 | 65 |
| 1871 | 5 15 | 3 40 | 1899 | 1 66 | 1 13 |
| 1872 | 4 10 | 3 00 | 1900 | 1 68 | 1 05 |
| 1873 | 3 05 | 1 00 | 1901 | 1 30 | 1 05 |
| 1874 | 1 90 | 45 | 1902 | 1 54 | 1 15 |
| 1875 | 1 65 | 90 | 1903 | 1 93 | 1 50 |
| 1876 | 4 23⅛ | 1 48¾ | 1904 | 1 85 | 1 50 |
| 1877 | 3 70 | 1 53¾ | 1905 | 1 61 | 1 27 |
| 1878 | 1 87½ | 78¾ | 1906 | 1 64 | 1 58 |
| 1879 | 1 28¾ | 63⅛ | 1907 | 1 78 | 1 58 |
| 1880 | 1 24⅜ | 71¼ | 1908 | 1 78 | 1 78 |
| 1881 | 1 01¼ | 72½ | 1909 | 1 78 | 1 43 |
| 1882 | 1 35 | 49¼ | 1910 | 1 43 | 1 30 |
| 1883 | 1 24¾ | 83¼ | 1911 | 1 35 | 1 35 |
| 1884 | 1 15⅝ | 51¼ | 1912 | 2 00 | 1 30 |
| 1885 | 1 12⅝ | 68 | 1913 | 2 50 | 2 00 |
| 1886 | 92¼ | 60 | 1914 | 2 50 | 1 45 |
| 1887 | $ 90 | $ 54 | 1915 | 2 25 | 1 35 |
| | | | 1916 | 2 60 | 2 25 |

*The Price of Petroleum From 1860 to 1916*
*Courtesy of The Oil and Gas Man's Magazine,*
*Butler, Pennsylvania, 1916*

With the advance of oil developments around its vicinity, Butler has taken a step in advance. New buildings are being erected, and all branches of business have become stimulated. A number of wells are now going down in the Butler district, upon which the further advance of the town for the present depends. Butler is a well-ordered town, possessing all the necessary adjuncts to civilization, such as a court house, jail, churches, school houses, two newspapers, etc.In 1870 the population of Butler was 1,935, by 1877 3,300 people lived there, and in the next ten years doubled.

Butler was never an oil boom town, but it supplied the oil region with goods and services needed to fuel the oil territory. (History of Petroleum, Vol. I and II, J. T. Henry, 1873)

## The Fourth Sand

We have already mentioned, briefly, the discovery of a fourth sand, or oil rock, in the vicinity of Karns City. It was first found by Mr. Charles Stewart, a native of Butler County. He purchased an abandoned well on the Scott farm, near Karns City, and after a month's vain effort to make it pay as a pumping well, resolved to sink it deeper.

He began this labor and after nearly eight weeks' drilling in daylight, struck the first flowing well in the fourth sand. This well is known as the "Bank's" or "Stewart" well, and averages 400 barrels daily. The entire oil community, including the shrewdest operators, scouted the idea of this being the "fourth sand," and claimed that the well had not originally been drilled deep enough.

Messrs. Tack and Morehead, however, on the adjoining farm, commenced to drill one of their abandoned wells, known as the McCleer No. 1, about four weeks ago, and struck the fourth sand on the 15th of September, with a good show of oil. They continued to drill until the 18th of September, when the well commenced flowing at the rate of 700 barrels per day! The most experienced operators claim that the fourth sand is only prolific at these points, where a spur branches out from the main belt, and this is evidently the spur of the Modoc belt. This theory will be very fully studied within the next 60 or 90 days, as there is a determination on the part of everybody owning an abandoned or non-paying well to try it.

The fourth sand, thus far developed, is from 65 to 75 feet below the third sand, and is of excellent quality. (The History of Petroleum, Vol. I and II, J. T. Henry, 1873)

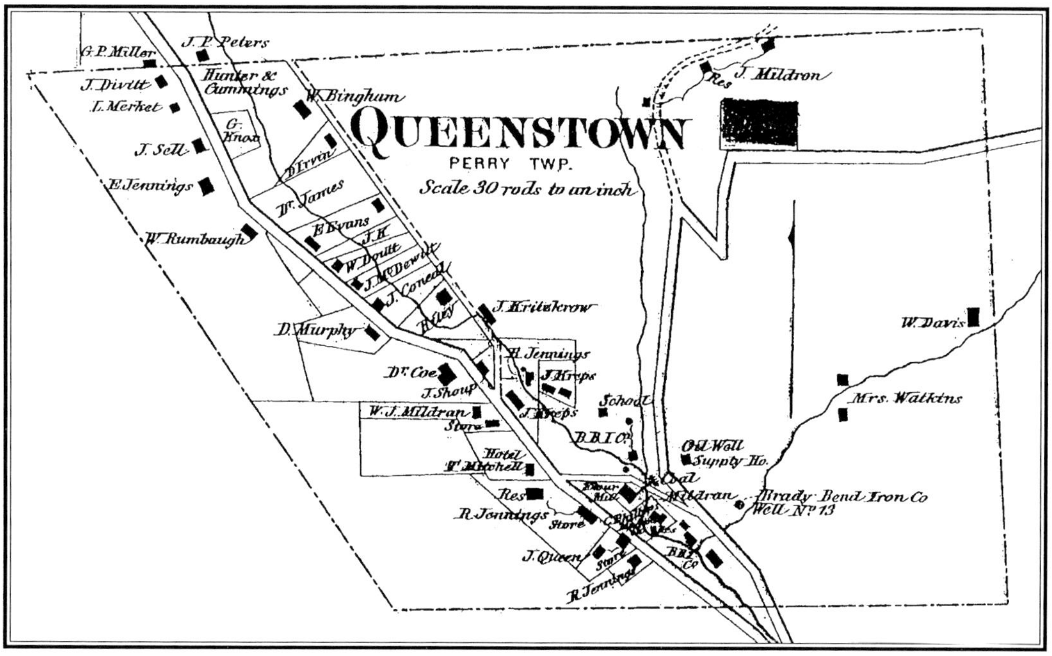


*Atlas of the County of Butler, 1874*

## Queenstown, Clarion County

The borough of Queenstown was incorporated in 1858. The town was named after John Queen, who located in the place in 1848. At that date the only persons residing within the present limits of the borough were Daniel Day and Abraham Teegard and their families.

No regular survey of lots was ever made, but pieces of land were sold to purchasers as they were wanted, by J. Queen, R. Jennings and Daniel Day, who owned the land now comprised in the borough. These lots were taken up by employees of the Brady's Bend Iron Works, and in a few years Queenstown became a small but flourishing village.

Oil production, while it did not greatly increase the population of Queenstown, materially aided its business interests. The Armstrong well, on the Meldron farm, was the first producing well struck in the neighborhood. This well began flowing April 17, 1870. It caught fire and burned three or four days. Good judges estimated the first day's flow at one thousand barrels. Other wells were soon completed in the vicinity of Queenstown, some of which are still producing. In 1872, Chas. Phillips began manufacturing all kinds of oil producers' implements at Queenstown. He employed from eight to ten men, and carried on a very successful business until 1881, when he moved away.

Queenstown is situated on a small tributary of Sugar Creek. The southern limit of the borough is the northern line of Brady's Bend Township. In 1860, the population was 127; in 1880, 217. (History of Butler County, Pennsylvania, 1883, Waterman, Watkins and Co., Chicago, Illinois)

*Brady's Bend Rolling Mills, 1873.*
*The Rolling Mills at Brady's Bend have been in existence for about fifty years. Their full capacity gives employment to 800 men. At the present time there are about 500 men at work. The town is situated on Sugar Creek, about a half mile from the Allegheny River, and has six churches built by the miners and mill men. (History of Petroleum, Vol. I and II, J. T. Henry, 1873)*

## East Brady

"East Brady has a population of nearly 1,500. There are five hotels and one church, one machine shop and foundry, a number of groceries and dry goods houses, and three schools. There is a large amount of freight business done here on the Allegheny Valley Railroad, consequent upon the opening up of the oil districts of Modoc and Millerstown.

"East Brady has a weekly paper, The Independent, which enjoys a large circulation, Col. Young, one of the live men of the profession of journalism, being its editor and proprietor.

"East Brady was never considered an oil boom town and was located outside the boundaries of the Lower Belt. This town supplied goods, services and shipping to the oil region. (History of Petroleum, Vol. I and II, J. T. Henry, 1873)

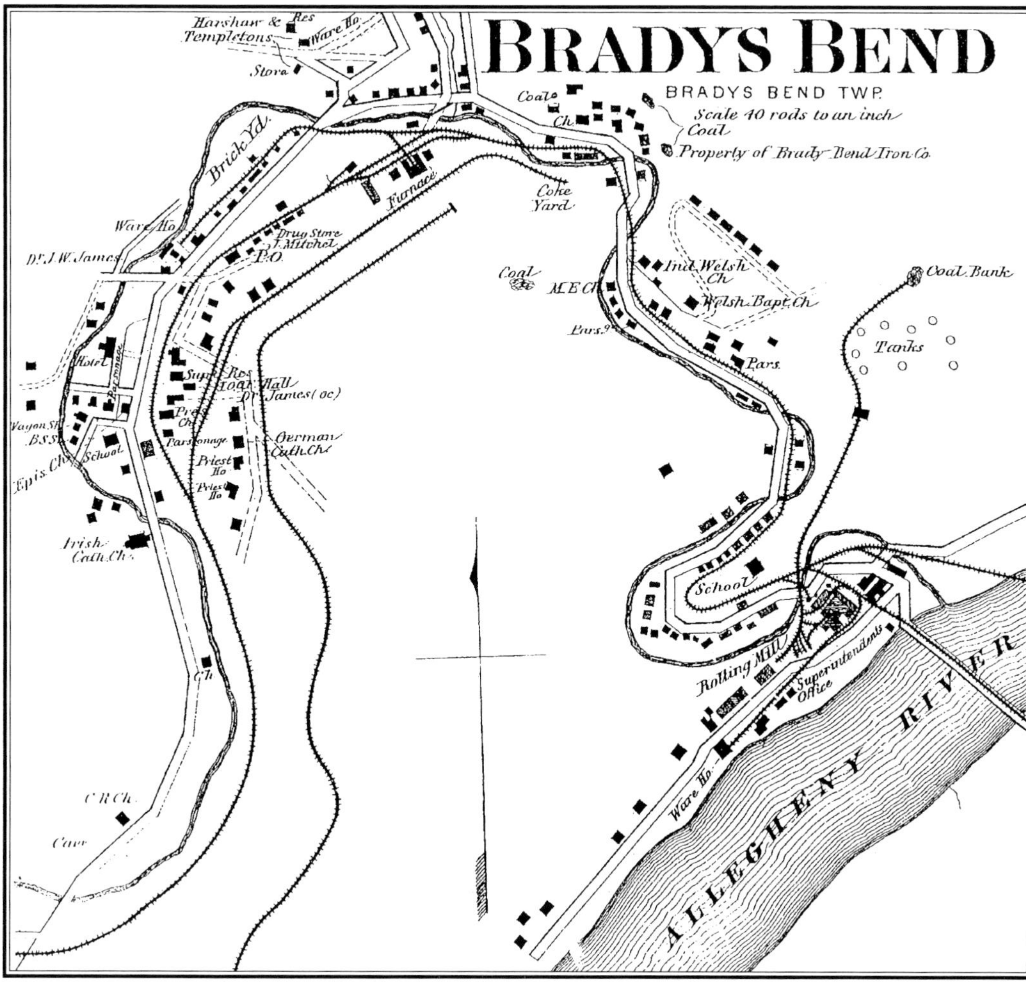


*Brady's Bend, Atlas of the County of Butler, 1874*

## The Derrick, June 15, 1880
## A Bad Cold or Distressing Cough

Dry, parched sore throat, pneumonia, bronchial and asthmatic attacks, weakened and debilitated state of the system, all these dangerous symptoms are cured by "Dr. Swayne's Compound Syrup of Wild Cherry." The sore lungs yield to its healing properties. An occasional dose of "Swayne's Pills" should be taken to keep the bowels free. They are excellent for torpid liver and other complaints.

## John Mather, Famous Oil Region Photographer

John Mather recorded the early days of oil discovery with his camera and Oh! how wonderful his photographs are. Thanks to John, we can shake hands with Colonel Edwin Drake, ride an oil barge down Oil Creek, or walk the long-forgotten streets of that famous Venango County oil ghost town, Pithole.

*John Mather, Oil Region Photographer, Titusville, Pennsylvania*
*Drake Well Museum, Titusville, Pennsylvania*

*John Mather memorial, Oakland Cemetery near Dempseytown, Pennsylvania, 1915*

John Mather lived in Titusville, Pennsylvania, and that is where he had his studio. In his latter years, life was hard and by the time he passed on, Mather was pretty much "down on his luck." John Mather's fame has grown over the years and today he is much more appreciated than at the time of his death.

John and his wife are buried in the Oakland Cemetery, an old country cemetery located off the beaten path between Dempseytown and Cherrytree, Pennsylvania, on Route 417 North. You, a student of early oil history, need to "motor" one-half mile past Dempseytown and turn left on Gresham Road. The cemetery is located one mile on the left in a small grove of oaks. Franklin, Oil City, and Titusville are nearby.

This photograph shows the memorial and final resting place of the famous Western Pennsylvania oil region photographer, John Mather, and his wife.

John rests in this oil country cemetery and is now "sleeping the sleep that knows no waking." He died August 23, 1915. "Peace to his ashes." (John J. McLaurin, Sketches in Crude Oil)

# Men Who Made Oil History
# St. Petersburg
# A Bad Fall, Businessman Injured
# Special to the Derrick, June 18, 1880

"Mr. William Z. Blakslee met with a serious accident this morning while laying gas pipe from the Taylor well below town. He fell from a rock and injured his spine so badly as to cause paralysis to the lower extremity. He is lying in a critical condition. His physician has hopes of his recovery."

## Well-Known Oil Man, W. Z. Blakslee, Hurt in Accident
## Oil City Derrick, June 19, 1880

Terrible accident befalls W. Z. Blakslee, well-known oil man and partner in Blakslee Bros. Machine Shop, formerly of Tarr Farm, Venango County, and current resident of St. Petersburg, Clarion County.

## June 25, 1880

"W. Z. Blakslee, well-known oil man, who received a serious injury last week, causing complete paralysis of the lower extremity, is still in a critical condition."

## June 28, 1880

"Mr. W. Z. Blakslee is feeling some better this morning, but his condition is yet of a most critical nature."

## July 6, 1880
## The Death of W. Z. Blakslee, a Sketch of His Life.
## From our Special Reporter, St. Petersburg

"Mr. William Z. Blakslee, who was injured by falling into Spruce Run some two weeks ago, an account of which was given in these columns, died from the effects of his injuries Monday morning. The deceased was in his forty-seventh year, and a native of Connecticut, having been born in Newton, that state, December 7, 1833. A machinist by trade, he early embarked in company with his brothers in the manufacturing business at his native town, removing from there to New York in 1853, where he was successfully engaged until 1864, when, attracted by business in the oil regions, he located at Tarr Farm. They engaged in the manufacture of drilling tools and producing, since which time he has been closely identified with the oil business. Following the tide of operations, he removed to St. Petersburg in the Spring of 1872.

"Early recognizing the necessities of trade, he became one of the incorporators of the Antwerp Pipe Company, a pioneer in the practical transportation of oil through pipes. Mr.

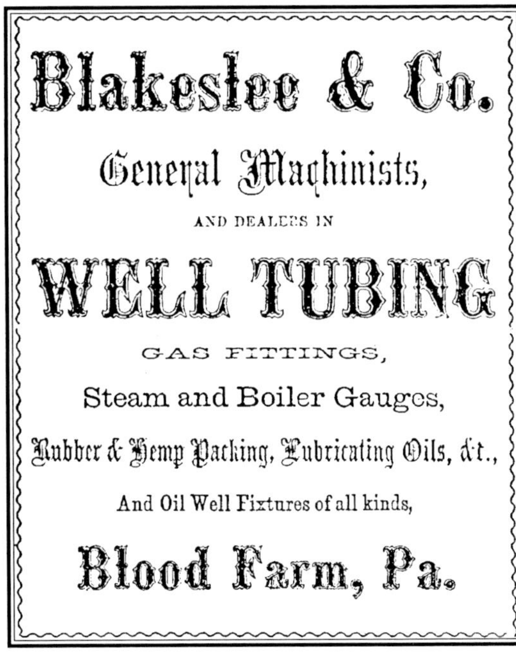


Blakslee was one of the earliest operators in the Bradford field, a partner in the producing firm of Bradley and Company, and a stockholder in the United lines.

"An active, energetic citizen, an honest, straightforward business man; a kind husband, father and brother, William was a man whose many genial qualities had gained hosts of friends throughout the entire oil region; cut off in the prime of life, his sad death will cast a gloom over a wide circle.

"He leaves a wife and three children, who have a heartfelt sympathy of the entire community. The funeral services were held at his residence on Main Street today at 11 a.m. His remains were taken to Albany, New York, for interment, accompanied by members of the Masonic fraternity of which order he was an esteemed member."

"Peace to his ashes." (John J. McLaurin)

The oil business in northwest Pennsylvania was fast paced and dangerous. Had William Z. Blakslee's life not been cut short at the age of 47 years, William would have no doubt ranked high on the list of successful, well-known Pennsylvania oil men. A lithograph on page 126 shows the Blakslee Bros. Machine Shop and the preceding information tells a far too common story of sorrow and disappointment in the "oil patch."

## Blakslee Bros. Machine Shop

*Charles F. Blakslee*
*1827-1913*

This information comes from the Pennsylvania Commemorative Biographical Record, 1898, page 982: William Z. Blakslee, partner in Blakslee Bros. Machine Shop, Main Street, St. Petersburg, was born December 7, 1833, and died July 5, 1880, of accidental injuries and buried in Albany, New York state. James B. Blakslee was born April 1823 and died in 1887 and buried in Newtowne, Connecticut. James became interested in oil wells in Pennsylvania and formed a partnership with his brother-in-law, John C. Beers, and from this nucleus formed an important industry. The third brother was Charles F. Blakslee. He was born September 26, 1827, and died at Franklin, Pennsylvania, on July 8, 1913, at the ripe old age of 85 years. Charles lived in St. Petersburg until the early 1890s before retiring to Franklin. At the age of 33 he enlisted in Co. B, 37th New York Infantry. His obituary stated he was a retired machinist and oil producer and one of the best known residents of Franklin. He is buried near the main gate of the Franklin cemetery.

Thanks men and "Peace to your ashes." (J. J. McLaurin)

## Men Who Made Oil History
## Northwest Pennsylvania Oil Man
## Gilmore C. Fink

Gilmore C. Fink was born in Petersburg, Pennsylvania, on November 23, 1852, and died September 28, 1888. Had his life not been cut short at the age of 35, Gilmore would have ranked high on the list of well-known Northwest Pennsylvania oil men. He was the son of Joel and Eve Fink, and learned the oil business from his father, a pioneer in the Lower Field. Many warm summer nights were spent drilling and pumping wells along the beautiful Allegheny and Clarion River hills.

Through the oil excitement of the 1870s in Clarion, Butler, and Armstrong Counties, business was profitable. In 1880 Gilmore was married to Birdie Estella Howe, daughter of Mrs. Harry Howe who managed a millinery and dress shop in Foxburg. In 1881, Gilmore C. Fink was listed as a member of the board of directors in the Foxburg Bank and owned stock in the Foxburg water works. In 1883, he started the Union Light and Heat Company, which supplied natural gas to Foxburg for lighting and heating. The company was sold to James Rowland and a group of investors in 1886. Fink also was associated with Thomas Healey in leasing and drilling in Butler County.

*Gilmore Fink*

On February 16, 1885, Fink formed a partnership with A. J. Zahniser of Foxburg. Under the name of Fink and Zahniser, they opened a large machine shop in Foxburg and along with machine work they sold boilers and engines. This machine shop was later operated by Philip Davis and son. About this time it was decided that it was time to move on to Washington, Pennsylvania. There the Washington Iron Works was formed under the name Fink and Zahniser. The business consisted of making and repairing oil well tools and machinery, the running of a foundry, and a general machine shop. They were dealers in boilers and engines and a hardware store sold implements and machinery. Business partners were G. C. Fink (three-eighths interest), A. J. Zahniser (three-eighths interest), S. H. Fink (one-eighth interest), and John N. Zahniser (one-eighth interest).

The Clarion Democrat reported in June 1885 that "Gilmore C. Fink is no longer associated with the Fox Estate," near Foxburg. He started as a common laborer, working the wells and left as superintendent of all oil well operations.

## Why Not Water?
## Another opportunity presented itself

In 1884, a gas well was drilled on the land of G. W. Hood on Hood Hill, west of Indiana, Pennsylvania. The well was drilled by Joel Fink of Livermore, formerly of St. Petersburg, and his son, Gilmore C. Fink, of Washington, Pennsylvania, to a depth of 2,655 feet. The well showed no trace of natural gas but at 325 feet produced a clean, clear, abundant amount of water. The town of Indiana had no good source of water and on March 12, 1886, Mr. Hood, Joel Fink, and Edward Rowe met at Mr. Hood's law office and the Indiana Water Company and the West Indiana Water Company were organized, each with $10,000 capital stock.

The original shares of stock were sold at $50 each to Joseph M. Fox of Philadelphia, Joel Fink, Gilmore C. Fink, Martin Earhart, G. W. Hood, Dr. N. F. Ehrenfield, Henry Hall, Edward Rowe, Harry W. Wilson and Samuel A. Smith. When it was discovered that a charter had been issued for another Indiana Water Company, the name was incorporated as the Clymer Water Co. of Indiana and Clymer Water Co. of West Indiana, on April 26, 1886. The first board of directors were G. W. Hood, president; Edward Rowe, secretary and superintendent; Harry W. Wilson, treasurer; Joel Fink, Joseph M. Fox and Gilmore C. Fink.

Edward Rowe designed the plans for the water system and Gilmore C. Fink was appointed managing director to proceed with construction. The system grew to two boiler iron holding tanks of 200,000 gallons and 100,000 gallons, five and one half miles of water main lines, 65 water plugs and 68 fire hydrants. The work began in 1886; however, the water was not turned on until June 1887.

## DISASTER STRIKES
## A Terrible Accident

The Indiana Democrat of June 7, 1888, has the following account of an accident that befell Mr. Gilmore C. Fink of Washington, Pennsylvania, and a former well-known resident of Clarion County.

A Serious Accident to a Former Resident of Indiana

"On Thursday last, Decoration Day, an accident occurred which for a time threatened the life of Gilmore Fink, who resided in this place for sometime. During the construction of the water works here in Indiana, Mr. Fink was on a visit to his brothers John M. and Steven H. in Washington, Pennsylvania, and was sitting in a buggy about to take a drive. Without warning, a run away team collided with the buggy. The horses ran off from two gentlemen who were behind Fink's vehicle. The tongue of their rig ran through the buggy striking Fink in the middle of the back below the shoulder blades. At the same time the large horses crushed the buggy he was sitting in, completely to the ground. The horses were lying one on each side of Mr. Fink when

*Gilmore C. Fink Memorial, Oakland Cemetery, Indiana, Pennsylvania*

the spectators appeared. He was immediately taken to his hotel and a physician called. It was found that he was completely paralyzed from the hips down and had sustained injuries which seriously threatened his life. A telegram was sent to his sister, Mrs. Melissa Vensel of this place, saying that feeling is gradually returning to the injured man's limbs and the physicians have some hope of his recovery. Mr. Charles H. Vensel is superintendent of construction at the new Indiana water works. Mr. Fink's father, Joel Fink of Livermore, Pennsylvania, and his sister Mrs. Vensel left on Monday and are now with him. We hope he may recover entirely. His wife, who resides in Allegheny City, was sent for. The injured man is the senior member of the firm Fink and Zahniser, who own an oil well machine shop here."

## "The roll is called up yonder"

Gilmore C. Fink was taken to his father's farm near Livermore in Blacklick Township, Indiana County, but recovery never came. Gilmore wrote his last will on September 11, 1888, and passed away on September 28, 1888. He left his wife, Birdie Estella, and his small son, Roy Howe Fink. Mr. Fink amassed a very sizeable estate and is buried in the Oakland Cemetery at Indiana, Pennsylvania. "Peace to his ashes." (John J. McLaurin)

## Men Who Made Oil History
## Robert Galey, Sr.
## Death Notice

Robert Galey, Sr., father of the Galey Brothers, well-known oil men, and Daniel Galey of Perry Township who lives at the former home and farm of the family, died at his residence in Bellevernon, Pennsylvania, on Thursday last at the advanced age of eighty-four years. During a long life of industry and frugality, Mr. Galey accumulated a fortune and his estate is estimated to be worth a half million dollars. His remains were laid to rest at Beaver Cemetery, Beaver, Pennsylvania, the home of his son, Thomas F. Galey.

The obituary above was published in the January 31, 1895, issue of the Clarion Democrat, Clarion, Pennsylvania. Robert Galey Sr. was John H. Galey's father. After much research, study, newspaper articles, books, letter and photographs, I must admit that the oilman I would most like to have met, the oilman who truly thought oil, slept oil, and lived oil, was good old John H. Galey. The riches of petroleum and the material wealth money could buy was not John's quest. W. L. Mellon, a nephew of Pittsburgh banker, Andrew W. Mellon, wrote in his autobiography his view of John H. Galey.

"He was an amazingly ingenious fellow. Although improvident, he was in my opinion beyond question the greatest oilman of my time. There are not many who will dispute this estimate. A quiet fellow, never boisterous, never a show off, he was unfailingly intrepid and aggressive in the hunt for oil. Always money interested him much less than finding oil. His greatest satisfaction came from unlocking some long unsuspected treasure house of nature; and a succession of dry holes discouraged him not at all. Petroleum had John Galey bewitched. Often rich, he was never free. Under his bright eyes there went on incessantly an almost feverish thinking about the hiding places of petroleum."

*The small flat grass marker in the center of this photograph is John H. Galey's memorial. In the end, little was left.*

John Galey's last effort to discover a virgin oil region was in 1911 when he drilled three wells in the San Juan Basin in Northwest New Mexico. John, the "oil man's oilman" passed away in Joplin, Missouri April 12, 1918. John now rests in the Beaver Cemetery, Beaver, Pennsylvania, along with his family in Section K, Lot 5. "Peace to his ashes." (John McLaurin)

## MEN WHO MADE OIL HISTORY
## Capt. Henry H. Cumings

Henry H. Cumings was born at Monmouth, Illinois on December 1, 1840. Capt. Cumings graduated with the class of 1862 from Oberlin College. In 1862 he enlisted in the 105th Ohio Volunteer Infantry and participated in all principal engagements. With the rank of Captain, he was mustered out in June 1865.

*Henry H. Cumings*

Soon after the close of the Civil War, Capt. Cumings visited the Pennsylvania oil region. At Tidioute, Pennsylvania, which is located along the Allegheny River, he took charge of the Day and Co. oil refinery in 1866.

In 1873, the refinery was dissolved and Capt. Cumings hooked up with Jahu Hunter and formed the firm of Hunter and Cumings. They operated extensively in Butler and Armstrong Counties and were among the heaviest and most successful operators in Butler County. They were owners of the famous "Lady Hunter" well, located on the Cross Belt between Petrolia and Queenstown, which had a production of 8,000 barrels a day when first struck.

Hunter and Cumings also opened up the old "Ponties" pool east of Millerstown and drilled many wells at Herman Station where they were rewarded with great production. They also operated on the lands of the Brady's Bend Iron Company and drilled extensively on the hill opposite the town of East Brady, Pennsylvania.

Captain Cumings invested his profit wisely and the Tidioute Savings Bank was principally owned by Messrs. Hunter and Cumings up until the death of Jahu Hunter in 1897. Mr. Cumings was truly a public-spirited, generous and enterprising citizen.

*Henry Harrison Cumings*
*Of the Firm Hunter and Cumings*
*Senator H. H. Cumings represented Warren and Venango districts in the Pennsylvania Senate from 1899 to 1906. Henry H. Cumings 3rd, grandson of Senator Cumings, served in France during World War I with the American Ambulance Field Service. While returning to the United States on the transport "Antilles," a German submarine destroyed the ship and all were lost at sea.*

## Henry Harrison Cumings

The Warren Evening Mirror dated May 15, 1913, reported that Tidioute's leading citizen, H. H. Cumings, passed away May 14th at his home. Mr. Cumings had been on the decline since March. Henry Cumings was President of the Northwestern Association of the Grand Army of the Republic and was buried with full military honors in the cemetery at Tidioute, Pennsylvania. "Peace to his ashes." (John J. McLaurin)

## 1913 Tragedy

Only thirty-nine days before, Mrs. Charlotte Cumings had laid her beloved Captain to his eternal rest. She was just taking up her new duties and looked bravely at the future with hope and courage. Mrs. Cumings traveled from her home to Baltimore to visit her little granddaughter Minerva Ellis, who had fallen down stairs and broke her arm. Her father Ralph Cumings promised her a ride in his new Ford Roadster as soon as she had sufficiently recovered. On the twenty-second of June 1913 little Minerva was given her promised ride. As the automobile reached Slade Avenue, the steering gear jammed causing the car to turn square across the road, overturned and pinned Ralph and his mother underneath. At this very moment a big touring car could not steer clear, ran over Mrs. Cumings and she died instantly. On a street in Baltimore she passed from earthly sight and caught her first "Vision of Glory".

*Cumings Memorial, Tidioute, Pennsylvania*

# Men Who Made Oil History
# Jahu Hunter

Mr. Jahu Hunter was born at the family homestead two miles below Tidioute, Warren County, Pennsylvania, on October 3, 1830. The Hunter family was of Scotch Irish ancestry, a group that embraces mental vigor, physical energy and warm heartedness. Jahu learned a lesson in his early life and at the age of 17, through great savings and shrewd management, became the owner of a yoke of good oxen. While skidding logs out of the woods one day, by accident one of his oxen was fatally injured. This terrible accident meant half his fortune was gone and no way of fulfilling his contract. In later life he was in the habit of putting aside other losses in business, even though amounting to many thousands of dollars by this remark, "Oh! that is nothing; it does not compare to the loss of my dear ox."

*H. H. Cumings (left), Jahu Hunter (right), Tidioute, Pennsylvania businessmen. Photo courtesy Sketches in Crude Oil, J. J. McLaurin*

When oil was discovered and oil interests became a prominent factor in business of the region, Mr. Hunter formed an association with Henry Cumings under the name of Hunter and Cumings in 1873. In the oil business, they operated successfully and very largely. The famous "Lady Hunter" well was drilled in Butler County and brought a large fortune to the firm. Hunter and Cumings were large investors in the Tidioute Savings Bank, wheat lands in North Dakota, and many mining and lumbering interests.

*Jahu Hunter Memorial in the Tidioute Cemetery, Tidioute, Pennsylvania, 1897*

Jahu Hunter, well-known Tidioute business man, passed away on March 15, 1897 after being struck with apoplexy, and was laid to rest in the cemetery at Tidioute, Pennsylvania.

It was a terrible shock to the people of Tidioute and surrounding areas. "The News" published in Tidioute on April 25, 1902, reported another family tragedy when Jahu's only surviving child and president of the Tidioute Bank, Livingston L. Hunter, and Burgess Lyman B. Magill were instantly killed. Both were on business when the "City of Pittsburgh," an Ohio River steamboat, blew up and burned to the water line near Cairo, Illinois. There was a terrible loss of life with upwards of 100 people killed. "Peace to his ashes." (John J. McLaurin)

## MEN WHO MADE OIL HISTORY
### Daniel O'Day

Daniel O'Day was born in Clare County, Ireland, on February 6, 1844, and at the age of one year his parents brought him to the United States. Mr. O'Day obtained the rudiments of education by being home taught.

Through hard work and knowing how to get the job done, Mr. O'Day was the Vice-President of the National Transit Company. He first came to the oil country in August 1865 and took his first job at Titusville, Pennsylvania, in solving problems in the transfer and shipping of petroleum. With the growth of the oil business, cheaper and more efficient means of transporting oil were needed and the pipe line was the answer. The completion of an elaborate system of pipe lines between the wells and the refineries freed the oil industry from the constant tightening grasp of the railroads.

*The grand "Lady Hunter" well, flowing 100 feet high. The firm of Hunter and Cumings struck this gigantic spouter on October 9, 1874, two and one-half miles from Petrolia City, Butler County. The "Lady Hunter" produces its oil from a fourth layer of sand rock which is found in that locality. The great vein of oil was struck after penetrating 10 feet into the fourth sand at a depth of 1,475 feet. The first day's production was 8,000 barrels. It produced 145,000 barrels of oil prior to March 1, 1876, and is still flowing 150 barrels daily. Nearly all flowing wells are intermittent, spouting at regular intervals. Every half hour the ground rumbles, and oil and gas belches forth with terrible force and power. (Photograph by Frank Robbins, Oil City, Pennsylvania*

Mr. O'Day's first pipe line venture was in 1873 when the base for crude oil supplies shifted permanently from the valley of Oil Creek to the Clarion and Butler oil fields. He built a line from Emlenton, Pennsylvania, to the Clarion County oil fields, which was later extended to Oil City. It was originally known as the Emlenton Pipe Line. The firm's name was subsequently changed to the American Transfer Company, with headquarters at St. Petersburg and Oil City. This pipe line made possible a constant supply of crude oil to the refineries.

Mr. O'Day was a prominent factor in the great consolidation of pipe line companies which took place in 1877. He became a director of the United Pipe Line and general manager. The great Clarion and Butler oil fields gave Daniel O'Day his start. Mr. O'Day lived at No. 128 West 72nd Street, New York City.

## Daniel O'Day

*Daniel O'Day*

Mr. Daniel O'Day was vice president and general manager of the National Transit Company. He was born in County Clare, Ireland, February 6, 1844, and died at Royan, Southern France, on Thursday afternoon, September 13, 1906. Information on his death was received in Titusville shortly after noon, coming in the form of a telegram from Joseph Seep, purchasing agent of the Standard Oil Company, who was in New York on business.

According to the Titusville Morning Herald dated Friday, September 14, 1906, Mr. O'Day had stomach and bladder problems and it is thought that a broken artery was the probably cause of death.

On October 16, 1906, the Buffalo News reported that burial was in the Holy Cross Cemetery at West Seneca, New York. In 1909, West Seneca was changed to Lackawanna and is near Buffalo, New York.

"Peace to his ashes." (John J. McLaurin)

## Secret Code of the Standard Oil Company

John D. Rockefeller and his Standard Oil Company kept in touch daily with the northwest Pennsylvania oil region by telegraph. Well information, banking information, pipe line tankage records and all other oil business was passed back and forth and the Standard did not want their business to be known to any competition. In order to keep information private and further confuse competitors, a top secret code was developed. Instead of talking about the Argyle Savings Bank, the code word used was "Pully." For instance, if the Argyle Savings Bank dropped their interest rate, word would be telegraphed to Standard Oil that "Pully" rate decreases by one percent. If a new well came in at Petrolia, the code was Wells producing, price per barrel increasing in "Opal" region. This code was used for towns, banks, wells and producers names with hundreds of codes used. (Lenn McCorry)

*Daniel O'Day Memorial, Holy Cross Cemetery*

# Men Who Made Oil History
# James C. McKinney

One of the most influential and respected figures, together with his brother, John L. McKinney, ever to operate in the Appalachian oil industry was James Curtis McKinney, oil producer, banker, financier, industrialist, philanthropist.

Son of James and Lydia Drury McKinney, James was born in Pittsfield, Pennsylvania, 28 miles from Titusville, in 1844. Finishing his formal education at Waterford Academy, Erie County, in 1861, he joined a railroad surveying crew and helped locate several rail lines in the area. In 1863 he resigned to enter the lumber business, soon establishing his own yard at Franklin, Pennsylvania.

*James Curtis McKinney*
*Courtesy of Drake Well Museum, Titusville, Pennsylvania*

J. C. McKinney's lifelong interest in oil began when he drilled his first well below Franklin in 1864 and operated it himself. He and his brother, John, joined forces in 1868 and from that time on engaged in producing activities that eventually included every field east of the Mississippi. As principals the McKinney brothers were interested in more than one hundred thousand acres of oil producing territory. They produced and sold, in twenty years of operation, as much petroleum as any one firm in the industry of that time.

*The Hon. James C. McKinney Mausoleum where James was laid to rest on December 6, 1924, at Woodlawn Cemetery, Titusville, Pennsylvania. Col. Edwin Drake and his wife are buried at Woodlawn. "Peace to his ashes." (John J. McLaurin)*

In 1889 all of the McKinney brothers' production acreage was sold to the Standard Oil Company and J. C. and his brother became large stockholders in Standard. J. C. McKinney maintained his touch with the oil industry for many years as general manager of South Penn Oil Company, as an astute investor, and as a banker.

Moving to Titusville in 1876, J. C. McKinney became a large stockholder in the Titusville Iron Works (now the Struthers Wells Corporation), and a director of the Commercial Bank. His civic interests were many and his contributions to charities, church and community advancement were of the largest order. Typical of his generosity was a gift at the time he founded the Titusville Trust Company in 1918. After all stock had been subscribed for and sold, and after the handsome granite and marble bank had opened its doors to the public, it was announced in the Titusville Herald that the $300,000 structure was a gift to the stockholders from Mr. and Mrs. James Curtis McKinney.

## McKinney Road

James C. McKinney donated $65,000 toward the construction of the new concrete highway from Titusville to Rouseville. The Honorable James C. McKinney and family made possible the construction of this road by making the largest personal contribution ever received by the State Highway Department of Pennsylvania. This highway was officially opened November 25, 1924. James C. McKinney died at his home on 314 Union Street on December 6, 1924, and was laid to rest in the Woodlawn Cemetery, Titusville, Pennsylvania.

## MEN WHO MADE OIL HISTORY
## Capt. Jacob Jay Vandergrift

Captain Jacob Jay Vandergrift was born in Pittsburgh, Pennsylvania, April 10, 1827. He was a well-known businessman and capitalist of Pittsburgh and one of the pioneers in the development of petroleum and natural gas interests in Western Pennsylvania.

Among other enterprises, Mr. Vandergrift was associated with the organization of pipe lines in Butler County, Pennsylvania. Under the control of Vandergrift, Foreman and Co., pipe line systems were built in Butler County including the Fairview Pipe Line, Raymilton Pipe Line, The Cleveland Pipe Line, and the Millerstown Pipe Line. When perfected, these lines were organized into one general system under the name of the United Pipe Lines of Vandergrift, Foreman and Company.

***Col. Edwin Drake Memorial***

***Woodlawn Cemetery, Titusville, Pennsylvania***
***Col. Edwin Drake and his wife Laura's final resting place, Woodlawn Cemetery, Titusville, Pennsylvania. Col. Drake died in Bethlehem, Pennsylvania, on November 9, 1880. His remains were moved to Titusville and reinterred on September 2, 1902. His wife Laura passed away on May 17, 1916. "Peace to his ashes." (John J. McLaurin)***

*Jacob Jay Vandergrift*

In 1872, Mr. Vandergrift and several other capitalists organized the firm of H. L. Taylor and Co. for the development of oil lands in Butler and Armstrong Counties, and also the famous Argyle Savings Bank at Petrolia, Butler County. Both of these enterprises were successful and the Argyle Savings Bank during its existence was honored by great public confidence. The bank was so trusted by the locals and oil men that even though its capital was only $50,000, the bank's deposits at times reached one and a half million dollars.

Mr. Vandergrift's hard work and kindness of heart gave him many dear friends.

Jacob J. Vandergrift was born in 1827 and passed away in 1899. He was laid to rest in the Allegheny Cemetery at Pittsburgh, Pennsylvania, section eight and one-half, lot twenty-four. "Peace to his ashes." (John J. McLaurin)

## MEN WHO MADE OIL HISTORY
## Joseph M. Black

J. M. Black was born in Mercer Township, Butler County, in 1836. In 1859 he went to Altoona, Pennsylvania, and was employed in the manufacture of artificial gas and was first to help supply this mode of lighting to that city.

J. M. Black came up the Allegheny River to Emlenton, Pennsylvania, on the old steamboat "Venango," and located on a farm near Emlenton in 1867. While not a resident of Emlenton, he was always interested in Emlenton's welfare and was connected with most of its enterprises in one way or another. Joseph M. Black and Dr. A. W. Crawford of Emlenton were connected with the production of oil in that vicinity. They were among the first operators in the Parker's Landing fields and also involved with early developments at Scrubgrass.

*Joseph M. Black*

In connection with other parties, Mr. Black discovered the Byron Centre pool and helped to develop that field. He has been identified with the development of most of the fields around Emlenton from their start. Mr. Black was a Republican Venango County Commissioner elected in 1896 and had the trust and satisfaction of all who knew him.

Dr. A. W. Crawford was appointed by President Abraham Lilncoln as United States Consul at Antwerp, Belgium. (Refer to page 28). Mr. Crawford and Joseph M. Black were wonderful friends and partners in the local oil fields. Joseph M. Black was born December 30, 1836, married Candace Kohlmeyer in 1862, and passed on March 4, 1923.

Venango County Commissioner Black was buried at the Scrubgrass Stone Church cemetery in Scrubgrass Township, Venango County. He was 86 years old and had hosts of friends who will miss his kind voice and pleasant smile. "Peace to his ashes." (J. J. McLaurin)

## Men Who Made Oil History
## JOHN J. McLAURIN
## Newspaperman, Historian, and Author

*John J. McLaurin. Photograph from "Sketches in Crude Oil," 1898*

No newspaperman was better known in the northwestern Pennsylvania oil fields than John J. McLaurin. For nearly fifty years, John was a writer and correspondent for the Oil City Derrick, a daily newspaper everyone in the oil patch read. The baker read it, the butcher and the candlestick maker read it, but the oil man studied the paper. Each day new wells, territories, accidents, and production figures were released.

After Col. Edwin Drake discovered petroleum at Titusville, Pennsylvania, John McLaurin followed the progression of new discoveries and few men were as close to the affairs of the oil country as John McLaurin.

In 1872, John was employed as a traveling correspondent with the entire petroleum region as his haunt. Adventure and tragedy, excitement and comedy were part of John's everyday workload with the words of his boss ringing in his head, "Get the news regard-

*John J. McLaurin Memorial, 1923*

*John J. McLaurin and his wife are laid to rest in the cemetery at Franklin, Pennsylvania, 1923*

less of man or devil." John traveled around the oil fields on foot, by horse and buggy, and by boat and train, and still holds the record of visiting eighteen wild and wooly oil boom towns in just one day. Not only did he visit these places and meet the people, his report was in print the next morning.

Our northwest Pennsylvania oil fields had a host of interesting and unique personalities, men and women who were drawn like a moth to a candle. For you and me, the students of early oil history, there is much to study, explore, and digest. John J. McLaurin was just one of those interesting and unique personalities. He was a man I have great respect for, even after 82 years since his passing. One of John's greatest and lasting gifts is his book, "Sketches in Crude Oil." This history of the early days of oil is a "must read" for all students of oil history.

Whenever an old oil man passed on and John J. McLaurin wrote the obituary, he always ended the tribute with "Peace to his ashes." Well, John, job well done and guess what I have to say, even though you have been gone all these years. "Peace to your ashes."

John J. McLaurin and his wife rest in the cemetery at Franklin, Pennsylvania. John was born November 29, 1841, and passed on August 28, 1923.

## Men Who Made Oil History
## John L. McKinney

Among the very first oil pioneers who began developing Oil Creek Valley of Pennsylvania, after Colonel Drake demonstrated that the substance lay concealed in the rock formations of earth, was John L. McKinney. Born at Pittsfield, Warren County, Pennsylvania, June 21, 1842, he began prospecting to discover petroleum in 1860. From that long-ago period to the present time he has steadily clung to the peculiarly fascinating business.

*John Luke McKinney*
*(Courtesy of Drake Well Museum, Titusville, Pennsylvania)*

In those early days he took hold of the drill and helped develop the oil fields of the Oil Creek Valley. In 1865 he was connected with the Bennehoff Run Petroleum Co., on the Warner farm above Petroleum Centre. In 1869-70, in partnership with his brother, J. C. McKinney, they drilled a number of wells in the Pleasantville and Franklin fields. From there they moved to the lower Allegheny River, Parker oil field, and assumed a leading part in developing the Butler oil region. While in that district the McKinney Brothers formed partnership with H. L. Taylor, John Satterfield, John Pitcairn, Jr., and F. S. McFarland, all engaged in the production of crude petroleum. In 1889 the firm sold its holdings to the Standard Oil Company and became a part of that organization.

John L. McKinney was next chosen as President of the Midland Division of the South Penn Oil Co. This company had 5,000 producing wells in various regions of the oil country, extending from Allegheny County, New York, to Allegheny County, Pennsylvania. Upon the organization of the Commercial Bank in Titusville in 1882, Mr. McKinney became President of the institution and served 32 years. He is a large stockholder in the Titusville Iron Co., and is also interested in the Second National Bank and the block it occupies. His home is located on North Perry Street in Titusville. In early life Mr. McKinney married Miss Ida D. Ford, and to the union was born two children, Glen Ford McKinney, now a lawyer in New York City, and Ida Ethelyn. The mother of these children died on May 11, 1894. For his second wife, Mr. McKinney married Miss Alliene Ford.

John Luke McKinney died on May 20, 1937, at his home on North Perry Street in Titusville at 94 years of age. He was laid to rest in the family plot near oilman Jonathon Watson in the Woodlawn Cemetery, Titusville. "Peace to his ashes." (The Titusville Morning Herald, Titusville, Pennsylvania; Genealogical and Personal History of Allegheny Valley, Pennsylvania, John W. Jordan, 1913)

***John Luke McKinney Memorial, Woodlawn Cemetery, Titusville, Pennsylvania. 1937***

## Obituary of S. D. Karns
## Recent Deaths, February 9, 1909

Samuel Duncan Karns, a widely known oil operator of the early days of the Butler-Clarion developments, died at his home in Nipton, California, Tuesday, February 9. Mr. Karns was born in 1842 near Natrona, Allegheny County, Pa., where his father was the owner of an extensive salt making plant and of many salt wells. When the Civil War broke out, the son went to the front as a lieutenant of a Pennsylvania regiment. In the latter part of 1866 or early in 1867, he made his appearance at Parker and engaged in the oil producing business, drilling several wells along the bank of the Allegheny River and, as the field enlarged, extended his operations to the Stone House section, where he opened up the McClymonds farm. His pioneer well was a short distance south of Petrolia, Butler County. A settlement grew up rapidly and soon assumed the proportions of a considerable village and was named Karns City.

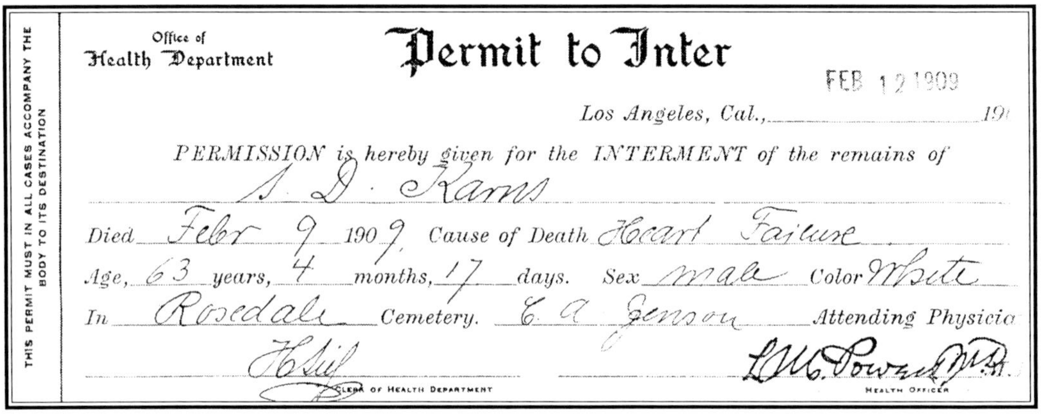

THIS PERMIT MUST IN ALL CASES ACCOMPANY THE BODY TO ITS DESTINATION

Office of Health Department

**Permit to Inter**

FEB 12 1909

Los Angeles, Cal., 19

PERMISSION is hereby given for the INTERMENT of the remains of S. D. Karns

Died Febr 9 1909 Cause of Death Heart Failure

Age, 63 years, 4 months, 17 days. Sex male Color White

In Rosedale Cemetery. E. A. Jenson Attending Physician

H Sief — Clerk of Health Department

L M Powers M.D. — Health Officer

***The Rosedale Cemetery in Los Angeles, California, is the resting place for "Dunc" Karns and his wife. (Courtesy Chris Fleming, Foxburg, Pennsylvania)***

He operated extensively throughout the area from its early activities up to the full period of the excitement. He organized the Karns Pipe Line Company and was the president of a bank, president of the Parkers Bridge Company, and organized the Parker & Karns City railroad. This railway, constructed in 1873, was the first narrow gauge road in the oil regions.

Mr. Karns became prominent in the various oil producers' movements and was especially active in the agitation which became general in October, 1872, of the 30-day shut down movement.

By 1876 he had amassed a great fortune and became the head of an organization formed to build a railroad from Harrisonburg, Va., to tidewater at Fredericksburg. This enterprise failing, the greater portion of his finances were absorbed and then the failure of the Butler County oil fields, in which his interests were centered, completed his financial ruin, which he was unable to retrieve. He continued oil operating until 1882 and then removed to the far western fields.

Although an aggressive and energetic man, Mr. Karns possessed many fine qualities that people admired.

(Courtesy of Chris Fleming, Foxburg, Pennsylvania. Obituary published in the Oil City Semi-Weekly, February 16, 1909)

## Men Who Made Oil History
## Cyrus Angell

*Cyrus D. Angell*

The oil town Angelica, named after C. D. Angell, "the lord of the soil," has an existence of only a few months, and yet it has attained considerable proportions. Active developments are being pushed on in the neighborhood, and no doubt this village will soon aspire to the title of city. Mr. C. D. Angell leased the J. Moore and J. Kepler farms on January 4, 1873. Then he sold at a profit five-eighths interest for the sum of $83,750. On the Kepler and Storey farms one-half mile west, a new town was built and named Angelica. Angell was founder of the "oil belt" theory.

Originally C. D. Angell was a native of Hanover, Chautauqua County, New York. He attended the Academy

***Edwin C. Bradley Memorial***

*Oilman E. C. Bradley buried in Woodlawn Cemetery at Wellsville, New York.* *Refer to page 285*

at Fredonia, New York, and the Genesee Wesleyan Seminary at Lima, New York. He was Chautauqua County school commissioner until 1856 and came into the oil territory empty handed in 1867. By 1873 his daily receipts of crude petroleum was upward of one thousand barrels a day. In 1871 Cyrus D. Angell ran an unsuccessful attempt to become State Senator of New York. An accident befell Cyrus's son, Elgin A. Angell, born 1849, and lost at sea July 4, 1898.

C. D. Angell was born April 24, 1826 and died near Chicago, Illinois, on June 20, 1905. He is buried at Forestville, Chautauqua County, New York, in the Forestville Pioneer Cemetery. "Peace to his ashes." (John J. McLaurin)

*Angell Memorial, Forestville Pioneer Cemetery, Forestville, New York, 1905. (Courtesy of Mildred L. Becker, Village Historian, Forestville, New York)*

*Joseph Seep*
*Joseph Seep was no stranger to fellow oilmen from the lower district. (Courtesy Drake Well Museum, Titusville, Pennsylvania)*

## Men Who Made Oil History
## Joseph Seep

Joseph Seep was born May 7, 1838, in Voerden, Hanover, Germany, and had a common school education. At the age of 11 he came with his parents to America and located in Richmond, Indiana. In 1869, Joseph Seep moved to Titusville, Pennsylvania and engaged in petroleum operations with his old friend, Mr. Bostwick, who had formed a partnership with J. B. Tilford under the name of Bostwick and Tilford. This firm joined the Standard Oil Company in 1871 and Mr. Seep became part of the concern which later was to become one of the greatest corporations in the world. Joseph Seep was to play one of the leading roles.

In 1886 he married Miss Kate Hillenmeyer and the couple had 11 children. Mrs. Seep passed away in Titusville on February 15, 1917.

As head of the Joseph Seep Purchasing Agency, he purchased crude oil used by Standard Oil Company. The amount of money which Mr. Seep has distributed in the Pennsylvania oil fields is tremendous in its proportions. Since the dissolution of Standard, the Seep Purchasing Agency has been an important part of the South Penn Oil Company and Mr. Seep was for many years president of that organization. In 1928, the Seep oil purchasing agency had 39 offices in the eastern and mid-western part of the United States. Seep purchased over one hundred million dollars of crude oil from the mid-1870s through 1918 for John D. Rockefeller.

Joseph Seep was one of the original officials of the Standard Oil Company and one of the best known residents of this section. Mr. Seep was the last of the group of original leaders in the Standard Oil Company to leave his working desk. All other outstanding oil men from the early and middle years of the Standard Oil Company had previously died or retired from active participation in the business of the great concern. Joseph Seep was chairman of the board of directors of the South Penn Oil Company and head of the Joseph Seep Purchasing Agency which purchased the largest amount of high grade crude oil from Pennsylvania, New York, West Virginia and Southeast Ohio. He was also chairman of the board of the Oil City Trust Company and a director of the Commercial Bank of Titusville. He was the last living man of the group of investors who founded the Seaboard Bank of New York City. The Seaboard Bank was designed at first simply as a help to the men interested in oil who lived or did business in New York City and has since become one of the greatest banking institutions in the world. On October 9, 1920, the Standard Oil Company of New Jersey christened the largest oil tanker in the world and named it "The Joseph Seep." The vessel was 440 feet long and held 80,250 barrels of oil.

*Joseph Seep Memorial, St. Catherine's Cemetery near Titusville, Pennsylvania*

Joseph Seep passed on at his home in Titusville, Pennsylvania, Sunday evening, April 1, 1928. The Titusville Herald dated Monday, April 2, reported that Titusville had lost its most highly esteemed citizen and burial will be in the family mausoleum at St. Catherine's Cemetery, near Titusville, Pennsylvania.

## Flying Machines

Sorrow, unfortunately, can be a part of any family and oil men were no exception. Their family members were business people who traveled for either business or vacation. In the summer of 1923, two of Joseph Seep's grandsons, Arthur and Joseph Seep, sons of Mr. and Mrs. Arthur Seep of Oil City, met their deaths in an airplane accident in France. The two young men had gone abroad and had enjoyed visits in England, Switzerland, and other places on the continent. On August 11, 1923, they engaged Gabriel Paulain, one of the foremost aviators of France, to pilot them on a trip. Over the village of Degmy the plane suddenly collapsed and crashed to the earth killing not only the two young men but the pilot as well, the three meeting death instantly.

The Titusville Herald dated July 29, 1932, carried the following story with the headline of "Plane Crash, Victim Former Oil City Girl." Mrs. Emily Bossom, daughter of the late Seaboard Bank of New York City president, Samuel G. Bayne, her son Bruce, and Prince Otto Erbach Furstenau of Germany were killed. The accident happened near Farnham, England. Witnesses said the wings seemed to crumple in air and explode as it crashed to the ground. Her son, Bruce Bossom, was an experienced pilot but it is thought that lightening hit the plane and caused it to fall. Samuel G. Bayne lived in Oil City and was associated with the Oil Well

Supply before moving his family to New York City. Mrs. Emily Bossom was married to Alfred C. Bossom, noted New York architect. They moved to England where he became a member of Parliament. In New York he designed skyscrapers for twenty-five years, designed the Titusville Trust Bank for J. C. McKinney, and in 1908 undertook the restoration of Fort Ticonderoga, New York. "Peace to his ashes." (John J. McLaurin) Refer to page 268.

## MEN WHO MADE OIL HISTORY
## Charles P. Hatch

*Charles P. Hatch*

Charles P. Hatch came to the Shaffer farm in 1865 as Buyer's Agent for the Empire Transportation Company. With the decline of the Shamburg and Pleasantville, Pennsylvania, oil fields, new territory had developed in Clarion, Armstrong and Butler Counties in Pennsylvania. The oil from these new fields found its way by pipeline to the Allegheny Valley Railroad at Foxburg and Parker by two independent pipelines. In 1872 Charles P. Hatch was General Manager of the Mutual Pipe Line which served the Clarion territory. The Empire Transportation Company purchased the line and took the business at hand.

This was soon thereafter followed by the purchase of the Union Pipe Line at Parker City, of which Charles Hatch also assumed the management. In order to meet the needs of the growing Butler field, Mr. Hatch constructed the Butler Pipe Line in 1876. Under the management of Charles P. Hatch, the Foxburg line, the Parker Line and the Butler Line were combined and reorganized under the name of the Union Pipe Company. Mr. Hatch later lived in the Philadelphia Bourse. Refer to page 418.

## Men Who Made Oil History
## T. W. Phillips

*T. W. Phillips*

Thomas Warton Phillips was a big operator in the firm Phillips Bros. T. W. Phillips was born February 23, 1835 and passed on July 21, 1912. He is buried in Oak Park Cemetery, New Castle, Lawrence County, Pennsylvania. (Courtesy of New Castle Public Library)

# Men Who Made Oil History
## George W. Brown

*George W. Brown*
*George W. Brown was the author of "Old Times In Oildom." George and his wife, Sarah, are buried in the cemetery at Youngsville, Warren County, Pennsylvania. George was born in 1828 and passed on in 1916. Thanks, George. "Peace to his ashes." (John J. McLaurin)*

*The Brown Memorial in the cemetery with a copy of his book resting against it. 2006*

# The Bullion Field, 1876
## The Phillips Oil and Gas Company

The Phillips family were among the earliest pioneers in the nation to enter the oil and gas industry. Isaac N. Phillips organized the New Castle Gas Company in 1856, three years before the Drake Well was drilled.

The four Phillips brothers purchased a large farm a half mile northeast of Petroleum Center along Oil Creek and half way between Oil City and Titusville, shortly after the Drake Well was drilled. It was bordered by the most prolific 1861-62 oil farms along the creek. The Espy, Caldwell, Boyd, and one known as the lower McAlheny were on this farm in May of 1861.

The Phillips brothers followed the oil industry south into Butler County with interests in Greece City in 1873, Modoc City in 1873, and then opened the Bullion Field on August 4, 1876, (this is north of Clintonville in Venango County). The Phillips brothers and Duncan Karns were instrumental in laying the Cleveland Pipe Line in mid-summer of 1873. This went from Greece City to the rail head at Forestville, which was a short cut to get oil to the Cleveland refineries.

In the spring of 1876 a company, of which Thomas W. Phillips was a part, leased several hundred acres in Clinton Township, Venango County. Prior to 1876 Jonathan Watson, the Crawford County oil king, had part of this property under lease but failed to find petroleum in paying quantities. At this time Mr. Watson had a lady spiritualist try to find petroleum but she failed to receive the much needed help from the unseen world and Jonathan Watson, Titusville's beloved oil man, declared the territory dry.

Little did Watson know but T. W. Phillips and Company had faith that could not be shaken and even after hole after hole was a dry, dry duster, Phillips Brothers continued to drill.

When the clouds of adversity seemed about to encompass them, John Taylor and R. C. Cundle sub-leased from the company a part of the William Gailey farm on Bullion Run and proceeded to test their holdings. Slowly down into Mother Earth sank the never-tiring drill, eating its way through rock and sand, and on August 9, 1876, the oil-bearing strata was reached. Among those who crowded the derrick floor that autumn morning was T. W. Phillips, waiting for the end of a chapter that meant much to him. When the drill had penetrated the pebbly rock several feet with but a faint showing of oil, Mr. Phillips turned to R. C. Cundle, saying, "I guess it's a forlorn hope, Bob. You might as well take down the tools." However, Cundle, loathe to admit defeat, said, "It may be, but I'll hit it again for luck." As the unexpected oft-times happens, it was so in this case. A few more strokes of the drill, a low, rumbling noise, and up through the casing, impelled by volumes of gas, rushed a stream of oil, and the long-kept secret was out. Thomas W. Phillips had proved his belief to be well founded. The news of the strike spread like wildfire and from city, town and hamlet there came hundreds of men eager for a chance to whirl the wheel of fortune. Well after well was struck, ranging in calibre from 50 to 300 barrels until Gailey No. 10 cut loose at a 500 barrel gait, which greatly augmented the interest in the field, and the advance guard of the oil brigade was reinforced by thousands.

W. H. H. Fithian, a United Oil Pipe Line man from the Foxburg and Parker City fields, had this to say about the new Bullion territory. "I accepted a position in the winter of '74 with the Grant Pipe Line Co., T. McConnell, Jr., principal owner and superintendent, and remained with that company until the winter of '77, when the Grant Pipe Line with a number of other small concerns were combined with the United Pipe Lines. I was transferred from Parker to Bullion, where quite a field was opened up, and assumed the agency for the company, succeeding Seymour Peck. Bullion did not last long, but it was a bolsterous place while it did last. Zack Bruner was the superintendent, and the principal operators were Phillips Bros., McCalmont Oil Co., Geo. H. Nesbitt, McKinney Bros., Geo. Conwall and W. T. Baum, of Franklin, generally called "Billy" from the fact of his being a big-hearted, kind, congenial soul, whom everybody liked. Mr. Baum was always a firm believer in the extension of the Bullion field to the west, and from what I can learn, his faith has been recently confirmed. Some of the men who used to come to Bullion, as I remember, were David Kirk, Gus Tack, Al Dilworth, Isaac "Farmer" Dean, all members of the McCalmont Oil Co. Although, as remarked before, the field soon played out, business was on the rush and the firm of Kerr, Gelder & Co. did its share. Some of the wells there, I understand, are still producing. After the collapse of Bullion I returned to Parker and shortly afterward was transferred to the Bradford office of the National Transit Co." (Oil Region Reminiscences, The Oil Men's Association, Butler, Pennsylvania, 1907)

# Summit City, Clinton Township
# 1877

On a commanding eminence one mile south of Gailey No. 1, the late Shadrick Simcox, a Clinton township farmer and operator, laid out a town. Although the history of Shamburg, Petroleum Center and Pithole was fresh in the minds of the people, there was a scramble for lots, and buildings costing thousands of dollars were erected by business men who dreamed not that history would repeat itself, that the subterranean wealth would ever cease to flow. A bank, hotels, stores, drinking dens and gambling halls sprung up in a day. Ben Hogan was there also, with his gilded palace of sin, luring men and women on to eternal ruin, and Summit City, a second to Pithole, the wickedest city of modern times, swung its banner to the breeze.

Past Summit City, on and off the scent like hounds upon the trail, operators fought their way, now deluged by success, again covered by defeat. The G. W. Berringer farm was the next to fall into line and proved to be a rich find. The town of Berringer now opened its eyes upon the oil-greased hills, followed by Dean City. From the latter place to Summit City, one mile distant, a gravelled walk was built six feet wide, called the "Devil's Lane," or "The Road to Ruin," a man would need a reserve force of self-will to run the gauntlet of a hundred and one gambling dens, speakeasies, etc., that lined the path. (Oil Region Reminiscences, 1907 Souvenir, The Oil Mans Association of Butler County)

*Main Street, Summit City*
*Sketches in Crude Oil*
*Oil Exchange Hotel 1877*
*(Drake Well Museum, Titusville, Pennsylvania; Personal notebook, Venango County Historical Society, John J. McLaurin)*

# June 18, 1877
# The "Big Injun," 3,000 Barrels

The high-water mark was reached when Crawford, Mitchell, Lee & Co. brought in the "Big Injun," on the Charles Henderson farm, which spouted 3,000 barrels the first twenty-four hours. To use the words of an old-timer who was on the ground at the time, "It was a holy terror," and the night following its initiation was the wildest in the history of the Bullion field. The people were frenzied by excitement; whiskey was carried around in buckets, all you could drink for ten cents, if you had not the "wherewith," drink anyway. The lockup, 12 by 22 feet, built of 2 by 4 scantling, was soon filled with drunken, cursing men and the morning sun revealed numbers handcuffed around trees and telegraph poles.

*The famous oil boom town of Summit City.*
*Summit House, Oil man's Hotel, 1877*

*This little ditty was written in 1907:*
*A wondrous change has taken place,*
*The grass and daisies grow*
*Where once the town of Summit stood*
*Just thirty years ago.*

As a "spouter" territory the "Big Injun" was the beginning of the end. The gushed ceased to gush. New fields were being opened up and all that now marks the spot where once stood Summit City, the Gotham of Venango County, is the old lockup, with its iron-barred windows. (Oil Region Reminiscences, 1907 souvenir, Butler County Historical Society.)

## Foster Farm, 1872

Cyrus D. Angell, whose research came up with the "Belt Theory," leased property from Foster Station to Scrub Grass, a distance of five miles. C. D. Angell also owned wells on Belle Island, a small strip of land in the Allegheny River just below Scrub Grass Station on the Allegheny Valley Railway about twenty-five miles by river below Oil City.

The famous "Burning Well," drilled in 1864, between Scrub Grass and Foster on a bend of the river, struck an immense gas vein. The tools got stuck in the well, the gas caught fire, and for seven years was not extinguished.

Mr. Frederick Prentice, a veteran "greaser" living in Franklin, Pennsylvania, was Angell's partner. A new station was built five miles below Oil City on the Allegheny Valley Railway and was named Prentice. A new "belt" was discovered and the rush was on. On June 18, 1877, the "Big Injun" well, drilled by Mitchell and Lee, flowed three thousand barrels the first day and a second well was drilled eight rods away and proved dry.

*Angell and Prentice Oil Co.*
*Foster farm, Allegheny River, 1872*

Many short-lived oil towns sprung up in this neck of the woods including Gas City, Freedom, Pittsville, Gregory, Kennerdell Station, Berringer City, Dean City and on the Simcox farm Summit City. Summit City's first house was built on December 8, 1876, and by June 1877, the oil town had over 200 buildings and one thousand five hundred people. Sadly, Abram Myers, its last resident, departed the ghost town in April 1889. The Bullion field extended seven miles in length and a narrow mile in width, swerving the tide from Bradford and ruling the petroleum roost for over a year. (John J. McLaurin, 1919)

## Butler County's Last Big Strikes
## Late 1870s through the early 1900s

"The fabulous Cross Belt, or Greece City line, did not follow a straight line from its western beginning at Greece City to its eastern terminus at Criswell. By projecting the Greece City line and extending the 22 degrees Angell Line southward, it was found that the two lines intersected south of Butler Borough near the village of Renfrew.

"Reiber and Huselton leased 720 acres in this area, known as Bald Ridge in 1880. An increase in activity in and around Bald Ridge became immediately apparent. In April of 1881 the Bald Ridge Oil Company was founded, followed in August of the same year by the organization of the Bald Ridge Oil and Transportation Company. The latter company numbered among its shareholders many prominent men of the community, among them J. D. McJunkin, John S. Campbell, Fred Reiber, W. D. Brandon, R. P. Scott, George Krug, B. C. Huselton, Henry Eitermiller and Jacob Reiber. (Bill C. McElvain)

"Operators were soon drifting into the new field. Wells of considerable size were brought in by Simcox and Meyers. A. Scheidemantle, drilling on the Weber farm in Penn Township, struck several fine wells.

## Phillips Brothers

"In August of 1882 the Bald Ridge Oil and Transportation Company sold its holdings to Phillips Brothers for the reputed sum of $160,000. This sale was followed by an extension of the pipe line from Petrolia to the new field and a definite increase in leasing over the Bald Ridge district. T. W. Phillips continued leasing in the district concentrating his activities largely on the land immediately bordering on Thorn and Connoquenessing creeks.

"Although his first wells were small, definite signs of large oil deposits in the vicinity were present. On August 16, 1884, the persistent exploration of the field by Mr. Phillips finally resulted in a 4,000 barrel per day strike on the Williamson Bartley farm in Penn Township. The Phillips well, at the height of its production, one of the largest if not the largest producer in Butler County down to that time, inevitably produced a stampede to the area comparable to the Millerstown boom ten years earlier. (Bill C. McElvain)

## Christie Brothers

"On October 11, 1884, Christie Brothers brought in the second large well on the Bartley farm. This well, on a lease immediately adjoining that of Phillips Brothers, came in at 150 barrels. By the 14th of the month, it was producing 250 barrels per hour or 6,000 barrels per day.

"At the time Christie Brothers completed their gigantic well on the Bartley farm, Boyd, Semple and Armstrong were drilling Armstrong Number 2 on the Marshall farm. Drilling in a part of the field not expected to produce heavily, tankage was not provided on a large scale.

When drilled in to the sand on October 25th, a flow of salt water resulted with no sign of oil. As the final act of an apparently ill-fated operation the well was shot on orders from Mr. Armstrong. The shooting, which took place on October 27 resulted in a flow of oil unparalleled in Butler County history. In one of the then-current oil works, the rush of oil from the well is described as follows:

*'In a few minutes, the ground was covered several inches deep with petroleum; the branches of the oak trees were like huge yellow plumes, and a stream as large as a man's body rushed down the hill to the road, where it filled the space beneath the small bridge at that place and, continuing down the hill, spread out upon the flats where the Johnson well is. In two hours these flats were covered with a flood of oil. The hillside was as if a yellow freshet had passed over it; heavy clouds of gas almost obscuring the derrick hung low in the woods, and still that mighty rush of oil continued. People living along Thorn Creek packed their household goods and fled to the hillsides; the pump station one and one-half miles down the creek had to extinguish its fires that night on account of the gas and all fires around the district were put out.'*

"No accurate gauge was ever taken of Armstrong Number Two. At the highest production point it was estimated that the well flowed 500 barrels an hour or at the rate of 12,000 barrels a day. It is reported that the owners received credit for 10,000 barrels a day from the pipe line company without any gauge being taken.

***Christie Well, Thorn Creek, October 1884***

## Thorn Creek

"McBride City and Phillips City sprang up on the heels of the Thorn Creek boom to flourish and fade. The fabulous productivity of the district was relatively short lived and the gushers and flowing wells soon deteriorated to pumping wells. The two boom towns, Phillips City and McBride City, did not long survive the production boom. McBride City still exists but Phillips City disappeared completely.

"The deterioration of the Thorn Creek field from a yield of 16,000 barrels a day in 1884 to 2,800 barrels at the end of July 1885, did not mark the end of Butler County's oil development. At the height of the Thorn Creek excitement in 1884, the 'Grandmother' well was drilled one mile west of Saxonburg. The well, a large producer, marked the founding of Golden City, and Hooks City in Parker township came into being in 1885 after successful drilling by Hooks Brothers on the Kelly farm.

"In 1886 the Extension Oil company drilled on the Welsh farm Jefferson township. This company was made up of R. B. Taylor, O. K. Waldron, Loyal S. McJunkin, W. P. Roessing, J. A. McMarlin and others. Their first well produced 100 barrels a day and was the beginning of extensive development in Jefferson township. (Bill C. McElvain)

## The Glade Run Field

"By the close of 1886 the Glade Run field had become the most interesting of the newer oil developments in the county. Looking for a southwest extension of the Thorn Creek field, T. W. Phillips in August of 1886, struck a 125 barrel per day well on the Critchlow farm in the Glade Run district. As the field was developed to the southwest, several wells yielding as high as 125 barrels an hour were drilled. Mr. Phillips production at Glade Mills eventually reached 6,000 barrels per day. During the year 1887, his oil production from this and other fields attained 1,100,000 barrels despite the fact that his wells were shut in for the last two months of the year.

"The Reibold field came into prominence in September of 1887 with the drilling of Beam Well No. 6. The well, when completed, produced 140 barrels an hour. At the height of its yield, the Reibold area yielded 9,000 barrels a day.

***Standard rig. Note the ladder from the ground to the top and the brave, or maybe crazy, driller hanging near the top. (Butler County Historical Society, Butler, Pennsylvania)***

## Connoquenessing Field Opened 1889

"Connoquenessing Township became the operational front in the spring of 1889 when John A. Steele drilled a well on the Anderson farm on Connoquenessing creek. The Connoquenessing Township field, known as the 'Hundred-Foot District,' reached the top of its production in 1890. The Humphrey, Brandon, Hazlett, Amberson, Rader and other nearby farms produced fabulously from the Hundred-Foot Sand. The Peter Rader farm alone produced 1,000,000 barrels of oil. After reaching the peak of production, the Hundred-Foot District declined rapidly.

"The Brownsdale field and the Harmony and Zelienople territories were developed during 1892. In 1893, a pool was opened south of Evans City on the Connelly farm. The extension of the Brownsdale field was the eventual means of opening the Cooperstown district. Reiber and Bradner opened this latter field in 1894 and its development continued through 1898. Producers flocked to this district. The Cooperstown development was the last major excitement of the county. (Bill C. McElvain)

## Washington and Concord Townships

"In the year 1900 Washington and Concord townships became the center of new developments when the Speechley sand was tapped. In the course of the next four or five years, over 3,000,000 barrels of petroleum were extracted from the Speechley field. While other production areas of a relatively minor nature have been found since, probably the last of Butler County's legendary wells came in 1905 when the 'Spotty McBride' well was drilled on the A. K. Waldron farm. The 'Spotty McBride' well was a 2,000 barrel gusher. The field, however, was limited in scope and the wells of short duration."

This article was published in the Butler Eagle's sesquicentennial edition in July 1950, which marked the 150th anniversary of the founding of Butler County. The article was researched and written by William C. McElvain. Butler Eagle, Butler, Pennsylvania, July 1950. Thanks to you, too, Bill C. McElvain. "Peace to his ashes." (John J. McLaurin)

*Spotty McBride's oil well, Butler County*
*The title reads, "Largest oil well in the U. S., flows over 2000 barrels per day." 1905*
*(Butler County Historical Society, Butler, Pennsylvania)*

## A Slip of the Tongue

After being converted by a higher power, Ben Hogan came back to the oil country as an evangelist. In an address to the good people at the Karns City Methodist Church, Ben remarked, "Where I used to raise Hell up there in Clarion County, they now raise corn." That got their attention. (Oil and Gas Man's Magazine, Butler, Pennsylvania, 1907-1918, Drake Well Museum)

## "I'm awfully sorry too, Dad."

In the gay days of Petrolia City, when everybody had money, there was among its citizens a leading oil producer and member of the Petrolia Oil Exchange. This gentleman was reported to be worth three hundred thousand dollars. He was a man of faultless dress, free with his money, gallant in bearing and withal a fit companion for the bravest.

The old proverb, "A fool and his money soon part," held good in this particular case, linked in arms with "Old John Barleycorn." He passed down life's highway at a merry clip.'

This writer saw him shaking dice with an equally rush spirit at fifty dollars a throw and a few years later saw him in the last stages of alcoholism, asking for a dime to get a drink. Speculation, wine, and women have wrecked many oil country lives and crumpled many fortunes. "I just wonder who this writer was talking about?" (The author)
("Flowing Gold" by Rex Beach, Harper and Bros., Publishers, New York and London, England. The Oil and Gas Man's Magazine, Published in Butler, Pennsylvania, 1913)

*"I'm awfully sorry too, Dad."*

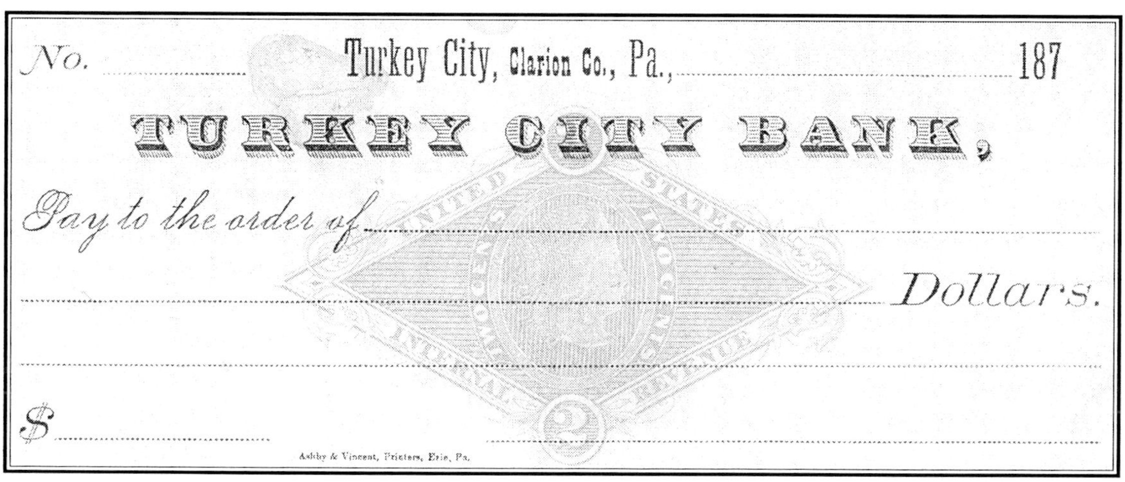

No. .................. Turkey City, Clarion Co., Pa., .................. 187

TURKEY CITY BANK,

Pay to the order of ..................

.................. Dollars.

..................

$ ..................

Ashby & Vincent, Printers, Erie, Pa.

*Karns City, 1873*
*A warm summer's day in oil country frozen forever in time. The derrick in the center is the "Shasta" well drilled by the Cooper Brothers in June 1872. It came in at 120 barrels and the Karns City "Boom" was on.*

*These old photographs are so important to the story and can actually take the reader back into time. Imagine, you working on the "Shasta", imagine, you driving your horse and buggy down a tree lined country road, imagine...*

# Final Words

John Mather was a photographer from Titusville who through his efforts captured the early days of Oildom. Mather's photographs span from Colonel Edwin Drake to Bradford and beyond.

Throughout my research I've asked myself, "If Mather photographs exist of Venango's Pithole and Red Hot, where is Clarion's Antwerp and Keating City or what became of many Butler County oil field photographs?" Altogether there are 3,274 Mather glass negatives in the Drake well museum. To my dismay, I may have found the answer.

In 1892 a terrible flood devastated Titusville and put over five feet of water in Mather's Studio. Destroyed were over 16,000 irreplaceable, historical glass negatives of people, places and things throughout the early oil fields.

Maybe, just maybe, the photographs I, as a oil history student dream of, were the victims of the cruel oil covered muddy waters of oil creek.

Much more can and should be researched and recorded about the early days of oil in North Western Pennsylvania. Many lost photographs and written histories still rest in forgotten places. This text is a small insight into the beginning of oil discovery in our "Lower Fields." It is my hope that this information will help you, the student of oil history, better understand those wonderful exciting days of both "Boom" and "Bust!" – Gary McKinney

*Gary and Helene McKinney*
*St. Petersburg, 1975*

# SOURCES


Allegheny Cemetery, Butler St., Pittsburgh, Penna.

Arter, John M., *Arter's Oil Field Directory for 1872,* 1872, Ashby & Vincent Publishing Co., Erie, Penna.

Asbury, Herbert, *The Golden Flood, An Informal History of America's First Oil Field,* 1942

Ashbaugh, John J., *Civil War Diary*, 1906, St. Petersburg, Penna.

"Atlas of Armstrong County, Pennsylvania," Illustrated, published by Pomeroy, Whitman and Co., Philadelphia, 1876

"Atlas of the County of Butler," G. M. Hopkins and Co., Philadelphia, 1874

"Atlas of the Oil Region," Beers and Co., 1864

Barcly, J. H., and Co., *The Progress,* St. Petersburg, Penna.

Bayne, Samuel G., *Derricks of Destiny,* 1924, Brentano's Publishers, New York, NY

"Beautiful, Bountiful Butler County," Butler County Historical Society, Lenn "Pithole" McCorry

Beers, J.H. and Co., *Commemorative Biographical Record of Central Pennsylvania,* 1898

Black, Brian, *Petrolia, The Landscape of America's First Oil Boom,* John Hopkins University Press

Botsford, Harry, *The Valley of Oil,* 1946, Hastings House, New York, NY

Bradford Landmark Society, Bradford, Penna.

Bradford Public Library, Bradford, Penna.

Brown, George W., *Old Times in Oildom,* 1909, Derrick Printing Co., Oil City, Penna.

Buffalo Bill Historical Center, 720 Sheridan Avenue, Cody, Wyoming

Buffalo Historical Society, Buffalo, New York

Buffalo Public Library, Buffalo, New York

*Butler County Citizen,* June, 1872, Butler, Penna. (found in Butler Library, Butler, Penna., microfilm)

Butler County Historical Society, National City Bank Building, Main Street, Butler, Penna.

"Butler County, the Second Hundred Years," Butler County Historical Society, Ralph Goldinger and Audrey Fetters

*Butler Eagle,* "J.V. Ritts Death Notice", June 24, 1937, Butler, Penna.

Caldwell, J.A., *Illustrated Historical Atlas of Clarion County, Pennsylvania,* 1877 (Reprint, Record Press, Rimersburg, Penna.)

Centennial History Committee, *St. Petersburg Centennial Book, 1872-1972,* St. Petersburg, Penna.

Centennial Souvenir of Butler and Butler County, Pennsylvania," Goodwin and Curry

"Centennial Souvenir of Butler County," Godwin and Curry

Clarion County Historical Committee, *Clarion County Centennial Book, 1840-1940,* Clarion, Penna.

*The Clarion Democrat,* Clarion, Penna. Dec. 1875-Jan. 1880 (Clarion Public Library microfilm)

Clark, James, *Spindletop,* 1952

"Cram's Descriptive Review Showing Development of the State of Pennsylvania," 1917

Cribbs, John W., *The Adventures of Jack Cribbs,* First Volume, 1909

Daniel O'Day Obituary, Buffalo News, Buffalo, New York, October 16, 1906

Davis, A. J., *History of Clarion County, Pennsylvania,* 1887

*Democratic Herald,* Butler, Pennsylvania, Jacob Ziegler, Editor, June 1870 (Butler Library on microfilm)

Derrick Publishing Co., *The Derricks, Handbook of Petroleum,* 1898, Oil City, Penna.

Derrick Publishing Co., *Romance of American Petroleum and Gas,* Vol. 1, 1911, Oil City, Penna.

Derrick Publishing Co., *Souvenir, Illustrating the City of Oil City,* 1896

"Destructive Floods in Pennsylvania," New York Times, February 5, 1883

"Directory of the Butler Oil Region," J. Wiggins and Co., 1873

Dolson, Hildegarde, *The Great Oilorado,* 1959

Donehoo, George P., *Pennsylvania, A History,* Lewis Historical Publishing Co., Inc., 1926

*East Brady Independent,* East Brady, Penna. Dec. 1869 - Mar. 1871, Editor Samuel Young (found on microfilm at Butler Library, Butler, Penna.)

Eaton, Rev. S.J.M., "Nature's Wonder Gift to Man", *Petroleum: A History of the Oil Region of Venango County, Pennsylvania,* 1866

*Emlenton Sun,* Emlenton, Penna.

Empire Gas & Fuel Ltd., *Empire Bradley, 75th Anniversary Booklet,* 1956, Wellsville, NY

Flaherty, Kathy J., *Hills, Dales and Oil Trails: Pittsburgh to Titusville Penna. 2003 Field Trip,* 2003

Forestville Cemetery, Forestville, New York

Franklin Cemetery, Franklin, Penna.

Freeport Public Library, Freeport, Penna.

"Genealogical and Personal History of the Allegheny Valley, Pennsylvania," John W. Jordan, 1913

Giddens, Paul H., "Contemporary Accounts 1859-1872, The Use of the Divining Rod in Locating Oil Wells", *Early Days of Oil,* 1947, Princeton University Press, Princeton, NJ

Henry H. Cumings Obituary, Warren County Mirror, May 15, 1913

Henry, James T., "The Robert's Torpedo", *The Early and Later History of Petroleum with Authentic Facts in Regard to its Development in Western Pennsylvania,* 1873, J. V. Roger Co., Philadelphia, Penna.

"Historic Landmarks and Places, Butler County, Pennsylvania," Lenn McCorry, Butler County Historical Society

History Committee, The, *Knox Centennial Book 1876-1976,* Knox, Penna.

"History of Armstrong County, Pennsylvania," Robert Walter Smith, 1883

"History of Butler County, Pennsylvania," Waterman and Watkins and Co., Chicago, Illinois, 1883

Holy Cross Cemetery, Lackawanna, New York

Hulings, L. W., Publisher, *March of Progress 1775-1933,* 1933, Corsica, Penna.

Hunter and Cumings Obituaries, The News, Tidioute, Penna.

"If These Hills Could Talk," Brady's Bend Historical Society, Brady's Bend, Penna.

"In Memoriam, Henry H. and Charlotte Cumings," by Rev. Fradenburgh, Derrick Publishing Co., 1913, Oil City, Penna.

J. Wiggins and Co., Butler County Oil Field Directory, 1873

Jordan, John W., *The Genealogical and Personal History of the Allegheny Valley,* 1913

*The Journal,* Parker City, Penna., March 1876, Clark Wilson, Proprietor (found on microfilm at Butler Library)

Keatley, Miss Hettie A., *Compendium of Edenburg, Clarion Jacksonian,* Print 1877, Clarion, Penna.

Knowles, Ruth Sheldon, *The Greatest Gamblers, The Epic of American Oil Explorations,* 1959, McGraw Hill Publishers, New York City, NY

McLaurin, John J., *Sketches in Crude Oil,* 1896, Horace McFarland Co.

McLaurin, John J., original "Sketches in Crude Oil" notebook (found at Venango County Historical Society, Franklin, Penna.)

Miller, Ernest C., *Early Daze in Oil,* 1974

Miller, Ernest C., *John Wilkes Booth in the Pennsylvania Oil Region,* 1987

Neely, R. H., *Oil Field Local,* St. Petersburg, Penna.

"The New York Industrial Recorder," Union Razor Company, Tidioute, Penna., 1904

North Cemetery, Butler, Penna.

O'Day, John Christopher M.D., *Oil Wells in The Woods,* 1905, Deposit, NY, *The Oquaga Press*

*Oil and Gas Man's Magazine,* 1907-1918, published in Butler, Penna.

Oilmen's Association of Butler County, Pennsylvania, *Oil Well Reminiscences,* 1907

"The Opening of the Great Wolf Run Pool, McKean County," Oil and Gas Man's Magazine, Vol. 7, No. 8, 1912

"Our First Thirty Years, Titusville Trust Company 1918-1948," Titusville, Penna.

*Parker City Daily,* Parker City, Penna., Aug. - Sep. 1875, G. A. Needle (found at Butler Public Library on microfilm)

*Parker City Oilmen's Journal,* Parker City, Penna.

Penn-Brad Oil Museum, Bradford, Penna.

"Pennsylvania, A Guide to the Keystone State," Oxford University Press, New York, 1940

"Pennsylvania Political Governmental, Military and Civil," by Frederic A. Godcharles

*Petroleum Age,* 1881-1887, Bradford, Penna.

*Petroleum Centre Daily Record,* Petroleum Centre, Penna.

*Petroleum Gazette,* 1897-1917, published Titusville, Penna. & Pittsburgh, Penna.

*The Petroleum Monthly,* Vol. 1, 1870-72, Oil City, Penna.

"The Petroleum Register," Directory of the Petroleum Industry, 1917

Reed, Sally Jordan, *Perry Township, Clarion County, Pennsylvania,* 1998, Pub. by Mechling Associates, Inc.

Riesenman, Joseph, Jr., *The History of Northwestern Pennsylvania Comprising the Counties of Erie, Crawford, Mercer, Venango, Warren, Forest, Clarion, McKean, Elk, Jefferson, Cameron and Clearfield; Vol. 1,* 1943, Lewis Historical Publishing Company, Inc., New York, NY

St. Catharine Catholic Cemetery, Titusville, Penna.

Smiley, A.W., Hon., *Interesting Chronological Events of Foxburg, Pennsylvania,* 1914, Clarion, Penna.

Tait, Samuel W., Jr., *The Wildcatters, Oil-Hunting in America,* 1946

"They Struck Oil," Hildegarde Dolson, 1959

'Thomas W. Phillips," Oil and Gas Man's Magazine, Vol. 7, No. 6, 1912, Drake Well Museum, Titusville, Penna.

Thompson, Eric. V., *A Brief History of Major Oil Companies in The Gulf Region,* University of Virginia

"Three Score Years, and More," W. Floyd Clinger, 1973

Titusville Benson Library, Titusville, Penna.
*Titusville Morning Herald,* Titusville, Penna.
Tozer, J. A., *Crude Local,* 1872, St. Petersburg, Penna.
Tozer, J. A., *The Daily Press,* St. Petersburg, Penna.
"20th Century History of Butler and Butler County, Pa.," James A. McKee, 1909
Vaux, Sarah Anson, and Shoup, Kenneth Lester, *Yearning to Breathe Free,* 1977.
"Venango County, Her Pioneers and People," John J. McLaurin, 1919
Venango County Historical Society, *History of Venango County, Pennsylvania, 1879-1976,* Franklin, Penna.
"Venango County, Pennsylvania," Vol. I and II, Charles A. Babcock, J. H. Beers & Co., 1919
*The Venango Spectator,* Published in Franklin, Venango County, Penna.
Village of Forestville, New York, town historian Mildred Becker, C. D. Angell
Warren County Historical Society, Warren, Penna.
Warren Public Library, Warren, Penna.
Wellsville Library, Wellsville, New York
Western Pennsylvania Historical Society, Pittsburgh, Penna.
"The Wickedest Man," Joseph Millard, Gold Metal Books, Fawcett Publications, NY, 1954
Woodlawn Cemetery, Titusville, Penna.
Woodlawn Cemetery, Wellsville, New York
Woodlawn Cemetery, Wilkensburg, Penna.
Wrigley, H.E., Lucas, D.J. and Lesley, J.P., *Second Geological Survey of Pennsylvania 1874.*

# CREDITS


A special thanks goes to the staff at the following locations for their assistance:

Butler County Library, Butler, Penna.
Clarion County Historical Society, Clarion, Penna.
Clarion Printing, Clarion, Penna.
Drake Well Museum Library at the Pennsylvania Historical and Museum Commission, Titusville, Penna., with special recognition to Susan Beates
Edenburg Public Library, Knox, Penna.
Foxburg Free Library, Foxburg, Penna.
Indiana County Historical Society, Indiana, Penna.
Oil City Library, History Room, Oil City, Penna.
Record Press, Rimersburg, Penna.
Venango County Genealogical Club, Oil City Library, Oil City, Penna.

And also:
Paul Ashbaugh of St. Petersburg, Penna.
Mildred L. Becker, Village Historian, Forestville, New York
Becky Crum Reinsel, Executive Director, Butler County Historical Society, Butler, Penna.
Clifford Dittman of St. Petersburg, Penna.
Roger Eddinger of Foxburg, Penna.
Barbara L. Hajel of Knox, Penna., for the Edenburg photographs and invoice headings
Sarah Lavadsky, Researcher, Receptionist, Butler County Historical Society, Butler, Penna.
Billy Master of St. Petersburg, Penna.
Homer Miller, Turkey City Historian, Bradford, Penna.
Lenn G. "Pithole" McCorry, Butler County Oil Historian
Karle Delbert McKinney, my dad, who shared his appreciation of history with me
Zelma and Nellie, my grandmother and great aunt — thank you for your love of history
Carol West Shaffer, "Rich Uncle Marcus," Hulings family research
Lawrence Shoup of St. Petersburg, Penna.
Cal Stevens, Karns City, Penna.
Harold "Red" Stoughton, Chicora, Penna. Thanks for the ride through the Butler County oil fields.
David L. Weber, Pleasantville, Penna., Oil Researcher
Ralph Vasey of St. Petersburg, Penna.

A special thanks goes to my two best animal friends. Fawn and Jeanna, my two little min-pins kept me company and warm. Their thoughts and opinions were always well taken.

Arf-arf
Gary S. McKinney

# INDEX

## A

Abrams, E. E., 401
Abrams, James H., 400
Accidents, 24-27, 147, 163,168, 188, 195, 248, 252, 255, 275, 301, 305, 373, 384, 387, 427, 440, 462, 465, 477
Adventures of Jack Cribbs, 71-72
Agnew, Wm., 98, 101
Aiken, Jerome, 15, 174, 178, 287
Aikens, J., 265
Akin, J., 223
Akin, L. D., 400
Aldinger, C. D., 417
Aldinger, C. F., 417
Allebach, Frank, 311
Allen, C. F., 304
Allen, R. B., 361, 441
Altoona, 471
Amm, James, 447
Anderson, Samuel, 310
Andersonville, 54, 56-58
Andersonville Prison Camp, 54, 58
Andrews, F., 107
Andrews, Frank W., 268
Andrews, J. C., 165
Angell, C. D., 156, 268, 320, 366, 370, 441, 475-476, 482
Angell, Cyrus, 475
Angell, Cyrus D., 156, 475-476, 482
Angell, Elgin A., 476
Antwerp City Hotel, 24, 39-40, 43, 50, 149, 192, 202, 204, 208, 213
Antwerp No. 1, 9
Antwerp Pipe, 49, 59-62, 64-66, 107, 116, 118, 167, 171, 174, 176, 181, 194, 196, 202, 208, 212-213, 361, 459
Anvil, Drake Well, 444-445
Archbold, John D., 291, 419
Argyle Bank, 366
Argyle City, 366-368, 382
Argyle Savings Bank, 366-367, 468, 471
Armstrong, S. P., 94
Armstrong County Map, 328
Asbury, Herbert, 383
Ashbaugh, Abraham, 75
Ashbaugh, Adam, 14-15, 75, 81, 99-100, 115-116
Ashbaugh, Daniel, 14-15, 51-52, 75, 81
Ashbaugh, Daniel Farm, 14, 51
Ashbaugh, Florence, 48, 58
Ashbaugh, H., 102
Ashbaugh, Henry, 38, 75
Ashbaugh, Howard, 15
Ashbaugh, J. J., 11, 14-15, 19, 21, 24, 39-40, 47-48, 54-58, 61, 64, 75, 80, 83, 95, 114, 146, 190-191, 202, 223, 275, 301-303
Ashbaugh, J. J. Farm, 15, 19, 39, 47, 61, 64, 83, 95, 114, 146, 190-191, 202, 301-303
Ashbaugh, Jacob, 14-15, 75
Ashbaugh, John, 14-15, 26, 75, 81, 100, 115, 120, 163
Ashbaugh, Nettie, 114
Ashbaugh, Susanna, 38
Ashland, 87, 145, 193-194

## B

Backus, D. C., 370
Barber, Hank, 72
Barbour, John K., 447
Barclay, F. H., 15, 175
Barkley, Augustus, 75
Barnhart, Andrew, 414
Barnsdall, Theodore, 293
Barnum, James, 36
Barr, Henry, 1
Barthold, C., 372
Bartley, Williamson, 484
Barton, E., 103
Bashline, Aryl Ann, 380
Bateman, F. E., 401
Battle of Williamsburg, 55
Battles, Joe, 209
Baum, W. T., 268, 480
Beals, Hiram, 67, 75, 87
Beals, Jacob, 75, 255
Beals, William, 75
Bear Creek, 105, 106, 327-328
Bear Creek Pump Station, 346
Beardsley, E. C., 15, 231
Beatty, Anson, 227
Beaver City, 138, 255
Beaver Township Map, 244, 255
Beaver Twp., 243-244
Beck, J., 404
Becker, Mildred L., 476
Beckwith, E. H., 261
Bedford, 42, 44, 289

Beers, D. W., 86
Beers, John C., 460
Beeson, A. C., 369-370
Bell, Dr. S. D., 418
Bell, Finley, 91
Benedict, M. C., 372
Bennett, James, 67, 308
Benson, Jennie, 385
Berg, John, 319
Berlin, George, 53, 243-244
Berlin Hotel, 193
Berringer, G. W., 481
Berringer City, 483
Best, Henry, 53
Best, J. I., 194, 248
Bethune, J. H., 41
Biery, C. W., 14
"Big Injun" Well, 482
Bishop, James, 59
Black, A. R., 67, 103, 142, 204-205
Black, J. C., 215
Black, J. G., 206, 210
Black, Jacob, 30, 244
Black, Joseph M., 471
Black, Paul, 267, 295
Blackman, C., 212
Blacksmiths, 44, 67, 232
Blair, Robert, 135
Blair's Corners, 134, 135
Blake, J. E., 104
Blakely, Orie J., 230
Blakeslee, Barton, 290
Blakeslee, Charles, 223
Blakeslee, Dr. Thomas, 290
Blakslee, C. F., 15, 62, 64, 67
Blakslee, Charles, 173
Blakslee, Charles F., 460
Blakslee, Irene C., 70
Blakslee, J. B., 15, 67, 126
Blakslee, James B., 62, 460
Blakslee, W. Z., 15, 67, 70, 126, 459
Blakslee, W. Z., 15, 67, 70, 459
Blakslee, William, 62, 179, 459
Blakslee, William Z., 62, 126, 459-460
Blakslee, Wm. Z., 64
Blakslee Bros., 15, 62, 64, 113, 126, 460
Blanchard, 13, 96, 117-118, 176, 184
Blaney, James, 368
Blaney Hotel, 450
Blufawitch, L., 214
Bohlen, P. B., 119
Bone, C. C., 209
Booher, Theodore, 202
Book, Dr. W. P., 415
Booth, John Wilkes, 60, 154-156
Borland, J., 402
Bossom, Alfred C., 478
Bossom, Bruce, 477
Bostaph, Daniel, 1, 75
Bostaph, Samuel, 75
Bostaph, Squire, 280
Bostwick, J. A., 209, 435
Bostwick, Jabez A., 297
Both, M. M., 186
Bottner, R. J., 372
Bowen, A. A., 204-205
Bowen, Robert, 117
Bowers, Kitty, 381-382
Boyd, James, 102
Boyle, P. C., 29, 373
Braden, G. T., 419
Bradford, 298, 299
Bradford Commercial Bank, 268, 299
Bradley, Edwin Colton, 62, 64, 174, 265, 285-286, 475
Bradys Bend, 170, 457
Brandon, W. D., 483
Brant, Harmon, 75
Bredin, Anne, 319
Bredin, J. S., 160
Brewer, S. M., 335
Brice, James, 75
Brodhead, J. J., 294
Bronsdall, William, 269
Brown, George, 3, 335, 347-348, 354, 479
Brown, James E., 348
Brown, N., 231, 401
Brown, R. L., 3, 79, 146
Brown, T. P., 400
Brown, W., 3, 231-232, 346-348, 355, 479
Brownson, Marcus, 441
Bruin, 319, 322-323, 363-364, 428
Bruner, Zack, 480
Bryan, Robert, 15, 26, 44, 163, 165-166, 175, 204-205, 231
Bryon, Lowen, 75
Bryon, Robert, 75
Buena Vista, 440-442
Buffalo Bill, 388-389
Bullion Field, 479
Bullman, Joseph O., 344
Burke, P. R., 401

Bushnell, Joseph, 209, 420, 435
Butler City, 307, 379, 399, 403, 429, 444, 467
Butler County Bank, 420
Butler County Map, 320
Butler County National Bank, 70
Butler County Oil Fields, 318
Butler County Oil Map, 320
Butler Pipe Line, 363, 478
Butt, G. W., 15, 67, 223
Byers, Willie, 193
Byrd, Edward, 290
Byrom, Captain, 210

## C

Calderen, H. D., 103
Callensburg, 98, 101, 108-109, 294
Campbell, A. L., 319, 323-324, 366-367
Campbell, Charles, 149
Campbell, F. M., 368
Campbell, James, 282, 368
Campbell, John, 427
Campbell, John S., 483
Carbon Center, 449-450, 453
Carpenter, Henry, 191, 235-236
Carpenter, S. W., 45
Central Hotel, 351, 368, 373, 379, 388, 413, 415-418, 420, 425
Chalfant and Graff, 118
Chambers, Wesley, 232, 243
Chapman, Honorable Oscar L., 444
Chappell, E. G., 45
Chechak, John, 61
Chicora, 319, 325, 428
Chrisman, Adam, 75
Christie, Peter, 368
Christie Bros., 484
Christy, J. A., 174, 180
Chustz, J. A., 181
Civil War Veteran, 36, 56, 344
Clapp, W. H., 268
Clarion County Bank, 230, 232, 258
Clarion County Courthouse, , 11, 15, 64
Clarion River, 6-8, 86, 87, 93
Clark, C., 253, 327
Clark, J. C., 253
Clark, W. E., 368
Clawson, C. M., 252
Clawson, Jamie, 57
Clayton, Charlie, 279, 283
Cleveland, 66, 103, 184, 363, 470, 479
Cleveland Pipe Line, 470, 479
Clintonville, 479
Coal Oil Johnny, 286
Cochran, Alan, 28
Cogley City, 108
Cole, Thomas, 295
Colgan, James L., 63-64
Collner, L., 172, 175, 180
Collner, Lewis, 22, 75, 84, 146, 167, 177, 179
Collner, Samuel, 75
Collner Farm, 12-13, 80, 84-86, 96-97, 114, 118, 146, 150-151
Columbia Conduit Pipe Line, 401, 420, 427, 428
Commercial Bank, 268, 469, 473, 477
Compton, Murat, 370
Connoquenessing, 428-429, 484, 487
Continental Hotel, 146
Conver, Aaron, 75
Conver, Jacob, 75
Cook, Smith, 16, 22, 187, 210
Cooke, Smith, 175
Cool, M., 118
Cooper, John, 253
Cooperhead Line, 279
Corbett, William, 282
Corbitt, Carrie, 226
Corell, Harry, 229, 232, 238
Craig, Samuel, 331
Cram, R. W., 368-369
Cramer, Charles H., 15, 19
Cramer Opera House, 23
Crawford, Dr. A. W., 28-30, 471
Crawford, E. G., 230, 232
Crawford, Ebon, 120
Crawford, H. J., 312
Crawford, Mary McMillen, 30
Crawford, R. P., 79
Crawford, Robert, 3
Crazy Nell, 48, 390
Crazy Nellie, 48
Cribbs, Jack, 70-72
Cribbs, John L., 253
Crime, 27-28, 48, 481
Criswell, David S., 325, 451
Criswell, J. E., 400
Criswell, R. W., 392
Criswell City, 450-452
Crosby, Pratt S., 447
Cross Belt, 324-325, 483
Cubbison, Dilly, 312
Cudahy, Michael, 290

Culver, H., 179
Cumings, Captain, 465, 466, 467
Cumings, Charlotte, 465, 494
Cumings, Henry H., 464-465, 494
Cumings, Ralph, 465
Cumings, Senator H. H., 465
Cummings, Daniel, 115
Cummings, Jim, 388
Cunchula, Bill, 15
Cunchula, Violet, 19
Cundle, R. C., 480
Cupler, Adam, 33
Curley, Barney, 186
Curley, Charles, 186
Curll, D. B., 294
Curly Green, 48, 199
Cypher, D. A., 370

# D

Dahle, R. J., 194
Dahte City, 108
Daughenbaugh, W. H., 258-259
Davis, A. J., 23, 70, 78, 159, 494
Davis, E., 154, 164, 233
Davis, Ebinger, 75
Davis, F. T., 187
Davis, J., 23, 70, 78, 159, 259
Davis, Philip, 461
Day, George, 209
Dean City, 481, 483
Dearolph, John, 71, 75
DeHaven, R. W., 22, 67
DeHavens, R., 175
Delo, Daniel, 294
Delo, George P., 162
Densmore, James, 60
DeRolph, John, 16, 195
Derousse, H., 169
Derousse, Harry, 95
Derousse, Henry, 97
Derrousse, Harry, 179
Devore, John, 75
Dewey, W. W., 102
Dieter, Michael, 414
Dilks, 374
Dilworth, Al, 480
Dilworth, Richard, 453
Dimick, George H., 268, 368, 379, 382
Dimmick, George H., 323
Dittman, Adam, 16, 75
Dittman, David, 75, 81
Dittman, David Jr., 75
Dittman, David Sr., 75
Dittman, John, 75
Divener Well, 420
Divining Rod, 31, 34-35, 273, 494
Dogs, 112, 113, 115
Dolson, Hildegarde, 494-495
Don, Leper, 75
Donaldson, J. B., 67, 86, 200-201
Dorn, Dr., 192, 195
Dorsey, Thomas, 418
Drake, Colonel Edwin L., 1, 10, 54-55, 155, 269, 292, 381, 442, 469-470, 472-473, 491
Drilling, 31-35, 239, 301-303
Duffy, Peter, 445
Duggan, Dave, 207
Dunlap, S. N., 27, 165
Durham, William, 449
Dutt, R., 67, 69, 198

# E

Earhart, Martin, 462
East Brady, 456
East Foxburg Map, 125
East Parker Map, 120
Eastwick, C. J., 209, 360
Eddinger, Samuel, 182
Eddinger, William, 153
Eden, 108, 221-222
Edenburg, 71, 139, 192, 221-238, 263
Edenburg Business Directory, 231-238
Edinger, Andrew, 75
Edinger, Charles, 75
Edinger, Daniel, 75
Edinger, George, 75
Edinger, Henry, 75
Ehrenfield, Dr. N. F., 462
Eitermiller, Henry, 483
Elk City, 144, 251, 253, 254, 265-268
Elliot, James, 206
Elliott, A. J., 88, 400
Ellis, Minerva, 465
Ellsworth, George B., 258
Emery, O. G., 399
Emlenton, 7-10, 28-30, 39, 49, 60-64, 81, 85, 88-89, 93, 98, 100-101, 104, 107, 116, 157, 159, 167, 169-170, 174-176, 178, 181, 190, 210, 215, 222-223, 230, 235, 237, 240-241, 255, 297, 304-314, 317, 336, 361, 365, 467, 471, 494

Emlenton Bank, 308
Emlenton Line, 170
Emlenton Pipe Line, 60, 175-176, 181, 297, 361, 467
England, 44, 278, 300, 319, 477-478, 489
Englets Bakery, 229
Eshnour, Andrew, 67
Ethelyn, Ida, 474
European Hotel, 109
Evans, A., 94, 99
Evans, Anna, 257
Evans, James, 257
Evans City, 487
Exchange Bank, 334
Exley Farm, 148, 162, 187, 252-253

## F

Fagundas City, 102
Fairview, 391-395
Fairview Hotel, 389
Fairview Pipe Line, 363, 392, 396-399, 439, 470
Female Seminary, 381, 388
Fern City, 108
Fertig, Charles, 75, 233
Fertig, Frank, 25, 67, 87, 103-104, 174
Fertig, John, 75, 79, 115, 268, 296-298
Fertig, Samuel S., 268
Fesenmyer, F. Wayne, 445
Fillman, Michael, 75, 95
Fillman, Mishal, 132
Findley, W. P., 176
Fink, Birdie Estella, 461, 463
Fink, Gilmore C., 54, 461-463
Fink, Joel, 127, 164, 180, 210
Fink, Joel, 2, 16, 54, 65-66, 75, 96, 104, 127, 151, 159, 164, 173, 180, 210, 256, 462-463
Fink, Johnas, 13
Fink, Jonas, 75, 98
Fink, Roy Howe, 463
Fink, S. H., 461
Finlay, Colonel J. B., 2
Finlay, H., 166
Finley, H., 106
Finley, Harris, 86
Finley, W. P., 16, 107
Fire, 20-26, 107, 110, 147, 151, 152, 157, 162, 163, 164, 165, 171-181, 185, 186, 188, 189, 191, 198-200, 203-205, 208, 226-230, 258, 276, 306-307, 327, 335, 336, 345, 356, 373, 374, 387, 391, 401, 415-417, 430, 435, 439, 440, 443, 447
First National Bank, 258, 374
Fithian, W. H. H., 214, 360, 480
Flagler, Henry M., 291
Fleming, Emmett, 153
Fletcher, Benjamin, 363
Foote, Dr. Paul D., 444
Ford, Alliene, 474
Ford, Ida D., 474
Fords Theatre, 155
Forest Home, 143
Forestville, 476, 479
Forst, Barney, 417
Fort Ticonderoga, 478
Foster Farm, 482-483
Four, Geo. H., 68
Fourth Sand, 324-325, 454
Foust, Barney, 160
Foust, Benjamin, 16, 75
Foust, Joseph, 16, 75, 80, 98, 107, 151
Foust, Philip, 75, 98
Foust Farm, 13, 80, 94, 97, 107, 116, 118, 120, 150-151, 163
Fowler, John W., 373
Fowler, T., 209
Fox, Hannah, 63, 310
Fox, Joseph M., 310, 462
Fox, Mary R., 187
Fox, Samuel, 2, 58-59, 63, 159, 256, 310
Fox, William, 223
Fox, William Logan, 60
Fox Farm, 2-3, 6, 60, 79, 159, 164, 181, 187, 191
Foxburg Bank, 97, 461
Foxburg Business Directory, 88-89
Foxburg Hotel, 88
Foxburg Map, 102, 122, 154, 158, 207
Foxburg Savings Bank, 170, 208-209
Fradenburgh, Rev., 494
France, 9, 26, 36, 44, 300, 465, 468, 477
France, Cyrus, 26
France, Dr., 36
Frank, Martin, 26, 75, 148
Frazier, George, 75
Frederick, Barnhart, 414-415, 417
Fredericksburg, 475
Freeport, 234, 325, 358, 408-409, 420, 449, 494
French Kate, 48, 111, 251, 267, 333, 380-382
French Mary, 380-381
French Mary Hill, 380-381
Frets, Solomon, 75
Frisbee, Fin, 351
Fry, Paul, 75, 100, 187

Fry, Polly, 187
Fuller, H., 322
Fuller, I. C., 79
Fullerton, H. R., 16, 345, 405
Fullertown, 62, 107-108
Fullmer, George, 75
Fulmer, Henry, 16, 68, 87, 175, 181
Funk, Captain A. B., 296

## G

Gailey, William, 480
Galey, J. H., 7-8, 79-80, 100-102, 244-245, 288-292, 349, 419, 441, 463-464
Galey, John, 7, 245, 288-289, 293, 349-350, 464
Galey, Mr. Buckwheat, 246
Galey, Robert Sr., 463
Galey, Thomas F., 463
Galey, Thomas M., 289, 293
Galeyville, 244
Garfield, James, 157
Garfield, President, 293
Garner, Henry, 75
Gas, 94, 149, 353-354, 452, 453
Gas City, 100, 335, 452, 483
Gates, Joseph, 162
Gaughly, John C., 214
George Bros., 68
German, 15, 24, 51, 56, 89, 207, 210, 212, 214-215, 233, 261, 416, 465
German National Bank, 416
Ghost Story, 50
Gibbs, F. H., 210
Giblers, Frank, 385
Giblet City, 249
Gibson, William, 369
Giddens, Paul H., 31, 111, 157, 299, 494
Gifford, J. M., 226, 233
Gilbert, David, 75
Gilger, B. F., 47
Gilgrist, John, 113
Gimlet Town, 165
Ginder, W. F., 57
Glade Run Field, 487
Globe Hotel, 351
Golden City, 486
Goldsmith, M., 68, 87
Goodwin, Tom, 207
Goodwin, W. W., 22
Goss, Alonzo, 248
Gosser, J. J., 311
Gotham, S. O., 373
Gottschalk, D., 21
Goughler, Amos, 63, 75
Goughler, Charles, 75
Goughler, Henry, 75
Goughler, Jacob, 75
Goughler, Lavina, 52
Goughler, Samuel, 75, 163
Grace, Peter M., 419
Graham, George H., 368, 370
Graham, Joseph, 446
Graham, Joseph Farm, 446
Graham, P., 87
Graham, Postmaster S. S., 258-259
Graham, W. T., 332
Grand Central Hotel, 373
Grant, Ulysses S., 248
Grant Pipe Line, 361, 420, 480
Grass Flat, 80, 147
Great Oil Belt City, 453
Greece City, 324, 428-433
Green, D. L., 204-205
Gregg Farm, 60
Gregory, Colonel W. S., 157
Gretter, Victor, 25-26, 44, 84, 117, 276
Guffey, J. M., 16, 83, 223, 233, 292, 419
Guffy, James M., 243, 265, 290, 291, 293
Gusher, Lucas, 292

## H

Hahn, Charles, 1
Hahn, Jacob, 1
Haines, John H., 399
Hale, George, 75
Hammond, John W., 79, 297-298
Hardison, James, 150
Harlansburg, 319
Harley, Henry, 250
Harley, S. W., 370
Harmony, 160, 487
Harms, G. R., 81
Harper, W. H., 36, 88
Harris, G. R., 12, 96
Harris, J. W., 205
Harrisburg, 8, 56
Hartley, William, 7, 289
Hatch, Charles P., 60, 121, 209, 418, 478
Hawk, William, 391

Hayden Hotel, 200
Hayes, Governor, 157
Hayes, President, 157
Hayes, Rutherford B., 157
Hays, Senator Thomas, 396
Hays, W. G., 396
Haysville, 392, 396
Hazelton, 233
Healey, Thomas, 461
Heart, J. D., 208
Hearts, J. D., 204
Heeter, Amos, 75
Heeter, D., 187
Heeter, Dan, 149
Heeter, Daniel, 75, 148
Heeter, William, 153
Heeter Farm, 119, 148-149, 153, 182, 187, 189-191, 208, 283
Heiner, J. H., 146
Heiner, John, 364
Henderson, Charles, 482
Henri, Flora, 385
Henry, J. H., 171
Henshaw, A., 417
Hepp, Fred, 22-23, 109, 171, 173
Herr, Charles E., 372
Herron, D., 22-23, 68
Hess, Dave, 253
Hess, John, 253
Hess, Michael, 150, 194
Hill, J., 237
Hillenmeyer, Kate, 476
Hinterlang, Alphonse, 448
Hoch, Martin, 415
Hogan, Ben, 110-111, 175, 251-252, 267, 322-323, 333, 368, 373,379-384, 384, 386, 388-390, 417, 481, 488
Hogan, Mayor, 382-383
Hood, G. W., 462
Hooks City, 486
Hovey Twp., 314
Howe, Birdie Estella, 461
Howe, Harry, 461
Hughes, James E., 369
Hugus, Simon, 16, 43, 202
Hulings, A. A., 8-9, 40, 46, 120, 162
Hulings, M., 8-10, 39-40, 62, 64, 109, 162, 218
Hulings, Michael, 9
Hummel Farm, 108-109, 187, 240, 242
Hunsicker, Martin, 63-64
Hunter, Jahu, 464-466
Hunter, Lady, 467
Hunter, Livingston L., 466
Huselton, B. C., 483
Huselton, Theodore, 428
Hutchinson, C. P., 160
Hyner, R. Farm, 156

# I

Ireland, L. N., 369
Iron City, 109, 418
Iron Furnace, 117, 131, 294, 310
Iron Pipe Line, 361, 363
Iron Tanks, 60, 61, 207, 354, 401, 404, 409,
Irwin, J. W., 26, 162
Isaac, Reecher, 69
Ishman, J., 186

# J

Jack, Tillman, 247
Jack, William, 153
Jackson, Andrew, 75, 313
Jackson, F., 79
Jackson, M., 360
Jackson, O. P., 418
James, Abram, 273
Jamison, John B., 368
Jefferson, 9, 28, 108, 117, 120, 137, 148, 187-188, 223-225, 229, 232, 234, 238, 240-241, 243, 253, 256, 294, 453, 486
Jefferson Furnace, 117
Jefferson Hotel, 253
Jennings, Richard, 325, 327, 379, 419
Johnson, Adam, 75
Johnson, B. G., 196
Johnson, C. E., 162
Johnson, Dr. F. J., 181
Johnson, F. I., 65, 99
Johnson, Michael, 210
Johnston, Moses, 107
Johnston, Tip, 248
Johnston, W. W., 214
Jones, David, 17, 182
Jones, George Jr., 75
Jones, J. T., 243, 265
Jones, William, 98
Jordan, Adam, 75
Jordan, John W., 474, 494

# K

Kane, John S., 400
Karns, Duncan, 3, 357, 358, 407, 409, 474, 479
Karns, Samuel D., 379, 405, 407
Karns, Samuel Duncan, 409, 474
Karns, Stephen Duncan, 407
Karns Bank, 358
Karns City, 399, 410
Karns Obituary, 474
Karns Pipe Line, 358, 404, 420, 475
Keating, John, 17, 75, 310
Keating, Judge, 83, 131, 289
Keating City, 45, 50, 108, 191, 491
Keating Farm, 13, 64, 95-97, 119-120, 146-149, 154
Keener, Nick, 327
Keir, S. M., 99
Kennedy, J. Q. A., 321
Kennedy, Jane, 53
Kerr, R. M., 360
Keystone Pipe Line, 420
Kibbe, W., 69
Kier, S. M., 80, 146, 244, 285
Kier, Samuel, 442
Kiley, J. D., 211
King, Tom, 357
Kirk, David, 453, 480
Kirley, Barney, 185
Kiser, Damal, 267
Kiser, Joseph, 265
Kistler, D. L., 80
Kittanning, 61, 118, 146, 348, 350, 384
Kline, Charles, 75
Klinger, Joel, 75
Klinger, Julius, 319
Klotz, Charles, 75
Knappenberger Oil Farm, 135
Knause, Samuel, 98
Knight, Daniel, 75, 194, 294
Knight Oil Farm, 140
Knoch, John A., 445
Kohlmeyer, Candace, 471
Kohlmeyer, Henry, 310
Kribbs, Captain George, 294
Kribbs, Christian, 294
Kribbs, George, 138, 294
Kribbs, P. F., 234, 238, 293-295
Kribbs Farm, 108, 138
Kribbs, George Farm, 138
Krug, George, 483
Kruger, Paul, 280
Kuhn, Leroy, 429
Kurtz, John M., 145

# L

Lambing, James M., 19
Lamburg, John, 366
Langworthy, F. H., 360
Larkin, William, 323
Laughner, James, 75
Lawrenceburg, 331, 356
Lawson, David, 6
Lee, John, 240-241, 420
Leedom and Patterson, 234, 265
Levans, Frank, 117
Levi, Alfred, 68
Levy, Howard M., 445
Lincoln, Abraham, 30, 156, 471
Litzenberg, Don, 429
Lockwood, Marcus. L., 81, 131, 288, 360, 419
Logue Oil Farm, 124
Loguetown, 303
Loomis, J. H., 174
Los Angeles, 278, 297, 474
Loughner, Sam, 184
Lovell, Jesse, 98
Lowers, Anthony, 75
Lowers, Jacob, 75
Lusk, William, 186-187
Lyons, J. C., 370

# M

Mackey, Ann, 202, 204
Magill, Lyman B., 466
Mahneke, Dr., 373
Malarkey, Thomas, 75
Malloney, John, 21
Malthorn, John, 26, 148
Manning, Daniel, 402
Mansion, Karns, 358, 409
Map of Blair, 134
Marshall, John, 312, 331
Marshall, Mary, 380
Marshall, S. S., 86
Martin, John, 148, 363
Martin, M. C., 58-59, 81
Martin, Samuel, 75
Martinsburg, 363
Masson, Frank L., 370

Mather, John, 457-458, 491
Mathorn, John, 75-76
Matson, Myron, 295
Matthew, Jack, 68
Mayer, Ludwig, 386
McArdle, Pete, 270
McBride, Spotty, 488
McBride City, 485-486
McCafferty, C. A., 45
McCain, Daniel, 311
McCalmont, G. B., 417
McCamon, Alexander, 76
McCarty, Wm., 21
McCaslin, Andrew, 309
McCleary, Thomas, 391
McClymonds, Hugh P., 399
McConnell, Thomas, 2
McCorkle, John, 396
McCracken, Ben, 161
McElvain, Bill C., 483-484, 486-488
McElvain, William C., 327, 488
McElwaine, J. B., 223
McElwee, James, 391
McFarland, F. S., 473
McFerson, Albert, 76
McGee, Alexander, 26, 148
McGinnis, J. J., 310
McGraw, Jim, 279-280
McJunkin, J. D., 483
McKee, John D., 375
McKee, R. W., 392
McKinney, Curt, 351, 419
McKinney, Gary, 491
McKinney, Glen Ford, 474
McKinney, J. C., 351, 441, 469, 473, 478
McKinney, James Curtis, 451, 469-470
McKinney, John Luke, 451, 469, 473-474
McKinney, Lydia Drury, 469
McKinney Well, 451
McKisson, John, 440
McLaughlin, John, 147, 351
McLaughlin, Thomas, 447
McLaurin, John, 27, 205, 248, 253, 257, 464, 472
McLaurin, John J., 51, 54, 221, 239, 241, 256-257, 285-286, 288, 295, 297, 379, 383, 385, 387, 404, 440, 443, 447, 458, 460, 463, 465, 467-473, 476, 478-479, 481, 483, 488
McMahon, Peter, 25
McMarlin, J. A., 486
McWillan, John, 186
Meadville, 9, 44, 61, 285
Mehrten, J., 197
Mellon, Andrew W., 463
Mellon, W. L., 291, 463
Mendenhall, J. G., 221
Miller, Earnest C., 264
Miller, Homer, 258
Miller Farm, 58, 162
Millers Farm, 60
Millerstown, 414-425
Mills, Major W. M., 289-290, 292
Mitchell, Alex, 8
Mobley, W. C., 223, 351, 405
Modoc, 433-439
Moffert, Bob, 450
Mongtown, 28, 108, 143, 225, 229, 265
Mongtown Map, 143
Monitor House, 127
Monroeville, 9, 28, 108, 117, 133-134, 159, 261-262
Monroeville Map, 134
Moore, Dr. C. R., 301
Moore, Johnny, 388
Moore, Wm. D., 89
Morehead, Alexander, 76
Morehead, S. W., 84
Morgan, Bill, 57
Morgan, John Jr., 7, 76
Morgan, Thomas, 95
Moriarty, D. D., 9, 308
Morrison, David, 323, 428
Moyer City, 108
Murphy, John D., 56
Murphy, Michael, 419
Murray, Edward, 1
Mutual Pipe Line, 59-60, 118, 153, 162, 297, 360, 478
Myer, C., 59
Myers, Abram, 483
Myers, J. T., 183
Myers, John T., 187

## N

National Hotel, 210
Neely, Alfred, 154
Neely, H., 81, 102, 160
Neely, Harison, 76
Neely, Hiram, 13, 76, 81, 84-85, 119, 147, 194
Neely, Hiran, 147
Neely, Isaac, 1, 76, 100, 102, 117, 119, 153-154, 159, 176

Neely, Paul, 119, 153-154
Neely, R., 17, 154
Neely, R. H., 17, 102
Neely, R. L., 100
Neely, Richard, 76, 101
Neely, William, 1, 76, 153, 159
Negley, John, 319
Nesbit, George, 379
Nesbitt, George H., 323
Nevil, Amos, 76
New Castle, 241, 478-479
Newton, A. R., 312
Norris, Josh, 360

# O

O'Day, Daniel, 455, 467-468
Oil City Pipe Line, 247, 286
Oil Exchange, 217, 362, 418
Oil Gauger, 375
Oil Price, 1860-1916, 454
Oil Times in Oildom, 347-361
Oil Well Supply, 42
Oliver, Charles H., 444
Outhouse, 453
Overy, Joseph, 420, 446
Owen, Richard, 28
Owens, Dick, 186
Owens, M., 187

# P

Page, John, 450
Painter, John, 368
Palmer, Charles, 121
Panton, James, 86
Parker, Fullerton, 350, 352, 405
Parker, James, 451
Parker, James P., 345
Parker, John, 327, 329
Parker, Judge, 329-331
Parker, Thomas M., 331
Parker, William, 329, 352, 399
Parker City, 3, 6, 8, 40, 111, 167, 182, 217, 248-249, 252, 305, 311, 314, 318, 321-322, 327, 329, 332-336, 344-359, 363, 379, 381-382, 386, 405, 408, 430, 443, 478, 480, 494-495
Parker City Map, 330
Parker City Savings Bank, 334
Parkers Landing, 3, 80, 84, 218, 322, 331
Patrick, J. B., 294
Patterson, Dr., 184
Patterson, J. C., 40, 88, 198, 202, 204-205
Patton, L. D., 20
Paulain, Gabriel, 477
Pauly, J. A., 103
Peck, Seymour, 480
Pedro, Dom, 299-303
Pennsylvania Transportation Co., 250-251
Perry, George, 289
Perryville, 108, 121, 314, 354
Perryville Map, 121
Pesevento, Ben, 15
Peters, Judge Richard, 13-14, 51, 53-54
Petersburg Hotel, 22, 67-70, 73, 82, 90-91, 172, 175, 191, 198, 285, 287
Petroleum Hotel, 237
Petrolia, 66, 157, 168-169, 219, 229, 236, 275, 283, 318, 323-326, 334, 346, 356-358, 363, 366-375, 376, 379-391, 401, 405, 426, 428-429, 439-440, 443, 447, 451, 464, 467-468, 471, 474, 484, 489
Petrolia City, 318, 334, 358, 368, 371-372, 374-375, 379-380, 383, 388-390, 451, 467, 489
Pettigrew, Dr. S. H., 400
Phillips, David, 17, 69
Phillips, O. C., 210
Phillips, T. W., 478, 480, 484, 487
Phillips, Thomas W., 478, 480
Phillips, Thomas Warton, 478
Phillips City, 485-486
Phillips Hotel, 348
Pickwick, 246
Pickwick Map, 134
Pierce, Dr., 149
Pierce, L. C., 205
Pike, Albert, 69
Pine, William, 36
Piney Map, 137
Pioneer, 51-54
Pipe Line, 58-64,121, 243, 327, 361, 397-399, 404, 418, 427
Pithole City, 48, 156, 381
Place, J. M., 61, 168
Plumer, Fred, 241, 420
Plumer, Frederick C., 240
Pollock, Charles, 181
Pollock, James, 3, 97
Pollock, John, 392
Porter, Frank M., 444
Porter, William, 13
Pratt, Charles W., 284

Prentice, Frederick, 482
Preston, John, 419
Primrose Hotel, 401
Pritner, Lindsey, 294

# Q

Queen, John, 455
Queen City, 319, 322
Queenstown, 325, 327, 455-456, 464

# R

Railroad, 49, 103, 123, 222, 240, 241, 329, 400, 405-408
Raleigh, Sir Walter, 149
Ralph, G. W., 89
Ratell, J., 20
Rathbun, Harvey, 440
Rea, Albert, 69
Reardon, Delilah, 258
Reardon, Edwin, 258-259
Red Hot, 21, 89, 171, 252, 491
Reed, Samuel R., 447
Refinery, 374
Reiber, Jacob, 483
Reiber Hotel, 416
Reichart, Michael, 109
Reighner, Franklin, 76
Reighner, Jeremiah, 76
Reynolds, J. B., 294
Richard, William, 22
Richison, Albert, 76
Richland Twp., 1, 53-54, 74-76
Richmond Map, 131
Riddle, Samuel L., 399
Riedel, Julius, 445
Rieder, Jake, 364
Rishel, John, 147
Ritchey, Alex, 76, 98
Ritchey, James, 53
Ritchey, Thomas, 305
Ritchey Farm, 305-306
Ritts, Charles, 18, 76, 81, 146
Ritts, Dan, 100, 115-116
Ritts, Daniel, 13, 76, 81, 96, 98, 100, 102, 118
Ritts, Daniel Farm, 13, 96, 98, 100, 102, 118
Ritts, Eli, 13, 81, 96, 100, 173
Ritts, Elias, 17, 70, 76, 78-79, 169, 243
Ritts, J. V., 17, 70, 79, 223
Ritts, John, 18, 76, 174
Ritts, John V., 70, 169
Ritts Farm, 13, 81-83, 96, 98-103, 107-108, 112, 115-120, 146, 148, 153, 159
Ritz Farm, 25, 190, 197
Ritzert, John, 448
Robbins, Frank, 191, 224, 239, 265, 267, 367, 408, 427, 428, 452, 467
Roberts Torpedo, 35-37
Robinson, Elisha, 2, 349, 352
Robinson, Nell, 390
Robinson, Nellie, 48
Robinson, William, 3, 69
Robinson, William D., 441
Robinsons, Elisha, 352
Rockefeller, John Davison, 65-66, 70
Rockwood, A. B., 25
Roebling, John A., 445
Roebling, Washington, 445
Roessing, W. P., 486
Rogers, Willie, 440
Roschy, Fred, 128
Ross, Joe, 312
Ross, Perry, 89
Ross, Young, 312
Rouse, Henry R., 269, 353, 382
Rowe, Edward, 462
Rowland, James, 461
Roy, Rob, 325, 352, 401
Rugh, M. J., 18, 204
Runkel, Anthony, 69
Rupert, Charlie, 50
Rupert, Jeremiah, 76
Rupert, Samuel, 65, 99
Rupert Farm, 12, 86, 96

# S

Salem Twp., 242
Samuel G., Bayne, 268-284, 477
Sand Pump Line, 81, 83, 93
Sanders, Margaret Webb, 317
Sargent, George, 214
Sawyer, Charlie, 390
Saxonburg, 445, 486
Seaboard Bank, 477
Second National Bank, 474
Seep, Joseph, 209, 476-477
Seldner, S. W., 386
Shady City, 108, 225
Shakely, Rosina, 52
Sharon Savings Bank, 258

Shaw Farm, 36
Sheehan, Mae, 444-445
Sheffer, Susan, 294
Sherman Hotel, 450
Shicker, Arnst, 76
Shingledecker, Florence Ashbaugh, 48
Shirey, Samuel, 76
Shirey, Solomon, 101
Shoup, Barney, 76
Shoup, David, 12, 21, 52-53, 57, 76, 80-81, 96, 146-147, 176, 194
Shoup, Elizabeth, 51
Shoup, Kenneth Lester, 51
Shoup, Michael, 26, 76, 148
Shoup, Sarah, 58
Shoup, Solomon, 52-53
Shoup, Willis, 313
Shoup Farm, 152-153, 176, 187, 190
Sible, George, 21
Sibley, Joseph, 419
Sigley, George, 76
Simpson, J. W., 221
Sitelr, Alex, 76
Slack, Colonel W. D., 327
Slarnes, R., 21
Small, S. J., 418
Smiley, A. W., 153-155, 162, 209, 223
Smiley, Alfred, 60, 153, 360
Smith, Billy, 55, 441-444
Smith, David, 373
Smith, Dr., 115
Smith, James H., 94
Smith, Jeremiah, 442
Smith, Jesse, 300-301
Smith, S. H., 371
Smith, Samuel A., 462
Smith, Samuel B., 442, 445
Smith, Uncle Billy, 55, 441-444
Smith, W. L., 236, 444
Smith, William, 148, 442-444
Smith, William Andrew, 442
Snow, James H., 58, 60, 247, 297
Snyder, Victorine, 114
Soliday, Abe, 283
Spear, H., 154
Spears, H., 80
St. James Hotel, 24, 185
St. Joe, 446-449
St. Peterburg, 20, 56, 82, 101, 119, 294, 297
St. Petersbug Hotel, 67
St. Petersburg Bank, 78, 79, 298
St. Petersburg Business Directory, 15-19, 66-70
St. Petersburg Map, 12, 125
St. Petersburg Savings Bank, 17, 70, 78, 79, 82, 175, 181, 296-298
Stanard, Daniel, 6
Steele, John, 286
Steele, John A., 487
Steffee, William, 69
Sterner, John, 76
Sterner, Reuben, 18, 76
Stevens, Cal, 398-399
Stewart, Adam, 414
Stewart, Charles, 454
Stewart, Charles C., 324
Stewart, Thomas, 113
Stiles, George G., 420
Stobble, Charles, 76
Stoble, Frederick, 76
Stocker, J., 214-215
Stockman, Mary, 390
Stokes, J. F., 210
Stores, 41-44, 311, 335
Storey, Alexander, 440
Stowell, J. C., 205
Stump Creek Island, 6-8
Sullivan, James, 255
Summit City, 388, 480-483
Sutherland, J. H., 26, 179
Swan, Bryan, 327
Swan, Ed, 327
Swayne, Dr., 457

# T

Tack, Gus, 480
Tarbell, Ida, 66
Tarr Farm, 64, 91, 94, 459
Taylor, Alonzo, 387
Taylor, E. A., 367, 370
Taylor, John, 473, 480
Taylor, Tom, 311
Teegard, Abraham, 455
Texas, 244, 259, 288, 291-293
Texter, George, 243
Thomas, Billy, 253
Thompson, Jesse, 1
Thompson, Robert, 449
Thompson, William, 399

Thorn Run Creek, 486
Tidioute Savings Bank, 465-466
Tilford, J. B., 476
Tippery, Andrew, 76
Titusville Trust Bank, 478
Tompkins, Lydia H., 288
Tonks, John, 335
Torpedo, 33, 35-37, 93, 130, 274-276, 387
Triangle City, 108, 196-197, 211-212, 240, 245, 247, 251
Triangle City Hotel, 196, 211, 245
Triangle Map, 136
Troutman, 324, 439
Truby, Simon, 3, 146
Trunkey, Judge, 282
Turkey City Bank, 258, 489
Turkey City Business Directory, 259
Turkey City Hotel, 259-260
Turkey City Map, 132
Turkey Run Bank, 262
Turner, John, 153-154, 193, 237

# U

Undertaker, 40-41
Unicum, Edward, 180
Union Pipe Line, 60, 213-214, 243, 260, 374, 420, 478
United Pipe Line, 60, 243, 286, 369-370, 372, 387, 398, 420, 467
United States Hotel, 155, 229, 260

# V

Van Syckel, Samuel, 427
Van Syckle, M. E., 153
Vandegrift, J. J., 419
Vandergrift, Jacob Jay, 470-471
Vandergrift, William, 435
Vasey, J. P., 166
Vaux, Sara Anson, 51
Vensel, Albert, 15
Vensel, Alfred, 76
Vensel, B., 100, 102
Vensel, Barney, 13, 18, 22-23, 54, 69, 76, 96, 116, 172, 175, 223
Vensel, C., 18, 54, 191, 223
Vensel, Charles, 54, 76, 223
Vensel, Hannah, 54
Vensel, Jacob, 18, 53
Vensel, Melissa, 463
Vensel, Nelson, 26, 54, 76, 95, 148
Vleck, Van, 237

# W

Waldron, A. K., 488
Waldron, O. K., 486
Walker, Billy, 195
Walker, Jennie G., 30
Walker, May, 317
Walley, John, 327
Warden, Will, 430
Warden, William G., 435
Washington, George, 28, 202
Washington City, 55
Watkins, H. W., 155
Watson, Colonel Thomas, 1
Watson, J., 268
Watson, Jonathon, 10, 254, 474
Watson, W. S., 223
Watt, Charlie, 283
Watt, Chas, 69
Watt and McElwain, 65
Weaver, Cooney, 317
Webb City, 289
Weeds, Sam, 172
Weller, Joseph, 310-311
Wellman, Dr., 37
Wellsville, 286, 475
Welsh, J. H., 22, 69
Werner, John, 290
West, A. A., 91, 212, 254, 290, 373, 475
Westermann, H. L., 414
Westinghouse, George, 292
Westmoreland, 43, 53, 149-150, 200, 204-205
Westmoreland Bakery, 200, 204-205
Westmoreland Hotel, 149
Wetter, Major Henry, 244
Wheeler, Fred, 447
White, Dr., 292
Whitehill, David, 1, 194
Whitely, R. D., 444
Whitling, Dan, 69
Whitling, George, 14, 76
Whittling, Daniel, 76
Whittling, George, 76, 98
Whittling, Wm. H., 87
Widel, Eugene, 56
Widle, Elias, 306
Wilcox Hotel, 355
Wildman, F. J., 237

Wilson, Alexander, 1
Wilson, Augustus, 248
Wilson, Chas, 70
Wilson, Clark, 334, 336
Wilson, E., 86, 237
Wilson, Harry W., 462
Wilson, Judge, 284
Wilson, Major, 373
Wilton, Chief, 283
Winner, Joseph E., v
Winsch, J. F., 310
Winsor, G. W., 174, 179, 206
Wire Rope, 445
Wolf, J. D., 248
Wright, E. G., 196

## Y

Young, David, 12, 96
Young, Samuel, 392
Youngsville, 347-348, 353, 383, 479
Youngsville Savings Bank, 348

## Z

Zahniser, A. J., 461
Zahniser, John N., 461
Zavinger, J. H., 174
Zehrung, W. S., 444
Zeigler, Jacob, 319, 321
Zelienople, 487